Peptides: Synthesis and Applications

Peptides: Synthesis and Applications

Editor: Owen Chase

R CALLISTO REFERENCE

www.callistoreference.com

Callisto Reference,
118-35 Queens Blvd., Suite 400,
Forest Hills, NY 11375, USA

Visit us on the World Wide Web at:
www.callistoreference.com

ISBN: 978-1-63239-986-1 (Hardback)

Cataloging-in-Publication Data

Peptides : synthesis and applications / edited by Owen Chase.
    p. cm.
Includes bibliographical references and index.
ISBN 978-1-63239-986-1
1. Peptides. 2. Amino acids. I. Chase, Owen.
QP552.P4 P47 2018
572.65--dc23

Table of Contents

Preface

Peptides are short chains of amino acids and are hence also called small proteins. Peptides are very important hormonal functions and regulate the body's metabolism. Peptide hormones trigger the release of fat and the creation of muscles. They also help the body to absorb more amino acids and manage the growth hormone. The research on peptides is very important for those seeking to study human fitness and muscle development. The extensive contents of this book provide the readers with a thorough understanding of the subject. It includes some of the vital pieces of work being conducted across the world, on various topics related to peptides.

The researches compiled throughout the book are authentic and of high quality, combining several disciplines and from very diverse regions from around the world. Drawing on the contributions of many researchers from diverse countries, the book's objective is to provide the readers with the latest achievements in the area of research. This book will surely be a source of knowledge to all interested and researching the field.

In the end, I would like to express my deep sense of gratitude to all the authors for meeting the set deadlines in completing and submitting their research chapters. I would also like to thank the publisher for the support offered to us throughout the course of the book. Finally, I extend my sincere thanks to my family for being a constant source of inspiration and encouragement.

Editor

Characterization of the Highly Variable Immune Response Gene Family, *He185/333*, in the Sea Urchin, *Heliocidaris erythrogramma*

Mattias O. Roth, Adam G. Wilkins, Georgina M. Cooke, David A. Raftos, Sham V. Nair*

Department of Biological Sciences, Macquarie University, North Ryde, NSW, Australia

Abstract

This study characterizes the highly variable *He185/333* genes, transcripts and proteins in coelomocytes of the sea urchin, *Heliocidaris erythrogramma*. Originally discovered in the purple sea urchin, *Strongylocentrotus purpuratus*, the products of this gene family participate in the anti-pathogen defenses of the host animals. Full-length *He185/333* genes and transcripts are identified. Complete open reading frames of *He185/333* homologues are analyzed as to their element structure, single nucleotide polymorphisms, indels and sequence repeats and are subjected to diversification analyses. The sequence elements that compose *He185/333* are different to those identified for *Sp185/333*. Differences between *Sp185/333* and *He185/333* genes are also evident in the complexity of the sequences of the introns. He185/333 proteins show a diverse range of molecular weights on Western blots. The observed sizes and pIs of the proteins differ from predicted values, suggesting post-translational modifications and oligomerization. Immunofluorescence microscopy shows that He185/333 proteins are mainly located on the surface of coelomocyte subpopulations. Our data demonstrate that *He185/333* bears the same substantial characteristics as their *S. purpuratus* homologues. However, we also identify several unique characteristics of *He185/333* (such as novel element patterns, sequence repeats, distribution of positively-selected codons and introns), suggesting species-specific adaptations. All sequences in this publication have been submitted to Genbank (accession numbers JQ780171-JQ780321) and are listed in table S1.

Editor: Sebastian D. Fugmann, Chang Gung University, Taiwan

Funding: The authors have no support or funding to report.

Competing Interests: The authors have declared that no competing interests exist.

* E-mail: sham.nair@mq.edu.au

Introduction

Many invertebrate immune systems studied to-date consist of highly complex repertoires of pattern recognition receptors (PRRs), regulatory and effector systems, but lack hypervariable recognition molecules that are homologous to vertebrate immunoglobulins. Recent studies also indicate the ability of some invertebrate immune systems, such as those of some arthropods, to specifically discriminate between different pathogens [1–3] and there is also evidence which suggest that some invertebrate immune systems may be capable of heightening responses to repeated challenge by the same type of pathogen [4,5], a phenomenon analogous to immunological memory [6]. However, the molecular bases of these immunological features have not been established. Some invertebrate immune genes, such as the scavenger receptor cysteine rich repeat (SRCR) genes of the sea urchins [7], are organized as large gene families that specify diverse repertoires of closely-related products [8,9]. This diversity is presumably brought about by gene duplication and divergence, gene conversion and gene rearrangement during PRR expression [10]. Another strategy involves post-transcriptional diversification of a small number of immune-response genes. An example of this is the down syndrome cell adhesion molecule (Dscam) gene family in *Drosophila melanogaster* and *Anopheles gambiae*, which can generate thousands of alternatively spliced transcripts from single copy genes [4,11]. Other examples of diversified families of genes in invertebrates include those encoding the fibrinogen related proteins (FREPs) in snails [12] and those encoding the variable chitin binding proteins (VCBPs) in cephalochordates [13].

Sea urchins are members of the echinoderm phylum, which is a sister group to the chordates. Thus, sea urchins are important for investigations of the immunology and the evolution of immune systems in the deuterostome lineage. The purple sea urchin, *Strongylocentrotus purpuratus*, has an elaborately equipped immune system [8]. The numbers of genes encoding putative PRRs (e.g. TLR, SRCR, NOD and NLR) in the *S. purpuratus* genome are much higher than those found in vertebrate genomes [14]. In addition, a unique class of highly variable immune-gene family, known as *Sp185/333*, also functions in immunity and is considerably upregulated upon immunological challenge [15–17]. More than 860 full-length *Sp185/333* cDNAs [16,17] and 171 genes [18] have been sequenced to-date. The diversity of *Sp185/333* s is based on the presence or absence of 25 to 27 sequence blocks, called elements, depending on how the sequences are aligned [15,17,19]. Elements do not randomly appear in the genes, but are present in distinct element patterns (i.e. combinations of elements; 15–18) and all *Sp185/333* sequences can be categorized according to their element patterns. Diversity is further enhanced by single nucleotide polymorphisms (SNPs), short insertions and deletions (indels) and a number of sequence repeats that appear throughout *Sp185/333* sequences [15,18] that enable two equally feasible alignments of the sequences. Previous studies indicated that *Sp185/333* with *E2*, *D1*, *01*, and *E3* element patterns were under positive selection for diversification (dn/ds>1; see Ref. 16). Furthermore, sequences from the first element of all Sp185/333 sequences were under positive selection. However, *Sp185/333* with *C1*, *A2*, and *E1* element patterns were under negative selection (dn/ds<1). The mechanisms that generate the high variability in *Sp185/333* sequences are unknown. However, gene conversion and DNA recombination driven by microsatellites that flank *Sp185/333* genes have been purported as possible mechanisms that promote gene diversification [19,20]. Post-transcriptional processing of *Sp185/333* mRNAs is also thought to contribute to diversity after their synthesis [21]. The deduced Sp185/333 polypeptides carry a predicted signal peptide, an N-terminal glycine-rich, a C-terminal histidine-rich region and patches of acidic residues [16]. The Sp185/333 proteins include up to six different types of tandem and interspersed repeats [19] and several conserved N- and O- linked glycosylation sites [16]. Almost all have an RGD motif and lack cysteine residues. *De novo* predictions point to the lack of discernible secondary and tertiary structures, including the absence of known functional domains. Sp185/333 proteins are localized to the cell surface of small phagocytes and are present in peri-nuclear vesicles of both small and polygonal phagocytes [22]. It has been speculated that Sp185/333 proteins may play a role in cell-cell interactions to form syncytia and initiate encapsulation of invading pathogens [22]. Although genome sequencing projects indicate the presence of *185/333* homologues in other sea urchins, only the *Sp185/333* gene family has been characterized to-date [15–20,22–25].

We report here the *He185/333* genes, transcripts and proteins in coelomocytes of the sea urchin, *Heliocidaris erythrogramma*. This family exhibits characteristic features of the *Sp185/333* family and the striking sequence diversity of the genes and proteins are common to both families. Full-length *He185/333* genes and transcripts possess element patterns, SNPs, short indels, as well as tandem and interspersed repeats. He185/333 proteins show a broad range of diversity in sizes and pIs and are expressed on the surfaces of some coelomocytes. Although *He185/333* and *Sp185/333* share many attributes, there are a few substantial differences between them. For example, *He185/333* sequences tend to be shorter than *Sp185/333* sequences. *He185/333* sequences also consist of element patterns that are not found amongst the *Sp185/333* sequences. There are also significant differences in intron structures, codon diversity, sequence repeats, amongst others.

Materials and Methods

All reagents were purchased from Sigma Aldrich or Amresco, unless otherwise indicated.

Sea urchins

All animals were collected according the rules indicated in the scientific collection permit (permit number F95/403–7.1) issued to Macquarie University by the NSW Department of Primary Industries, Australia. *Heliocidaris erythrogramma* specimens were collected at Clifton Gardens in Sydney Harbour and maintained at 22°C in the Macquarie University sea water facility in 50-liter tubs with continuous recycling of sea water from Sydney Harbor. Sea urchins were fed once a week with fresh algae that was collected from Sydney Harbour. All necessary permits were obtained for the described field studies.

Immune challenge and extraction of coelomic fluid

As described previously [24], between 0.5 ml and 25 ml of coelomic fluid was harvested from each animal and mixed with an equal volume of ice-cold calcium- and magnesium-free sea water (CMFSW-EI; 460 mM sodium chloride, 10.73 mM potassium chloride, 7.04 mM disodium sulfate, 2.38 mM sodium hydrogen carbonate) with 30 mM EDTA and 50 mM imidazole, pH 7.4. Coelomocytes were pelleted (1200–5000×g for 4 min at 4°C) and washed in CMFSW-EI before further processing. When required, sea urchins were challenged with 1.2×10^8 heat-killed *E. coli* cells (inoculum volume of 0.2 ml) per animal for 24 h or 48 h (for subsequent protein extraction and microscopy) prior to harvesting coelomic fluid.

Total RNA and DNA isolation

Coelomocytes collected from sea urchins 24 hr after challenge with *E. coli* were pelleted and extracted using TRI Reagent® (Molecular Research Center Inc.) for RNA and DNA isolation according to the manufacturer's instructions. The extracted RNA and DNA were each dissolved in 50 μl of nuclease free water. The quality of RNA was determined by formaldehyde agarose gel electrophoresis [26], while DNA quality was assessed by standard agarose gel electrophoresis. RNA quality was considered good when formaldehyde agarose gels showed 18 S and 28 S ribosomal RNA bands, without smears. DNA preparations that contained high molecular weight (MW) DNA, without smearing, were considered to be of good quality.

RT-PCR

RT-PCR was carried out with 0.05–3 μg of total RNA, using either Superscript III (Invitrogen) or PowerScript reverse transcriptase (Clontech) in conjunction with SpAncRT or SMARTIV and CDIII (Clontech) oligonucleotides (Table 1). All RT-PCR procedures were carried out in accordance with manufacturers' protocols. Both the 5′ and 3′ oligonucleotides used in reverse transcription reactions provided binding sites for primers used in PCR amplifications.

He185/333 PCR amplification optimization

PCR amplification of 185/333 sequences was carried out in 50 μl reactions using the Advantage 2 PCR system (Clontech) or Phusion DNA polymerase (Finnzymes). Each reaction consisted of 0.2 μM dNTPs, 0.4 μM of each primer (Table 1) and 1 μl of RT-PCR reaction or gDNA as template. The cycling parameters were as follows; for Advantage 2, an initial denaturation at 94°C for 2 min, followed by 35 cycles of denaturation at 94°C for 30 s, annealing at 60°C for 30 s and extension at 72°C for 3 min followed by a final extension at 72°C for 5 min. For amplification using Phusion DNA polymerase, the following conditions were used: an initial denaturation at 98°C for 1.5 min, followed by 35 cycles of denaturation at 98°C for 15 s, annealing at 60°C for 20 s and extension at 72°C for 20–40 s, followed by a final extension at 72°C for 3 min. Amplification products were analyzed by agarose gel electrophoresis.

Amplification strategy

Attempts to amplify *He185/333* sequences from *H. erythrogramma* using *Sp185/333* primers [16] were not successful. This was not unexpected, as the diversity of *185/333* sequences creates formidable barriers to amplification, especially when applied across species. Thus, a modified rapid amplification of cDNA ends (RACE) strategy was conducted (Fig. 1): Linker sequences at the 5′ (SMARTIV oligo) and 3′ (CDSIII) ends of the double stranded

Table 1. Oligonucleotides used in RT-PCR, PCR and sequencing.

Letter	Name	Function	Sequence (5'– 3')
A	SMARTIV Oligo	5' linker used in RT-PCR	AAGCAGTGGTATCAACGCAGAGTGGCCATTACGGCCGGG
B	CDSIII 3' PCR primer	3' linker used in RT-PCR and as generic 3' PCR primer	ATTCTAGAGGCCGAGGCGGCCGACATGd(T)$_{30}$N$_{-1}$N
C	SpAncRT	3' linker used in RT-PCR	ACTATCTAGAGCGGCCGC(T)$_{16}$V
D	SMARTIVPCRF	5' PCR primer, binds to SMARTIV Oligo	AAGCAGTGGTATCAACGCAGAGT
E	SpR1	3' PCR primer, binds to SpAncRT	ACTATCTAGAGCGGCCGCTT
F	185F5	*Sp185/333* forward primer	GGAACYGARGAMGGATCTC
G	185R6	*Sp185/333* reverse primer	GCAGCATCAGTTTCTTCKTCTC
H	MRHE1855'UTRF1	*He185/333* 5' UTR primer	GCTAGTTCTCTCTTGGAAGCTGGACGA
I	MRHE1855'UTRF4	*He185/333* 5' UTR primer	TCTCTTGGAAGCTGGACGAA
J	MRHE1855'UTRF2	*He185/333* 5' UTR primer	TGGAAGCTGGACGAAGGGAAAGGA
K	MRHE185R6	*He185/333* 3' UTR primer	TGGAAAAGACATCAGTGACA
L	MRHE185R9	*He185/333* 3' UTR primer	CTTTCAGGAATGTAATTGTCTTGAT
M	MRHE185R10	*He185/333* 3' UTR primer	CTGCAATTTTTCTACAAACTCA
	MRPGEMTM13F	Forward sequencing primer	CGCCAGGGTTTTCCCAGTCACGAC
	MRPGEMTM13R	Reverse sequencing primer	TCACACAGGAAACAGCTATGAC

The letters in the first column refer to the primer binding sites indicated in Fig. S1.

He185/333 cDNAs provided annealing sites for PCR primers. Primers (5' SMARTIV primer and 3' CDSIII or 3' SpR1) specific to the linker regions were used in combination with *Sp185/333* primers to generate partial *He185/333* sequences (5' and 3' ends) that in turn enabled primers specific to *He185/333* UTRs to be designed (MRHE1855UTRF1, MRHE1855UTRF2, MRHE5 UTRF4, MRHE185R6, MRHE185R9, MRHE185R10; see Table 1). By using these primers in PCR reactions, full length (i.e. complete ORFs) *He185/333* cDNA and gDNA sequences were amplified from *H. erythrogramma*. Those PCR primers were designed to bind to highly conserved nucleotide stretches in the UTRs and the leader/element 1 regions. Although this approach allowed us to identify more than 100 unique *He185/333* sequences (see Results), *He185/333* sequences that contain variant UTRs, if present, will not be identified. Despite this, the population of unique *He185/333* sequences that we identified was sufficiently large for this study.

Cloning and sequencing of PCR products

Amplicons from PCR reaction mixes or amplicons separated by agarose gel electrophoresis were purified with the PCR purification or QIAquick gel extraction kits according to the manufacturer's instructions (Qiagen) and eluted in 30 μl or 50 μl of 10 mM Tris-HCl, pH 8.0. The DNA was quantified by spectrophotometry. Purified DNA were ligated to pGEM T-Easy Vectors (Promega) overnight at 4°C and subsequently used to transform chemically competent One Shot® Top Ten (Invitrogen) or JM109 (Promega) *E. coli* cells according to standard protocols. Transformed bacterial culture (200 μl) were plated on pre-warmed LB-Agar plates with 100 μg/ml ampicillin, 50 μg/ml X-Gal and 0.5 mM IPTG, and incubated overnight at 37°C.

Plasmids were isolated from overnight cultures of *E. coli* using the QIAprep miniprep kit (Qiagen) according to the manufacturer's instructions. Purified plasmids were eluted in 50 μl of 10 mM Tris-HCl, pH 8.0 and evaluated by spectrophotometry and agarose gel electrophoresis. DNA inserts were sequenced at the Australian Genome Research Facility (AGRF, University of Brisbane, Queensland). Inserts were sequenced in both the forward and reverse directions with MRPGEMTM13F and MRPGEMTM13R sequencing primers (Table 1).

Analysis of DNA sequences

After removing vector and primer sequences computationally, the forward and reverse sequences for each cloned insert were aligned using the BioEdit sequence alignment editor [27]. Homology searches were performed using nucleotide BLAST [28]. Sequences homologous to *Sp185/333* were aligned with ClustalW2 (default parameters, nucleotide sequences: gap initiation penalty = 3, gap extension penalty = 1, base match score = 2, base mismatch penalty = 1; amino acid sequences: gap initiation penalty = 8, gap extension penalty = 2; see Refs. [29], [30]) and the alignments were further edited manually in BioEdit.

Immunofluorescence labeling and confocal microscopy

Coelomocytes were collected from sea urchins 48 hrs after challenge with E. coli. The density of coelomocytes in the final suspension was determined using a Neubauer Cell haemocytometer and adjusted to 1×10^6 cells/ml with CMFSW-EI (on ice). The cell suspension (100 μl) was pipetted onto glass slides and cells were allowed to settle and adhere to the slide surface for 5 min. Cells were fixed at 17°C in 0.5%–1% paraformaldehyde (PFA) in CMFSW-EI in a two-step procedure: (i) in 0.5% PFA for 15 min, (ii) 1% PFA for 15 min. After three washes in CMFSW-I, cells were permeabilized with 0.025% Triton X100 in CMFSW-I for 3 min followed by three washes for 5 min each. Non-specific epitopes were blocked with heat inactivated horse serum for 30 min at 17°C before the primary rabbit antisera mix was added for 1 h at 17°C. The primary antisera mix contained three polyclonal antibodies, anti-Sp185–66, anti-Sp185–68 and anti-Sp185–71 (each diluted 1:10,000 in CMFSW-I, kindly provided by L. Courtney Smith), which targeted the N-terminal-, central- and C-terminal- 185/333 regions [22]. The antibodies were raised in rabbits against synthetic peptides corresponding to those regions. The anti-Sp185/333 antibodies cross react with the He185/333

A

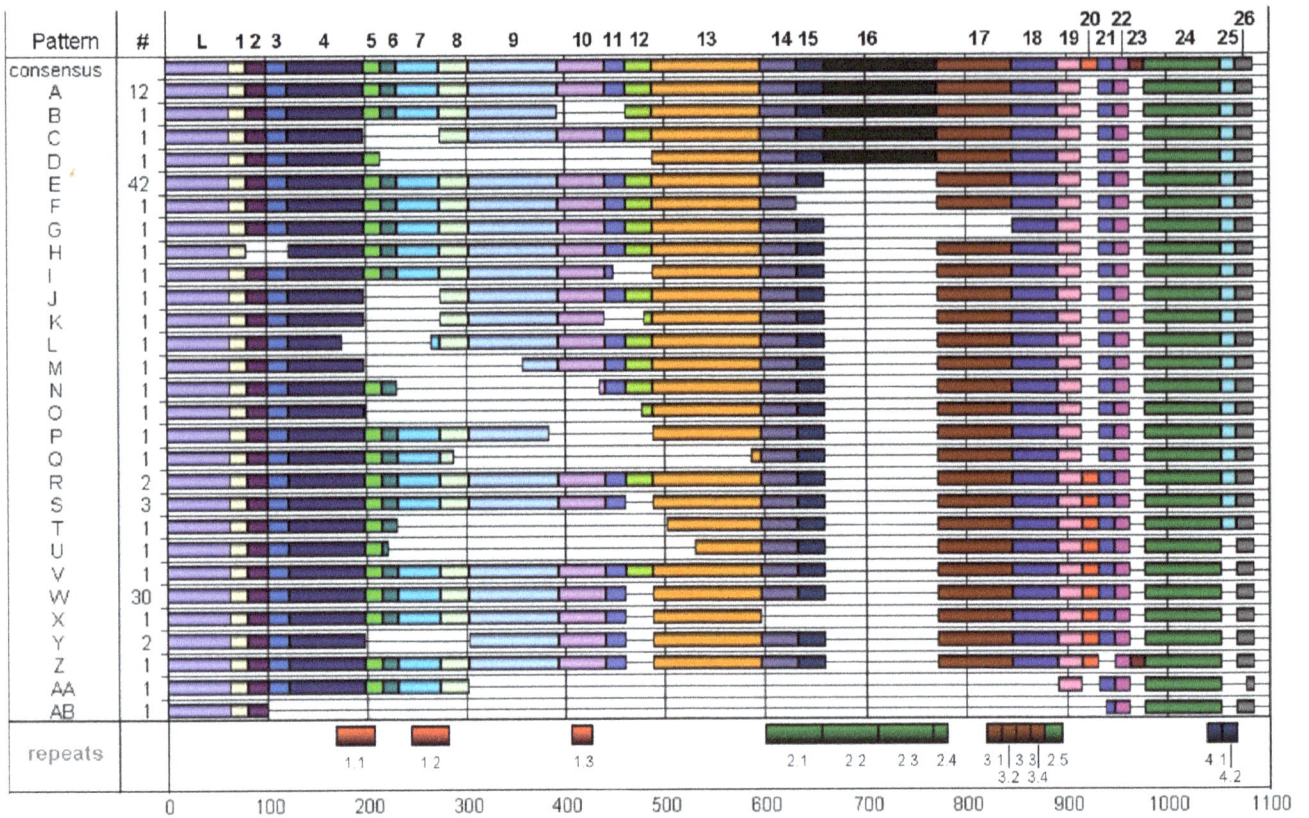

B

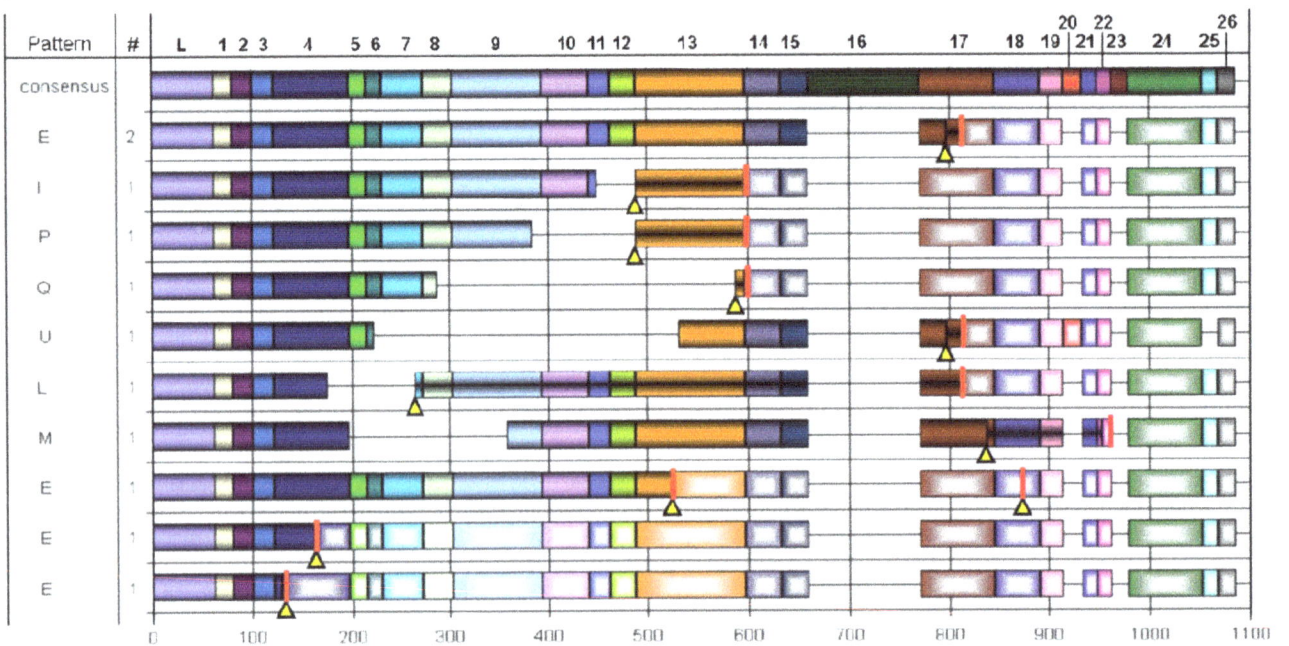

Figure 1. Element patterns of *He185/333* **cDNAs. A.** A total of 26 distinct elements (1–26 and leader (L) – see consensus pattern in the top row of the alignment) have been identified amongst He*185/333* sequences, and these are arranged into 28 unique element patterns. A horizontal line indicates elements that are "missing". The locations of sequence repeats are shown at the bottom of the figure. The numbers at the bottom of the alignments indicate nucleotide positions. **B.** Element patterns of *He185/333* cDNAs with mutations. Sequences with the element patterns E, I, P, Q, U, L, M contained insertions, deletions or point mutations, resulting in frame shifts and/or early stop codons. Mutations are marked by black lined yellow arrowheads, early stop codons as red bars across an element. Arrowheads falling together with bars are cases of stop codons caused by point mutations, as seen in the three lower element patterns. Arrowheads and bars that are located apart from each other are cases of insertions or deletions, resulting in frame shifts and downstream early stop codons. Elements between mutations and early stop codons contain black central lines

to point out missense translations, due to the frame shift. Elements after early stops are patterned with white centres to highlight non-translated regions of the cDNAs.

proteins as described [24]. After 3 washes, slides were incubated for 1 h in the dark at 17°C in a cocktail consisting of secondary antibody (mouse anti rabbit IgG-Alexafluor 546 conjugate), actin counterstain (phalloidin-Alexafluor 488 conjugate) and/or nuclear counterstain (Toto; all dyes from Invitrogen). Slides were washed three times in CMFSW-I. Finally, Biomeda Gel Mount (ProSci-Tech) was added, a coverslip overlaid and sealed with nail polish, and the slides stored at 4°C until analyzed on the Fluoview 300 laser scanning confocal microscope (Olympus).

Protein extraction

Total protein from coelomocytes was extracted and purified for subsequent SDS-polyacrylamide gel electrophoresis (SDS-PAGE). Total coelomic fluid (12.5–25 ml depending on the size of the sea urchin) was collected after removal of Aristotle's Lantern and coelomocytes were pelleted at 3000×g for 5 min at 4°C. The cell pellet was lysed in lysis buffer (5 ml; 7 M urea, 2 M thiourea, 1% 3-[4-heptyl]phenyl-3-hydroxypropyldimethylammoniopropane-sulfo-nate [ASB-C7BzO], 40 mM Tris base, 15 mM acrylamide [Bio-Rad], 10 mM tributyl phosphine [TBP], and 1x complete protease inhibitors [Roche]) using a French press (Thermo Spectronic). The lysate was incubated for 1 h at room temperature to allow complete reduction and alkylation of cysteines. Cell debris was pelleted by centrifugation for 10 min at 5000×g, the supernatant was transferred to a fresh tube and proteins were precipitated for 30 min at room temperature with five volumes of acetone and centrifuged for 10 min at 5000×g. The acetone supernatant was decanted, the pellet was air dried for 5 min and dissolved in 0.5–1.5 ml of sample buffer (7 M urea, 2 M thiourea, 1% ASB-C7BzO). Finally, the protein solutions were desalted using Micro Biospin tubes (Bio-Rad) and the eluates stored at −20°C.

Protein assay

Proteins were quantified using the Non Interfering™ Protein Assay kit (G-Biosciences) according to the manufacturer's instructions. BSA was used as the protein standard reference.

SDS polyacrylamide gel electrophoresis (SDS-PAGE)

SDS-polyacrylamide gels (10%) were loaded with 20 μg protein/lane and separated for an initial 10 min at 100 V followed by 80 min at 150 V and stained with Coomassie blue silver stain (20% methanol, 10% (w/v) ammonium sulfate, 1.6% orthophosphoric acid, 0.15% (w/v) Coomassie G250). Other gels were loaded with 10 μg proteins per lane followed by Western blotting (see below). Either Broad Range or Precision Plus protein markers (Bio-Rad) were used as size standards.

Western blotting and immunodetection

Proteins were blotted onto polyvinylidene fluoride membranes for 1 h at a constant current of 2 mA/cm² in a Transblot semidry transfer cell (Bio-Rad) followed by two washes in Tris buffered saline (TBS buffer, 20 mM Tris-HCl, pH 7.5, 150 mM sodium chloride) for 10 min each at room temperature, blocking of the membrane at 4°C overnight in TBS buffer containing 10% skim milk powder (block buffer), three washes at room temperature in TBS-Tween/Triton buffer (TBS buffer, 0.05% (v/v) Tween 20, 0.2% (v/v) Triton X-100) and incubation with the primary antibody (equal mix of anti-185/333 antisera diluted 1:10000 in

block buffer). This was followed by three washes in TBS-Tween/Triton buffer for 10 min each at room temperature and incubation in the secondary antibody (goat anti rabbit IgG Alkaline Phosphatase conjugate, 1:20000 in block buffer), followed by three final washes in TBS-Tween/Triton buffer. Protein-antibody complexes were visualized by colour reaction with 3 ml BCIP/NBT per 150 cm² membrane area for 2–5 min at room temperature and photographed with a digital camera (Canon EOS 40 d).

Phylogenetic analyses

Twenty-five randomly chosen 185/333 cDNA sequences from each species, H. erythrogramma and S. purpuratus, were aligned in ClustalW with default parameters and the alignment was further refined manually in BioEdit. Phylogenetic analyses were performed in PAUP*4.0b10 [31] using character based, distance-based and model-based (Maximum Parsimony, MP; Neighbor-joining, NJ and Maximum Likelihood, ML) methods of analysis. For MP analysis, a heuristic search strategy was employed to identify the most parsimonious tree. All characters were treated as unordered and un-weighted, while gaps were treated as missing data. Bootstrap re-sampling based on 1000 replicates was used to assess the support of relationships for the majority-rule consensus tree. For the NJ and ML phylogenetic analyses MODELTEST v.3.06 [32] was used to estimate the most likely model of sequence evolution for the sequence data. Based on the Akaike Information Criterion (AIC), Tamura-Nei (+G) was selected as the most likely model of sequence evolution for 185/333. Corrected genetic distances based on 2025 positions in the alignment were calculated in PAUP*4.0b10. NJ and ML trees were obtained in PAUP*4.0b10 using model parameters specified by MODEL-TEST and NJ was also assessed with 1000 bootstrap replicates.

Diversity analysis

The diversity of He185/333 sequences was determined from their alignments using the HyPhy suite of algorithms that were accessed via Datamonkey [33,34]. Unique, full-length He185/333 cDNA sequences were processed to remove 5′ and 3′ untranslated regions (UTRs). As there were 112 He185/333 sequences in our dataset, we customized our analytical approach to circumvent data processing restrictions on Datamonkey, which will only process a maximum of 100 sequences at a time. We developed a custom script to randomly select 100 He185/333 sequences from our dataset for the analysis. These sequences were aligned using ClustalW [29] as described above and uploaded to Datamonkey for diversity analysis. Each alignment was subjected to automatic nucleotide substitution model detection, generation of NJ trees and then SLAC (Single Likelihood Ancestor Counting, see Ref. [35]), FEL (Fixed Effects Likelihood, see Ref. [35]), IFEL (Internal Fixed Effects Likelihood, see Ref. [36]) analyses. The diversity scores were considered to be significant at a confidence interval of p≤0.1. The final diversity score for He185/333 sequences was the consensus of the data output from all three analytical algorithms (SLAC, FEL and IFEL). This was repeated a further nine times (i.e. a total of ten sets of sequences, each containing 100 sequences, were analyzed) and a consensus diversity score was generated. A similar analysis was conducted on 231 Sp185/333 cDNA sequences so that the data from the two families of 185/333 sequences could be compared.

Results

He185/333 cDNA sequences

A total of 112 unique full-length cDNAs were obtained. BLAST searches revealed significant homology of all sequences only to *Sp185/333* mRNAs and genes, with sequence identity of 68% to 74%. All cDNAs contained an open reading frame (ORF) of 219 to 1050 nucleotides, a Kozak consensus sequence (5′-CAGA-CATGG-3′; see Ref. [37]) and an in-frame stop codon.

Optimal alignment of the 112 *He185/333* revealed conserved sequence blocks or sequence elements (Fig. 1A and S2), a feature associated with *185/333* sequences. Each element has a conserved sequence and amongst the 112 He*185/333* sequences that were characterised in this study, a total of 31 different elements were identified. The shortest elements were 15 bp in length (elements 7, 11, 14, 16, 26, 27, 28 and 30), while the largest one was 111 bp long (element 21). Each *185/333* sequence is composed of a mosaic of elements, which is referred to as an element pattern. 29 distinctive element patterns were evident amongst the He*185/333* sequences (alphabetically labelled A–AC, Fig. 1A). In our library, all except for six element patterns were singletons; A (12 clones), E (41), R (2), S (3), W (30) and Y (2). Depending on the primers that were used, regions of up to 33 bp of the 5′ UTRs and up to 105 bp of the 3′UTRs were amplified (note: the location of the 5′UTR was based on the alignment of the *He185/333* sequences with *Sp185/333* sequences, especially with the region surrounding the initiating codon). One partial *He185/333* cDNA contained the 3′end of the ORF and an entire 3′UTR including the polyadenylation signal sequence (PAS, 5′-ATTAAA-3′) was located 185 bp downstream of the stop codon and 21 bp upstream of the poly A-tail (data not shown). Of the 112 cDNA sequences, 11 had early stop codons and/or frame shifts resulting from indels or point mutations (Fig. 1B) and encode truncated polypeptides, some with missense sequence.

He185/333 gDNA sequences

To date, 39 unique genes were sequenced, and these varied in length between 1261 bp and 2301 bp (Fig. 2 and S3). They consisted of two exons, the first and shorter of which was 55 bp long (excluding the putative 5′ UTR, of which 33 bp were amplified). The size of the larger second exon ranged from 749 bp to 905 bp (excluding the 3′ UTR, of which either 30 bp or 105 bp were amplified, depending on the primers used in PCRs). The intron ranged in size from 457 bp to 1392 bp. The element patterns in the second exon matched the cDNA element patterns A, E, I, W and Y (Fig. 2B). An additional element pattern, AC, which was not identified in the library of cDNA sequences was found among the gDNA sequences. Element patterns were also evident in the introns (Fig. 2B). A total of 10 intron elements (i1–i10) were identified and these were in four recognizable intron element patterns (alpha (α), beta (β), gamma (γ) and delta (δ)). The intron patterns, when combined with exon patterns, form nine gene element patterns; E-α, AC-α, W-α, Y-α, E-β, R-β, W-β, E-γ and W-δ.

Deduced polypeptide sequences

The deduced polypeptides had between 72 and 349 aa with predicted MW ranging from 8 to 39 kDa and the predicted isoelectric points (pIs) ranged from 4.63 to 6.99. The first 21 aa represented a predicted signal sequence (see Ref. [38] and Fig. 3), suggesting an extracellular destination of the He185/333 proteins. At the same time the He185/333 proteins were predicted to lack both transmembrane regions [39] and canonical signatures of glycosylphosphatidylinositol-anchors [40]. The deduced polypep-

tides also contained up to ten potential N- and one O-linked glycosylation sites [41] and a total of 16 potential serine-, two threonine- and one tyrosine-phosphorylation sites were predicted [42]. The translated sequences included an N-terminal glycine-rich and a central histidine-rich region. The predicted polypeptides were also rich in arginine, which were evenly distributed along the polypeptide and typically constituted 11%–12% of total number of amino acids. The histidine rich region contained a poly-histidine stretch of at least six, but usually more (8 to 13) consecutive residues. Common web-based programs did not predict extensive secondary structures or folding patterns.

Repeats

We found four types of tandem and interspersed sequence repeats in 104 of the 112 185/333 deduced polypeptide sequence, named types 1 to 4 (Fig. S4 and S5) and were comparable to five repeat types in *S. purpuratus* (Fig. S6). Eight sequences that contained non-synonymous substitutions and frameshift mutations (sequences He_185/333_cDNA_105–112) were omitted in the analysis of sequence repeats because major parts of their deduced amino acid sequence were not homologous to 185/333 proteins. Most of the He185/333 repeats were homologous to repeats found in *S. purpuratus* but varied in the maximum repeat copy numbers and in their length.

Phylogenetic analysis

As *185/333* sequences are members of a diversifying gene family [15–19,21], the identification of these sequences in a second species (*H. erythrogramma*) enabled us to compare the sequence similarity amongst *185/333* sequences from these two animal groups. Analysis of the phylogenetic relationships between unique *He185/333* and *Sp185/333* cDNAs indicates that the sequences clustered according to the species from which they derived (Fig. 4). Of the 2025 bp of the aligned sequences, 418 characters were variable and 287 were parsimony informative. The Tamura-Nei corrected genetic distances ranged from 0.00216 to 0.07705 for *He185/333* and from 0.00115 to 0.12781 for *Sp185/333* (Table S2). The genetic distances between *He185/333* and *Sp185/333* were significantly greater and ranged from 0.27949 to 0.38071. Moreover, groups of *185/333* within each species clustered separately from one another in branches with well-supported bootstrap values, indicating presence of subfamilies that may have originated by duplication and divergence from a common founding member.

Within the *H. erythrogramma* clade, there was a correlation between the sequence clusters and the element patterns of the *He185/333* sequences within those clusters. Three distinct clusters were evident: one cluster consisted exclusively of sequences with exon element pattern W (sequences HE085, HE045, HE041, HE037 and HE033), while the second contained sequences with element patterns E, F, M and O (sequences HE061, HE054, HE081, HE078, HE049, HE029, HE017, HE013, HE005, HE021, HE009, HE001 and HE057). The third cluster, which was most diverse in terms of the element patterns, contained the patterns A, C, G and S (sequences HE097, HE093, HE089, HE073, HE069, HE065 and HE025).

Diversity analysis

A total of 112 unique *He185/333* cDNAs were analyzed using the HyPhy suite via Datamonkey to detect diversity and selection pressure at the level of individual codons (Fig. 5, and table S3). The consensus output from SLAC, FEL and IFEL identified 17 codons (4.6%) that were under negative selection, while nine codons (2.5%) were under positive selection ($p < 0.1$). In compar-

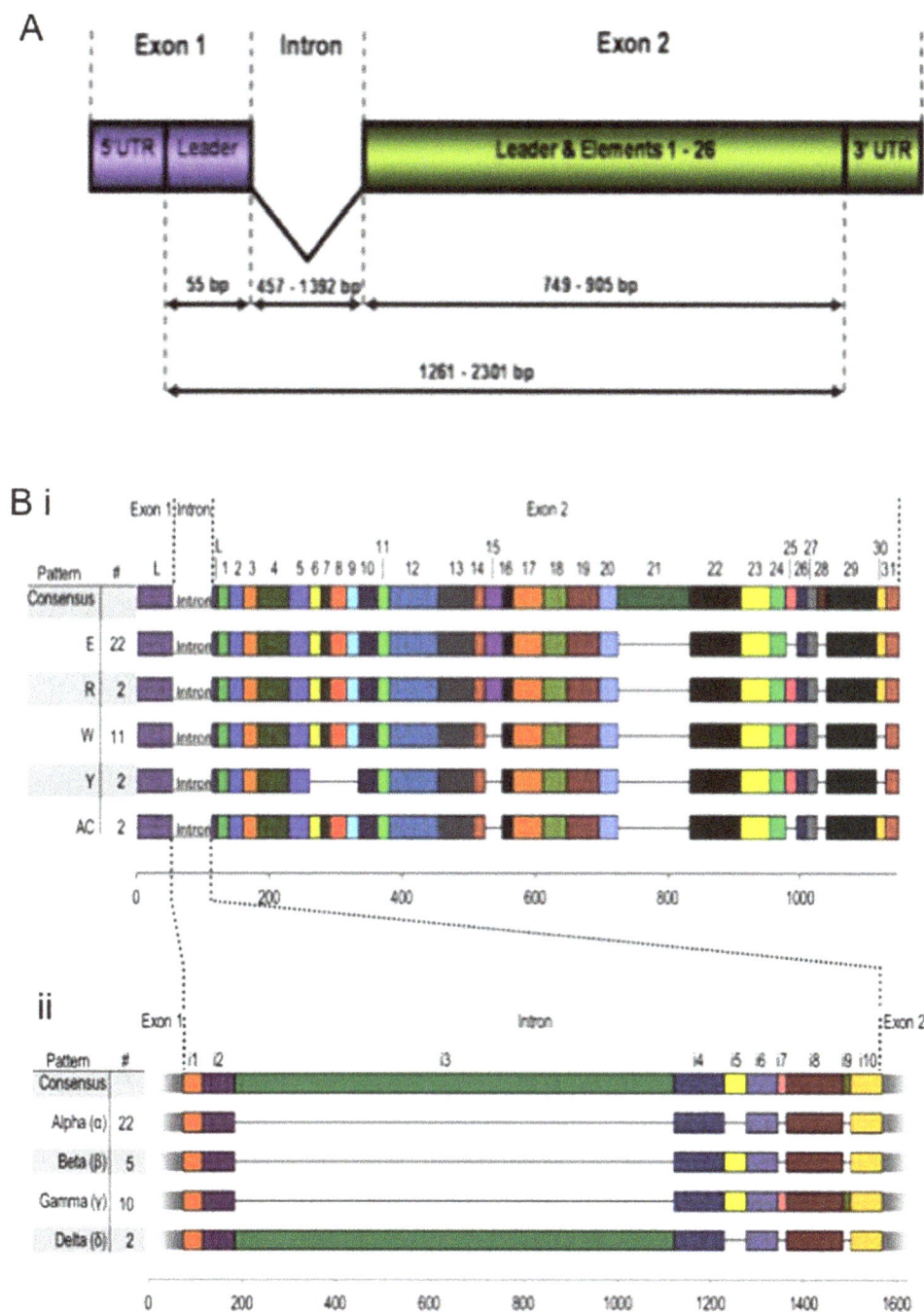

Figure 2. *He185/333* **gene structure. A.** Genes consist of a short exon 1 and a longer exon 2, separated by one intron. The leader is 63 bp in length and is interspersed between positions 55/56 by the Intron. Thus, the leader represents exon 1 (excluding the 5′UTR) but also forms the first eight nucleotides of the second exon. The intron varies in size between 457 bp and 1392 bp. The second, long exon ranges from 749 bp to 905 bp (excluding the 3′UTR) and contains the mosaic organisation of elements. The 5′ and 3′ UTRs of *He185/333* genes have only been partially sequenced. **B. Element patterns of 39 manually aligned** *He185/333* **gDNA sequences. i.** The coding regions (exons 1 and 2) are interspersed by one intron, dividing the leader in two parts, as indicated at the top of the diagram. Genes identified to-date show the five exon element patterns E, I, R, W and Y as described for cDNAs in figure 1A. Introns are simplified by interrupted, checkered boxes. The consensus at the top of the diagram is based on all element patterns, including cDNA element patterns as shown in figure 1A. **ii.** Similar to exons, introns align optimally with insertion of large gaps, resulting in ten intron elements and four intron element patterns, designated alpha (α), beta (β), gamma (γ) and delta (δ). Individual intron elements are named i1 to i10 and are shown as differently shaded gray boxes. Exons are represented as fading extensions to the left and right of intron element patterns. **i & ii.** Combinations of exon and intron element patterns define gene element patterns of which the following nine have been identified among the 39 gene sequences (number of individual sequences with according gene element pattern in brackets: E-α (11), I-α (2), W-α (8), Y-a (1), E-β (2), R-β (2), W-β (1), E-γ (10), W-δ (2). The first element pattern is the consensus.

Amino acid position	0		100			200			300		
Polypeptide	N-signal		glycine-rich			histidine-rich					-C
Pot. phosphorylation	TS	SS	S	S	S	SSSS	S	S	S TY S	S	S
Pot. glycosylation						N	N	N N N N	N	N	N O
Repeats		1.1	1.2	1.3		2.1	2.2	2.3 2.4	3.1 3.3 2.5 / 3.2 3.4		4.1 / 4.2

Figure 3. Structural overview of deduced consensus of He185/333 polypeptides. Approximate amino acid positions are given at the top, amino/carboxy-terimini are labeled with N and C to the left and right of the consensus, respectively. Polypeptides carry a signal sequence at the N-terminus, followed by glycine-rich and histidine-rich regions. Potential phosphorylated serine, threonine and tyrosine residues are marked with S, T and Y respectively and potential N- and O-linked glycosylations with N and O, respectively. Repeats, as shown for cDNA element patterns in figure 2A and in table S2, are shown at the bottom and are labeled with repeat type followed by copy number.

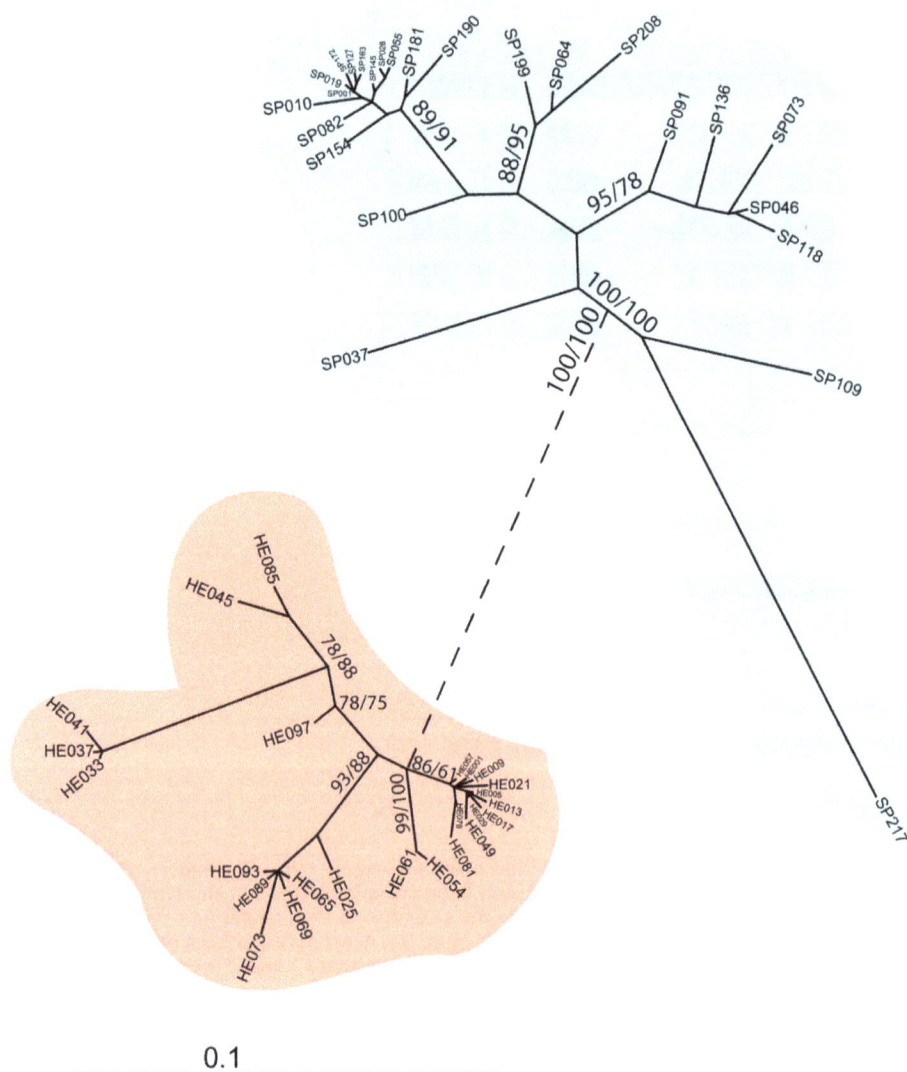

Figure 4. Phylogenetic relationship between 25 *He185/333* and *Sp185/333* cDNAs. *He185/333* clades are highlighted in a darker shade than *Sp185/333* clades. The phylogenetic tree shows clearly that 185/333 sequences cluster according the species they originated from. The main branch separating the two clades was shortened for display purposes (dashed line). It is representative of the corrected genetic distance of 0.27949 and is supported by 100% of bootstraps. Hence, all sequences within one species had lower genetic distances to each other (highest value 0.07705 for *He185/333* and 0.12781 for *Sp185/333*, respectively) than to any sequence from the opposite species. Boot Strap values are shown next to the branches with the ordering NJ/MP. Values below 50 are not displayed.

ison, the *Sp185/333* sequences had 64 codons (11%) under negative selection and 15 codons (2.6%) under positive selection. While positively and negatively selected sites were evenly distributed along the length of the *He185/333* alignments, 39 of the 64 negatively selected sites were located within the first 200 codons of the 581 codon *Sp185/333* alignment, while only six sites were located in the last 200 codons. Positively selected sites also appeared to cluster within the first 200 codons, with eleven of the 15 positively selected sites present.

He185/333 protein expression profile

Western blots identified a broad range of MW for He185/333 proteins, ranging from expected values for monomers deduced from cDNA sequences to MW greater than 206 kDa (Fig. 6A). The majority of He185/333 positive bands were greater than 75 kDa, and their MW were not altered by the strong reducing agent TBP. There were considerable variations in the repertoires (relative sizes and intensities on Western blots) of He185/333 proteins between different individual sea urchins (Fig. 6A). Such variations in He185/333 protein repertoires were also evident within individuals after they were injected with heat-killed bacteria, filtered sea water or sterile needle injury (Fig. 6A).

Interestingly, He185/333-positive bands on 1DE Western blots could not be associated with specific bands on Coomassie-stained gels. In fact, strong He185/333 signals on Western blots were associated with non-stained areas on Coomassie gels (Fig. 6A and 6B). Dheilly et al. also noted similar observations of Sp185/333 proteins [24]. This is thought to be due to the chemistry of protein-dye interactions: the post-translational modifications of the 185/333 proteins are believed to interfere with Coomassie blue interacting with those proteins, resulting in a poor alignment of Coomassie stained 185/333 proteins and the corresponding bands on Western blots (see Ref. [24] and asterisks in Fig. 6A and 6B).

2DE Western blots showed a diverse range of MW and isoelectric points (data not shown), and were very similar to those of Sp185/333 proteins [24]. Disparities between observed MW and pI and those predicted from He*185/333* sequences were observed. This suggests that He*185/333* proteins are post-translationally modified. Furthermore, the protein spots were arranged in trains, an effect often associated with differential glycosylation [24].

Localization of He185/333 proteins in coelomocytes

Immunofluorescence confocal microscopy showed that He185/333 expression was present in two distinct coelomocyte subpopulations: filopodial and lamellipodial amoebocytes (Fig. 7). These coelomocytes are morphologically and functionally (phagocytosis)

equivalent to the corresponding cell types in *S. purpuratus* [25,43]. Z-stack analysis of the images indicated that the majority of the He185/333 proteins were located on the plasma membranes (data available on request). In some instances, He185/333 proteins were detected in perinuclear areas of large filopodial cells (Fig. 7C), which may reflect their distribution within the organelles involved in protein biosynthesis (e.g. endoplasmic reticulum, Golgi apparatus and transport vesicles). He185/333-specific immunofluorescence on amoebocyte surfaces appears to be patchy (Fig. 7A–D), suggesting that the distribution of He185/333 proteins may be tightly clustered. This was particularly evident for filopodial amoebocytes, where distinct knobs of He185/333-associated fluorescence were observed (Fig. 7A).

Discussion

We describe the discovery and characterization of the *He185/333* gene family in the sea urchin, *Heliocidaris erythrogramma*, and present characterization of this gene family in a second group of sea urchins, after its original discovery in *S. purpuratus* [15,23]. Sequences that are homologous to *Sp185/333* have been identified in other sea urchin genome sequencing projects, including *S. franciscanus* and *A. fragilis* (see Ref. [23] and http://www.ncbi.nlm.nih.gov/genome/?term=strongylocentrotus+purpuratus). All evidence to-date indicate that the *185/333* gene families are unique to the sea urchins, as no homologues have been identified in other animal groups.

Similarities between *He185/333* and *Sp185/333*

Our analysis shows that *He185/333* sequences bear many similarities to those from *S. purpuratus*. These include (i) the organization of *185/333* cDNA and gene sequences into elements, (ii) sequence variations due to indels, SNPs and repeats, (iii) structure of *185/333* genes (two exons separated by an intron), (iv) structural features of the predicted 185/333 polypeptides (a hydrophobic leader, a glycine rich and a histidine rich region and several potential N- and O-linked glycosylation sites), (v) a large repertoire of 185/333 polypeptide sizes (from ~39 kDa to >206 kDa), as well as differences in the repertoires of 185/333 polypeptides between individual animals, (vi) changes in the repertoire of 185/333 polypeptides upon immunological insult or injury, and (vii) cell surface and peri-nuclear expression of 185/333 proteins in amoebocytes. These similarities suggest that He185/333 and Sp185/333 are closely related molecular families that bear substantial structural and functional similarities.

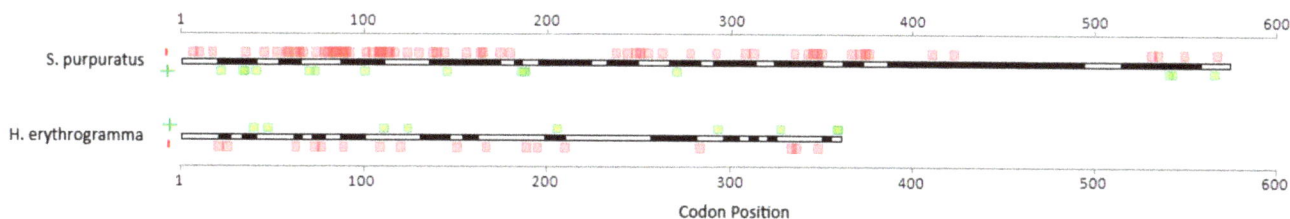

Figure 5. A comparison of codon diversification between *S. purpuratus* and *H. erythrogramma* 185/333 cDNA alignments. Sequence alignments of 231 *S. purpuratus* and 112 *H. erythrogramma* cDNAs were used to investigate recombination and selective pressure at the codon level. Element patterns are shown as black and white banding, while positively- and negatively-selected codons are shown as green and red boxes above and below each alignment, respectively. Negative selection (against codon diversification) appears to be prevalent across *185/333* sequences from both species. Significant (p<0.1) negative selection was detected in 64 codons across the 581 codon-long *S. purpuratus* alignment, while 15 codons were positively-selected. The 360 codon-long *H. erythrogramma* alignment contained 17 codons under significant negative selective pressure and nine codons with positive selective pressure. *S. purpuratus* element patterns adapted from [16].

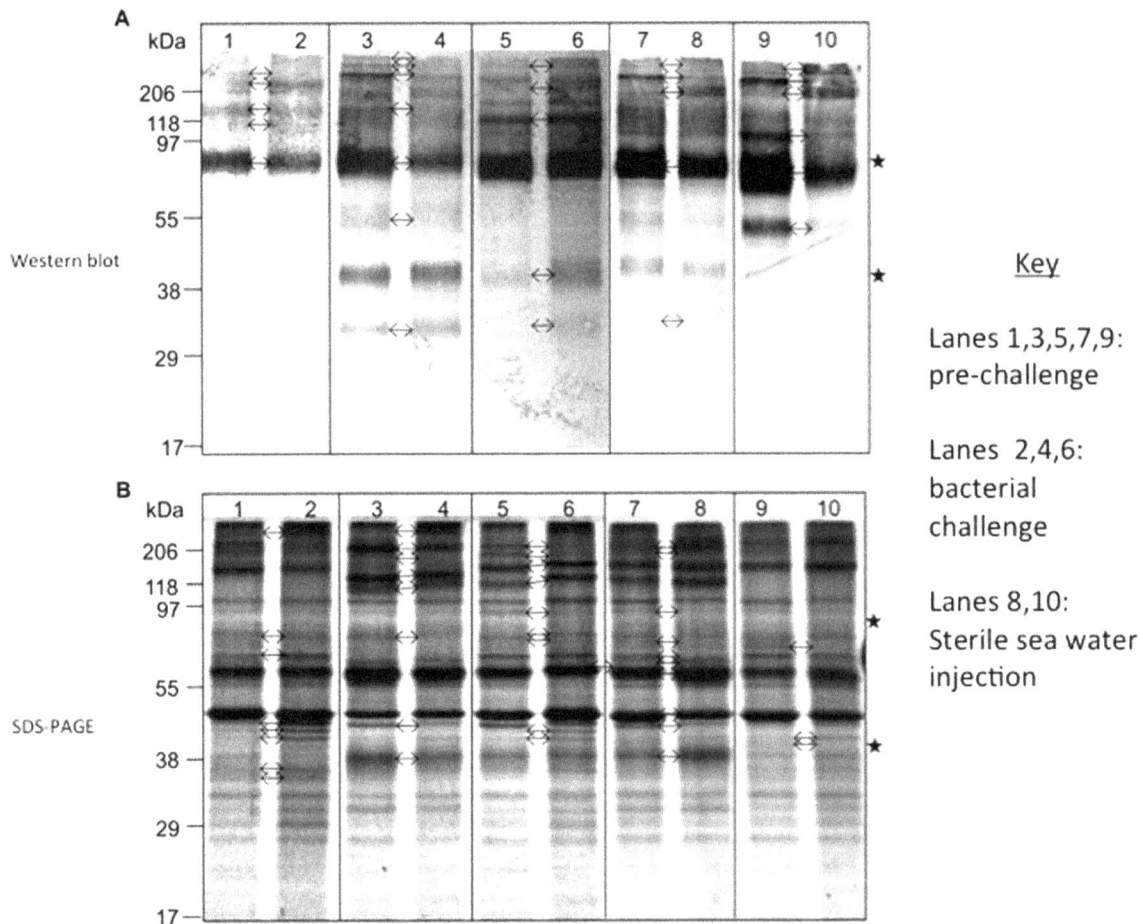

Figure 6. Total coelomocyte proteins from five sea urchins were analysed by Western blotting (A) and SDS-PAGE (B). Coelomocyte proteins were extracted from animals either before (pre-) or after (post-) immunological challenge with heat killed bacteria. Other treatments included injections with filtered sea water (FSW) and injury (pricked with a sterile needle). In both panels A and B, lanes with odd numbers (1, 3, 5, 7 and 9) show pre-challenged protein profiles while lanes with even numbers (2, 4, and 6) show post-challenge protein profiles. Lanes 8 and 10 show the profiles after filtered sea water injection and Injury, respectively. Asterisks to the right of the figures indicate regions that are not stained with Coomassie Blue (B), which contain He185/333+ bands in the Western blot. **A.** The Western blot shows a diverse pattern of He185/333 proteins between animals but also changes within individuals before and after immunological challenge, FSW injection and injury. Bands on the blot are not as discrete and sharp as their corresponding bands on the Coomassie Blue stained gel, but appear to be rather diffuse and large. Arrows between pre- and post-challenged samples indicate bands that change in intensity or are present/absent as a result of the experimental treatment. **B.** The Coomassie Blue stained gel shows discreet, sharp protein bands, some of which differ in size and intensity between animals and within individuals before and after immune challenge. None of the bands, however, could unambiguously be identified as He185/333+ band when compared to the Western blot.

Differences between He185/333 and Sp185/333

Although both Sp185/333 [15,16,18] and He185/333 sequences are composed of elements (conserved sequence blocks), the nucleotide sequences of elements, as well as the number of elements in each gene family, are very different. Hence, an independent element pattern system had to be developed for He185/333, which resulted in 29 cDNA element patterns with 31 elements as opposed to 38 patterns with 25 elements in Sp185/333 [16,17]. Also, although all four repeat types in H. erythrogramma find their homologous counterparts in S. purpuratus, the overall repeat lengths and maximum copy numbers vary between the two species [19].

Phylogenetic analysis shows that despite all similarities, the sequences cluster into two groups that are defined by the species from which they originate. Hence, although there are clear homologies between Sp185/333 and He185/333 the distinct clustering of the two groups suggests independent evolution of the two gene families after the divergence of the host animals. Although extensive divergence is evident within He185/333 sequences and Sp185/333 sequences, there is also a clear distinction between these homologous sequences in the two groups of animals. Hence, although there are clear homologies between Sp185/333 and He185/333 the distinct clustering of the two groups suggests independent evolution of the two gene families after the divergence of the host animals. It is estimated that the Strongylid sea urchins diverged from Heliocidaris approximately 35 MYA [47]. These groups of sea urchins are geographically isolated: H. erythrogramma is predominantly found in the southern hemisphere, while Strongylocentrotus' habitats are located in the northern hemisphere [46–47]. Although it is not clear if the differences in their developmental life histories contribute to the differences that exist in the 185/333 homologues from these species (S. purpuratus is an indirect developer, while H. erythrogramma is a direct developer), it is reasonable to speculate that the

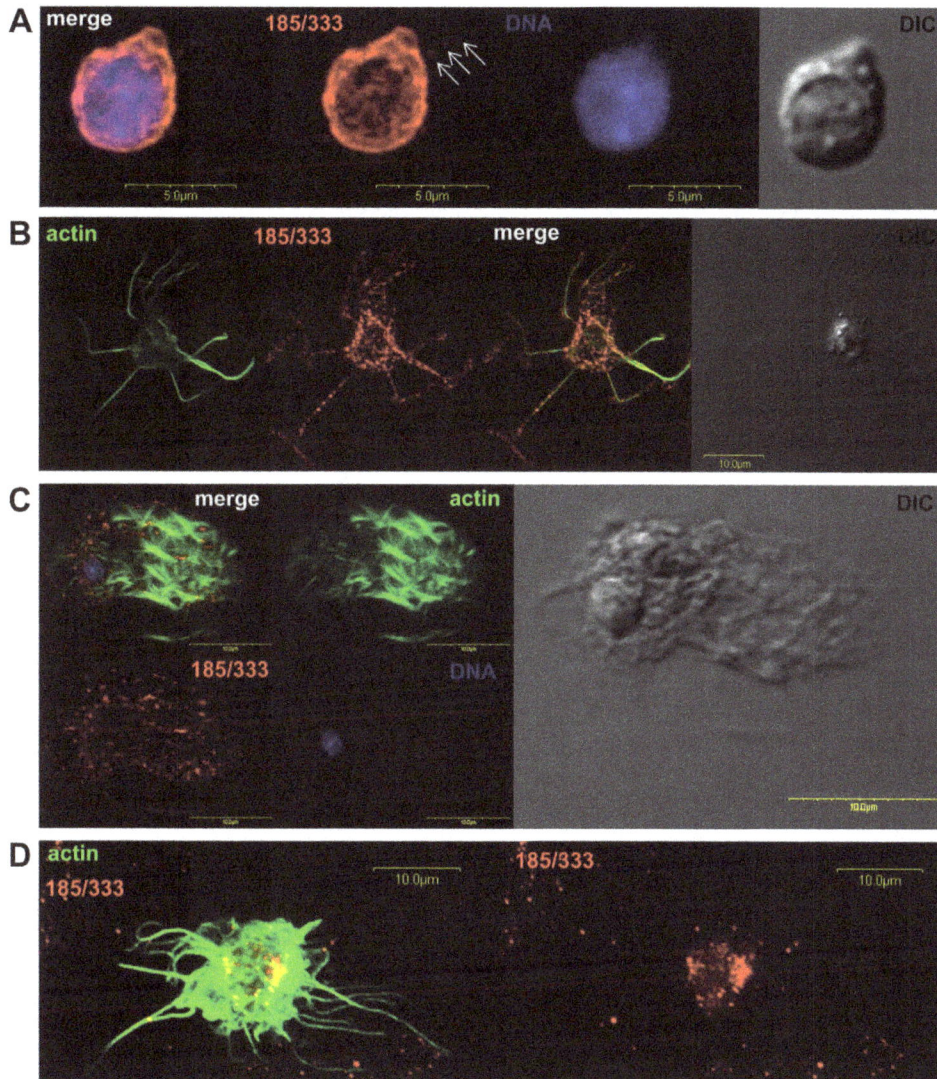

Figure 7. Cellular expression of He185/333 proteins. A–D. Immunofluorescence and differential intereference contrast (DIC) microscopic images of different coelomocyte types expressing He185/333 proteins (red) and actin (green). Nuclear DNA appears in blue. **A.** A small filopodial amoebocyte expressing *He185/333* proteins. *He185/333* staining is found on the cell surface and the clustering of *He185/333*-associated fluorescence in knobs is evident in a filopodium (white arrows). **B.** A large filopodial amoebocyte expressing *He185/333* proteins in dense knobs. *He185/333* signals are not uniformly distributed but are found in patches. **C.** A large filopodial amoebocyte expressing *He185/333* within the cell body in perinuclear areas. **D.** A lamellipodial amoebocyte expressing *He185/333* as knobs on the cell surface.

pathogen pressures that are present in their respective habitats may have driven the evolution of *185/333* sequences in these two groups of sea urchins. Experimentally, this was also evident from the difficulties in our initial attempts to amplify He185/333 sequences in H. erythrogramma based on Sp185/333 sequences. Experimentally, this was also evident from the difficulties in our initial attempts to amplify *He185/333* sequences in *H. erythrogramma* based on *Sp185/333* sequences. The analysis of diversity within *He185/333* and *Sp185/333* cDNA sequences revealed a number of similarities as well as several differences between the two families. In both families, a greater proportion of codons were under negative selection than under positive selection. Negatively-selected codons represent those sites that are invariant and are likely to encode amino acids that mediate critical structural or functional roles [36]. In both families, negatively-selected codons occurred throughout the lengths of the sequences. In *He185/333*, positively-selected sites are also distributed throughout the

sequence, whilst in *Sp185/333*, the majority of the positively-selected codons are located in the first half of the sequence. The significance of this is not clear. In concurrence with previous studies, our data support the notion that 185/333 polypeptides lack distinct 'hypervariable' regions. Our data, the combined result of three separate diversity analyses, indicate that pressure to diversify (positive selection) or conserve (negative selection) is not evident at the level of individual sequence elements (p<0.1). In their paper, Buckley et al. have suggested that RNA editing (deamination) may be a potential source of sequence diversification [21]. While we have not looked for evidence of RNA editing in our study, our demonstration of codon-level selection supports this notion.

Taken together, the *185/333* sequences from both *S. purpuratus* and *H. erythrogramma* appear to be under positive selection for diversification, although the selective pressures that drive this diversification are not known [15–21,23]. Another, albeit unrelated, gene family in the sea urchins also undergoes positive selection and diversification. The bindin proteins on sea urchin sperm cells are involved in species-specific fertilisation. Molecular analyses indicate that the regions flanking the highly conserved core region of the *bindin* genes accumulate point mutations and indels. Phylogenetic analysis of the evolutionary rates of the bindin genes from several sea urchin species indicates that *Strongylocentrotus* and *Heliocidaris bindin* alleles have undergone rapid evolutionary divergence, compared to the other sea urchins (approximately four times greater in these species, compared to the others) [47–48]. It will be interesting to compare the evolutionary diversification rates between *185/333* genes with *bindin* genes, as well as those of the other immune response gene families.

Overall, our data indicated that *He185/333* cDNA sequences are less diverse than their *Sp185/333* counterparts; however, this may be because the set of unique *He185/333* cDNA sequences was smaller than the set of *Sp185/333* sequences. The *He185/333* cDNA sequences were obtained solely through RT-PCR experiments, implying that biases arising from the use of PCR primers may have reduced our ability to amplify diverse repertoires of *He185/333* cDNA sequences. *Sp185/333* sequences were initially obtained from screening of cDNA libraries, which enabled primer design for the subsequent production of RT-PCR amplicons that were cloned and sequenced [15–17].

While *Sp185/333* [18] and *He185/333* genes share structural similarities, there are major differences in their intron sequences. Firstly, the conservation of sequences between *Sp185/333* and *He185/333* introns is low (identity <50%) and BLAST searches using *He185/333* intron sequences, as queries do not match to *Sp185/333* gene sequences. Secondly, intron-types in *He185/333* genes are structurally different from those in *Sp185/333* and consist of elements, which enable their classification into intron-element patterns (designated as α, β, γ, and δ). In contrast, *S. purpuratus* introns (α, β, γ, δ, and ε) are defined by sequence dissimilarities and SNPs as based on phylogenetic analysis because the alignment of the *Sp185/333* introns is not improved by the insertion of gaps to identify elements [18]. Last, some *He185/333* genes carry unusually long introns of more than 1300 bp. This is approximately 2–3 times the average length of most *He185/333* and all *Sp185/333* introns and exceeds the combined lengths of *He185/333 coding regions of* exons 1 and 2 by several hundred base pairs. Introns as large as this have not, as yet, been identified in *S. purpuratus*.

In general, homologous stretches of sequence between *Sp185/333* and *He185/333* were not well conserved. On average, sequence identity of corresponding sequence stretches was about 70–80%, depending on the length of sequence that was compared. However, there were regions that were highly conserved: the leader sequences (first ~55 bp of the open reading frame) were relatively well conserved between *Sp185/333* and *He185/333* (approximately 90% sequence identity at the amino acid level). Some well-conserved blocks of sequence, ranging from 10 to 15 amino acids showed identities of up to 100% between the two species. These sequence blocks often corresponded to regions of repeated sequences.

None of the predicted He185/333 polypeptide sequences contain RGD motifs. In Sp185/333 sequences, most of the predicted polypeptides have a single RGD motif [15,16] but several Sp185/333 sequences that lack this motif have also been identified [17]. RGD motifs serve as binding sites for integrins [44], a family of plasma membrane anchor proteins which interact with cytoskeleton–connecting proteins and are involved in cell adhesion and signal transduction [45]. Assuming that both Sp185/333 and He185/333 subserve the same general immunological

functions, it seems unlikely that RGD motifs in 185/333 proteins have fundamental functional significance. Similarly, the fact that cysteine residues are found in He185/333 proteins but are absent in all full-length Sp185/333 proteins (they have been predicted in missense sequences, see Ref. [24]), implies that disulphide-mediated tertiary or quaternary interactions in He185/333, if they do occur at all, may be functionally irrelevant. However, it is possible that such differences represent species-specific adaptations as a consequence of immunological or evolutionary pressures that are specific to each host. Our data indicate that disulphide bonding does not stabilize the high MW forms of He185/333. The presence of high MW *He185/333* proteins even in the presence of the strong reducing agent TBP suggests that they are likely to be non-disulphide stabilized, but covalently-linked oligomers. Similar discrepancies in the sizes of recombinantly-expressed Sp185/333 proteins were also identified [22], again suggesting that oligomerization occurs between Sp185/333 proteins. Since other sea urchin proteins were not present in the recombinant expression system, it is plausible that 185/333 proteins self-oligomerize to form higher-order structures.

In summary, *He185/333* and *Sp185/333* are homologous immune gene families that share many common features. They also demonstrate sufficient differences to suggest that the gene families have undergone independent evolution after the divergence of the host species. The data provided in the manuscript will contribute to our understanding of the evolution of this gene family that appears to be unique to the echinoderms.

Supporting Information

Figure S1 RACE-PCR amplification of *He185/333* sequences. Oligonucleotides are represented as arrows and are labeled A-M. Messenger RNA was reverse transcribed (RT) into double stranded cDNA with addition of oligonucleotide linkers (A, B or C) to the 5′ and 3′ ends. Subsequent PCRs with the cDNA used linker primers (D, B or E) in combination with Sp185/333 specific primers (F, G) to produce partial *He185/333* sequences corresponding to 5′ and 3′ ends, including parts of UTRs to which *He185/333* specific primers were designed. PCRs using these *He185/333* UTR-primers (H-M) resulted in the amplification of full length *He185/333* sequences from cDNA and gDNA templates.

Figure S2 Nucleotide sequence alignment for 112 *He185/333* cDNAs generated in Clustal W and BioEdit. The best alignment was obtained by insertion of large gaps, resulting in 26 sequence blocks (elements). The 26 elements are numbered along the top and separated by vertical black lines. The first and last three nucleotides represent the start and stop codon, respectively. The alignment shows full-length cDNAs irrespective of mutations that lead to missense and early termination upon translation. The cDNAs with such mutations are translated accordingly in Fig. S4.

Figure S3 Nucleotide sequence alignment for 39 *He185/333* gDNAs generated in Clustal W and BioEdit. The untranslated regions (5′UTR and 3′UTR), leader, 26 exon elements and ten intron elements (with suffix "i") are labeled along the top and separated by vertical black lines. The start and stop codons are shaded in green and red coloured boxes, respectively and labeled along the top. The exon elements have been numbered according to the categorization system based on the 112 cDNA sequences shown in Fig. S2. The leader, which is identical to the predicted signal sequence, counts 63 nucleotides but is separated by the intron between nucleotide positions 55 and 56.

Figure S4   Alignment of 112 deduced He185/333 polypeptides generated in Clustal W and BioEdit. The polypeptide sequences were deduced from the cDNA sequences shown in Fig S2. The 26 elements are numbered along the top and separated by vertical black lines. Four types of repeats with tandem or interspersed-incomplete structures are highlighted in differently coloured boxes and labeled along the top. Also indicated along the top are glycine- and histidine-rich regions (orange and magenta arrows, respectively), predicted O-linked and N-linked glycosylation sites (red and blue triangles, respectively), as well as predicted serine, threonine and tyrosine phosphorylations (red, blue and green lightning bolts, respectively).

Figure S5   Repeats found in 104 translated *He185/333* cDNA sequences, presented as sequence logos and linear sequences. The figure shows repeat types and maximum copy number (left column), as well as sequence variations within repeats (central column) and the structure of each repeat type (right column). For example, there are up to three copies of the type 1 repeat (1.1, 1.2 and 1.3) in He185/333 deduced polypeptides. Sequence variations within repeats are depicted as sequence logos and as plain text. Sequence logos were generated using the software, Geneious (Geneious v5.4, http://www. geneious.com). The size of each letter within the sequence logos is proportional to the frequency of that residue at the specific position in the He185/333 alignment. For example, the first position of repeat 1.1 is G (glycine), with a value of 1, because it is invariant at that position amongst the 104 He185/333 sequences. In the plain text below the logos variant amino acids at specific positions are indicated in brackets.

Figure S6   Comparison of the four types of repeats from *He185/333* translated sequences with five *Sp185/333* repeat types (15). The number of copies of each type of repeat varies between the two species and among sequences within one species. For example, type 1 repeat is present as two complete tandem repeats and one interspersed incomplete repeat in *H. erythrogramma* but as up to four tandem repeats in *S. purpuratus*. In *He185/333*, type 3 repeat appears as four tandem copies and is homologous to a portion of Sp185/333 type 5 repeat. Repeat type 5 is present up to three times in Sp185/333. Similarly, the last four residues of *S. purpuratus* repeat type 4 (GDQD) are not part of the homologous repeat sequence in He185/333. Most, but not all, He185/333 repeats had homologues amongst Sp185/333 sequences, as some repeats were unique to each species. For

example, although the type 4 repeat sequence in He185/333 was homologous to a sequence stretch in Sp185/333, the latter was not repeated in Sp185/333. Finally, repeat type 2 of *H. erythrogramma* is present up to five times (four complete in tandem plus one incomplete repeats). The homologous sequences in Sp185/333 are composed of types 2, 3 and 4.

Table S1   Genbank accession numbers of He185/333 and Sp185/333 sequences.

Table S2   Tamura-Nei genetic distance matrix of phylogenetic comparison between 25 He185/333 and 25 Sp185/333 cDNA sequences. The rates are assumed to follow gamma distribution with shape parameter $=0.8094$. The distances in light and dark gray shadings represent intra-species comparisons, while those without background shading represent comparisons between He185/333 and Sp185/333 sequences.

Table S3   Summary of diversity analysis carried out on He185/333 and Sp185/333 sequences. Only those codons (codon#) that are positively or negatively selected are indicated in this table. The diversity analysis was carried out using three separate algorithms (SLAC, FEL and iFEL) and the table indicates whether codons are positively ($+$) or negatively ($-$) selected according to each of the algorithms. Codons are considered to be under selection ($+$ or $-$) if two or more of the analytical algorithms indicate significant selection pressure (columns entitled 'consensus'). Blank columns specify codons that are not considered to be under significant selection by an algorithm.

Acknowledgments

The authors would like to acknowledge significant contributions made by Prof L. Courtney Smith (George Washington University) for providing the antisera used in this project, as well as her suggestions for improving this manuscript. We also acknowledge the assistance and advice provided by Cameron Hill and Dr Ben Herbert (Macquarie University, Sydney) for optimizing 2DE protocols. Dr Ante Jerkovich and Dr Artur Sawicki provided assistance with protein extractions using the French press.

Author Contributions

Conceived and designed the experiments: SVN. Performed the experiments: MOR. Analyzed the data: MOR SVN. Wrote the paper: MOR SVN. Provided overall guidance and advice: DAR. Performed sequence diversity analysis: AGW. Performed phylogenetic analyses of He185/333 sequences: GMC. Developed the sequence diversity display tool described in this manuscript: AGW. Analysed phylogenetic relationships between 185/333 sequences using PAUP: GMC.

References

1. Kurtz J (2005) Specific memory within innate immune systems. Trends in Immunology 26: 186–192.
2. Roth O, Kurtz J (2009) Phagocytosis mediates specificity in the immune defence of an invertebrate, the woodlouse Porcellio scaber (Crustacea: Isopoda). Developmental and Comparative Immunology 33: 1151–1155.
3. Kurtz J, Armitage S (2006) Alternative adaptive immunity in invertebrates. Trends in Immunology 27: 493–496.
4. Dong Y, Taylor HE, Dimopoulos G (2006) AgDscam, a hypervariable immunoglobulin domain-containing receptor of the Anopheles gambiae innate immune system. PLoS Biology 4: e229.
5. Sadd BM, Schmid-Hempel P (2006) Insect immunity shows specificity in protection upon secondary pathogen exposure. Current Biology 16: 1206–1210.

6. Little TJ, Kraaijeveld AR (2004) Ecological and evolutionary implications of immunological priming in invertebrates. Trends in Ecology and Evolution 19: 58–60.
7. Pancer Z (2000) Dynamic expression of multiple scavenger receptor cysteine-rich genes in coelomocytes of the purple sea urchin. Proceedings of the National Academy of Sciences of the United States of America 97: 13156–13161.
8. Hibino T, Loza-Coll M, Messier C, Majeske AJ, Cohen AH, et al. (2006) The immune gene repertoire encoded in the purple sea urchin genome. Developmental Biology 300: 349–365.
9. Rast JP, Smith LC, Loza-Coll M, Hibino T, Litman GW (2006) Genomic insights into the immune system of the sea urchin. Science 314: 952–956.
10. Pasquier LD (2006) Germline and somatic diversification of immune recognition elements in Metazoa. Immunology Letters 104: 2–17.

11. Watson FL, Püttmann-Holgado R, Thomas F, Lamar DL, Hughes M, et al. (2005) Extensive diversity of Ig-superfamily proteins in the immune system of insects. Science 309: 1874–1878.

12. Zhang SM, Loker ES (2003) The FREP gene family in the snail Biomphalaria glabrata: additional members, and evidence consistent with alternative splicing and FREP retrosequences. Fibrinogen-related proteins. Developmental and Comparative Immunology 27: 175–187.

13. Cannon JP, Haire RN, Litman GW (2002) Identification of diversified genes that contain immunoglobulin-like variable regions in a protochordate. Nature Immunology 3: 1200–1207.

14. Södergren E, Weinstock GM, Davidson EH, Cameron RA, Gibbs R a, et al. (2006) The genome of the sea urchin Strongylocentrotus purpuratus. Science 314: 941–952.

15. Nair SV, Del Valle H, Gross PS, Terwilliger DP, Smith LC (2005) Macroarray analysis of coelomocyte gene expression in response to LPS in the sea urchin. Identification of unexpected immune diversity in an invertebrate. Physiological Genomics 22: 33–47.

16. Terwilliger DP, Buckley KM, Mehta D, Moorjani PG, Smith LC (2006) Unexpected diversity displayed in cDNAs expressed by the immune cells of the purple sea urchin, Strongylocentrotus purpuratus. Physiological Genomics 26: 134–144.

17. Terwilliger DP, Buckley KM, Brockton V, Ritter NJ, Smith LC (2007) Distinctive expression patterns of 185/333 genes in the purple sea urchin, Strongylocentrotus purpuratus: an unexpectedly diverse family of transcripts in response to LPS, beta-1, 3-glucan, and dsRNA. BMC Molecular Biology 8: 16.

18. Buckley KM, Smith LC (2007) Extraordinary diversity among members of the large gene family, 185/333, from the purple sea urchin, Strongylocentrotus purpuratus. BMC Molecular Biology 8: 68.

19. Buckley KM, Munshaw S, Kepler TB, Smith LC (2008) The 185/333 gene family is a rapidly diversifying host-defense gene cluster in the purple sea urchin Strongylocentrotus purpuratus. Journal of Molecular Biology 379: 912–928.

20. Miller C, Buckley KM, Easley RL, Smith LC (2010) An Sp185/333 gene cluster from the purple sea urchin and putative microsatellite-mediated gene diversification. BMC Genomics 11: 575.

21. Buckley KM, Terwilliger DP, Smith LC (2008) Sequence Variations in 185/333 Messages from the Purple Sea Urchin Suggest Posttranscriptional Modifications to Increase Immune Diversity. The Journal of Immunology 181: 8585–8594.

22. Brockton V, Henson JH, Raftos DA, Majeske AJ, Kim YO, et al. (2008) Localization and diversity of 185/333 proteins from the purple sea urchin – unexpected protein-size range and protein expression in a new coelomocyte type. Journal of Cell Science 121: 339–348.

23. Ghosh J, Buckley KM, Nair SV, Raftos DA, Miller C, et al. (2010) Sp185/333: a novel family of genes and proteins involved in the purple sea urchin immune response. Developmental and Comparative Immunology 34: 235–245.

24. Dheilly NM, Nair SV, Smith LC, Raftos DA (2009) Highly variable immune-response proteins (185/333) from the sea urchin, Strongylocentrotus purpuratus: proteomic analysis identifies diversity within and between individuals. The Journal of Immunology 182: 2203–2212.

25. Dheilly NM, Birch D, Nair SV, Raftos DA (2011) Ultrastructural localization of highly variable 185/333 immune response proteins in the coelomocytes of the sea urchin, Heliocidaris erythrogramma. Immunology and Cell Biology: 1–9.

26. Lehrach H, Diamond D, Wozney JM, Boedtker H (1977) RNA Molecular Weight Determinations by Gel Electrophoresis under Denaturing Conditions, a Critical Reexamination. Biochemistry 16: 4743–4751.

27. Hall TA (1999) A user-friendly biological sequence alignment editor and analysis program for Windows 95/98/NT. Nucl Acids Symp Ser 41: 95–98.

28. Altschul SF, Madden TL, Schäffer AA, Zhang J, Zhang Z, et al. (1997) Gapped BLAST and PSI-BLAST: a new generation of protein database search programs. Nucleic Acids Research 25: 3389–3402.

29. Thompson JD, Higgins DG, Gibson TJ (1994) CLUSTAL W: improving the sensitivity of progressive multiple sequence alignment through sequence weighting, position-specific gap penalties and weight matrix choice. Nucleic acids research 22: 4673–4680.

30. Larkin MA, Blackshields G, Brown NP, Chenna R, McGettigan PA, et al. (2007) Clustal W and Clustal X version 2.0. Bioinformatics 23: 2947–2948.

31. Swofford DL (2003) PAUP*. Phylogenetic Analysis Using Parsimony (*and Other Methods). Version 4. Sunderland, Massachusetts: Sinauer Associates. p.

32. Posada D, Crandall KA (1998) MODELTEST: testing the model of DNA substitution. BIOINFORMATICS APPLICATIONS NOTE 14: 817–818.

33. Delport W, Poon AFY, Frost SDW, Kosakovsky Pond SL (2010) Datamonkey 2010: a suite of phylogenetic analysis tools for evolutionary biology. Bioinformatics 26: 2455–2457.

34. Pond SLK, Frost SDW, Muse SV (2005) HyPhy: hypothesis testing using phylogenies. Bioinformatics 21: 676–679.

35. Kosakovsky Pond SL, Frost SDW (2005) Not so different after all: a comparison of methods for detecting amino acid sites under selection. Molecular Biology and Evolution 22: 1208–1222.

36. Pond SLK, Frost SDW, Grossman Z, Gravenor MB, Richman DD, et al. (2006) Adaptation to different human populations by HIV-1 revealed by codon-based analyses. PLoS Computational Biology 2: e62.

37. Cavener DR, Ray SC (1991) Eukaryotic start and stop translation sites. Nucleic Acids Research 19: 3185–3192.

38. Bendtsen JD, Nielsen H, von Heijne G, Brunak S (2004) Improved prediction of signal peptides: SignalP 3.0. Journal of Molecular Biology 340: 783–795.

39. Krogh A, Larsson B, von Heijne G, Sonnhammer EL (2001) Predicting transmembrane protein topology with a hidden Markov model: application to complete genomes. Journal of molecular biology 305: 567–580.

40. Pierleoni A, Martelli PL, Casadio R (2008) PredGPI: a GPI-anchor predictor. BMC Bioinformatics 9: 392.

41. Caragea C, Sinapov J, Silvescu A, Dobbs D, Honavar V (2007) Glycosylation site prediction using ensembles of Support Vector Machine classifiers. BMC Bioinformatics 8: 438.

42. Blom N, Gammeltoft S, Brunak S (1999) Sequence and structure-based prediction of eukaryotic protein phosphorylation sites. Journal of Molecular Biology 294: 1351–1362.

43. Smith LC, Rast JP, Brockton V, Terwilliger DP, Nair SV, et al. (2006) The sea urchin immune system. Invertebrate Survival Journal 3: 25–39.

44. Giancotti FG, Ruoslahti E (1999) Integrin Signaling. Science 285: 1028–1033.

45. Miranti CK, Brugge JS (2002) Sensing the environment: a historical perspective on integrin signal transduction. Nature Cell Biology 4: E83–90.

46. Pederson HG, Johnson CR (2007) Growth and age structure of sea urchins (Heliocidaris erythrogramma) in complex barrens and native macroalgal beds in eastern Tasmania. ICES Journal of Marine Science 65: 1–11.

47. Palumbi SR, Lessios HA (2005) Evolutionary animation: How do molecular phylogenies compare to Mayr's reconstruction of speciation patterns in the sea? PNAS. 102: 6566–6572.

48. Lessios HA, Zigler KS (2012) Rates of sea urchin bindin evolution. In Rapidly Evolving Genes and Genetic Systems. Rama S. Singh, Jianping Xu, and Rob J. Kulathinal (eds). Oxford University Press.

2

Accurate Assignment of Significance to Neuropeptide Identifications Using Monte Carlo K-Permuted Decoy Databases

Malik N. Akhtar[1,9], **Bruce R. Southey**[1,9], **Per E. Andrén**[2], **Jonathan V. Sweedler**[3], **Sandra L. Rodriguez-Zas**[1,4,5]*

1 Department of Animal Sciences, University of Illinois at Urbana-Champaign, Urbana, Illinois, United States of America, **2** Department of Pharmaceutical Biosciences, Uppsala University, Uppsala, Sweden, **3** Department of Chemistry, University of Illinois at Urbana-Champaign, Urbana, Illinois, United States of America, **4** Department of Statistics, University of Illinois at Urbana-Champaign, Urbana, Illinois, United States of America, **5** Institute for Genomic Biology, University of Illinois at Urbana-Champaign, Urbana, Illinois, United States of America

Abstract

In support of accurate neuropeptide identification in mass spectrometry experiments, novel Monte Carlo permutation testing was used to compute significance values. Testing was based on k-permuted decoy databases, where k denotes the number of permutations. These databases were integrated with a range of peptide identification indicators from three popular open-source database search software (OMSSA, Crux, and X! Tandem) to assess the statistical significance of neuropeptide spectra matches. Significance *p-values* were computed as the fraction of the sequences in the database with match indicator value better than or equal to the true target spectra. When applied to a test-bed of all known manually annotated mouse neuropeptides, permutation tests with k-permuted decoy databases identified up to 100% of the neuropeptides at *p-value* $< 10^{-5}$. The permutation test *p-values* using hyperscore (X! Tandem), *E-value* (OMSSA) and Sp score (Crux) match indicators outperformed all other match indicators. The robust performance to detect peptides of the intuitive indicator "number of matched ions between the experimental and theoretical spectra" highlights the importance of considering this indicator when the *p-value* was borderline significant. Our findings suggest permutation decoy databases of size 1×10^5 are adequate to accurately detect neuropeptides and this can be exploited to increase the speed of the search. The straightforward Monte Carlo permutation testing (comparable to a zero order Markov model) can be easily combined with existing peptide identification software to enable accurate and effective neuropeptide detection. The source code is available at http://stagbeetle.animal.uiuc.edu/pepshop/MSMSpermutationtesting.

Editor: Frederique Lisacek, Swiss Institute of Bioinformatics, Switzerland

Funding: This work was supported by NIH R21 DA027548 and P30 DA 018310. The funders had no role in study design, data collection and analysis, decision to publish, or preparation of the manuscript.

Competing Interests: The authors have declared that no competing interests exist.

* Email: rodrgzzs@illinois.edu

9 These authors contributed equally to this work.

Introduction

Neuropeptides participate in cell to cell communication and regulate many biological processes such as behavior, learning, and metabolism [1]. Mass spectrometry has revolutionized neuropeptide characterization and quantification [2–7]. However, detection is complicated by the neuropeptide size (typically 3 to 40 amino acids long) and by the complex post-translational processing that includes cleavage, and amino acid modifications of prohormones into neuropeptides [1,8].

Database search programs are commonly used to identify peptides from tandem mass spectrometry experiments [9]. These programs generate *in silico* theoretical spectra from target databases of known peptide sequences that have masses within a range (tolerance) of the observed peptide mass. The *in silico* spectra are then compared to the observed experimental spectra and indicator scores that signify the closeness of the match are computed. To assess the statistical significance of these matches, the observed-target match indicator is compared to the distribution of indicator values under the null hypothesis of no match using various methods. In the popular target-decoy approach, the experimental spectra are compared to spectra from a decoy database consisting of peptides sequences that were generated by reverting or reshuffling the amino acids in the sequences of the target database [9–11].

For neuropeptide identification, the target-decoy approach can result in false negatives because the small size of many neuropeptides leads to low observed-target match indicator values and consequently low significance levels [11]. Furthermore, the small size of many neuropeptide leads to few decoy reshuffled sequences and the resulting granularity of the null distribution of decoy scores further lowers the significance levels [11–15]. At the protein level, alternative identification approaches have attempted to address the challenge of assessing statistical significance [16,17].

However, the implementations of the previous approaches do not work with widely used database search programs, do not use all the information resulting from the mass spectrometry experiment, and are biased by peptide length or assume one-direction progressive processing. Approaches that rely on fewer limiting assumptions and that use all the information available need to be evaluated.

Permutation tests are well-suited for neuropeptide database searches by helping to overcome the finite combination of amino acids from small neuropeptides and do not rely on directional assumptions. Furthermore, permutation testing provides strong control of Type I errors thus minimizing the incidence of false positive results [18]. Under the null hypothesis of no match, the experimental spectrum of a peptide is the result of a random sequence of amino acids provided that the total mass is close to the experimental mass. This requirement stems from the database search program strategy that only accepts sequences within a user determined range of the experimental spectra. Computation of the permutation statistical significance requires the distribution of the peptide-spectrum match scores generated by the database search program under the null hypothesis that there is no correct match. This distribution is then generated by searching the experimental spectrum against a decoy database considering all possible amino acid sequences within the predetermined range of the experimental spectra. The permutation $p\text{-}value$ is then the proportion of peptide-spectrum scores obtained from the decoy database that are equal to greater to the score obtained using the target database. Under the null hypothesis any amino acid can be present at any position of the sequence, thus, addressing the exchangeable assumption required by the permutation test [18].

Monte Carlo sampling is used to reduce the number of possible sequences while providing an unbiased estimate of the $p\text{-}value$. Furthermore, the loss in statistical efficiency when estimating the $p\text{-}value$ decreases with increasing number of random samples [18]. The previously demonstrated advantage of the Monte Carlo permutation approach proposed over existing decoy generation based on sequence reversion or reshuffling of the target sequence is the improved definition of the null distribution [18]. The larger number of decoy sequences results in lower granularity and, thus, more precise assessment of the statistical significance of the observed matches. Two major advantages of the Monte Carlo permutation approach proposed over existing dynamic programming approaches [16,17] is the simplicity of integration to existing database search programs, the use of all spectra information available and consideration of all possible spectra matching processes.

This study demonstrates the use Monte Carlo permutation testing to overcome the limitations of current protein identification approaches to accurately assess neuropeptide statistical significance. This approach combines and extends the model-free property of current decoy databases with the more extensive search of dynamic programming approaches. The aims are: (1) to develop permutation resampling methodology that can be easily integrated with existing peptide database search software, and (2) to demonstrate the advantages of this approach to provide accurate measures of neuropeptide match significance using ideal and real experimental neuropeptide spectra. Supporting objectives were: (1) to develop and implement complementary novel permuted databases; (2) to determine the number of permutations required for accurate significance levels; and (3) to identify the neuropeptide match indicators within and across programs that are better suited to provide accurate statistical significance.

Materials and Methods

Tandem Spectral Dataset and Target Database

Tandem mass spectra from a comprehensive list of 103 experimentally-obtained and manually annotated mouse neuropeptide were obtained from the SwePep database (http://www.swepep.org). These spectra were obtained using linear ion trap mass spectrometer coupled with liquid chromatography and electrospray ionization source [19]. Neuropeptides were manually validated after identification using the X! Tandem database search program [20]. The independent manual annotation step also ensured that the subsequent software comparison would not be biased in favor of the X! Tandem database search program. Of these, 80 neuropeptides were unmodified and the remaining 23 encompassed post-translational modifications (PTMs) including C-terminal amidation, N-terminal acetylation, phosphorylation, pyroglutamination and oxidation. The spectra corresponded to 5, 68, 25, and 5 peptides that had precursor charge states +1, +2, +3 and +4, respectively, and all charge states were observed in modified and unmodified peptides.

Ideal uniform spectra of all possible $b\text{-}$ and $y\text{-}$ions with +1 product charge state were simulated for 103 annotated experimental spectra. The ideal spectra also included all the PTMs identified in the corresponding experimental spectra. The neutral mass loss peaks due to loss of single water or ammonia molecules from the $b\text{-}$ and $y\text{-}$ions were simulated regardless of their position in the ions sequence. These ideal spectra are expected to be correctly identified at an extremely high significance level because these spectra are equivalent to the theoretical spectra internally generated by the database search engine.

A comprehensive target database of 618 mouse neuropeptides was obtained from the PepShop database (http://stagbeetle.animal.uiuc.edu/pepshop; [21]). This target database encompassed the neuropeptides corresponding to the 103 tandem spectra studied. The neuropeptides in the PepShop were assembled from the known 95 mouse prohormones present in SwePep [19] and UniProt [22] complemented with NeuroPred [23] predictions. The neuropeptides in the target database ranged from 2 to 223 amino acids in length because this included all known experimentally confirmed mouse neuropeptides as well as all possible intermediate and other peptides produced during the processing of prohormones. The target database of neuropeptides is available at http://stagbeetle.animal.uiuc.edu/pepshop/MSMSpermutationtesting.

Database Search Programs and Database Searching

Three open source database search programs were used in this study: Crux [24] (version 1.37), OMSSA [25] (version 2.1.8), and X! Tandem [20] (version 2013.02.01.1). These commonly used open source programs were selected because the code could be modified to ensure comparable search parameter specification and enabled to retrieve intermediate indicators of the strength of the match between the observed and target or decoy spectra. The observed-target or observed-decoy spectra match indicators extracted from OMSSA were: number of matched fragment ions, lambda or Poisson mean match indicator, Poisson probability of the lambda match indicator, and corresponding $E\text{-}value$ of the match (Poisson probability multiplied by the effective database size). The spectra match indicators extracted from X! Tandem were: number of matched fragment ions, intermediate convolution score (product of the intensities of the shared $b\text{-}$ and $y\text{-}$fragment ions between experimental and theoretical spectra), hyperscore (factorial of the number of matching $b\text{-}$ and $y\text{-}$ions multiplied by the convolution score), and $E\text{-}value$ (calculated from the distribution of hyperscores scores). The spectra match indicators extracted from Crux were: number of matched fragment ions, Sequest Sp score (Sp), cross-correlation score (XCorr), deltaCn score (ΔCn) and *p-value* that is calculated from the Weibull distribution fitted to the XCorr scores of observed-theoretical spectra matches [26].

For comparable neuropeptide identification across the three programs the following search parameters from our prior research [11] were used: (1) precursor ion tolerance: 1.5 Da; (2) fragment ion tolerance: 0.3 Da (OMSSA and X! Tandem); mz-bin-width: 0.3 (Crux) (3) searches were performed with and without PTMs. The PTMs evaluated were: amidation, phosphorylation, N-terminal acetylation, acetylation of lysine, pyroglutamination of glutamine, methylation of lysine and arginine residues, sulfation of tyrosine residue, and oxidation of methionine; (4) "protein" (OMSSA) or "enzyme: custom cleavage site" (X! Tandem and Crux) to prevent peptide cleavage since the detection of neuropeptides does not involve protease digestion; (5) fragment ion charge: default values; (6) OMSSA "ht" option was set to eight to filter database peptides that have at-least one theoretical fragment ion match to one of the top eight most intense peaks in the observed spectra; and (7) peptide mass: monoisotopic; 8) Crux *p-values* were computed using 1000 Weibull points because this information provides more accurate *p-values* than the default 40 Weibull points [11].

Permutation Approach and K-Permuted Decoy Databases

A Monte Carlo permutation test approach based on biological, computational and statistical considerations was used to generate decoy sequence databases. The resulting decoy database accommodates limited changes in the specified search parameters and can be used by all database search programs without the need to modify the original program code. A decoy database considering all possible amino acid sequences can be used to generate the distribution of the peptide-spectrum match scores under the null hypothesis of no correct match. This distribution is required by database search programs to assess the statistical significance of the match. This requirement results in an extremely large number of sequence since a decoy database of only 10-amino acid long peptides consists of 6.13×10^{12} sequences. This number would be further increased to account for different possible lengths of the target neuropeptides. Due to the potential size of a database encompassing all possible sequences, a Monte Carlo permutation approach based on a set of candidate peptides was used to generate a random sample of all possible sequences. Applying the same strategy used by the database search programs, 236 mouse neuropeptides within 12 Da of the precursor mass of the 103 studied neuropeptides were considered as candidate peptides. The 12 Da arbitrary threshold enabled the creation of single flexible peptide catalog that could be used with different database programs while permitting different mass tolerance specifications and precursor charge states. The arbitrary threshold does not influence the results because the database search programs ignore candidate sequences outside the settings and all peptide-spectrum scores involving target database size will be equally affected. Figure 1 depicts the correspondence between the lengths of neuropeptides in the target database, the 103 experimental neuropeptides and the neuropeptides that fall within 12 Da of the 103 peptides. A new decoy peptide database was obtained by generating a set of random sequences from each of the candidate peptides. Random sequences were generated by sequentially replacing each amino acid in the sequence of each candidate peptide by a randomly selected amino acid from the 19 amino acids from the candidate peptide list (leucine and isoleucine were

considered the same amino acid due to the same neutral masses). This was repeated until the predetermined number of permuted sequences per candidate sequence was obtained. The resulting permuted sequences are comparable to those generated from a Markov model of order zero such that the 19 amino acids are equally likely at all positions. These permuted sequences were collected into a single database after removal of duplicate peptides and sequences present in the target database. This procedure was used to generate k-permuted decoy sequence databases where the numbers of unique permuted sequences per candidate peptide (k) were: 10^3 ($K10^3$ with 236,000 decoy peptide sequences), 10^4 ($K10^4$ with 2,360,000 decoy peptide sequences), 10^5 ($K10^5$ with 23,600,000 decoy peptide sequences), and 10^6 ($K10^6$ with 236,000,000 decoy peptide sequences). The target database was appended to each of the four k-permuted databases to create a combined target-k-permuted decoy database. The combined database search is more accurate than separate database searches and to avoid zero *p-value* [27–29]. This strategy also removed potential database size dependency of the match indicators between target and permuted sequences because the correct match was evaluated under the same database sizes as the permuted databases.

The search of spectra against the k-permuted decoy databases produced many matches that were indistinguishable from each other based on the indicators reported by the programs (e.g., number of matched ions, hyperscore, convolution score, and *E-value* for the X! Tandem). Matches were considered "homeometric" [18] when the matches had the same indicator values across programs and the matched peptides masses were within ± 1.5 Da from each other. Figure 2 depicts the number of peptides with homeometric matches ranging from 1 to 10 for the $K10^6$ k-permuted decoy database across the three databases search programs. Homeometric matches were counted only once while calculating the number of random peptides that have an indicator value equal or better than the true target peptide. This strategy resolved the challenge that database search programs were not able to differentiate between such matches that are technically redundant and ensured the calculation of permutation *p-values* that were unbiased by these effects.

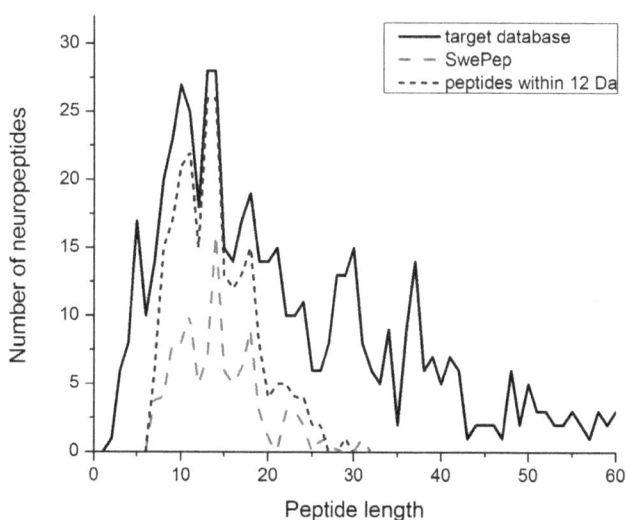

Figure 1. Distribution of neuropeptides length in target database peptides (less than 60 amino acid in length are shown), 103 studied peptides, and 236 peptides that fall within ±12 Da of the 103 peptides.

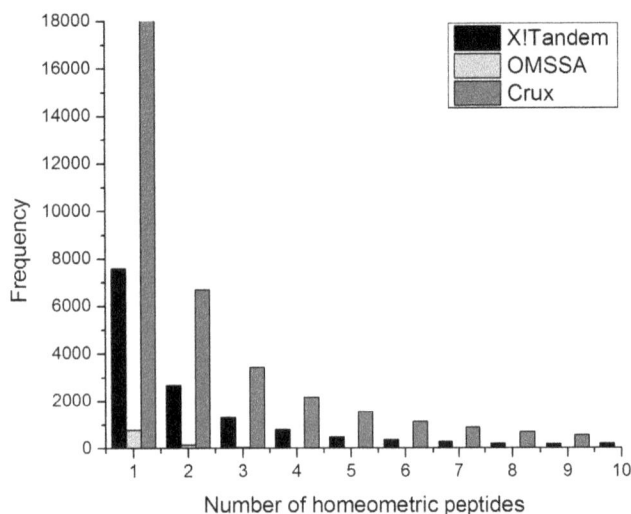

Figure 2. Frequency (number) of spectra with 1 to 10 homeometric matches for K10⁶ k-permuted decoy databases across the three database search programs (X! Tandem, OMSSA, and Crux).

For each database search program and target sequence, the observed tandem spectra were searched for matches within each combined target-k-permuted decoy spectra. The permutation *p-values* were estimated as the fraction of combined target-k-permuted decoy peptides, excluding any homeometric matches that have a matching indicator score equal or better than the score of target peptide.

A comprehensive evaluation of the k-permuted decoy approaches, programs, and peptide match indicators was undertaken including: (a) Search for ideal uniform simulated spectra against the target database using all three database search programs; (b) Search for real tandem spectra against the target database using all three database search programs; (c) Search for the 80 tandem spectra containing no PTMs against the $K10^3$, $K10^4$, $K10^5$, and $K10^6$ target-k-permuted decoy databases without PTM specification using all three database search programs; (d) Search for the 80 tandem spectra containing no PTMs against the $K10^5$ k-permuted database with PTM specification using all three database search programs; and (e) Search for the 23 tandem spectra containing PTMs against the $K10^5$ k-permuted database with PTM specification using OMSSA and X! Tandem. Crux was excluded from this last comparison due to considerable amount of search time required.

Results and Discussion

A k-permuted decoy database approach that resolves limitations of existing methods to assess the significance of peptides is presented. The proposed approach can be used with any database search program, especially when the program lacks of a statistical approach for *p-value* calculation, and can be integrated to target and target-decoy databases. The k-permutation strategy proposed supersedes a decoy database of randomly generated decoy peptides that underestimates the false discovery rate because, by definition, the vast majority of the random peptides will be true negatives. Non-random approaches such as reversing peptide sequences are more appropriate than random approaches in generating decoy databases [9–11].

Results from a three step benchmarking strategy were used to evaluate the performance to detect neuropeptides using target-k-permuted decoy databases. First, a baseline performance was obtained by comparing ideal simulated spectra against a standard "target database" using the three database search programs. Then, observed tandem spectra were matched to a target database. Lastly, the observed tandem spectra were matched to different target-k-permuted decoy databases. Evaluation of results among modified and non-modified spectra enabled understanding of the influence of PTMs on the search results. The source code to generate k-permuted decoy databases is available at http://stagbeetle.animal.uiuc.edu/pepshop/MSMSpermutationtesting.

Peptide Detection using Ideal Simulated Spectra and a Target Database

Table 1 summarizes the results from the three database search programs when 103 ideal uniform spectra were simulated with all *b*- and *y*-ions including neutral mass losses and searched against the target database. The search of ideal simulated spectra demonstrated the ability of the database search methods to assign an E-*value* or *p-value* to each peptide-spectrum match in the absence of technical or biological noise [30]. Although the E-*values* and *p-values* reported by these programs have different interpretations and are computed differently between programs, match counts based on the same threshold are reported to facilitate the identification of trends.

The three programs matched all unmodified neuropeptides correctly at an E-*value* or *p-value* $< 2 \times 10^{-1}$. At an E-*value* or *p-value* $< 1 \times 10^{-2}$, OMSSA, X! Tandem, and Crux identified 80 (100% of unmodified neuropeptides), 80 (100% of unmodified neuropeptides), and 73 (91.25% of unmodified neuropeptides) peptides, respectively. This trend was consistent with previous study that compared Crux, OMSSA and X! Tandem [11]. Our study confirmed the lower significance values that Crux computes for peptides less than 45 amino acids in length [11]. OMSSA E-*values* averaged more significant matches than X! Tandem for the 32 peptides that were less than 13 amino acids in length. However, for the 48 peptides longer than 12 amino acids in length, the difference in significance levels of X! Tandem and OMSSA decreased on the average with 8, 18, and 22 peptides getting lower, equal, and better significance levels for the X! Tandem than OMSSA, respectively.

For the 23 neuropeptides with PTMs and an E-*value or p-value* $< 1 \times 10^{-1}$, OMSSA, X! Tandem, and Crux correctly detected 23 (100% of modified neuropeptides), 18 (78.26% of modified neuropeptides), and 23 (100% of modified neuropeptides) peptides, respectively. X! Tandem failed to correctly match five peptides with N-terminal acetylation modification instead these five peptides were matched with incorrect internal acetylation modification at 9th lysine residue. The failure in the peptide detection of X! Tandem was only observed when multiple PTMs were specified in the search specification. The five peptides were correctly detected when only N-terminal acetylation was used in the search specification. At an E-*value or p-value* $< 1 \times 10^{-2}$, 23 (100% of modified neuropeptides), 18 (78.26% of modified neuropeptides), and 22 (95.66% of modified neuropeptides) peptides were detected by OMSSA, X! Tandem, and Crux, respectively. The three peptides that were not significant for OMSSA at E-*value* $< 1 \times 10^{-4}$ all had a pyroglutamination (Q residue) modification. Two of these peptides, somatostatin (gene symbol: SMS) [87–100] (QRSANSNPAMAPRE; charge state +2) and secretogranin-2 (gene symbol: SCG2) [205–216] (QELGKLTGPSNQ; charge state +1), were significant for the X! Tandem and Crux at an E-*value or p-value* $< 1 \times 10^{-4}$. A nine

Table 1. Peptide detection significance levels using ideal simulated spectra of the 103 peptides with and without any post-translational modifications (PTMs) and all *b*- and *y*-ions including neutral mass losses against a standard target database across database search programs (OMSSA, X! Tandem, and Crux).

Program	PTMs	Significance[a]							$P \leq 10^{-2}$ [b]
		0	1	2	3	4	5	≥6	
X! Tandem	None	0	0	4	4	2	6	58	74
	Amidation	0	0	0	0	0	0	9	9
	Oxidation	0	0	0	0	0	0	1	1
	Pyroglutamination	0	0	1	0	1	1	1	4
	Phosphorylation	0	0	0	0	0	1	3	4
	N-terminal acetylation	0	0	0	0	0	0	0	0
OMSSA	None	0	0	0	0	0	0	79	79
	Amidation	0	0	0	0	0	0	9	9
	Oxidation	0	0	0	0	0	0	1	1
	Pyroglutamination	0	0	1	2	1	0	0	4
	Phosphorylation	0	0	0	0	0	0	4	4
	N-terminal acetylation	0	0	0	0	0	0	5	5
Crux	None	2	5	12	52	3	1	2	70
	Amidation	0	0	2	3	2	0	2	9
	Oxidation	0	0	0	1	0	0	0	1
	Pyroglutamination	0	1	1	0	1	0	1	3
	Phosphorylation	0	0	1	2	0	0	1	4
	N-terminal acetylation	0	0	0	3	1	0	1	5

[a]Significance threshold (t) for matched to be considered significant at an E-*value* or *p-value* < 1×10^{-t} (t= 0 to >=6).
[b]Cumulative number of peptides with an E-*value* or *p-value* < 1×10^{-2}.

amino acid long peptide secretogranin-1 (gene symbol: SCG1) [667–675] (QKIAEKFSQ; charge state +2) was not significant for all three programs at an *E-value* or *p-value* $< 1 \times 10^{-4}$, while the same peptide was missed by the Crux at *p-value* $< 1 \times 10^{-2}$.

Peptide Detection using Observed Spectra and a Target Database

Table 2 summarizes the performance of the three database search programs when the 80 experimental tandem spectra containing no PTMs were searched against the target database. All peptide assignments by the three database search methods were correct at an *E-value* or *p-value* $< 5 \times 10^{-1}$. At an *E-value* or *p-value* $< 1 \times 10^{-2}$, OMSSA, X! Tandem and Crux detected 80 (100% of unmodified neuropeptides), 71 (88.75% of unmodified neuropeptides), and 63 (78.75% of unmodified neuropeptides) peptides, respectively. The higher number of significant peptide detections by OMSSA relative to Crux was consistent with the prior reports [11]. The three search methods were less accurate on 23 observed spectra with PTMs when searched against the standard target database (Table 2). From the correctly matched peptides for each program, at an *E-value* or *p-value* $< 1 \times 10^{-2}$, OMSSA, X! Tandem and Crux detected 20 (86.95% of modified neuropeptides), 15 (65.21% of modified neuropeptides), and 17 (73.91% of modified neuropeptides) peptides, respectively.

The 80 spectra without PTMs were searched against the target database using three database search programs and with PTM specifications. X! Tandem peptide detection significance levels for the 76, 3, and 1 target peptide remained unchanged, decreased, and increased, respectively, relative to the searches involving no PTMs. The changes in the significance levels of the four peptides were due to higher number of candidate peptides available in the PTM searches which in turn changed the estimation parameters used in the *E-value* computation. The OMSSA peptide detection significance levels decreased for the majority of the previous peptides (75 out of 80 peptides) or remained unchanged (5 out of 80 peptides) when searches included PTMs, respectively. Crux peptide detection significance levels were improved when searches included PTMs with 65 and 29 peptide detections at *p-value* $< 1 \times 10^{-2}$ and $< 1 \times 10^{-4}$, respectively. Comparison of peptide detections across PTM scenarios indicated that at *p-value* $< 1 \times 10^{-2}$, 54 peptides were detected by both scenarios, 11 peptides were detected in the PTM scenario, 9 peptides were detected in the no PTMs scenario, and 6 peptides were not detected by either scenario. The target peptides with low XCorr scores remained undetected either across both scenarios or with PTM search. The clear positive correlation between significance level and XCorr score for the PTM searches relative to the searches without PTMs could be due to the higher number of low scoring matches in the searches with PTMs than without PTMs. The Crux resampling from the low scoring matches might have resulted in a shift on the distribution of XCorr scores towards lower scores than the target peptides XCorr scores.

X! Tandem Peptide Identification using a K-Permuted Decoy Database

Table 3 summarizes the $\log_{10}$ transformed of the *E-values* to the target database and permutation *p-values* computed for the X! Tandem indicators: number of matched ions, hyperscore, *E-value*, and convolution score using the 80 spectra without PTMs across the four target-k-permuted decoy databases studied. The permutation *p-values* from number of matched ions, hyperscore and *E-value* showed that the X! Tandem *E-values* from the target database were dramatically underestimated (less significant) for

most target peptides. Detection and significance level using the number of ions matched, hyperscore and *E-value* were almost the same across all target-k-permuted decoy databases. Only at the 10^6 permutations did the *p-values* for number of ions matched started to differ from the *p-values* from the hyperscore and *E-value* match indicators. This trend was expected as the hyperscore is a function of the product of factorial of the number of matched ions and the ion intensity values and *E-value* is a function of the hyperscore.

The convolution score resulted in fewer target peptide identifications with higher number of sequence permutations due to relative increase in the number of decoy matches with equal or better scores. From the $K10^3$, $K10^4$, $K10^5$, and $K10^6$ target-k-permuted decoy databases, 72 (90% of unmodified neuropeptides), 31 (39% of unmodified neuropeptides), 9 (11% of unmodified neuropeptides), and 10 (13% of unmodified neuropeptides) peptides were identified at *p-value* $< 1 \times 10^{-2}$, $< 1 \times 10^{-3}$, $< 1 \times 10^{-4}$, and $< 1 \times 10^{-4}$, respectively. These results showed that the convolution score alone was less suitable to discriminate between true target and decoy matches than the hyperscore and *E-value*.

Comparison of the *p-values* obtained from the target-k-permuted decoy number of matched ions, hyperscores and convolution scores suggested that roughly 10^5 permutations were required for significant *p-value* computations using the convolution scores. Higher number of sequence permutations provided better separation between the significance levels of the three indicators. There were 7 peptides with *E-values*$<10^{-7}$ from the target database indicating that the lower bound of *p-values* appeared to be far smaller than the limit provided by the $K10^6$ permuted database. Comparable performance (significance level) using number of matched ions and hyperscore were observed with fewer permutations or lower significance thresholds. This novel finding suggests that more significant detections can be obtained by permuting the X! Tandem hyperscore and number of matched ions indicators, even with a relatively small k-permuted decoy database size.

Crux Peptide Identification using a K-Permuted Decoy Database

Table 4 summarizes the $\log_{10}$ transformed permutation *p-values* computed for the Crux match indicators: number of matched ions, XCorr, ΔCn, and Sp using the 80 spectra without PTMs across the four target-k-permuted decoy databases. Higher number of sequence permutations increased the significance values using the number of matched ions and Sp. This trend was due to the lower number of matched ions and Sp scores of the decoy peptide matches relative to the target peptides. The two non-detected peptides could be attributed to the low number of decoy candidates for those peptides rather than to an increase in the number of decoy peptides with equal or better scores. The hindering effect on the match significance of better or equal decoy matches on Sp was more evident with the large decoy databases at *p-value* $< 1 \times 10^{-5}$.

Peptide detection was less significant when using XCorr relative to Sp and number of matching ions. The drop in significance level with increase in threshold and database size was due to the higher number of decoy peptides reaching XCorr levels better or equal than the target peptides. The detection and significance computation using XCorr and ΔCn (the difference in XCorr between candidates) was similar across all target-k-permuted databases which reflects that the range of these match indicators stabilized. The range of possible XCorr values was limited by the number of observed spectrum peaks because the background adjustment is

Table 2. Peptide detection significance levels using experimental spectra of the 103 peptides with and without any post-translational modifications (PTMs) against a standard target database across database search programs (OMSSA, X! Tandem, and Crux).

Program	PTMs	Miss[c]	Inc[d]	Significance[a]							Cum N[b]
				0	1	2	3	4	5	≥6	P≤10^{-2}
X! Tandem	None	0	0	1	8	11	15	16	11	18	71
	Amidation	0	0	0	0	0	0	0	0	9	9
	Oxidation	0	0	0	0	0	0	0	0	1	1
	Pyroglutamination	0	0	0	0	1	0	1	1	1	4
	Phosphorylation	0	0	0	0	0	0	0	1	3	4
	N-terminal acetylation	0	5	0	0	0	0	0	0	0	0
OMSSA	None	0	0	0	0	1	2	1	3	73	80
	Amidation	1	0	1	0	0	0	1	0	6	7
	Oxidation	0	0	0	0	0	0	0	0	1	1
	Pyroglutamination	0	0	0	0	0	0	1	0	3	4
	Phosphorylation	0	0	0	0	0	0	0	0	4	4
	N-terminal acetylation	0	1	0	0	0	0	0	0	4	4
Crux	None	0	0	9	8	9	44	1	0	9	63
	Amidation	0	0	0	1	5	1	1	0	1	8
	Oxidation	0	0	0	0	0	1	0	0	0	1
	Pyroglutamination	0	0	1	1	0	2	0	0	0	2
	Phosphorylation	0	0	0	2	2	0	0	0	0	2
	N-terminal acetylation	0	0	0	1	1	2	0	1	0	4

[a] Significance threshold (t) for matched to be considered significant at an E-*value* or p-*value* < 1×10^{-t} (t=0 to >=6).
[b] Cumulative number of peptides with an E-*value* or p-*value* < 1×10^{-2}.
[c] Number of peptides missed by program.
[d] Number of peptides with incorrect post-translational modification assignment.

Table 3. Performance of the target and alternative k-permuted decoy databases used with the X! Tandem database search program using spectra from 80 unmodified neuropeptides.

Database[a]	Indicator	Significance Levels of the Permutation p-values[b]							Cum. Num. of Peptides[c]	
		0	1	2	3	4	5	≥6	≥10^{-2}	≥10^{-4}
Target	E-value	1	8	11	15	16	12	17	71	45
K10^3	# ions	0	0	76	4	0	0	0	80	0
	Hyperscore	0	0	76	4	0	0	0	80	0
	Convolution	0	8	70	2	0	0	0	72	0
	E-value	0	0	76	4	0	0	0	80	0
K10^4	# ions	0	0	0	80	0	0	0	80	0
	Hyperscore	0	0	0	80	0	0	0	80	0
	Convolution	0	5	44	31	0	0	0	75	0
	E-value	0	0	0	80	0	0	0	80	0
K10^5	# ions	0	0	0	0	80	0	0	80	80
	Hyperscore	0	0	0	0	80	0	0	80	80
	Convolution	0	3	36	32	9	0	0	77	9
	E-value	0	0	0	0	80	0	0	80	80
K10^6	# ions	0	0	0	0	1	79	0	80	80
	Hyperscore	0	0	0	0	0	80	0	80	80
	Convolution	0	4	30	36	5	5	0	76	10
	E-value	0	0	0	0	0	80	0	80	80

[a]Target: database of 236 neuropeptide sequences; K10^3: k-permuted decoy database size of 236,000 peptides; K10^4: k-permuted decoy database size = 2,360,000 peptides; K10^5: k-permuted decoy database size = 23,600,000 peptides; K10^6: k-permuted decoy database size = 236,000,000 peptides.
[b]Significance threshold (t) for target spectrum to be considered significant at significance thresholds <1×10^{-1} (t=0 to >=6).
[c]The cumulative number of peptides at 1×10^{-2} and 1×10^{-4} thresholds.

Table 4. Performance of the target and alternative k-permuted decoy databases used with the Crux database search program using spectra from 80 unmodified neuropeptides.

Database[a]	Indicator[b]	Significance Levels of the Permutation p-values[c]							Cum. Num. of peptides[d]	
		0	1	2	3	4	5	≥6	≥10^{-2}	≥10^{-4}
Target	p-value	9	8	9	44	1	0	9	63	10
K10^3	# ions	0	2	78	0	0	0	0	78	0
	XCorr	3	11	66	0	0	0	0	66	0
	Sp	0	2	78	0	0	0	0	78	0
	ΔCn	3	11	66	0	0	0	0	66	0
K10^4	# ions	0	0	1	79	0	0	0	80	0
	XCorr	3	10	14	53	0	0	0	67	0
	Sp	0	0	1	79	0	0	0	80	0
	ΔCn	3	10	14	53	0	0	0	67	0
K10^5	# ions	0	0	0	1	79	0	0	80	79
	XCorr	3	10	8	23	36	0	0	67	36
	Sp	0	0	0	1	79	0	0	80	79
	ΔCn	3	10	8	23	36	0	0	67	36
K10^6	# ions	0	0	0	0	2	78	0	80	80
	XCorr	3	10	9	19	22	17	0	67	39
	Sp	0	0	0	0	4	76	0	80	80
	ΔCn	3	10	9	19	22	17	0	67	39

[a]Target: database of 236 neuropeptide sequences; K10^3: k-permuted decoy database size of 236,000 peptides; K10^4: k-permuted decoy database size = 2,360,000 peptides; K10^5: k-permuted decoy database size = 23,600,000 peptides; K10^6: k-permuted decoy database size = 236,000,000 peptides.

[b]# ions: permutation p-values computed for the number of matched b- and y-ions. XCorr: permutation p-values computed from the XCorr scores of the matches. Sp: permutation p-values computed from the Sp scores of the matches. ΔCn: permutation p-values computed using X! Tandem ΔCn.

[c]Significance threshold (t) for matched to be considered significant at p-value $< 1 \times 10^{-t}$.

[d]Cumulative number of peptides with p-values thresholds of 1×10^{-2} and 1×10^{-4}.

expected to be constant across permuted database sizes. This result indicates that only a relatively few permuted sequences are required to cover the range of XCorr values and that higher number of permutations offer greater precision to detect match differences.

OMSSA Peptide Identification using a K-Permuted Decoy Database

Table 5 summarizes the $\log_{10}$ transformed permutation *p-values* calculated for the OMSSA match indicators: number of matched ions, lambda match indicator, *p-value*, and *E-value* using the 80 spectra without PTMs across the target-k-permuted decoy databases. Comparison between the target database and the permutation *p-values* indicated that most peptides were accurately estimated by OMSSA suggesting that the k-permuted database size was unimportant. Examination of the few peptides with underestimated *E-values* suggested that these peptides had fewer intense MS/MS ion peaks resulting in lower 75% quartile values than peptides of similar size with lower *E-values*. This result indicates that OMSSA *E-values* may be less reliable in the presence of multiple low intensity spectra peaks.

Detection and significance computation using the number of matched ions, OMSSA *p-value* and *E-value* indicators was identical across all k-permuted decoy databases. However, the lambda parameter was less suitable than the other OMSSA match indicator to discriminate matches than the other match indicators. Differences in the lambda indicator for the same observed spectrum were mainly determined by the total number of theoretical *m/z* values for product ions and hence by the length of the decoy peptide sequence. After a relatively few permutations, the range of possible sequences is determined such that fewer permutations are required to determine the distribution of the lambda parameter than other match indicators.

Impact of PTM on Peptide Identification using a K-Permuted Decoy Database

Searches of 80 peptides with no PTMs including the specification of common neuropeptide PTMs improved the significance of the detection in target-k-permuted decoy databases. Using X! Tandem, all 80 observed peptides were identified at *p-value* $< 1 \times 10^{-5}$ using the number of matched ions and hyperscore indicators in the $K10^5$ permuted database, while convolution score indicator detected only 7 (8.75% of unmodified neuropeptides) peptides. Consistent with searches without PTMs using the OMSSA program, when the searches included PTMs the number of matched ions and *E-value* indicators provided more significant permutation *p-values* than the lambda indicator. For Crux, specification of PTMs reduced the performance (significance levels) of the number of matched ions, XCorr, and Sp indicators in the $K10^5$ database. The lower significances was due to corresponding increase in the decoy peptides with equal or better scores than the target peptides with increase in decoy database size when PTMs are considered in the search. Using the $K10^5$ permuted database, OMSSA and X! Tandem correctly identified the 20 and 17 of spectrum with PTMs as the first match, respectively. Both programs correctly identified the same 16 peptides, 6 peptides were identified by only one program and 1 peptide was not detected by either program. There were 4 peptides unmatched by X! Tandem only and the unmodified forms were matched outside the top 20 matches. The unmatched peptide, acetyl-YGGFMTSEKSQTPLVT, was undetected by OMSSA both in the target or k-permuted databases. X! Tandem was able to match the correct sequence, however the match has an additional amidation. Manual evaluation would have corrected the match as the amidation was on an unexpected amino acid and the non-amidated form was closer to the precursor mass then the amidated form.

The remaining 2 peptides that were unmatched by OMSSA were both amidated. One peptide, SYSMEHFRWGKPV-amide, was correctly identified as the 15th best match by OMSSA with the unamidated form providing the best match. The difference in monoisotopic mass between modified and unmodified was less than 1 Da. The experimental spectrum had a precursor *m/z* value of 541.70 with an assigned a 3+ charge state. At a 3+ charge state the predicted *m/z* values were 541.9294 and 541.6014 for the unmodified form and amidated forms, respectively. Biologically the unmodified form would be identified as a probable match since this sequence is an intermediate in the amidation process and the unmodified sequence is uncommon among neuropeptides because this form lacks the terminal G-residue after cleavage [11]. Consequently this unmodified peptide could be considered a match for OMSSA.

Comparison of Peptide Database Search Programs

Overall the k-permuted decoy databases allowed the detection of more peptides based on real spectra than the use of the standard target database regardless of the database search program. The search of spectra against the k-permuted decoy databases produced many matches that were indistinguishable from each other based on the indicators reported by the programs (e.g., number of matched ions, hyperscore, convolution score, and *E-value* for the X! Tandem). Permutation testing is computational demanding even with Monte Carlo sampling (Table 6). The increase in time across permutated database sizes is a consequence of the exponential increase in the number of sequences evaluated. However, the $K10^5$ database provided adequate results and all programs completed the search within 35 CPU minutes using a single process Intel Core i7-3770 CPU @ 3.40 GHz. This timing is the result of single-processor searches that ignored possible parallel processing of individual spectra. The advantages of Monte Carlo permutation approaches to assess the statistical significance of neuropeptide matches could be further advanced by simultaneously running groups of observed spectra using parallel processing.

An alternative approach to generate a permutated database is to perform targeted permutation of specific regions such as the terminal amino acids to disrupt *b*- and *y*-ion series. While other regions can be permuted, the advantage of permuting only the terminal peptides is that this strategy is independent of peptide size. The size of the required database quickly increases from 84,960 sequences per target peptide when one terminal position was permuted to 47,045,880 sequences per target peptide when 3 terminal positions were permuted. Evaluation of terminal permuted databases demonstrated that this approach offered similar yet less significant matches than the whole sequence permuted database approach. Also, this permutation approach had the disadvantage of providing a large number of homeometric matches since experimental ions near the termini are required to differentiate the order of amino acids. Thus, results from this approach are not reported.

With the goal of accurate significance evaluation of protein matches, dynamic programming-related approaches have been proposed [31]. However, dynamic programming assumes that a problem (i.e., spectra matching) can be divided into independent components. In the context of tandem spectra, any division based on sequence location creates dependent components because changing an amino acid in any location will change both the *b*-

Table 5. Performance of the target alternative k-permuted decoy databases used with the OMSSA database search program using spectra from 80 unmodified neuropeptides.

Database[a]	Indicator[b]	Significance Levels of the Permutation p-values[c]							Cum. Num. of Peptides[d]	
		0	1	2	3	4	5	≥6	≥10^{-2}	≥10^{-4}
Target	E-value	0	0	1	2	1	3	73	80	77
K10^3	# ions	0	2	78	0	0	0	0	78	0
	Lambda	0	9	71	0	0	0	0	71	0
	p-value	0	2	78	0	0	0	0	78	0
	E-value	0	2	78	0	0	0	0	78	0
K10^4	# ions	0	0	1	79	0	0	0	80	0
	Lambda	0	5	11	64	0	0	0	75	0
	p-value	0	0	1	79	0	0	0	80	0
	E-value	0	0	1	79	0	0	0	80	0
K10^5	# ions	0	0	0	0	80	0	0	80	80
	Lambda	0	5	8	24	43	0	0	75	43
	p-value	0	0	0	0	80	0	0	80	80
	E-value	0	0	0	0	80	0	0	80	80
K10^6	# ions	0	0	0	0	2	78	0	80	80
	Lambda	0	5	8	17	18	32	0	75	50
	p-value	0	0	0	0	0	80	0	80	80
	E-value	0	0	0	0	0	80	0	80	80

[a]Target: database of 236 neuropeptide sequences; K10^3: k-permuted decoy database size of 236,000 peptides; K10^4: k-permuted decoy database size = 2,360,000 peptides; K10^5: k-permuted decoy database size = 23,600,000 peptides; K10^6: k-permuted decoy database size = 236,000,000 peptides.

[b]# ions: permutation p-values computed for the number of matched b- and y-ions. Lambda: permutation p-values computed from the Poisson mean of matches. p-value: permutation p-values computed from the p-value reported by the OMSSA for the matches. E-value: permutation p-values computed using OMSSA E-values.

[c]Significance threshold (t) for matched to be considered significant at p-value < 1×10^{-t}.

[d]Incorrect: the program provided an incorrect match.

[e]Cumulative number of peptides with p-value < 1×10^{-2}.

[f]Cumulative number of peptides with p-value < 1×10^{-4}.

Table 6. Computation times in seconds for search of 80 unmodified spectra against different databases using a single process Intel Core i7-3770 CPU @ 3.40 GHz.

Database[a]	Database Search Program		
	Crux	OMSSA	X! Tandem
Target	5	11	1
K10³	7	56	41
K10⁴	61	915	476
K10⁵	200	1220	467
K10⁶	2162	24475	5196

[a]Target: database of 236 neuropeptide sequences; K10³: k-permuted decoy database size of 236,000 peptides; K10⁴: k-permuted decoy database size = 2,360,000 peptides; K10⁵: k-permuted decoy database size = 23,600,000 peptides; K10⁶: k-permuted decoy database size = 236,000,000 peptides.

and y-ion fragment series. Further any mass change must be balanced by a corresponding change in another part of the sequence such that the overall mass is within the specified tolerance of the original mass. Also, the implementation of these approaches limit high computational requirements by limiting the information considered or through analytical assumptions. These strategies resulted in non-exhaustive libraries that could lead to biased statistical significance assessment. In one case, the algorithm used is location based such that the only one ion series can be used [16,17] due to interrelationship between ion series and that precursor must remain within the preset tolerances. However, using only one series is not as effective as using both ion series and that one ion series can be more informative than the other series [17]. In the other case, the score for a given number of matched peaks is assumed to encompass the score from fewer matched peaks [11]. This assumption fails when different sets of peaks are being matched from the same peptide and the number of peaks in common changes. Both strategies do not consider the optimal starting location such that a peptide will be dropped from consideration when a region of the spectrum has a poor match score despite the higher score in other unevaluated regions. The published algorithms appear to lack error corrections for common problems of incorrect peak assigned due to charge state, presence of chimeric peptides, and missing peaks. Also, both dynamic programming strategies do not have a clear approach to account for peptide length that has been proven to bias the statistical significance of neuropeptides identifications [16]. Lastly, both approaches cannot be directly applied to the open source X! Tandem, Crux and OMSSA unlike the straightforward permutation approach proposed in this study. Although the lack of comparable basis challenges the benchmarking of strategies, the Monte Carlo permuted database approach proposed addresses the previous limitations while enabling simple integration to database search programs and prompt results.

Conclusions

The present study demonstrated that the k-permuted decoy database is an effective and computationally feasible approach to accurately calculate the statistics of neuropeptide matches from complex tandem MS datasets. Unlike other proposed methods to control multiple testing such as target-decoy approaches, permutation testing provided strong control of Type I error such that

neuropeptides are detected at high confidence of significance. The implication of this finding is that an extensive decoy database is not required to accurately detect neuropeptides and this can be exploited to increase the speed of the search.

This study demonstrated the relative superiority of specific detection indicators for database search programs. The indicators E-value, hyperscore, and Sp score from the OMSSA, X! Tandem, and Crux programs, respectively, performed better than other indicators. The results indicated that 10^5 permutations per peptide were sufficient to provide significant peptide identifications. Indication of the suitability of the Monte Carlo permutation approach using 10^5 permutations was the capability of all three database search programs to detect all or nearly all neuropeptides at p-$value < 10^{-4}$ and the absence of a trend for lower statistical significance with higher permutation number. A promising finding is the robust performance of the simple indicator, number of matched ions between the experimental and theoretical spectra to detect peptides. This intuitive indicator identified the vast majority of the peptides also identified by other indicators such as hyperscore, Sp and E-value that rely on assumptions or parametric specifications. This result also highlights the importance of considering the number of matched ions when a match is borderline significant. The results have shown that, in conjunction with database search programs, the k-permuted sequence databases allowed the detection of more peptides and exhibited high consensus among the various indicators and database search programs.

Permutation testing approached developed here can easily be integrated into standard database search programs to compute spectrum specific p-$values$ for any indicator reported by the program. Through the generation of decoy peptides, the permutation approach could offer insights into unknown or unexpected neuropeptides (including those resulting from PTMs or polymorphisms or chimeras) not present in the target database. Further, the k-permuted databases can be generated once and shared between programs and the community.

Author Contributions

Conceived and designed the experiments: BRS SRZ JVS. Performed the experiments: MNA BRS. Analyzed the data: MNA BRS. Contributed reagents/materials/analysis tools: PEA. Contributed to the writing of the manuscript: SRZ JVS BRS MNA PEA.

References

1. Hook V, Funkelstein L, Lu D, Bark S, Wegrzyn J, et al. (2008) Proteases for processing proneuropeptides into peptide neurotransmitters and hormones. Annu Rev Pharmacol Toxicol 48: 393–423.
2. Hummon AB, Amare A, Sweedler JV (2006) Discovering new invertebrate neuropeptides using mass spectrometry. Mass Spectrom Rev 25: 77–98.
3. Zamdborg L, LeDuc RD, Glowacz KJ, Kim YB, Viswanathan V, et al. (2007) ProSight PTM 2.0: Improved protein identification and characterization for top down mass spectrometry. Nucleic Acids Res 35: W701–6.
4. Xie F, London SE, Southey BR, Annangudi SP, Amare A, et al. (2010) The zebra finch neuropeptidome: Prediction, detection and expression. BMC Biol 8: 28-7007-8-28.
5. Zhang X, Petruzziello F, Zani F, Fouillen L, Andren PE, et al. (2012) High identification rates of endogenous neuropeptides from mouse brain. J Proteome Res 11: 2819–2827.
6. Jia C, Lietz CB, Ye H, Hui L, Yu Q, et al. (2013) A multi-scale strategy for discovery of novel endogenous neuropeptides in the crustacean nervous system. J Proteomics 91: 1–12.
7. Southey BR, Lee JE, Zamdborg L, Atkins N, Jr, Mitchell JW, et al. (2014) Comparing label-free quantitative peptidomics approaches to characterize diurnal variation of peptides in the rat suprachiasmatic nucleus. Anal Chem 86: 443–452.
8. Svensson M, Skold K, Svenningsson P, Andren PE (2003) Peptidomics-based discovery of novel neuropeptides. J Proteome Res 2: 213–219.
9. Nesvizhskii AI (2010) A survey of computational methods and error rate estimation procedures for peptide and protein identification in shotgun proteomics. J Proteomics 73: 2092–2123.
10. Perkins DN, Pappin DJ, Creasy DM, Cottrell JS (1999) Probability-based protein identification by searching sequence databases using mass spectrometry data. Electrophoresis 20: 3551–3567.
11. Akhtar MN, Southey BR, Andren PE, Sweedler JV, Rodriguez-Zas SL (2012) Evaluation of database search programs for accurate detection of neuropeptides in tandem mass spectrometry experiments. J Proteome Res 11: 6044–6055.
12. Sadygov RG, Yates JR, 3rd (2003) A hypergeometric probability model for protein identification and validation using tandem mass spectral data and protein sequence databases. Anal Chem 75: 3792–3798.
13. Carr S, Aebersold R, Baldwin M, Burlingame A, Clauser K, et al. (2004) The need for guidelines in publication of peptide and protein identification data: Working group on publication guidelines for peptide and protein identification data. Mol Cell Proteomics 3: 531–533.
14. Kapp EA, Schutz F, Connolly LM, Chakel JA, Meza JE, et al. (2005) An evaluation, comparison, and accurate benchmarking of several publicly available MS/MS search algorithms: Sensitivity and specificity analysis. Proteomics 5: 3475–3490.
15. Frese CK, Boender AJ, Mohammed S, Heck AJ, Adan RA, et al. (2013) Profiling of diet-induced neuropeptide changes in rat brain by quantitative mass spectrometry. Anal Chem 85: 4594–4604.
16. Kim S, Mischerikow N, Bandeira N, Navarro JD, Wich L, et al. (2010) The generating function of CID, ETD, and CID/ETD pairs of tandem mass spectra: Applications to database search. Mol Cell Proteomics 9: 2840–2852.
17. Alves G, Ogurtsov AY, Yu YK (2010) RAId_aPS: MS/MS analysis with multiple scoring functions and spectrum-specific statistics. PLoS One 5: e15438.
18. Ernst MD (2004) Permutation methods: A basis for exact inference. Statistical Science 19: 676–685.
19. Falth M, Skold K, Norrman M, Svensson M, Fenyo D, et al. (2006) SwePep, a database designed for endogenous peptides and mass spectrometry. Mol Cell Proteomics 5: 998–1005.
20. Craig R, Beavis RC (2004) TANDEM: Matching proteins with tandem mass spectra. Bioinformatics 20: 1466–1467.
21. Southey BR, Akhtar MN, Andrén PE, Sweedler JV, and Rodriguez-Zas SL (2013) A comprehensive resource in support of sequence-based studies of neuropeptides. 6: 144.
22. UniProt Consortium (2010) The universal protein resource (UniProt) in 2010. Nucleic Acids Res 38: D142–8.
23. Southey BR, Amare A, Zimmerman TA, Rodriguez-Zas SL, Sweedler JV (2006) NeuroPred: A tool to predict cleavage sites in neuropeptide precursors and provide the masses of the resulting peptides. Nucleic Acids Res 34: W267–72.
24. Park CY, Klammer AA, Kall L, MacCoss MJ, Noble WS (2008) Rapid and accurate peptide identification from tandem mass spectra. J Proteome Res 7: 3022–3027.
25. Geer LY, Markey SP, Kowalak JA, Wagner L, Xu M, et al. (2004) Open mass spectrometry search algorithm. J Proteome Res 3: 958–964.
26. Klammer AA, Park CY, Noble WS (2009) Statistical calibration of the SEQUEST XCorr function. J Proteome Res 8: 2106–2113.
27. Higdon R, Hogan JM, Van Belle G, Kolker E (2005) Randomized sequence databases for tandem mass spectrometry peptide and protein identification. OMICS 9: 364–379.
28. Elias JE, Gygi SP (2007) Target-decoy search strategy for increased confidence in large-scale protein identifications by mass spectrometry. Nat Methods 4: 207–214.
29. Knijnenburg TA, Wessels LF, Reinders MJ, Shmulevich I (2009) Fewer permutations, more accurate P-values. Bioinformatics 25: i161–8.
30. Frank AM, Savitski MM, Nielsen ML, Zubarev RA, Pevzner PA (2007) De novo peptide sequencing and identification with precision mass spectrometry. J Proteome Res 6: 114–123.
31. Yin P, Bousquet-Moore D, Annangudi SP, Southey BR, Mains RE, et al. (2011) Probing the production of amidated peptides following genetic and dietary copper manipulations. PLoS One 6: e28679.

3

The C-Terminal Domain of Nrf1 Negatively Regulates the Full-Length CNC-bZIP Factor and Its Shorter Isoform LCR-F1/Nrf1β; Both Are Also Inhibited by the Small Dominant-Negative Nrf1γ/δ Isoforms that Down-Regulate ARE-Battery Gene Expression

Yiguo Zhang*, Lu Qiu, Shaojun Li, Yuancai Xiang, Jiayu Chen, Yonggang Ren

The Laboratory of Cell Biochemistry and Gene Regulation, College of Medical Bioengineering and Faculty of Life Sciences, Chongqing University, Shapingba District, Chongqing, China

Abstract

The C-terminal domain (CTD, aa 686–741) of nuclear factor-erythroid 2 p45-related factor 1 (Nrf1) shares 53% amino acid sequence identity with the equivalent Neh3 domain of Nrf2, a homologous transcription factor. The Neh3 positively regulates Nrf2, but whether the Neh3-like (Neh3L) CTD of Nrf1 has a similar role in regulating Nrf1-target gene expression is unknown. Herein, we report that CTD negatively regulates the full-length Nrf1 (i.e. 120-kDa glycoprotein and 95-kDa deglycoprotein) and its shorter isoform LCR-F1/Nrf1β (55-kDa). Attachment of its CTD-adjoining 112-aa to the C-terminus of Nrf2 yields the chimaeric Nrf2-C112^{Nrf1} factor with a markedly decreased activity. Live-cell imaging of GFP-CTD reveals that the extra-nuclear portion of the fusion protein is allowed to associate with the endoplasmic reticulum (ER) membrane through the amphipathic Neh3L region of Nrf1 and its basic c-tail. Thus removal of either the entire CTD or the essential Neh3L portion within CTD from Nrf1, LCR-F1/Nrf1β and Nrf2-C112^{Nrf1}, results in an increase in their transcriptional ability to regulate antioxidant response element (ARE)-driven reporter genes. Further examinations unravel that two smaller isoforms, 36-kDa Nrf1γ and 25-kDa Nrf1δ, act as dominant-negative inhibitors to compete against Nrf1, LCR-F1/Nrf1β and Nrf2. Relative to Nrf1, LCR-F1/Nrf1β is a weak activator, that is positively regulated by its Asn/Ser/Thr-rich (NST) domain and acidic domain 2 (AD2). Like AD1 of Nrf1, both AD2 and NST domain of LCR-F1/Nrf1β fused within two different chimaeric contexts to yield Gal4D:Nrf1β^{607} and Nrf1β:C270^{Nrf2}, positively regulate their transactivation activity of cognate Gal4- and Nrf2-target reporter genes. More importantly, differential expression of endogenous ARE-battery genes is attributable to up-regulation by Nrf1 and LCR-F1/Nrf1β and down-regulation by Nrf1γ and Nrf1δ.

Editor: Yoshiaki Tsuji, North Carolina State University, United States of America

Funding: This work was supported by the National Natural Science Foundation of China (key program 91129703 and project 31270879) awarded to Prof. Yiguo Zhang (University of Chongqing, China). The authors state that the funders had no role in study design, data collection and analysis, decision to publish, or preparation of the manuscript.

Competing Interests: The authors declare no competing financial and other interests exist.

* Email: yiguozhang@cqu.edu.cn

Introduction

Collectively, the cap'n'collar (CNC) family of transcription factors controls a variety of critical homeostatic and developmental pathways through regulating the expression of antioxidant response element (ARE)-driven genes, encoding antioxidant proteins, detoxification enzymes, metabolic enzymes and 26S proteosomal subunits [1,2]. The CNC family comprises the founding member *Drosophila* Cnc protein, the *Caenorhabditis elegans* protein skinhead-1 (Skn-1), the vertebrate activator nuclear factor-erythroid 2 (NF-E2) p45 subunit, NF-E2-related factor 1 (Nrf1, including its long form transcription factor 11 (TCF11) and its short form Locus control region-factor 1 (LCR-F1, also called Nrf1β)), Nrf2 and Nrf3, as well as the transcription

repressors Bach1 (BTB and CNC homolog 1) and Bach2 [1,3–6]. All other family members except Skn-1 form a functional heterodimer with a small Maf factor or other basic region-leucine zipper (bZIP) proteins, before binding to ARE sequences in their target gene promoters [7–9].

Amongst mammalian CNC-bZIP proteins, Nrf1 and Nrf2 are two principal factors to regulate ARE-driven cytoprotective genes against cellular stress [10–12]. Surprisingly, most of researches have focused on Nrf2, that is considered to be a master regulator of adaptive responses to oxidative stressors and electrophiles [13,14]. However, Nrf2 is dispensable for development because global knockout of its gene in mice yields viable animals [15]. Specifically, $Nrf2^{-/-}$ mice do not develop any spontaneous cancer, although they are more susceptible than wild-type mice to carcinogens [16]. By contrast, relatively less is known about Nrf1, although global knockout of the LCR-F1/Nrf1β-encoding sequence from its gene locus in the mouse causes embryonic lethality and severe oxidative stress [17–20]. Moreover, conditional knockout of *Nrf1* (also called *nfe2l1*) in the liver, brain and bone results in non-alcoholic steatohepatitis (NASH) and hepatoma [21,22], neurodegeneration [23,24], and reduced bone size [25]. These studies demonstrate that Nrf1 fulfils a unique and indispensable function responsible for maintaining cellular homeostasis and organ integrity.

It is axiomatic that the distinct biological functions of Nrf1 and Nrf2 are determined by differences in their primary structures. It is clearly reported that the function of Nrf2 is dictated by its seven structural domains, namely Nrf2-ECH homology 1 (Neh1) to Neh7 (Fig. 1A), which are conserved amongst different metazoan species [26,27]. The central Neh1 domain comprises both the CNC and bZIP regions, which mediate its DNA-binding to ARE and heterodimerization with other bZIP proteins [8,9,28,29]. The N-terminal Neh2 domain contains a redox-sensitive Kelch-like ECH-associated protein 1 (Keap1)-binding degron, that targets the normal homeostatic Nrf2 protein to the ubiquitin ligase cullin 3/ Rbx1-dependent proteasomal degradation pathway [30–32]; a two-site substrate recognition model was proposed that the forked-stem homodimer of Keap1 binds to the DLG and ETGE motifs within this domain [33–36]. The C-terminal Neh3 domain is required for Nrf2 activity through interaction with chromo-ATPase/helicase DNA-binding protein 6 (CHD6) [37]. The central Neh4 and Neh5 are two transactivation domains (TADs) that contribute to positive regulation of Nrf2 by recruiting CRE-binding protein (CREB)-binding protein (CBP) [38,39]; the activity is controlled through a tight interaction with nuclear receptor-associated coactivator 3 (RAC3, also called steroid receptor coactivator-3 (SRC-3)) [40]. The Neh6 domain, adjacent to the N-terminus of the Neh1 domain, acts as a redox-insensitive β-transducin repeat-containing protein (β-TrCP)-binding degron that negatively regulates Nrf2 [41,42]. The newly-defined Neh7 domain can directly bind to retinoic X receptor α (RXRα), allowing inhibition of Nrf2 activity [27]. In the light of the above information, similar but different structural domains of Nrf1 are identified on the basis of the amino acid sequence similarity with other CNC-bZIP proteins [43].

Significantly, the unique biological function of Nrf1 is dictated by its specific biophysio-chemical properties, such as those provided by the N-terminal domain (NTD, that negatively regulates this CNC-bZIP factor) and the central Asn/Ser/Thr-rich (NST) glycodomain, which are absent from Nrf2. Our previous work showed that negative regulation of Nrf1 by its NTD is associated with the endoplasmic reticulum (ER) through its N-terminal homology box 1 (NHB1) signal sequence, which lacks a signal peptidase cleavage site [43,44]. Upon translation of Nrf1, its NHB1 sequence enables the newly-synthesized polypeptide to be integrated in a specific topological orientation within and around the membranes [45,46]. During topological folding of Nrf1, it is anchored within the ER through its NHB1-associated transmembrane-1 (TM1) region. The connecting TAD sequences, that comprise acidic domain 1 (AD1), NST glycodomain, AD2 and possibly serine-repeat (SR) domain, are transiently translocated into the ER lumen. The NST domain is therein glycosylated to allow Nrf1 to represent an inactive 120-kDa glycoprotein, because the luminal TAD is unlikely to transactivate the nuclear target genes, although its DNA-binding CNC-bZIP domain is positioned on the cyto/nucleoplasmic side of membranes. When stimulated by biological cues, the TAD sequences are partially partitioned out of the ER and retrotranslocated across membranes into the cyto/

nucleoplasmic compartments, where Nrf1 is deglycosylated to yield an active 95-kDa factor because its TADs can gain access to the general transcriptional machinery, enabling transactivation of its target genes. Therefore, the membrane-topological organization of Nrf1 dictates selective post-translational processing of this CNC-bZIP protein to generate a cleaved activator (i.e. 85-kDa), a short weak activator (e.g. 55-kDa LCR-F1/Nrf1β), and other small dominant-negative isoforms (e.g. 36-kDa Nrf1γ and 25-kDa Nrf1δ). However, it is not clear how these Nrf1 isoforms together control the overall transcriptional ability to fine-tune target gene expression to an extent to which is required for cellular homeostasis and organ integrity.

Within AD1, the Neh2-like (Neh2L) subdomain (aa 156–242) does not negatively regulate Nrf1 *via* a Keap1-dependent ubiquitin proteasomal degradation pathway [3,43], although the interaction was examined in total cell lysates [47]. Conversely, Neh2L contributes to the stability of Nrf1, particularly the 120-kDa glycoprotein [48]. The Neh4-like (Neh4L) subdomain is lost in Nrf1 because it is removed by alternative splicing of the transcript encoding the longer isoform TCF11 [49,50]. The Neh5-like (Neh5L) subdomain (aa 280–298) is essential for AD1-mediated transactivation by Nrf1 [12,45]. Similarly, AD2 (and possibly SR domain) appears to positively regulate Nrf1, but its contribution to the short LCR-F1/Nrf1β is unclear. By contrast, the Neh6-like (Neh6L) domain (situated between the TADs and DNA-binding domain) negatively regulate Nrf1 through the β-TrCP-mediated and/or PEST (Pro/Glu/Ser/Thr-enriched degron)-dependent proteolytic degradation pathways [48,51], but its negative effect exerted on LCR-F1/Nrf1β is unknown. Lastly, the C-terminal domain (CTD, aa 686–741) of Nrf1 comprises the major Neh3-like (Neh3L, including a CRAC5 motif and TMc) region and a basic c-tail (BCT) (Fig. 1, B and C), but whether CTD has an effect on Nrf1 similar to the Neh3 domain in Nrf2 is not elucidated.

In the present study we have examined whether: (i) the CTD of Nrf1 exerts a positive or negative effect on its ability to regulate ARE-battery genes; (ii) attachment of the CTD to Nrf2 alters the transactivation activity of the resulting chimaeric factor; (iii) the 36-kDa Nrf1γ or 25-kDa Nrf1δ (both lacking all TAD regions) act as a *bona fide* dominant-negative form that competitively inhibits wild-type Nrf1 and Nrf2; (iv) the 55-kDa Nrf1β acts as a weak activator and is positively regulated by its AD2 and NST domains; and (v) expression of endogenous ARE-driven genes is differentially regulated by distinct Nrf1 isoforms.

Experimental

Chemicals and antibodies

All chemicals were of the highest quality commercially available. Proteinase K (PK) was obtained from New England Biolabs. Mouse monoclonal antibodies against the epitope V5 and Xpress were from Invitrogen Ltd. Antisera against Nrf1 or Nrf2 were produced in rabbits by commercial companies.

Expression constructs

Expression constructs for full-length mouse Nrf1, Nrf1β, Nrf1γ, Nrf1δ, Nrf2 and its mutants, along with a variety of other expression constructs for GFP-CTD fusion protein, and their mutants lacking various lengths of CTD in Nrf1, have been created as described previously [12,43]. A series of expression constructs for Gal4D-Nrf1 and Gal4D-Nrf1 were generated by ligating cDNA fragments that encode various lengths of Nrf1 to the 3′-end of cDNA sequence encoding the Gal4 DNA-binding domain (called Gal4D), within pcDNA3.1Gal4D-V5, through the

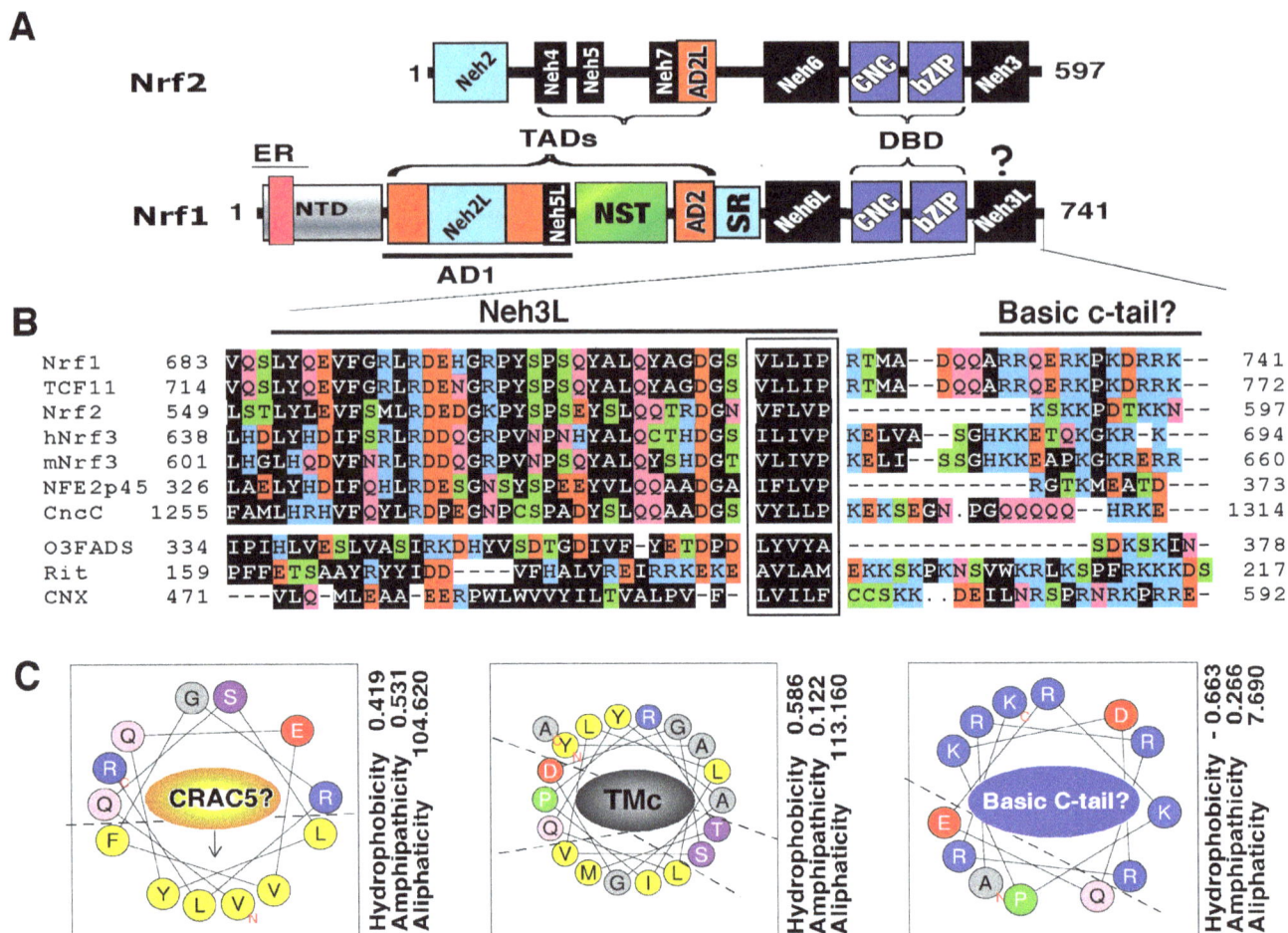

Figure 1. The Neh3L-containing CTD of Nrf1 is conserved in the CNC-bZIP family. (A) Schematic representation of discrete domains of Nrf1 and Nrf2. Locations of the ER signal, transactivation domains (TADs, including AD1, NST and AD2), DNA-binding domain (DBD, including CNC and bZIP) are indicated within Nrf1. The Neh3L region is situated within the C-terminal domain (CTD) of Nrf1. The positive regulation of Nrf2 by its Neh3 domain occurs through direct interaction with CHD6 [37], but it is not identified as one of Nrf1-interacting proteins [71]. **(B)** An alignment of amino acids covering CTD in Nrf1 and other CNC-bZIP factors with ER-resident proteins. The CNC family comprises both water-soluble members (i.e. NF-E2p45 and Nrf2) and membrane-bound NHB1-CNC members (including Nrf1, TCF11, Nrf3, CncC, and Skn-1, albeit the latter lacks both the corresponding ZIP and Neh3L regions). The distinction between Neh3L and Neh3 from Nrf1 and Nrf2 is attributable to different positioning relatively to membranes. Amongst the NHB1-CNC proteins, the core Neh3L is conserved with an ER-resident protein, omeg-3 fatty acid desaturase (O3FADS). Its N-terminally flanking CRAC motif is present in Nrf1 (numbered as CRAC5), TCF11 and Nrf3, but is absent from other members. The C-terminal basic cluster is predicted to possess an ER-retention signal (K/RxK/R), which ensembles to those in calnexin (CNX), O3FADS and Rit (Ras-like protein in all tissues). The conversed hydrophobic pentapeptide is boxed due to the representative in Nrf2 that is essential for its interaction with CHD6 [37]. **(C)** Bioinformatic prediction of three discrete regions within CTD of Nrf1. It is proposed that both CRAC5 and TMc sequences could be wheeled into two relative stable amphipathic helices only upon interaction with amphipathic membranes, whilst a positively-charged helix folded by the basic C-terminal peptide could interact electrically with the putative negatively-charged head group of membrane lipids. Three physico-chemical parameters related with the helical folding (i.e. aliphaticity, hydropathicity and amphipathicity) were calculated using the ProParam tool (http://web.expasy.org/protparam/).

BamHI/EcoRI cloning site [43]. Two different types of chimaeras of Nrf1 with Nrf2 were created, one of which were engineered by replacing the *BamHI/EcoRI* fragment (encoding aa 1–328 of Nrf2) with various length cDNA encoding the N-terminal aa 1–607 of Nrf1 (N607) to yield N607:C270^{Nrf2}, Nrf1^{N607}:C270^{Nrf2} and their mutants (in which C270^{Nrf2} represents the C-terminal 270 aa of Nrf2 being retained); another were made by inserting additional three fragments encoding the C-terminal 112 aa of Nrf1 (C112^{Nrf1}) into the *XhoI/XbaI* sites situated immediately downstream of the entire Nrf2-coding pcDNA3.1/V5 Bis B vector to produce Nrf2:C112^{Nrf1} and its mutants. The fidelity of all cDNA products was confirmed by sequencing.

Cell culture, transfection, and reporter gene assays

Unless otherwise indicated, monkey kidney COS-1 cells (3×10^5, which had been previously purchased from ATCC and maintained in our laboratory) were seeded in 6-well plates and grown for 24 h in Dulbecco's Modified Eagle Medium (DMEM) contain 25 mM glucose and 10% foetal bovine serum (FBS). After reaching 70% confluence, the cells were transfected with a Lipofectamine 2000 (Invitrogen) mixture that contained an expression construct for wild-type Nrf1 or a mutant protein, together with each of $P_{-1061}Nqo1$-Luc, $P_{SV40}Nqo1$-ARE-Luc, $P_{SV40}GSTA2$-6×ARE-Luc or $P_{TK}UAS$×4-Luc [12,43], along with pcDNA4 HisMax/*lacZ* encoding β-galactosidase (β-gal) that was used as a control for transfection efficiency. Additional reporter genes lacking the ARE sequence were used as negative controls. Luciferase activity was measured approximately 36 h after transfection with an expression vector for Nrf1, Nrf1-Nrf2, Gal4-Nrf1, or their mutants and was calculated as fold change (mean ± S.D) relative to 1.0 (of the background activity, i.e. obtained following co-transfection of an empty pcDNA3.1/V5 His B vector and an ARE-driven reporter after subtraction of the non-specific value from co-transfecting an empty pcDNA3.1/V5 His B vector and a non-ARE-containing reporter). The data presented each represent at least three independent experiments undertaken on separate occasions that were each performed in triplicate. Differences in their activities were subjected to statistical analyses.

Real-time qPCR analysis of endogenous genes in cells expressing distinct Nrf1 isoforms or cognate small-interfering RNA (siRNA)

The human embryonic kidney (HEK)-293T cells (that had been previously purchased from ATCC and maintained in our laboratory) were cultured in DMEM supplemented with 5 mmol/L glutamine, 10% (v/v) FBS, 100 units/ ml of either of penicillin and streptomycin) in the 37°C incubator with 5% CO2. The cells had been tranfected with a Lipofectamine 2000 mixture that contained each of expression construct for distinct Nrf1 isoforms, along with an empty pcDNA3 vector (as a blank control) or an Nrf1-target siRNA (5′-CCCAGCAAUUCUACCAGCCU-CAACU-3′) to knock down the endogenous expression) alongside with a scramble siRNA (5′-UUCUCCGAACGUGUCACG-3′); both siRNA sequences were synthesized by Genepharma (Shang-hai China). At 48 h after transfection, these cells were harvested and then subjected to extraction of total RNAs, which were subsequently reverse-transcribed into a single-stranded cDNAs as temples of real-time qPCR reactions using the Perfect Real-Time and SYBR Premix Ex-Taq kit), according to manufacturers' protocols (TaKaRa, Dalian, China). The expression levels of Nrf1 and its target genes in the cells that had been transfected with the vehicle plasmid or the scramble siRNA were given the value of 1.0, and other data were calculated as a ratio of this value and shown as a fold change (mean ± S.D). Notably, mouse $Nrf1^{-/-}$ embryonic fibroblasts (a gift from Dr. Akira Kobayashi with Doshisha University, Japan) had been grown in DMEM containing 10% (v/v) foetal calf serum, 10 μg/ml of insulin, 5.6 μg/ml transferrin, 6.7 ng/ml of sodium selenite, 0.25% (w/v) NaHCO3 and 100 units/ ml of penicillin/streptomycin), and then were allowed to restore distinct Nrf1 isoforms by transfecting their expression constructs alone or in various combinations, followed by real-time qPCR analysis of Nrf1-target gene expression. The data presented each represent at least three independent experiments, each of which were performed in triplicate, followed by statistical analysis of significant differences.

Live-cell imaging combined with in vivo membrane protease protection assays

COS-1 cells expressing each of green fluorescent protein (GFP) fusion proteins with the CTD of Nrf1 or its mutants and ER/DsRed were subjected to in vivo membrane protease protection assays, along with live-cell imaging as reported previously [52]. Briefly, COS-1 cells were permeabilized by digitonin (20 μg/ml) for 10 min, and were then allowed the in vivo membrane protection reactions against digestion by PK (50 μg/ml) for various lengths of time before addition of 0.1% (v/v) Triton X-100 (TX). In the experimental course, the live-cell images were acquired every one min under a 40× objective lens mounted on the Leica DMI 6000 green and red fluorescence microscopes

equipped with a high-sensitivity HAMAMATSH ORCA-ER camera, cell environment control units (at 37°C in 5% CO_2 culture conditions) and a definitive focus module. Relative fluorescence units were measured with Simulator SP5 Multi-Detection system for GFP with 488-nm excitation and 507-nm emission and for DsRed with 570-nm excitation and 50-nm emission. The intensity of the fluorescent signals lined is quantified automatically according to the instrumental software.

Immunocytochemistry followed by confocal imaging

COS-1 cells had been transfected with with expression constructs for V5-tagged Nrf1, GFP-CTD or their mutants (1.3 μg DNA of each) and allowed to recover for 24 h as described elsewhere [43]. Thereafter, subcellular location of proteins was examined by immunocytochemistry followed by confocal imaging. Fluorescein isothiocyanate (FITC)-labelled second antibody was used to locate V5-tagged proteins. Nuclear DNA was stained by 4′,6-Diamidino-2-Phenylindole (DAPI). The ER/DsRed (an ER-localized red fluorescent protein marker) gave a red image in the ER. The merge signal represents the results obtained when the three images were superimposed. The corresponding quantitative data shown here were calculated by determining the percentage of cells (at least 200 cells counted) in which the extra-nuclear stain, i.e. cytoplasmic plus ER (called simply C) was greater than or equal to the nuclear stain (called N), as opposed to the percentage of cells in which the extra-nuclear stain was less than the nuclear stain. Bar = 20 μm.

Western blotting

Equal amounts of protein prepared from cell lysates loaded into each electrophoretic well were monitored by β-gal activity, and visualized by immunoblotting with distinct antibodies against V5, Xpress or Nrf2. Glyceraldehyde-3-phosphate dehydrogenase (GADPH) or β-Actin served as an internal control to verify amounts of proteins that were loaded in each well. On some occasions, nitrocellulose membranes that had already been blotted with an antibody were washed for 30 min with stripping buffer before being re-probed with an additional primary antibody [53].

Bioinformatic analysis

The membrane-topology of Nrf1 was predicted using several bioinformatic algorithms, including the TopPred (http://mobyle.pasteur.fr/cgi-bin/portal.py?form=toppred), HeliQuest (http://heliquest.ipmc.cnrs.fr/) and AmphipaSeek (http://npsa-pbil.ibcp.fr/cgi-bin/npsaautomat.pl?page=/NPSA/npsaamphipaseek.html) programmes. The T-Coffee program was employed to align Nrf1 amino acid sequences with those of its orthologues.

Statistical analysis

The statistical significance of changes in Nrf1 activity and its target gene expression was determined using the Student's t test or Multiple Analysis of Variations (MANOVA). The data presented herein are shown as a fold change (mean ± S.D), each of which represents at least 3 independent experiments undertaken on separate occasions that were each performed in triplicate.

Results

The transactivation activity of Nrf1 is negatively regulated by its CTD

The CTD of Nrf1 is highly conserved amongst different vertebrate species (Fig. S1), and shares 50% sequence identity and 70% similarity with the Neh3 domain of Nrf2, as well as equivalents of other CNC-bZIP factors (Fig. 1B). Although the Neh3 domain was suggested to be required for Nrf2 activity [37], our ARE-driven reporter assay of COS-1 cells (that had been transfected with an expression construct for wild-type Nrf1 or its mutants, illustrated in Fig. 2A) showed evidence to the contrary, that the Neh3L region within CTD contributes to negative regulation of Nrf1 activity (Fig. 2B). Indeed, Nrf1-mediated gene transactivation was significantly increased by deletion of the entire CTD (to yield Nrf1$^{\Delta CTD}$) or its portions (to yield Nrf1$^{\Delta Neh3L}$ and Nrf1$^{\Delta BCT}$, lacking aa 686–733 or 723–741, respectively), to an extent close to that obtained from an NTD-deleted mutant (i.e. Nrf1$^{\Delta NTD}$ lacking aa 1–124 within its NTD), and even near to that of the wild-type Nrf2. Western blotting revealed that Nrf1$^{\Delta NTD}$ has an electrophoretic mobility similar to that of Nrf2 (80-kDa on NuPAGE gels) (Fig. 2C, *upper panel, lanes 3 vs 8*). When compared with wild-type Nrf1, slight faster migrations of mutant glycoproteins (~120-kDa), deglycoproteins (~95-kDa) and cleaved polypeptides (~85-kDa) were observed in Nrf1$^{\Delta CTD}$, Nrf1$^{\Delta BCT}$, Nrf1$^{\Delta Neh3L}$ and Nrf1$^{\Delta CTD}$ (Fig. 2C, *upper panel, cf. lanes 5–7 with 2*). In addition, these three mutants also appeared to reduce generation of the cleaved 85-kDa Nrf1 polypeptide (Fig. 2C, *upper panel*) and/or the shorter Nrf1β/LCR-F1 of 55-kDa (Fig. 2C, *middle panel*), as compared to their wild-type equivalents (*cf. lanes 5–7 with 2*).

Intriguingly, no marked changes in the subcellular location of Nrf1$^{\Delta CTD}$ and Nrf1$^{\Delta BCT}$ were visualised by confocal microscopy, when compared with the wild-type Nrf1 showing primarily extra-nuclear staining that superimposed with the red fluorescent signal from the ER/DsRed maker (Fig. 2, D and E), although previous subcellular fractionation data of the 120-kDa Nrf1$^{\Delta CTD}$ (i.e. Nrf1$^{\Delta 686-741}$) mutant protein revealed a modest enhancement of its recovery in ER fractions relative to wild-type glycoprotein [12] (indicating that integration of Nrf1 within ER is unaffected by loss of its CTD). By contrast, the Nrf1$^{\Delta Neh3L}$ mutant (in which a classic bipartite nuclear localization signal (NLS) sequence ^{625}RDIRRRGKNKMAAQNCRKRKL^{645} is retained) enabled a significant increase in the nuclear staining of ~55% cells observed (Fig. 2, D and E). By contrast, deletion of the basic region encompassing the NLS (to yield the Nrf1$^{\Delta BR}$ mutant) caused a significant decrease in the nuclear staining of ~3% cells, as revealed by a relative enhancement in either the cytoplasmic staining of ~90% cells (Fig. 2, D and E) or the abundance of Nrf1$^{\Delta BR}$ mutant proteins between 120-kDa and 85-kDa (Fig. 2C, *upper panel*). Such a decrease in the nuclear localization of Nrf1$^{\Delta BR}$ resulted in a reduction in the reporter gene activity to approximately 40% of that obtained from the wild-type Nrf1 (Fig. 2B). Therefore, the transactivation activity of Nrf1 (and possibly its protein stability) is monitored by its NLS-containing basic region, enabling the CNC-bZIP protein to translocate the nucleus before binding ARE sequences in target genes. By contrast, the basic c-tail ^{730}RRQERKPKDRRK^{741} within CTD is not required for the nuclear localization of Nrf1 rather than for its primary cytoplasmic location.

Attachment of the CTD of Nrf1 to GFP enables the chimaeras to associate with the ER

To address whether the basic c-tail ^{730}RRQERKPKDRRK^{741} adjacent to the amphipathic Neh3L region of Nrf1 is required for its association with the ER or its localization in the nucleus, we attached the entire CTD and its portions to the C-terminus of GFP (giving GFP-CTD and mutant constructs, Fig. 3A) and then cotransfected them with an ER-resident marker into COS-1 cells, before being subjected to fixed-cell microscopy experiments. When compared with the primary nuclear imaging of GFP alone

(as theoretically a protein with molecule mass of less than 40 kDa can easily translocate the nucleus), an increase in the extra-nuclear subcellular distribution of the fusion protein GFP-CTD appeared to be similar to that of GFP-CTD$^{\Delta 701-741}$ (Fig. 2B). By contrast, CTD$^{\Delta 731-741}$, CTD$^{\Delta 723-741}$, CTD$^{\Delta 714-722}$ or CTD$^{\Delta 705-713}$ gave a slight increase in the nuclear green staining of GFP fusion protein, relative to the extra-nuclear staining (Fig. 3B). This observation indicates that the basic c-tail of Nrf1 does not act as functional NLS, and give rise to a possibility that the basic c-tail adjoining the amphipathic TMc region enables the GFP-CTD fusion protein to be retained in putative extra-nuclear subcellular compartments.

To test this hypothesis, we performed live-cell imaging of GFP-CTD and its mutants combined with *in vivo* membrane protease protection assays, in order to determine whether the extra-nuclear-localized proteins is capable of being either associated with the cytoplasmic side of the ER or dislocated from the lumen to the cytoplasmic side of the membrane. In these experiments, COS-1 cells that had been transfected with expression constructs for CTD-GFP or its mutants together with ER/DsRed were first pre-treated for 10 min with digitonin to pemeabilize cellular membranes before being challenged with PK for 3–90 min (in the presence of digitonin) to digest cytoplasmic proteins. Thus we envisaged that if extra-nuclear GFP fusion protein was not tethered to membranes, it would diffuse rapidly into the extracellular compartments while the plasma membrane was disrupted by digitonin; otherwise, if the GFP fusion protein was associated with the ER membranes or transferred from the lumen to the cytoplasmic side of the membrane, it would become vulnerable to digestion by PK. As anticipated, Figure 3 (C and D) showed the green fluorescent signal from extra-nuclear CTD-GFP and GFP:CTD$^{\Delta 731-741}$ appeared to be partially superimposed upon red fluorescent images presented by ER/DsRed. No apparent change in the intensity of their green signals was observed within 10 min following treatment of the cells with digitonin (Fig. 3, C and D), but subsequently exposure to PK allowed the GFP fusion protein to be digested gradually from 3 min to 20 min before being destroyed completely. Similarly, live-cell imaging of GFP-CTD$^{\Delta 723-741}$ and GFP-CTD$^{\Delta 714-722}$ revealed that the extra-nuclear subcellular proteins appeared to be unaffected by digitonin, and most of them were completely digested by PK within 10 min (Fig. 4A) or ~20 min (Fig. 4B). By contrast, GFP-CTD$^{\Delta 701-741}$ (retaining a cholesterol recognition/amino acid consensus (CRAC5) motif only) gave strong staining in both the nuclear and extra-nuclear compartments (Fig. S2A), in which it appeared insensitive to PK until the protease incubation time was extended to 45 min, and then it was digested to disappear by 48 min (Fig. S2B). Taken together, these results indicate that an extra-nuclear subcellular fraction of GFP-CTD could be tethered to the cytoplasmic side of ER membranes or be dynamically dislocated into cytoplasmic subcellular compartments.

LCR-F1/Nrf1β is a weak activator, whilst Nrf1γ (or Nrf1δ) acts as a dominant-negative inhibitor

To identify the biological significance of the CTD within distinct Nrf1 isoforms, we created a series of expression constructs for five distinct isoforms (Fig. 5A) into COS-1 cells, in order to determine which isoforms are activators or dominant-negative isoforms. The $P_{SV40}Nqo1$-ARE-Luc reporter gene assay showed that the shorter LCR-F1/Nrf1β exhibited a ~1.6-fold transactivation activity against the background value of 1 (which was measured from transfection of cells with an empty pcDNA3 vector; it is notable that the baseline reporter activity is theoretically deduced to be mediated by endogenous Nrf1, Nrf2 and/or other transcription factors that enable to bind the ARE-

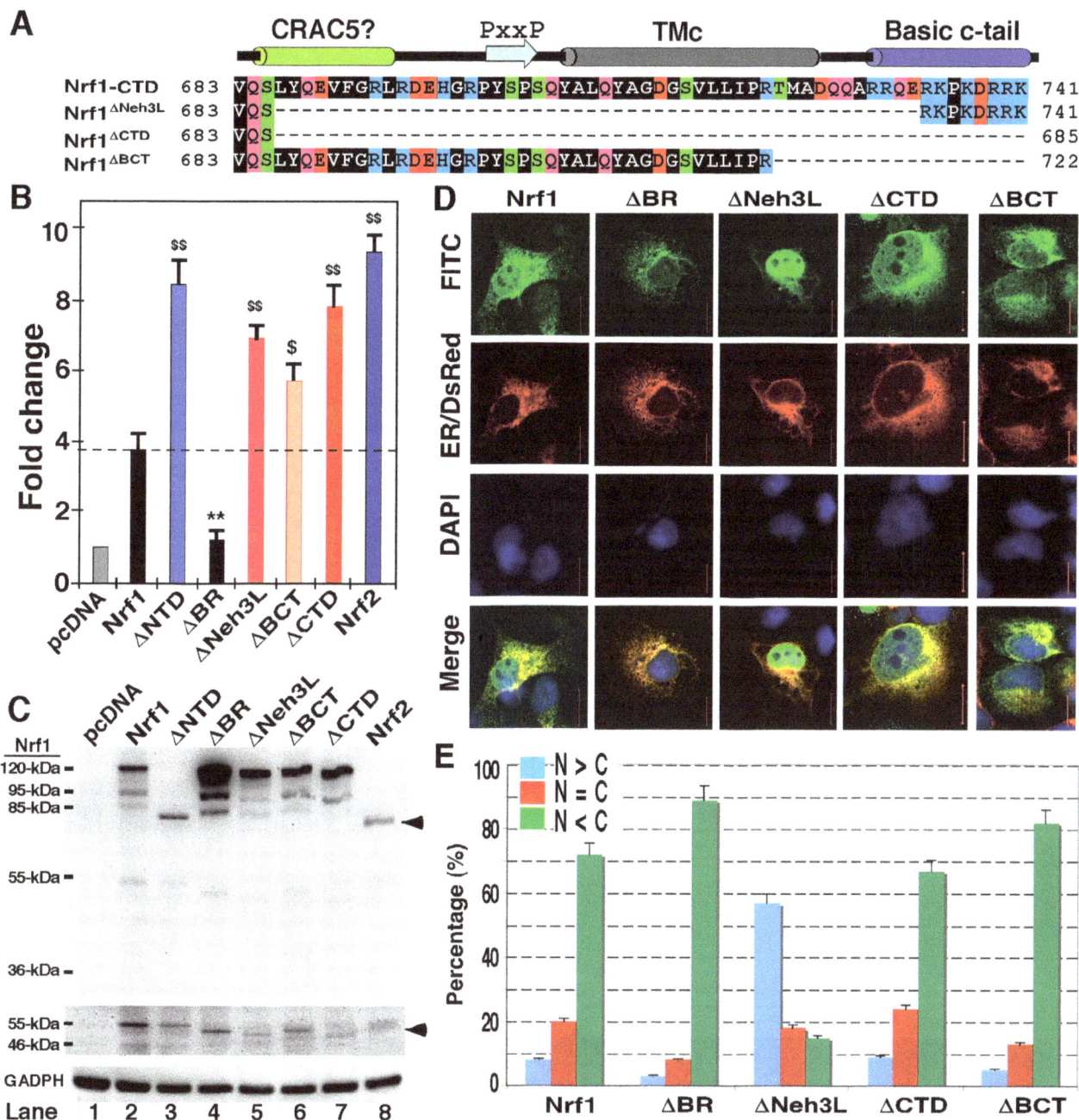

Figure 2. Nrf1 is negatively regulated by its CTD. (**A**) Diagrammatic representation of various lengths of CTD in Nrf1 and its mutants. The putative secondary structure of discrete regions within CTD is shown (*upper cartoon*). (**B**) Luciferase activity was measured from COS-1 cells had been transfected with each of expression constructs for Nrf1 or its mutants (1.2 μg), together with $P_{SV40}Nqo1$-ARE-Luc (0.6 μg) and β-gal plasmid (0.2 μg), and allowed to recover in fresh media for an additional 24 h before lysis. The data were calculated as a fold change (mean ± S.D) of transactivation by Nrf1 or its mutants. Significant increases ($, p<0.05 and $$, p<0.001, n=9) and decreases (**, p<0.001, n=9) in activity relatively to wild-type Nrf1 are indicated. (**C**) The above-prepared cell lysates (30 μg of protein) were resolved by gradient LDS/NuPAGE containing 4–12% polyacrylamide in a Bis-Tris buffer system and visualized by western blotting with antibody against the V5 epitope. The amount of proteins loaded into each electrophoresis sample well was adjusted to ensure equal loading of β-gal activity. An arrow indicates Nrf2 with a molecular mass of ~80-kDa estimated (*upper panel*), whereas another arrow points to the brightly-contrasted band of ~55-kDa Nrf1β (*middle panel*) that was cropped from the same gel as shown in the upper panel. GAPDH served as a protein-loading control (*lower panel*). It is notable that the same protein exhibits distinct mobility on different electrophoretic gels in different running buffer systems (*cf.* Figs. 2*C* with 5*C*). (**D**) COS-1 cells were co-transfected with 1.3 μg DNA of each of the above-described expression constructs and 0.2 μg of the ER/DsRed plasmid, and then allowed to recover from transfection for 24 h before being fixed. Subcellular location of proteins was examined by immunocytochemistry followed by confocal imaging. FITC-labelled second antibody was used to locate V5-tagged proteins. Nuclear DNA was stained by DAPI. The ER/DsRed gave a red image in the ER. The merge signal represents the results obtained when the three images were superimposed. (**E**) The quantitative data of imaging (corresponding to those shown in panel **D**) were calculated by determining the percentage of cells (at least 200 cells counted) in which the extra-nuclear stain, i.e. cytoplasmic plus ER (called simply C) was greater than or equal to the nuclear stain (called N), as opposed to the percentage of cells in which the extra-nuclear stain was less than the nuclear stain. Bar=20 μm.

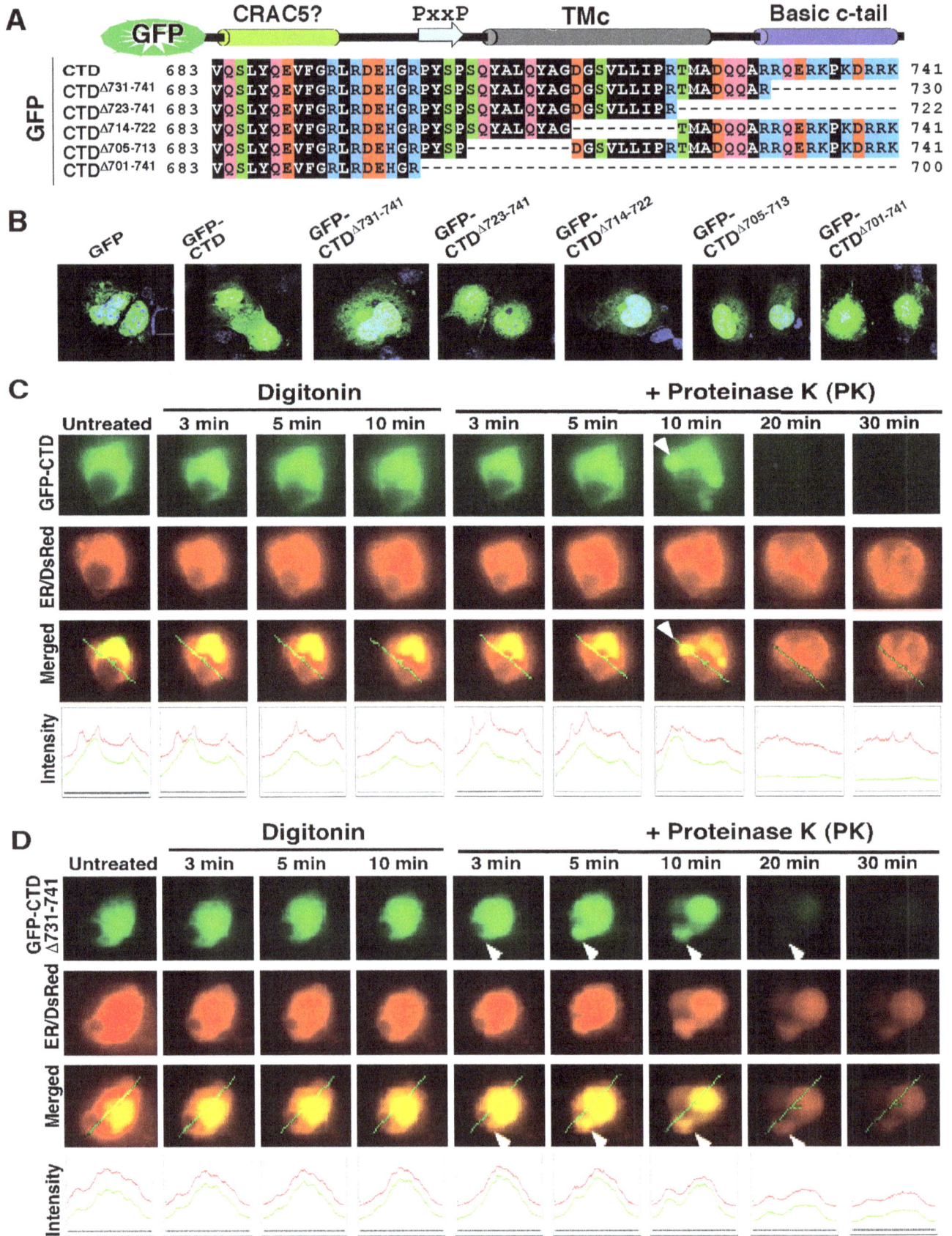

Figure 3. Imaging of fixed and live cells expressing GFP fusion protein with CTD of Nrf1 or its mutants. (A) Schematic of Six expression constructs for the GFP-CTD fusion protein and its mutants; these fusion proteins have been created by attachment of various lengths of CTD of Nrf1 to the C-terminus of GFP. (B) These indicated expression constructs each were transfected into COS-1 cells for 6 h. The cells were then allowed to

recover from transfection in fresh medium for 18 h before being fixed by 4% paraformaldehyde and stained for the nuclear DNA by DAPI. The green signals from GFP were observed under confocal microscope and merged with the DNA-staining images. (C and D) Live-cell imaging of GFP-CTD and its mutant GFP-CTD$^{\Delta731-741}$(lacking its basic c-tail). COS-1 cells had been transfected with expression constructs for either GFP-CTD (C) or GFP-CTD$^{\Delta731-741}$ (D), together with the ER/DsRed marker, before being subjected to real-time live-cell imaging combined with the *in vivo* membrane protease protection assay. The cells were permeabilized by digitonin 20 μg/ml) for 10 min, before being co-incubated with PK (50 μg/ml) for 30 min. In the time course, real-time images were acquired using the Leica DMI-6000 microscopy system. The merged images of GFP with ER/DsRed are placed (on *the third raw of panels*), whereas changes in the intensity of their signals are shown graphically (*bottom*). Overall, the images shown herein are a representative of at least three independent experiments undertaken on separate occasions that were each performed in triplicate (n = 9). The *arrow* indicates a 'hernia-like' vesicle protruded from the cytoplasm.

driven gene) (Fig. 5B), as it was expressed as a major single band of 55 kDa (Fig. 5C, *lane 2*). By contrast, over-expression of Nrf1β$_2$ (aa 424–741), Nrf1γ (aa 548–741) and Nrf1δ (aa 590–741) significantly diminished ARE-driven reporter gene activity from 40% to 20% of the background level (Fig. 5B). All the three small isoforms of Nrf1 were expressed as major proteins with molecular masses of between 38-kDa and 25-kDa, as migrated on the general electrophoretic gels containing 12% polyacrymide in an SDS Bis-Tris buffer system (pH 8.9), which were smeared with high mass polypeptide ladders from 46-kDa to 38-kDa (Fig, 5C). Together, these results indicate that the 55-kDa Nrf1β is *per se* a weaker activator than the full-length Nrf1, whereas smaller isoforms Nrf1β$_2$, Nrf1γ and Nrf1δ exert differentially dominant-negative effects on its activity.

Further examination of co-transfected cells by $P_{-1061}nqo1$-*Luc* and $P_{SV40}Nqo1$-*ARE-Luc* reporter gene assays revealed that forced expression of Nrf1γ caused a significant decrease in the transactivation activity of the full-length Nrf1 to ~60% of the background level (Fig. 5D1, *column 5*), whilst the activity of Nrf1 appeared to be unaffected by expression of LCR-F1/Nrf1β (as shown in Fig. 6B1, *left panel*). In the parallel experiments, Nrf1γ also dominantly repressed Nrf2-mediated reporter gene activity to ~25% of its original transactivation activity obtained from cells that had been transfected with an Nrf2 expression construct alone (Fig. 5D2, *column 10*). In addition, western blotting of these proteins that had been separated by 4–12% NuPAGE in an LDS Bis-Tris buffer system (pH 7.0) revealed no changes the abundance of Nrf1 and Nrf2 when co-transfected with Nrf1γ (Fig. 5E, *upper two panels*). By contrast, the transactivation activity of Nrf2 was neither suppressed by LCR-F1/Nrf1β (Fig. 6B2, *right panel*) nor influenced by the full-length Nrf1 (Fig. 5D2, *column 9*). Moreover, expression of Nrf1 and Nrf2 is also unaffected by LCR-F1/Nrf1β (Fig. 6C). Collectively, these findings demonstrate that Nrf1γ, but not LCR-F1/Nrf1β, is a *bona fide* dominant-negative form competitively against the intact wild-type factors Nrf1 and Nrf2.

LCR-F1/Nrf1β is positively and negatively regulated by its AD2 and CTD, respectively

Since CTD has the negative effect on the intact full-length Nrf1, we examine whether this domain auto-inhibits LCR-F1/Nrf1β. As expected, an approximately 1.5~2.0-fold increase in the transactivation activity of LCR-F1/Nrf1β resulted from removal of its essential Neh3L region (in Nrf1β$^{\Delta Neh3L}$) or the entire CTD (in Nrf1β$^{\Delta CTD}$) (Fig. 6A, *cf. columns 6 & 7 with 2*). These two CTD-deficient mutants of LCR-F1/Nrf1β also caused an enhancement in the reporter gene transactivation by Nrf1 or Nrf2 (Fig. 6B, *columns 6 & 7*), when compared to the indicated controls. However, removal of either CTD or its portions from Nrf1γ did neither disinhibit its dominant negative effect on ARE-driven gene activity (Fig. 6A, *cf. columns 9–10 with 8*), and nor influence the nuclear location of the smaller isoform (Fig. 7A). These data indicate that CTD is not ascribed to the dominant negation by Nrf1γ, albeit it enables negative regulation of both Nrf1 and LCR-

F1/Nrf1β. Furthermore, the activity of LCR-F1/Nrf1β was abolished by loss of the DNA-binding basic region in the Nrf1β$^{\Delta BR}$ mutant, but it did not act as a potential dominant-negative competitor against either endogenous or exogenous activities of Nrf1 and Nrf2 (Fig. 6, A and B, *columns 5 vs 2*). This is likely to be associated with a decrease in its nuclear staining to ~5% of cells examined (Fig. 7A).

To gain an insight into the biological significance of LCR-F1/Nrf1β, we next examined which regions of this isoform contribute to its transactivation activity to positively regulate ARE-driven genes. As shown in Figure 6A (*cf. columns 3 with 2*), the Nrf1β$^{\Delta413-448}$ mutant (lacking a major acidic-hydrophobic amphipathic region within AD2) significantly reduced Nrf1β-mediated transactivation activity to ~40% of the background level. Similarly, the Nrf1β$^{\Delta403-506}$ mutant (lacking the entire AD2 together with both SDS1 and SR/PEST2 degrons) exhibited ~60% of the background activity, though its dominant-negative effect was modestly alleviated as compared with that of Nrf1$^{\Delta413-448}$ (Fig. 6A, *columns 4 vs 3*). Moreover, both Nrf1β$^{\Delta413-448}$ and Nrf1β$^{\Delta403-506}$ enabled dominant-negative competition against the wild-type Nrf1 and/or Nrf2 activity to be blunted markedly (Fig. 6B, *columns 3 & 4*). In turn, these results demonstrate that AD2, but not the SR domain, contributes to the positive regulation of Nrf1β/LCR-F1. Upon Nrf1β/LCR-F1 co-expression, the abundance of Nrf1 and Nrf2 is unaffected (Fig. 6C).

Blockage of Nrf1γ expression results in an increase in the activity of Nrf1 and LCR-F1/Nrf1β

Distinct Nrf1 isoforms are proposed to arise from the in-frame translation from an internal ATG codon or the selective post-translational processing of a larger Nrf1 isoform [46,48]. As shown in Figure 7B, generation of the 55-kDa LCR-F1/Nrf1β was completely abolished by the Nrf1$^{4xM/L}$ mutant (in which the internal translation start codons of Nrf1 at Met$^{289/292/294/297}$ were mutated into leucines), when compared to the wild-type Nrf1 expression (*lanes 3 vs 2*). This mutant also caused an increase in its activity to transactivate $P_{sv40}nqo1$-*ARE*-luc reporter gene by approximately 50% more than that of the wild-type Nrf1 activity [46]. Further examinations revealed that expression of additional two polypeptides of 46-kDa and 43-kDa, besides the 55-kDa LCR-F1/Nrf1β was also, to a lesser extent, prevented by the Nrf1$^{4xM/L}$ mutant (Fig. 7B, *lane 3*). Similarly, two polypeptides of 46-kDa and 43-kDa appeared to be expressed in COS-1 cells that had been transfected with the Nrf1β$^{\Delta403-506}$ mutant (which was constitutively tagged by the Xpress or V5 epitopes at its N-terminus and C-terminus, respectively); the 46-kDa protein was detected by western blotting with both antibodies against V5 and Xpress, whilst the 43-kDa protein was only immunoblotted by antibodies against V5 rather than Xpress (Fig. 6A, *lane 4 in panel a1 vs a2*). These results rise to a possibility that proteolytic processing of the 55-kDa LCR-F1/Nrf1β occurs within and around its NST domain to yield 46-kDa and 43-kDa polypeptides.

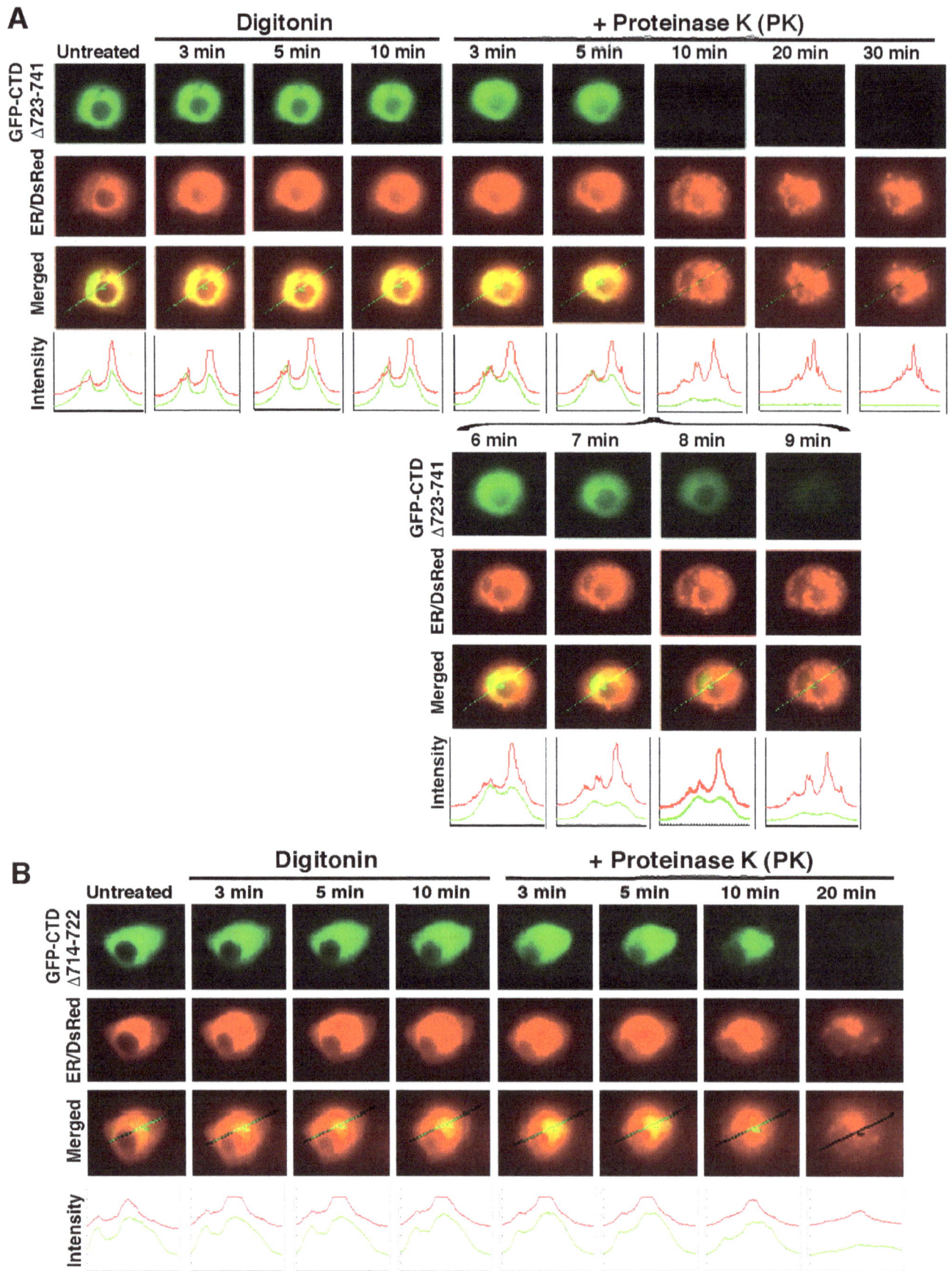

The abundance of the dominant-negative Nrf1γ isoforms (i.e. 36-kDa or possibly 38-kDa in Fig. 5C, *lane 4*) appeared to be unaffected by expression of Nrf1$^{4xM/L}$ (Fig. 7B, *lanes 3 vs 2*), but it was obviously diminished by the Nrf1β$^{M/L}$ mutant (in which an additional potential translation start codon at Met^{548} within Nrf1β was mutated into leucine) (Fig. 7B, *lanes 6 vs 7*), implying that Nrf1γ also arises from the internal translation. The failure to produce the dominant-negative Nrf1γ resulted in a ~2.0-fold increase in the reporter gene transactivation when compared to LCR-F1/Nrf1β-mediated activity (Fig. 7C, *columns 8 vs 7*). Interestingly, a further increase in the transactivation activity of LCR-F1/Nrf1β to ~4.0-fold activation was determined by deletion of the CTD to yield Nrf1β$^{M/LDCTD}$ (Fig. 7C, *columns 10 vs 8*). Similarly, LCR-F1/Nrf1β activity was also modestly enhanced by the Nrf1β$^{M/LDBCT}$ mutant (*columns 9 vs 8*). Furthermore, a combinative mutant Nrf1$^{5xM/L}$ (in which all five potential ATG start codons at Met$^{289/292/294/297}$ and Met^{548} were mutated into leucines) exhibited a higher activity of ~6.0-fold induction of $P_{SV40}Nqo1$-*ARE-Luc* reporter gene (Fig. 7C, *column 3*). Excitingly, the activity of Nrf1$^{5xM/L}$ was further increased to approximately 8.0- to 10.0-fold transactivation by loss of its N-terminal ER-targeting sequence and its CTD in Nrf1$^{5xM/L;\Delta ER+CTD}$ (Fig. 7C, *column 6*). Together, these results demonstrate that LCR-F1/Nrf1β is a weak activator that transactivate Nrf1-target gene expression, because its activity, as well as that of wild-type Nrf1, is negatively regulated by CTD and also is predominantly inhibited by smaller molecular weight dominant-negative forms of between 46-kDa and 25-kDa.

Chimaeric Gal4/Nrf1 and Gal4/Nrf1β factors are positively regulated by AD2 and NST domains, but are also negatively regulated by the Neh6 domain and its adjoining region

To determine that AD2 contributes to the transactivation activity of LCR-F1/Nrf1β, we created a series of chimaeric Gal4/Nrf1β factors by fusing Gal4 DNA-binding domain (Gal4D) with the N-termini of various portions of LCR-F1/Nrf1β and its mutants (Fig. 8A, *left panel*). P_{TK} *Gal4-UAS*-Luc reporter assay displayed a ~510 fold activity mediated by Gal4D/Nrf1β^{607} (containing the NST, AD2, SR/PEST2 and Neh6 regions, *column 11*). Lack of the acidic-hydrophobic amphipathic aa 413–448 of AD2 (in Gal4D/Nrf1β$^{607\Delta413\_448}$) caused a significant decrease in the reporter gene activity to ~30-fold (Fig. 8A, *column 12*). However, loss of both the entire AD2 and SR/PEST2 regions (in Gal4D/Nrf1β$^{607\Delta403-506}$) did not further decrease its transactivation activity (*column 13*). These data indicate that AD2 (and possibly the NST domain) is required for Nrf1β-mediated transactivation, whereas the SR/PEST2 sequence appears to be dispensable for the transactivation activity of Gal4D/Nrf1β^{607}. By comparison with Gal4D/Nrf1β^{607}, Gal4D/Nrf1β^{460} (retaining NST and AD2) exerted a higher activity to enable ~790-fold activation of the target reporter gene (Fig. 8A, *column 9*). Gal4D/Nrf1β^{526} (retaining NST, AD2 and SR/PEST2) also showed a ~630-fold activity, which was more than the ~510-fold induction by Gal4D/Nrf1β^{607} (Fig. 8A, *columns 10 vs 11*). These observations indicate that Nrf1β^{607}-mediated transactivation activity is positively regulated by both its AD2 and NTD regions, but is negatively regulated by its adjacent SR/PEST2 and Neh6L

regions. Further comparison of Gal4D/Nrf1β^{460} and Gal4D/$^{\Delta120}$N395$^{\Delta173-286}$ mutants (only retaining most of the NST domain) revealed that the latter Gal4D/$^{\Delta120}$N395$^{\Delta173-286}$ still exhibited a ~320-fold activity to mediate P_{TK} *Gal4-UAS*-Luc reporter gene expression, implying that besides AD2, the NST domain contributes to the positive regulation of Gal4D/Nrf1β^{460} (Fig. 8A, *columns 9 vs 8*).

The molecular basis for the positive regulation of the full-length Nrf1 by its AD2 was herein investigated by measuring differences in between the transcription activities of a series of chimaeric Gal4/Nrf1 factors (in which Gal4D was fused N-terminally with various portions of Nrf1 and its mutants). Gal4D/N460 (in which N460 represents the N-terminal 460 aa of Nrf1, containing AD1, the NST domain, and AD2) exhibited a ~810-fold transactivation activity, whereas loss of AD2 in Gal4D/N395 (i.e. the N395 portion of Nrf1 contains AD1 and the NST domain) decreased the reporter gene transactivation to ~590 fold (Fig. 8A, *columns 5 vs 6*). Conversely, a significant increase in the transcriptional activity to a maximal ~1,000-fold induction was observed by the removal of NTD from Gal4D/N395 (to yield the Gal4D/$^{\Delta120}$N395 mutant), and in turn the transactivation activity was markedly decreased to ~320 fold by deletion of most of AD1 (to yield the Gal4D/$^{\Delta120}$N395$^{\Delta173-286}$ mutant) (*columns 6 to 8*). These results demonstrate that Gal4/Nrf1 chimaeric factor is positively regulated by its AD1, NST and AD2 regions, and is negatively regulated by its NTD.

Further examinations of Gal4D/N526 (i.e. the N526 portion of Nrf1 contains NTD, AD1, NST, AD2 and SR/PEST2), its mutants Gal4D/N526$^{\Delta413-448}$ (lacking the major amphipathic portion of AD2 within Nrf1) and Gal4D/N526$^{\Delta403-506}$ (lacking both the entire AD2 and SR/PEST2 regions) (Fig. 8A, *columns 2 to 4*) revealed that AD2, rather than SR/PEST2, is required for transactivation by Gal4/N526. Comparison of Gal4D/N460, Gal4D/N526 and Gal4D/N607 (i.e. the N607 portion of Nrf1 covers its NTD, AD1, NST, AD2, SR/PEST2 and Neh6L regions) showed that ~810-fold activity of Gal4-driven reporter gene induced by Gal4D/N460 was significantly decreased from ~620-fold to ~400-fold changes measured from Gal4D/N526 and Gal4D/N607, respectively (Fig. 8A, *columns 1 to 3*). In addition, our previous publications had reported that Gal4D/N607 exhibited a similar transactivation activity to that of the wild-type full-length Gal4D/Nrf1 [12,46] (and thus Gal4D/N607 is herein used as a reference basis to create the above-described series of mutants illustrated in Fig. 8A, *left panel*). Collectively, these results demonstrate that both SR/PEST2 and Neh6L regions contribute to negative regulation of chimaeric Gal4D/N607 (or Gal4/Nrf1) factor.

Both AD2 and NST domains contribute to up-regulation of Nrf2-target gene by its chimaeras N607:C270^{Nrf2} and Nrf1β^{N607}:C270^{Nrf2}

Herein, whether AD2-containing TAD elements of Nrf1 make differential contributions to its chimaeric factors N607:C270^{Nrf2} and Nrf1β^{607}:C270^{Nrf2} (Fig. 9A) are investigated by measuring $P_{SV40}Nqo1$-*ARE-Luc* reporter gene assay. As shown in Figure 9B, N607:C270^{Nrf2} was created by fusing the N607 portion of Nrf1 with the C-terminal 270-aa portion of Nrf2 (designated C270^{Nrf2}, which retains its Neh6, CNC, bZIP and Neh3 domains).

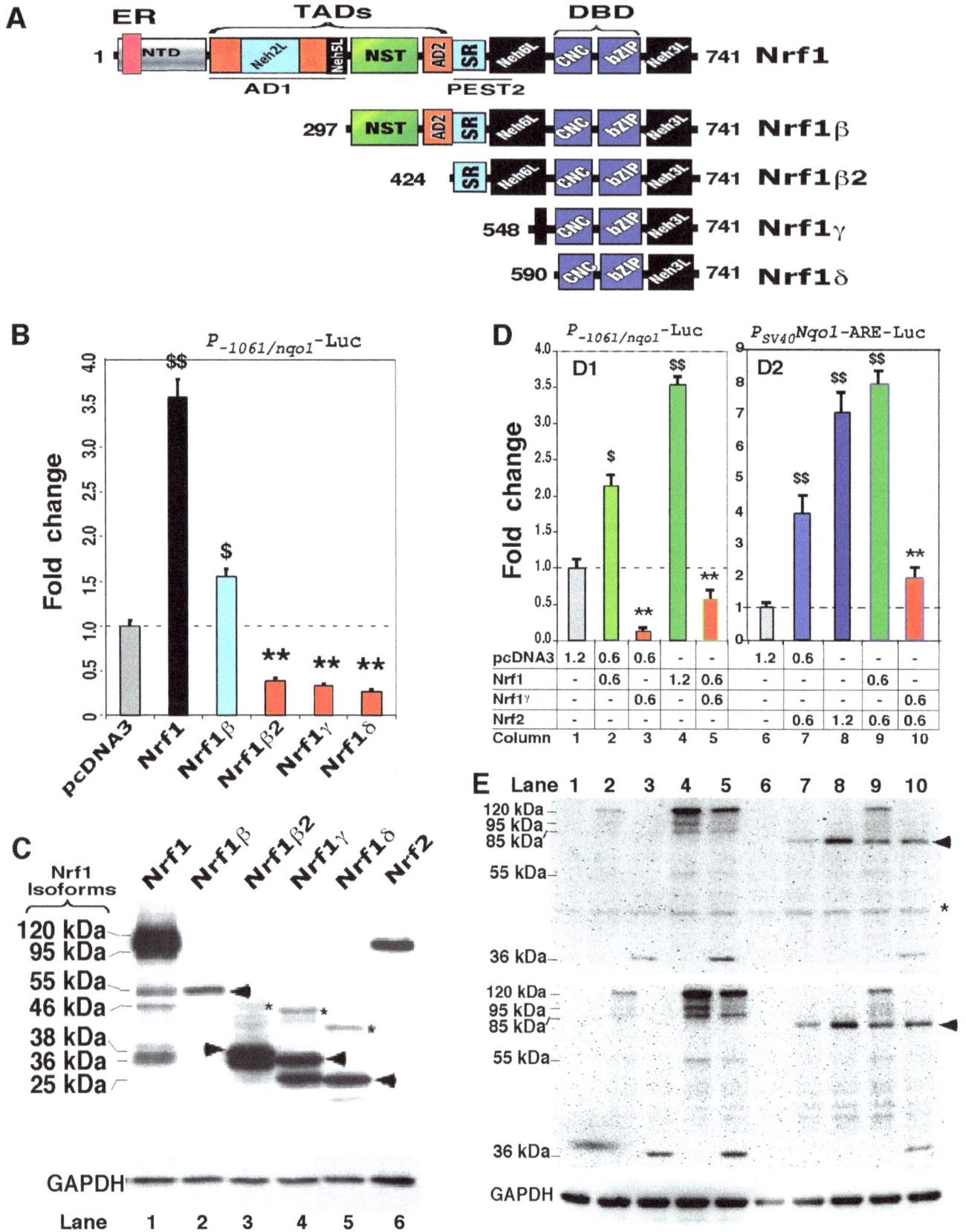

Figure 5. Opposing regulation of ARE-driven reporter genes by distinct Nrf1 isoforms. (**A**) Schematic shows structural domains of five different isoforms of Nrf1. Locations of ER-targeting signal, AD1 and PEST2 are also indicated within distinct domains. (**B**) Shows luciferase reporter gene activity measured from COS-1 cells that had been co-transfected with 1.2 μg of each expression construct for Nrf1 isoforms, together with 0.6 μg of $P_{-1061/nqo1}$-Luc (that is driven by the 1061-bp promoter of *Nqo1*) and 0.2 μg of β-gal plasmid. The data were calculated as a fold change

(mean ± S.D) of transactivation by distinct Nrf1 isoforms. Significant increases (\$, p<0.05 and \$\$, p<0.001, n = 9) and decreases (**, p<0.001, n = 9) in activity were calculated relatively to the background activity (obtained from transfection of cells with an empty pcDNA3 with reporter plasmids). (**C**) Total lysates of COS-1 cells expressing each of Nrf1 isoforms or Nrf2 were resolved by 12% SDS-PAGE in a Bis-Tris buffer system and visualized by immunoblotting with the V5 antibody. The position of migration of the V5-tagged polypeptide was estimated to be 120, 95, 55, 46, 38, 36 and 25 kDa, and GAPDH was used as an internal control to verify amounts of proteins loaded into each electrophoretic well. (**D**) Nrf1γ inhibits transactivation of ARE-driven genes by Nrf1 or Nrf2. COS-1 cells were co-transfected with indicated amounts of expression constructs for Nrf1, Nrf1γ and/or Nrf2, together with 0.6 µg of $P_{-1061}nqo1$-Luc (**D1**) or $P_{SV40}Nqo1$-ARE-Luc (**D2**) and 0.2 µg of β-gal plasmid. Thereafter, luciferase activity was measured and is shown as a fold change (mean ± S.D). Significant increases (\$, p<0.05 and \$\$, p<0.001, n = 9) and decreases (**, p<0.001, n = 9) in activity relatively to the background activity are indicated. (**E**) Total lysates of COS-1 cells co-transfected with expression constructs for Nrf1, Nrf1γ and/or Nrf2 alone or in combination (as indicated corresponding to those in panel D) was subject to separation by 4–12% LDS/NuPAGE in a Bis-Tris buffer system. The *upper two panels* represent similar images from different independent gels, on which location of Nrf2 migration is *arrowed*, whilst a non-specific protein band is *starred* (*). The position of the V5-tagged Nrf1 polypeptides of 120, 95, 85, 55, and 36 kDa is indicated. It is notable that the same proteins exhibit distinct mobilities on different electrophoric gels in different running buffer systems (cf. **C** with **E**).

Luciferase assay showed a ~4.3-fold transactivation activity of N607:C270^{Nrf2} (Fig. 9C, *column 1*), which was modestly greater than the wild-type Nrf1 activity (i.e. ~3.2-fold), but was much lesser than the wild-type Nrf2 activity (i.e. ~8.2-fold). Removal of most of NTD from the N607 portion yielded the $^{\Delta120}$N607:C270^{Nrf2} mutant, allowing its activity to be significantly increased to ~6.5-fold (Fig. 9C, *columns 2*), while the transactivation activity was decreased to ~1.6-fold by loss of most of AD1 (in the $^{\Delta120}$N607$^{\Delta173-286}$:C270^{Nrf2} mutant) (Fig. 9C, *column 3*). These data indicate that N607:C270^{Nrf2} is positively regulated by the AD1 of Nrf1 and is negatively regulated by its NTD. By comparison with N607:C270^{Nrf2}, the chimaeras N289:C270^{Nrf2} (i.e. the N289 portion contains AD1 and NTD) showed a lower activity (i.e. ~2.0-fold), similar to that of $^{\Delta120}$N607$^{\Delta173-286}$:C270^{Nrf2} (Fig. 9C, *columns 3 vs 4*), whilst the mutant chimaeras $^{\Delta120}$N289:C270^{Nrf2} (in which AD1 of Nrf1 was retained with a loss of its NTD) exhibited a relatively higher activity (i.e. ~4.4-fold), that was similar to that obtained from the entire N607:C270^{Nrf2} (Fig. 9C, *columns 5 vs 1*). Together, these data indicate that AD1-mediated transactivation is suppressed by NTD, whilst AD2 and the NST domain positively regulate reporter gene expression.

By comparison with N289:C270^{Nrf2}, approximately 1.9- to 2.8-fold increases in the reporter gene activity were respectively observed in N395:C270^{Nrf2} (in which N395 contains its NTD, AD1 and NST domain) and its mutant $^{\Delta120}$N395:C270^{Nrf2} (retaining AD1 and the NST domain) (Fig. 9C, *columns 6,7 vs 4*). Conversely, lack of most of AD1 (in the $^{\Delta120}$N395$^{\Delta173-286}$:C270^{Nrf2} mutant) significantly decreased its transactivation activity to a ~1.9-fold extent similar to those obtained from N289:C270^{Nrf2} or $^{\Delta120}$N607$^{\Delta173-286}$:C270^{Nrf2} (Fig. 9C, *columns 8 vs 3, 4*). These results demonstrate that besides AD1, the NST domain of Nrf1 positively regulates N395:C270^{Nrf2}.

Thereafter, we examined that both the NST domain and its adjoining AD2 contribute to the transcriptional activity of Nrf1β^{607}:C270^{Nrf2} chimaeric factor (in which Nrf1β^{607} contains its NST, AD2, SR/PEST2 and Neh6L regions, see Fig. 9B). Luciferase assay showed that Nrf1β^{607}:C270^{Nrf2} exerted a transactivation activity of ~2.0-fold, that was similar to that of N289:C270^{Nrf2} (Fig. 9C, *columns 9 vs 4*). By contrast, the $^{\Delta383}$Nrf1β^{607}:C270^{Nrf2} mutant (lacking most of the NST domain) showed a decreased activity (Fig. 9C, *column 12*), which is similar to the background value measured from COS-1 cells that had been transfected with an empty pcDNA4 vector. Both Nrf1$\beta^{607;\Delta413-448}$:C270^{Nrf2} (lacking the major acidic-amphipathic portion of AD2) and Nrf1$\beta^{607;\Delta403-506}$:C270^{Nrf2} (lacking both the entire AD2 and SR/PEST2 regions) showed similar reduction of ~1.4-fold activity (Fig. 9C, *columns 10 & 11*). These findings indicate that Nrf1β^{607}:C270^{Nrf2} is positively regulated by AD2 and the NST domain of LCR-F1/Nrf1β, but is negatively regulated by its SR/PEST2 or Neh6L regions.

Inhibition of Nrf2-target gene expression by attachment of the CTD to Nrf2

To explore whether attachment of the CTD from Nrf1 to Nrf2 exert a negative effect on target gene expression, we created a chimaeric factor Nrf2:C112^{Nrf1} by fusing the C-terminal 112-aa portion of Nrf1 (designated C112^{Nrf1}, that contains its ZIP and CTD) to the C-terminus of wild-type Nrf2. Luciferase assays of COS-1 and RL34 cells showed that the chimaeras Nrf2:C112^{Nrf1} exerted a ~2.2-fold activity, that was much lower than the wild-type Nrf2 (Fig. 9, D and E, *column 13*), implying that attachment of C112^{Nrf1} to Nrf2 suppresses its target gene activity. Conversely, removal of the CTD from the C112^{Nrf1} portion (to yield the Nrf2:C112$^{Nrf1;\Delta CTD}$ mutant) resulted in the major recovery of Nrf2-target reporter gene to be re-activated by ~6.5-fold (in COS-1 cells) or by ~8.5-fold (in RL-34 cells), indicating that, within the mutant chimaeras, the Nrf2 portion is liberated from the suppression by attached CTD from Nrf1 (Fig. 9, D and E, *column 15*). Removal of BCT also caused a modest increase in the reporter gene transactivation mediated by Nrf2:C112$^{Nrf1;\Delta BCT}$ (*column 14*), suggesting partial inhibition of Nrf2:C112^{Nrf1} by the basic c-tail ^{730}RRQERKPKDRRK^{741} of Nrf1. The difference between these chimaeric activities is not associated with their protein amounts, because no changes in the abundance of these chimaeric proteins were observed by western blotting of COS-1 cells (Fig. 9F). Overall, these data indicate that the chimaeric Nrf2:C112^{Nrf1} factor is endowed with the CTD within the C112^{Nrf1} portion to down-regulate Nrf2-target gene expression.

Endogenous ARE-driven genes are up-regulated by Nrf1 and LCR-F1/Nrf1β whilst down-regulated by Nrf1γ and Nrf1δ

Several previous studies [17,21,22,54,55] have showed that Nrf1 regulate the constitutive expression of a subset of ARE-driven genes responsible for maintaining cellular redox protein and lipid homeostasis. Real-time qPCR analysis of HEK-293T cells that had been transfected with *Nrf1*-targeting siRNA revealed that knockdown of this factor resulted in a significant decrease in the endogenous expression of mRNAs of *GCLC* (glutamate-cysterne ligase catalytic subunit), *GCLM* (glutamate-cysteine ligase catalytic modifier subunit), *NQO1* (NAD(P)H:quinone oxidoreductase 1) and *PSMB6* (the 26S proteasomal subunit B6) to ~20–25% as compared to their basal levels measured from the cells transfected with a scrambled control siRNA (Fig. 10A).

Recently, it has been showed that Nrf1 is essential for the hepatic lipid homeostasis through regulating both transcriptional coactivator genes *Lipin1* (LPIN1, also identified as a phosphatidate phosphatase) and *PGC-1β* (peroxisome proliferator-activated receptor-γ (PPAR-γ) coactivator-1β) [56]. As expected, our data showed that expression of *PGC-1β* was almost completely abolished upon siRNA knockdown of Nrf1 to ~42% of its basal

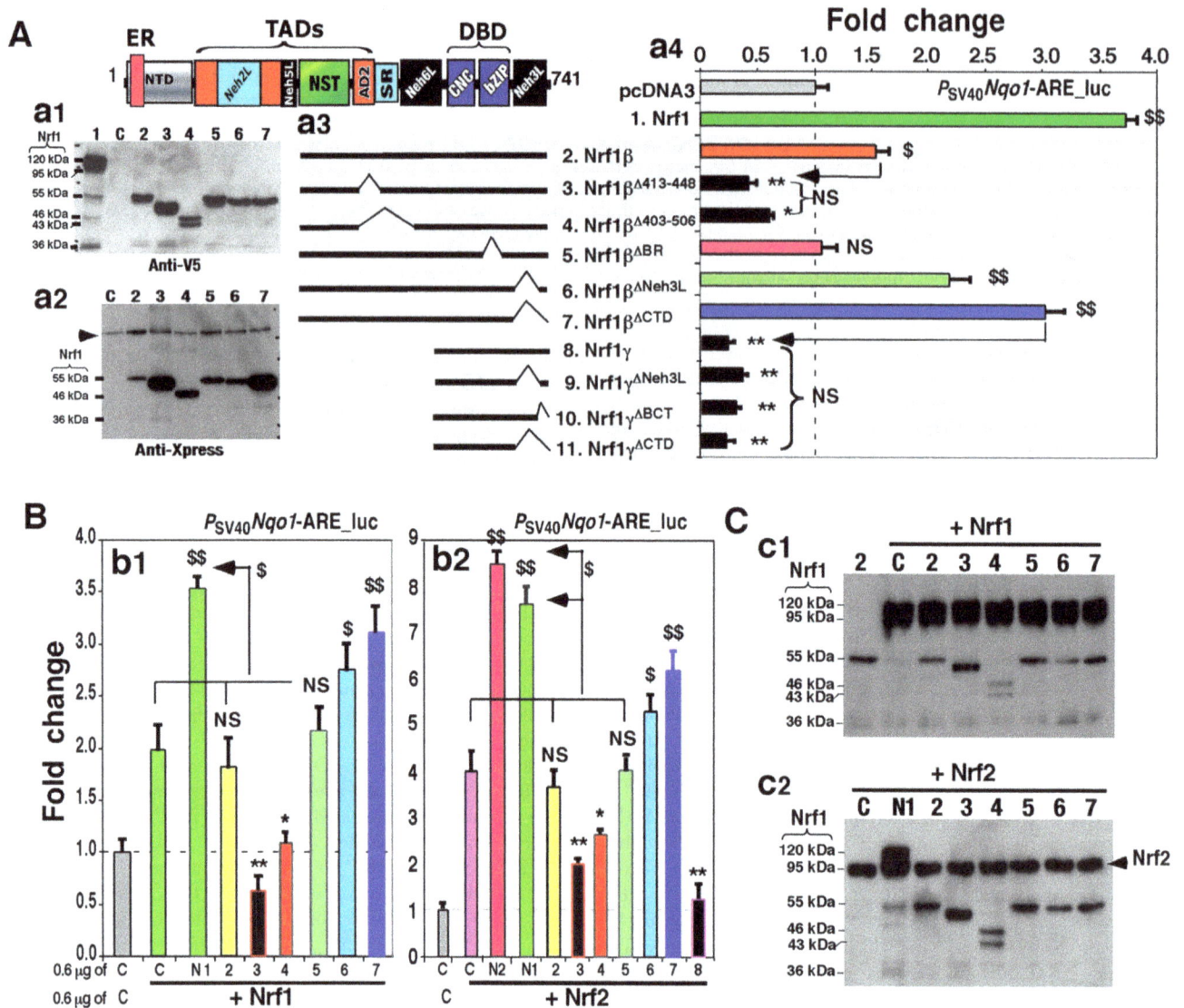

Figure 6. The weak activator Nrf1β/LCR-F1 is negatively regulated by its CTD. (A) The *middle* schematic representation of Nrf1β, Nrf1γ and their deletion mutants lacking various lengths of aa 297–741 of Nrf1 (*a3*). The contributions of the deleted regions to changes in the activity of Nrf1β and Nrf1γ, when compared with the background value, were examined using the $P_{SV40}Nqo1$-ARE-Luc reporter assay as described above. The *right panel* shows ARE-driven luciferase activity (*a4*) that was measured from COS-1 cells that had been co-transfected with each of numbered expression constructs and reporter plasmids. The data are shown as a fold change (mean ± S.D), and significant increases ($, p<0.05 and $$, p<0.001, n = 9) and decreases (*p<0.05, **p<0.001, n = 9) are indicated, relatively to the background value from transfection with an empty pcDNA3 control vector alone (C). The *left two panels* show western blotting of some of the above-transfected cell lysates with antibodies against either V5 (*a1*) or Xpress (*a2*). In addition, a non-specific protein-band is indicated (by *arrow*). The amount of protein applied to each polyacrylamide gel sample well was adjusted to ensure equal loading of β-gal activity. **(B)** COS-1 cells were co-transfected with each of the above-numbered expression constructs for Nrf1β, Nrf1γ and their mutants, together with an expression vector for wild-type Nrf1 (**N1**) or Nrf2 (**N2**), $P_{SV40}Nqo1$-ARE-Luc and β-gal plasmids. The cells were allowed to recover from transfection for 24 h before luciferase activity was measured. The data are shown as a fold change (mean ± S.D) of ARE-driven gene activity when compared with the background (value of 1.0). Significant increases ($, p<0.05 and $$, p<0.001, n = 9) and decreases (*p<0.05, **p<0.001, n = 9) are indicated. **(C)** The above-prepared cell lysates (*b1* and *b2*) were resolved using 4–12% LDS/NuPAGE and visualized by western blotting with V5 antibody (*c1* and *c2*). The electrophoresis band representing Nrf2 is indicated ((by *arrow*). The amount of protein loaded to each electrophoretic well was adjusted to ensure equal loading of β-gal activity.

level, accompanied with a similar reduction in expression of *Lipin1* (Fig. 10A).

Over-expression of the full-length Nrf1 that had been transfected in HEK-293T cells (Fig. 10B) caused a ~1.5–2.5-fold increase in expression of endogenous *GCLC*, *GCLM*, *NQO1* and *PSMB6*. The parallel experiments also showed that expression of *PGC-1β* was markedly up-regulated to ~10-fold by Nrf1, whilst no significant change in expression of *Lipin1* was observed. More

interestingly, similar expression patterns of such genes, with an exception of *PGC-1β* which was only induced by ~2.5-fold, as determined in HEK-293T cells that had been transfected with LCR-F1/Nrf1β (Fig. 10C). However, over-expression of either Nrf1γ or Nrf1δ resulted in a significant decrease in expression of the genes examined (Fig. 10D).

Collectively, the above results demonstrate that differential expression of endogenous ARE-driven genes is up-regulated by

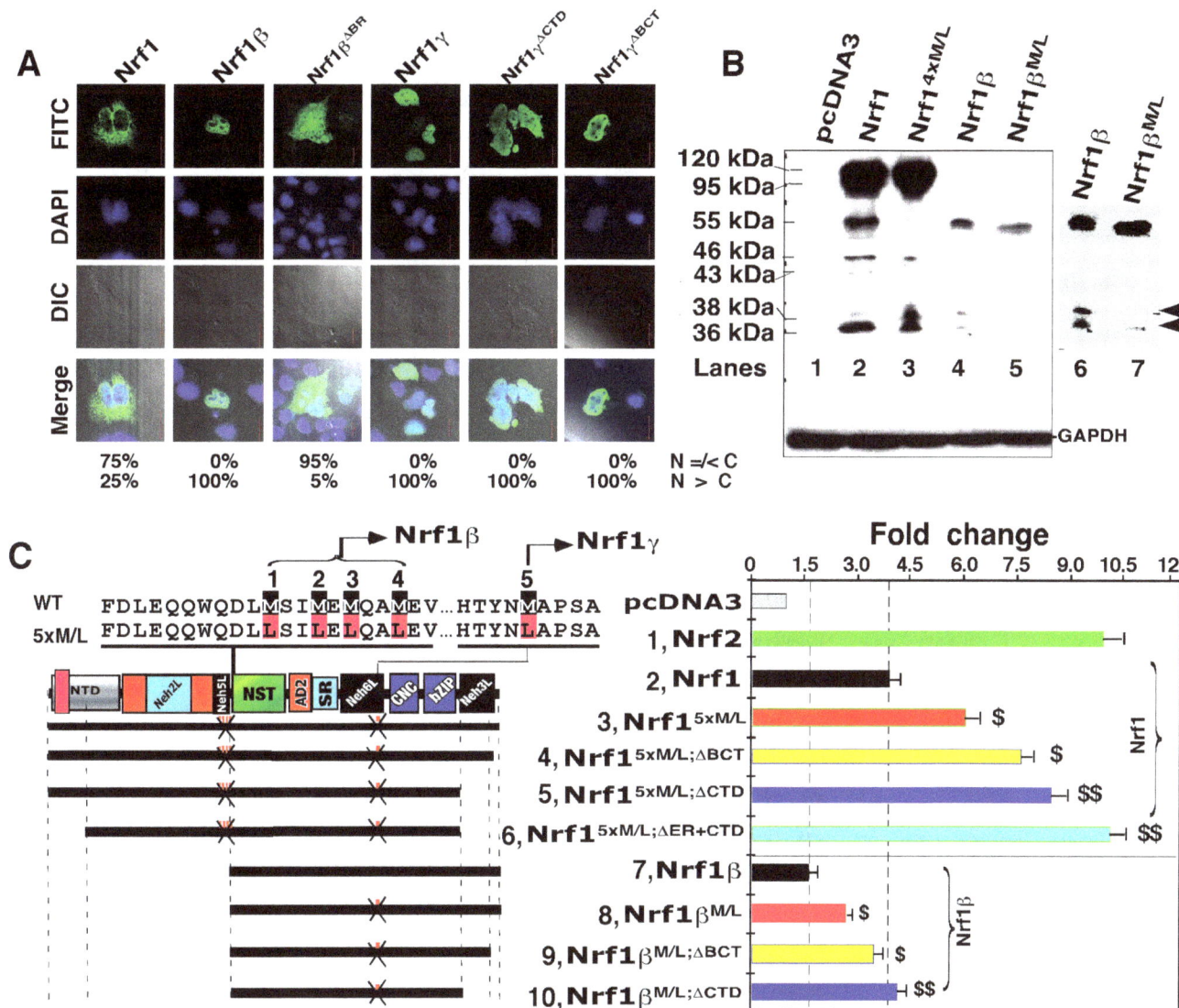

Figure 7. Blockage of Nrf1γ results an increase in the transactivation activity of Nrf1 and Nrf1β/LCR-F1. (A) Confocal imaging of COS-1 cells that had been transfected with 1.3 µg DNA of each expression construct for Nrf1, Nrf1β and Nrf1γ or mutants, before their subcellular locations were then examined by immunocytochemistry with FITC-labelled second antibody in order to locate V5-tagged proteins. Nuclear DNA was stained by DAPI. The merge signal represents the results obtained when the two images were superimposed with DIC from normal light microscopy. Bar = 20 µm. The quantitative data (*bottom*) were calculated as described in Figure 2D. (B) Western blotting of COS-1 cells that had been transfected with the indicated expression constructs for V5-tagged Nrf1, Nrf1β and Nrf1γ and their point mutants (Met into Leu, *below*). The *right panel* shows that the same gel as *the left panel* was exposed to X-ray for a little longer time. Two bands representing the 36-kDa Nrf1γ and a 38-kDa polypeptide are indicated (*arrows*). GAPDH served as an internal control to verify the amount of proteins applied to each electrophoresis well. (C) Schematic representation of Nrf1, Nrf1β, and their Met-to-Leu mutants with various deletions. The *upper left panel* shows amino acids adjoining five numbered Met residues; their mRNA codons can be recognized by ribosome for the internal initiation to translate Nrf1β or Nrf1γ. The first four or all five Met-to-Leu mutants were made respectively to yield Nrf1⁴ˣᴹ/ᴸ and Nrf1⁵ˣᴹ/ᴸ, whilst Nrf1β ᴹ/ᴸ contains the fifth Met-to-Leu mutant. Additional deletion mutants were created on the base of Nrf1⁵ˣᴹ/ᴸ and Nrf1βᴹ/ᴸ. The *right panel* shows luciferase reporter activity of COS-1 cells that had been transfected with 1.2 µg of each of indicated expression constructs, together with 0.6 µg of P_SV40nqo1-ARE-Luc and 0.2 µg of β-gal plasmids. The data are shown graphically as fold changes (mean ± S.D.) of transactivation by indicated factors. Significant increases ($, p<0.05 and $$, p<0.001, n = 9) in the activity are compared to the activity of the intact Nrf1 or Nrf1β.

Nrf1 and LCR-F1/Nrf1β, but is down-regulated by Nrf1γ and Nrf1δ. Differential contributions of these distinct isoforms to gene regulation were further determined by restoring ectopic Nrf1, LCR-F1/Nrf1β, Nrf1γ or Nrf1δ into Nrf1⁻/⁻ MEFs (in which aa 296–741 of endogenous Nrf1 had been genetically deleted), followed by real-time qPCR analysis of the resulting expression of GCLM (Fig. 10E) and PSMB6 (Fig. 10F). When compared with the basal GCLM and PSMB6 levels measured from transfection of

Nrf1⁻/⁻ MEFs with an empty pcDNA3 vector, their transcriptional expression was increased by restoration of Nrf1 and Nrf1β; the increased expression of GCLM and PSMB6 mediated by Nrf1 and Nrf1β was significantly prevented by Nrf1γ or Nrf1δ (Fig. 10, E and F), though no significant difference between the overall abundances of restored Nrf1 isoforms was measured (Fig. S3). Intriguingly, co-transfection experiments showed that Nrf1-mediated transcriptional expression of GCLM and PSMB6 seemed to

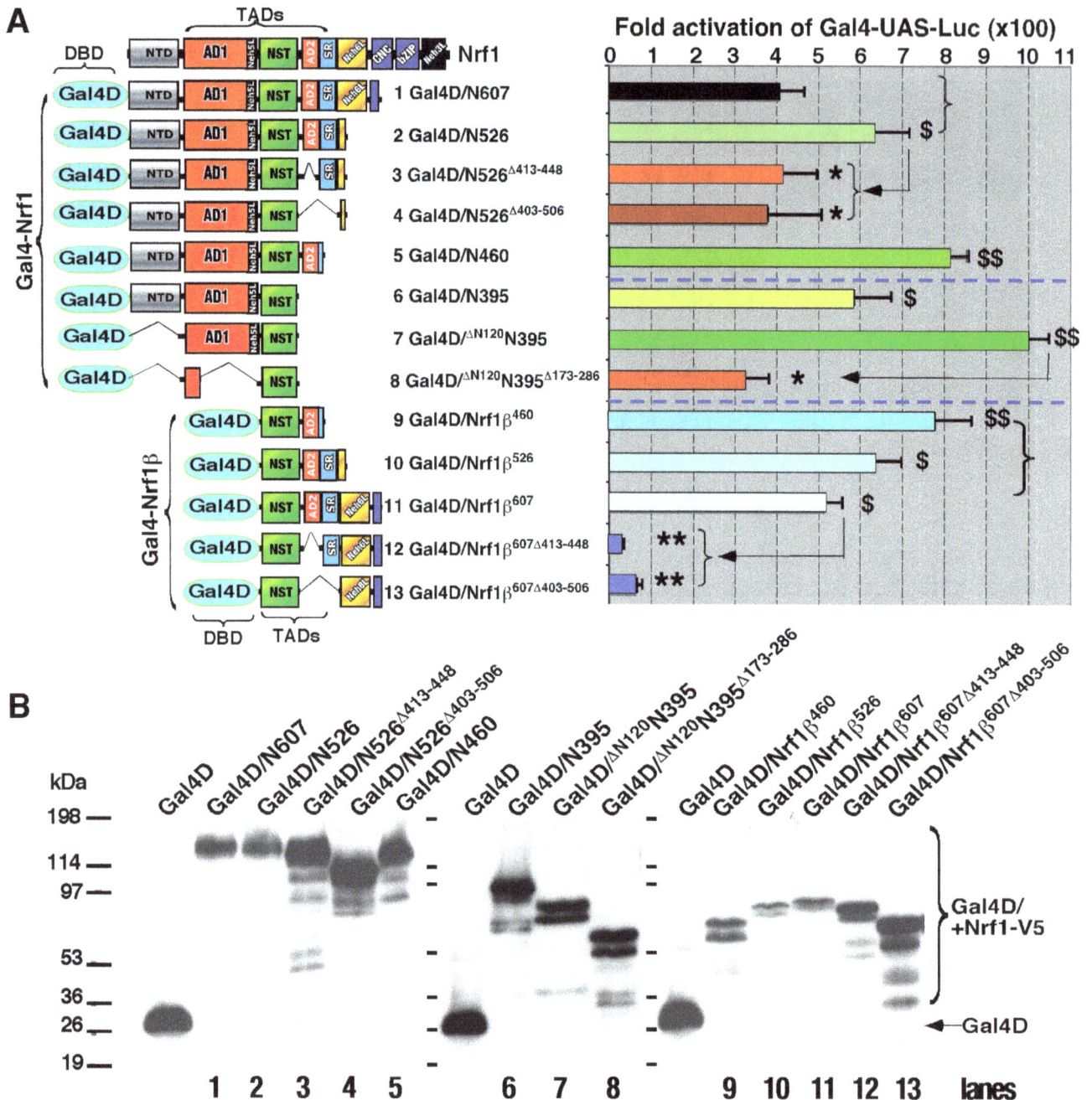

Figure 8. Both AD2 and NST domains positively regulates chimaeric Gal4-Nrf1 and Gal4-Nrf1β factors. (A) Schematic representation of expression constructs for Gal4D (Gal4 DNA-binding domain) fusion proteins containing various portions of Nrf1 or Nrf1β (*left panel*). They were created by ligation of their encoding cDNA fragments into the *BamHI/EcoRI* sites of the pcDNA3/Gal4-V5 vector. The *left panel* shows Gal4D-directed reporter activity that was measured from COS-1 cells had been cotransfected with each of indicated expression constructs for the various Gal4D/Nrf1 fusion proteins (1.2 µg), together with $P_{TK}UAS \times 4$-Luc (0.6 µg) and b-gal (0.2 µg) plasmids. The data are shown graphically as fold changes (mean ± S.D.) of transactivation by indicated Gal4-fusion factors when compared with the background (value of 1.0). Significant increases ($, p<0.05 and $$, p<0.001, n=9) and decreases (*p<0.05, **p<0.001, n=9) in activity relatively to the referenced activity are indicated (*arrows*). (B) The above-prepared cell lysates were resolved using 4–12% LDS/NuPAGE and examined by western blotting with V5 antibody. The electrophoretic bands representing free Gal4D and Gal4-Nrf1 fusion proteins are indicated. Samples loaded on each well were calculated to contain equal amounts of β-gal activity.

be modestly blunted by the relatively weak Nrf1β to a similar extent to that of Nrf1β restored alone (Fig. 10, E and F). Together with the data (as shown in Fig. 7B) and our previous results [12,48], these findings give rise to a possibility that the nuclear-localized LCR-F1/Nrf1β (i.e. 55-kDa) and its further degraded polypeptides (between 38-kDa and 25-kDa, acting as dominant-negative regulators) may gain more access to endogenous genes than the membrane-associated Nrf1 protein such that they occupationally bind target genes competitively against the intact Nrf1 factor.

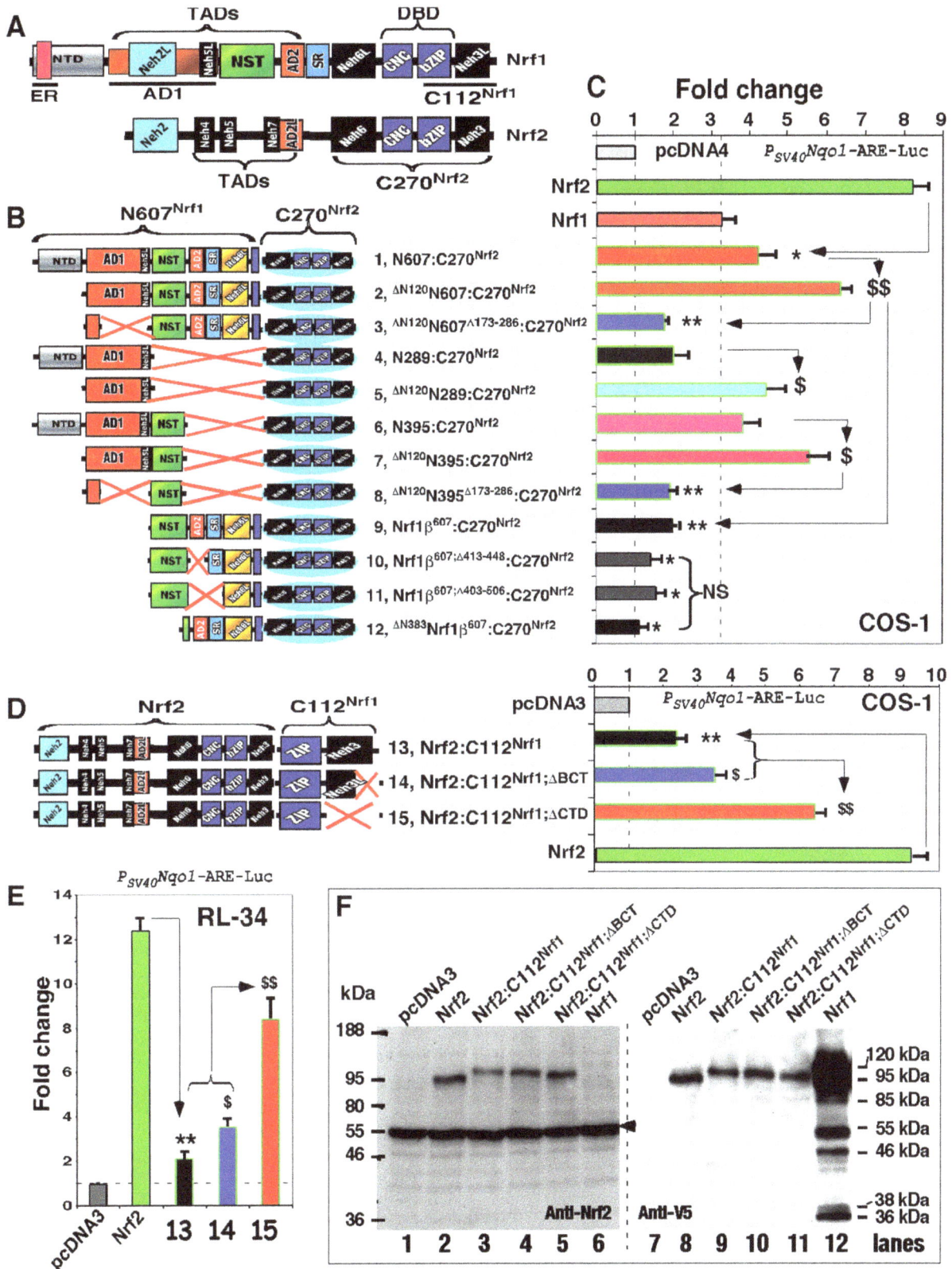

Figure 9. Nrf2-target gene expression is up-regulated by AD2 and NST domains of Nrf1 within chimaeras N607:C270^{Nrf2} and Nrf1β^{N607}:C270^{Nrf2}, but is also down-regulated by CTD of Nrf1 within additional chimaeras Nrf2:C112 ^{Nrf1}. (A) The *cartoon* shows structural domains of Nrf1 and Nrf2. The C-terminal residues 629–741 of Nrf1 (i.e. C112^{Nrf1}) cover both its bZIP and CTD regions (*upper*). In Nrf2, the C270^{Nrf2} represents its C-terminal 270 aa between positions 328–597 that cover its Neh6, CNC, bZIP and Neh3 domains (*lower*). **(B)** Diagrammatic representation of chimaeras that were composed of various portions Nrf1 and Nrf2. The cDNA fragments encoding different portions of the N-terminal aa 1–607 of Nrf1 (e.g. N607^{Nrf1}) and various lengths of the central aa 292–607 (i.e. Nrf1β^{N607}) were ligated into the BamHI/EcoR1 sites of the Nrf2/pcDNA4His/Max B construct. Thus a series of chimaeric proteins were created by fusing different regions of either N607^{Nrf1} or Nrf1β^{N607} to the N-

terminus of C270$^{\text{Nrf2}}$. (**C**) The transactivation activity of the above-described chimaeric factors as well as wild-type Nrf1 and Nrf2. This was determined by using P$_{SV40}$Nqo1-ARE-Luc and β-gal reporters that had been co-transfected with each of indicated expression constructs into COS-1 cells. The data are shown as fold changes (mean ± S.D.) of the transactivation activity when compared with the background (value of 1.0) that was measured from the blank co-transfection of cells with an empty pcDNA4 vector and the above two reporters. Thereafter, significant increases ($, p<0.05 and $$, p< 0.001, n = 9) and decreases (*p<0.05, **p<0.001, n = 9) in activity relatively to the referenced activity are indicated (*arrows*). (**D**) Additional three chimaeras are schematically shown (*left panel*), which were created by fusing the full-length Nrf2 to the N-terminus of either C112$^{\text{Nrf1}}$ or its mutants. These expression constructs for Nrf2 and its chimaeric proteins, together with P$_{SV40}$Nqo1-ARE-Luc and β-gal reporters, were co-transfected into either COS-1 cells (*right panel* of **D**) or RL-34 cells (**E**), before luciferase activity was assayed and the data are presented as fold changes (mean ± S.D.). Significant increases ($, p<0.05 and $$, p<0.001, n = 9) and decreases (**p<0.001, n = 9) in activity were calculated relatively to the activity arrowed. (**F**) Total lysates of COS-1 cells that had been co-transfected with expression constructs for Nrf1, Nrf2 and its three chimaeric proteins (shown in **D**) were resolved using 4–12% LDS/NuPAGE and then visualized by western blotting with antibodies against either Nrf2 or the V5 epitope (*left and right panels*, both blotting in the same gel-transferred nitrocellulose membranes). Amounts of protein loaded to each electrophoretic well were adjusted to ensure equal loading of β-gal activity. In addition, a non-specific protein band recognized by anti-Nrf2 antibody is arrowed (*right panel*).

Discussion

In the present study we found that: (i) both Nrf1 and its shorter isoform LCR-F1/Nrf1β are negatively regulated by their CTD and Neh6L regions, but both are positively regulated by their AD2 and NST domains; however a dual opposing effect of SR/PEST2 on Nrf1 is not exerted on LCR-F1/Nrf1β; (ii) the latter LCR-F1/Nrf1β is a relatively weak activator to the full-length Nrf1 factor, whilst the small molecular-weight Nrf1γ and Nrf1δ act as two dominant-negative forms that competitively inhibit the wild-type Nrf1 and Nrf2, as well as LCR-F1/Nrf1β; and (iii) differential expression of endogenous ARE-driven genes is up-regulated by Nrf1 and LCR-F1/Nrf1β, besides Nrf2, but is down-regulated by Nrf1γ and Nrf1δ. These together control the overall activity of Nrf1 to fine-tune the steady-state expression of target genes for maintaining cellular homeostasis.

LCR-F1/Nrf1β is a weak activator relative to the full-length Nrf1; both are inhibited by the small dominant-negative Nrf1γ and Nrf1δ

Accumulating evidence reveals that over eleven Nrf1 isoforms [46] are produced from the single *nfe2l1* gene, though they are differentially expressed in different mammalian species [1,57–60]. Upon translation, the mouse Nrf1 is biosynthesized as an unglycosylated 95-kDa polypeptide. During the membrane-topogenic vectorial process, the nascent 95-kDa polypeptide is co-translationally integrated within the ER, where it is glycosylated to become a 120-kDa glycoprotein [45,46]. Subsequently, the 120-kDa glycoprotein is partially repartitioned out of ER into the cyto/nucleoplasmic compartments, whereupon Nrf1 is deglycosylated to generate an active 95-kDa transcription factor [45]. Thereafter, the membrane-bound 95-kDa Nrf1 proteins are subjected to selective proteolytic processing to yield multiple cleaved polypeptides of between 85-kDa and 25-kDa [48]. In addition to the post-translational processing of Nrf1, its isoforms ranging between 65-kDa and 25-kDa can also be synthesized by translation through distinct in-frame start ATG codons embedded in various lengths of mRNA sequences, some of which arise from alternative splicing of variant transcript species [9,49,50,57,58,61].

Amongst these Nrf1 isoforms, its polypeptides of ~55-kDa, 36-kDa and 25-kDa are originally designated LCR-F1/Nrf1β [12,59,61], Nrf1γ and Nrf1δ [46,59], respectively. LCR-F1/Nrf1β migrates at a mass of 55-kDa in the pH 7.0 LDS/NuPAGE gel ([12,44] and herein), but the pH 8.9 Laemmli SDS-PAGE allows its mobility to be exhibited at an estimated size of 60-kDa [19] or 65-kDa (called p65Nrf1 [62]). To provide a clear explanation of the Nrf1 nomenclature in the literature, we also note a recent report on another confused Nrf1β [63]. In fact, this spliced variant was originally designated as Nrf1 clone Δ767 in 1998 [57] (GenBank accession No. NM_001130453.1); it was translated as

an N-terminally-truncated mutant lacking 181-aa of Nrf1 (spanning the entire NTD and one-third of the AD1) and thus was simply designated Nrf1^{ΔN} in 2009 [46,59], although additional 12 aa (i.e. MGWESRLTAASA) replaced the original 181-aa of Nrf1.

By comparison with the full-length Nrf1, the 55-kDa Nrf1β/LCR-F1 lacks both the NTD and the essential transactivation domain AD1, but still retains AD2 and NST domain required for its transactivation activity. It is thus postulated that Nrf1β/LCR-F1 is a nuclear activator with a poor transactivation activity; this has been confirmed by us and others [12,57,58,61]. Herein, our experiments confirm that endogenous Nrf1-target genes and ARE-driven luciferase reporters are up-regulated by LCR-F1/Nrf1β, as it acts as a *bona fide* activator with a weak activity relative to the wild-type Nrf1. On the other side, co-transfection of LCR-F1/Nrf1β- and Nrf1-expressed constructs appears to slightly decrease Nrf1-mediated transactivation of endogenous genes, but not ectopic reporter genes, to a similar extent to that of transactivation mediated by LCR-F1/Nrf1β alone. Together with previous publications [12,48,60,61,64], these findings indicate that the ER membrane-bound Nrf1 has tempo-spatial behaviour, that is distinct from the nuclear water-soluble LCR-F1/Nrf1β, to bind ARE sequences in Nrf1-target genes within different transcriptional assembly being built in different cell types. However, an exception was reported that LCR-F1/Nrf1β was thought to be a significant dominant-negative inhibitor of ARE-driven gene transactivation against the wild-type Nrf1 and Nrf2 [62]. The dispute on LCR-F1/Nrf1β suggests that it is an unstable protein that might be proteolytically processed to yield several small poplypeptides of 36-kDa and 25-kDa; this is supported by the fact that the activity of Nrf1β/LCR-F1 is significantly increased by blocking generation of either 36-kDa Nrf1γ or 25-kDa Nrf1δ polypeptides. Induction of Nrf1β/LCR-F1 activity may also be dependent on distinct stressors in different cell lines [57,58,64].

By definition, a dominant-negative mutant usually disrupts certain functions of the intact protein (i.e. *negative*) and also out-compete the endogenous protein in some way (i.e. *dominant*) [65,66], it is deduced that the 36-kDa Nrf1γ and the 25-kDa Nrf1δ, but not the 55-kDa Nrf1β/LCR-F1, act as *bona fide* dominant-negative isoforms. This is supported by lack of all the potential transactivation elements (i.e. AD1, AD2 and NST domain) in Nrf1γ and Nrf1δ [12,57,67]. Our evidence presented demonstrates that over-expression of either Nrf1γ or Nrf1δ almost abolishes transcriptional expression of endogenous ARE-battery genes and ectopic luciferase reporters against the wild-type Nrf1 or Nrf2. Further co-transfection experiments confirm that both Nrf1γ and Nrf1δ dominant-negatively inhibit the transactivation activity of Nrf1 and Nrf2. This occurs possibly through competitive interference with the intact functional assembly of the active transcription factor complex that binds ARE sequences in target genes.

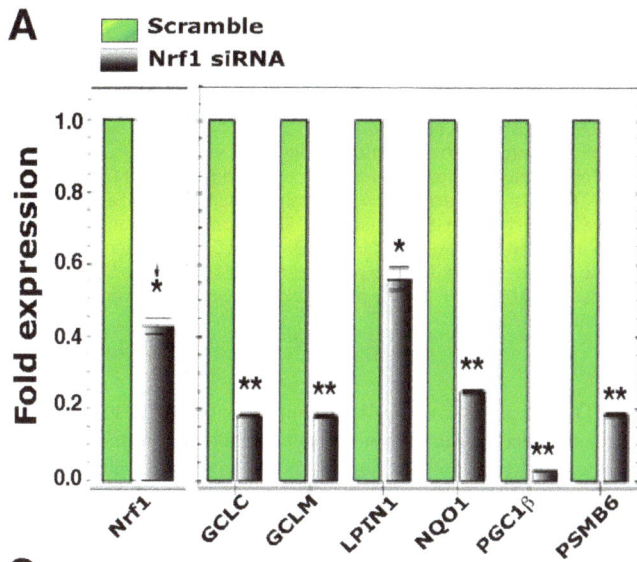

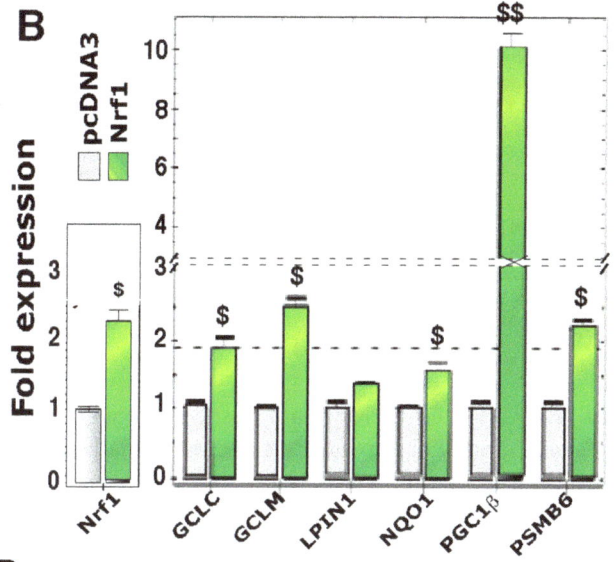

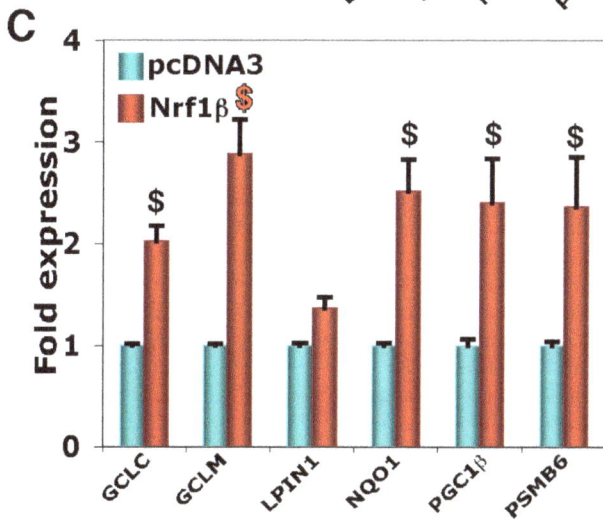

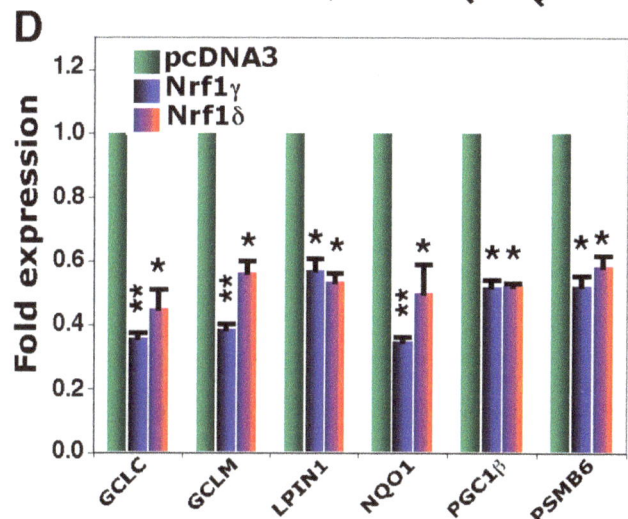

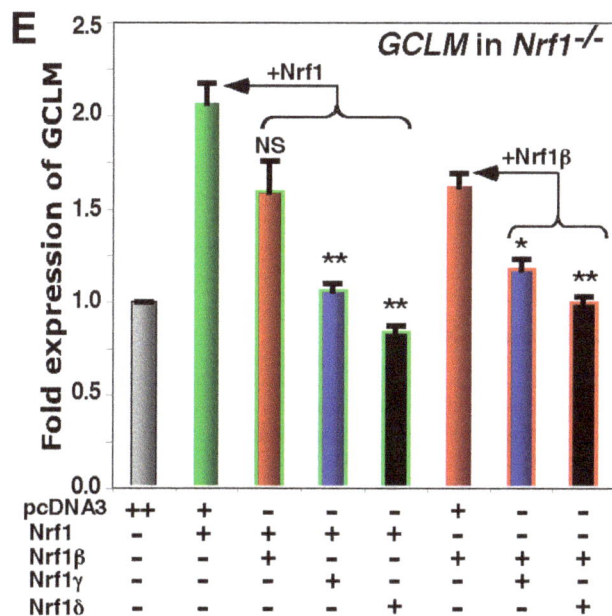

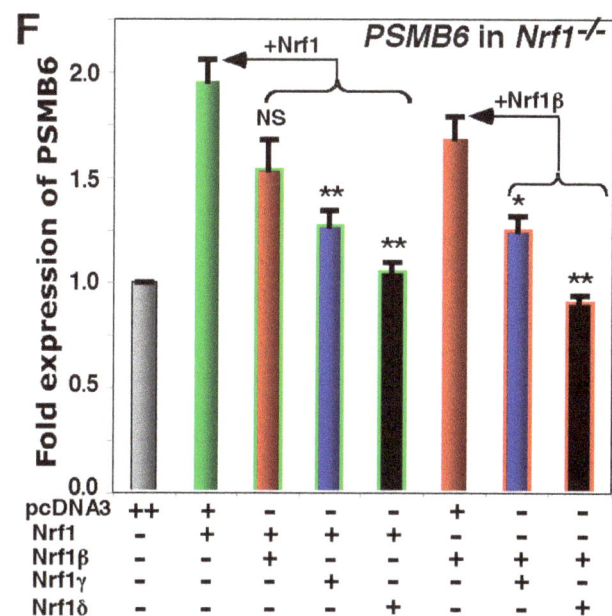

Figure 10. Endogenous genes are up-regulated by Nrf1 and Nrf1β/LCR-F1 but also down-regulated by Nrf1γ and Nrf1δ. (A) Knockdown of Nrf1 by its targeting siRNA, which, along with a scramble siRNA (as an internal control), was transfected into HEK 293T cells as described previously [12] (and maintained in our laboratory). Subsequently, changes in the mRNA expression of both the endogenous *Nrf1 per se* and Nrf1-target genes were analyzed by real-time qPCR. The data are shown as fold changes (mean ± S.D) in gene knockdown by Nrf1-siRNA relatively compared to the scramble value (1.0 set). Significant decreases (*p<0.005, **p<0.001, n = 9) in gene expression relatively to the basal level are indicated. **(B to D)** Expression constructs for Nrf1 (*B*), Nrf1β (*C*), Nrf1γ and Nrf1δ (*D*) (2 μg of cDNA each, along with an empty pcDNA3 control vector) were transfected into HEK 293T cells. Thereafter, alterations in the expression of Nrf1-target genes were determined by real-time qPCR, and were calculated as fold changes (mean ± S.D) in gene regulation by distinct Nrf1 isoforms when compared to the background (value of 1.0). Significant increases ($, p<0.05 and $$, p<0.001, n = 9) and decreases (*p<0.005, **p<0.001, n = 9) in gene expression relatively to the basal level are indicated. **(E and F)** Nrf1 and Nrf1β, Nrf1γ and Nrf1δ were restored into *Nrf1$^{-/-}$* MEFs, in which Nrf1 has been lost (see Fig. S3) before being transfected with expression constructs for distinct isoforms alone or in combination, which were indicated (+, 1 μg of cDNA; ++, 2 μg of cDNA). Subsequently, real-time qPCR was performed to determine changes in the expression of *GCLM* (*E*) and *PSMB6* (*F*). The data are presented as folds (mean ± S.D) relatively to the blank transfection with pcDNA3 alone (value of 1.0). Significant decreases (*p<0.005, **p<0.001, n = 9) in gene expression were calculated when compared to the level of genes regulated by Nrf1 or Nrf1β (*arrows*).

Both Nrf1 and LCR-F1/Nrf1β are positively regulated by their AD2 and NST domains but also are negatively regulated by their CTD and Neh6L regions

Our previous work showed that the full-length Nrf1 is a modular protein containing nine discrete domains, and both its NTD and AD1 are lost in the short LCR-F1/Nrf1β [43,45]. Besides AD1, both AD2 and NST domain also contribute to the full activity of Nrf1 [12,59]. In addition, the SR domain (aa 455–488) was early considered to function as a transactivation domain in the human Nrf1 and its long form TCF11 [68]. We recently found that the core SR region (aa 466–488) contributes to the basal and stimulated activity of Nrf1 [45]. However, removal of aa 403–506 covering the entire AD2 and most of SR/PEST2 causes a marked augmentation in the basal and stimulated activity of Nrf1 to a similar extent to that obtained from deletion of aa 413–448 covering a short core AD2 sequence [45]. In combination with our recent findings [45,48], these data indicate that both AD2 and SR also encompass negative regulatory degrons to inhibit the full-length Nrf1; this effect depends on distinct membrane-topological conformation (that enables repositioning of their adjoining degrons to target the protein for proteolytic degradation). However, removal of aa 403–506 from the water-soluble LCR-F1/Nrf1β does not cause a further decrease in its activity, when compared to deletion of aa 413–448 (in this study), suggesting that AD2, but not the SR domain, is required for its transactivation activity. Moreover, the Gal4-based reporter experiments demonstrate that AD2 and NST rather than SR domains contribute to transactivation mediated by LCR-F1/Nrf1β; this notion is supported by evidence obtained from the chimaeric N607^{Nrf1}:C270^{Nrf2} factor.

Previous studies have reported that Nrf1 is negatively regulated by its NTD and Neh6L regions from within the molecule through different mechanisms [43,48]. Our present work uncovers that beside NTD and Neh6L, CTD of Nrf1 also contributes to the negative regulation of both the full-length CNC-bZIP factor and its shorter isoform Nrf1β/LCR-F1, whilst the dominant-negative effect of Nrf1γ or Nrf1δ is unaffected by its CTD. The inhibitory effect of CTD is also elicited within its chimaeric factor resulting from attachment of CTD to Nrf2 that enables its transactivation activity to be diminished. However, the detailed mechanism(s) underlying the negative regulation by CTD of Nrf1, particularly Nrf1β/LCR-F1, remains to be determined. During topogenesis of the full-length Nrf1, topological folding of its TMc-containing CTD around membranes is dictated by the N-terminal topogon [12,46], and thus removal of CTD could enables the CNC-bZIP protein to be dislocated to the cyto/nucleoplasmic side of the ER membrane such that CTD-deficient mutant would gain more access to Nrf1-target genes than the full-length protein, thereby increasing its activity. Huang and colleagues reported that the

CTD and its adjacent bZIP domain enable LCR-F1/Nrf1β to directly interact with either the cytomegalovirus immediate-early protein 2 (IE2) [69] or a cell cycle-dependent microspherule protein 2 (MCRS2, that is involved in telomere shortening by inhibiting telomerase activity [70]). As a consequence, the transactivation activity of LCR-F1/Nrf1β is repressed by IE2 and MCRS2, although both its heterodimerization with a small Maf protein and its ability to bind ARE sequences in Nrf1-target genes are unaffected by these two interacting partners.

Our previous studies showed that LCR-F1/Nrf1β and Nrf1γ are primarily located in the nucleus, but another small portion of these two proteins are recovered in the membrane fraction, that are not protected by membranes and hence are rapidly digested by proteases in membrane protection assays [12,45,48]. In this study, confocal microscopy revealed that the basic c-tail ^{730}RRQERKPKDRRK^{741} does not appear to act as a functional nuclear localization signal, because removal of the basic peptide causes the resultant mutant protein to be modestly accumulated in the nucleus of some cells. Live-cell imaging experiments unravel that most of the extra-nuclear proteins of GFP-CTD or its mutants lacking various lengths of aa 714–741 from CTD are rapidly digested by PK in around 10–20 min. These indicate that the extra-nuclear fraction of GFP-CTD fusion protein could be associated with the cytoplasmic leaflet of membrane possibly through amphipathic or electrical interactions. In addition, attachment of CRAC5 from CTD to the C-terminus of GFP (to yields GFP-CTD$^{Δ701-741}$) allowed the fusion protein to be insensitive to PK digestion, relative to GFP-CTD. Collectively, these findings suggest that CRAC5, along with the amphipathic TMc region, enables the protein to be associated with the cholesterol-rich membrane microdomain (e.g. rafts). Together with the fact that CRAC5 is present in the Neh3L-containing CTD of Nrf1 but is absent from the Neh3 domain of Nrf2, therefore we envisage that the motif is attributed to the difference of Neh3L from Neh3.

Concluding Comments

Collectively, Nrf1, LCR-F1/Nrf1β and Nrf1γ are three major representatives of various isoforms arising from both post-transcriptional and/or post-translational processing of the *nfe2l1* gene products. Together with previous publications [12,57,58,61,64], the evidence that has been provided herein demonstrates that Nrf1β/LCR-F1 is *per se* a weak activator relative to the intact full-length Nrf1, and that the shorter isofrom does not functions as a *de facto* dominant-negative inhibitor of Nrf1 (and Nrf2), albeit it can modestly blunt wild-type Nrf1 activity to mediate endogenous ARE-driven gene expression. By contrast, Nrf1γ and Nrf1δ act as *bona fide* dominant-negative isoforms that competitively inhibit Nrf1 and LCR-F1/Nrf1β. We have also presented further evidence showing that both Nrf1 and LCR-F1/Nrf1β are positively regulated by their AD2 and NST domains, but also are negatively regulated by their Neh6L and CTD regions. However, CTD does not alter the dominant-negative effect of Nrf1γ. Intriguingly, it is to note that CTD of Nrf1 is homologous to the equivalent Neh3 domain of Nrf2; the latter Nrf2 is positively regulated by its Neh3 domain through direct interaction with CHD6 [37], but CHD6 is not detected as one of Nrf1-interacting proteins by the liquid chromatography-mass spectrometry (LC-MS)/MS [71]. Our previous studies indicated that the negative regulation of Nrf1 by CTD is associated with its topological folding within and around membranes [12,46], and the association could be enhanced by CRAC5 through possible interaction with the cholesterol-rich microdomain. Moreover, LCR-F1/Nrf1β is repressed by IE2 [69] and MCRS2 [70] through interaction with its CTD-adjoining region, whilst both its heterodimerization with small Maf protein and its activity of DNA-binding to target genes are unaffected. However, the detailed mechanism(s) underlying negative regulation of Nrf1 by its Neh3L-containing CTD remains to be deeply studied in order to elucidate distinction between Neh3L (particularly in LCR-F1/Nrf1β and Nrf1$^{\Delta N}$) and Neh3 of Nrf2.

Supporting Information

Figure S1 Alignment of amino acids covering the CTD of Nrf1 from 30 different species. The CTD of Nrf1 is highly conserved amongst different vertebrate species, particularly mammalian animals. This domain is composed of CRAC5, PYxP, TMc and Basic C-tail, and its secondary structure is predicated and shown in the main text.

Figure S2 Live-cell imaging of GFP-CTD$^{\Delta701-741}$ retaining CRAC5. COS-1 cells co-expressing GFP-CTD$^{\Delta701-741}$ (that lacks most of the CTD of Nrf1, but retains its CRAC5 only), together with the ER/DsRed marker, were subjected to live-cell imaging combined with the *in vivo* membrane protease protection assay. The cells were first permeabilized by 20-μg/ml digitonin for 10 min, before being co-incubated with 50-μg/ml PK for 90 min. The imaging data are shown that were obtained from 90 min digestion by PK (**A**), and the images obtained from 30 min to 48 min incubation with PK are presented (**B**).

Figure S3 Restoration of distinct Nrf1 isoforms in Nrf1$^{-/-}$ cells. Nrf1$^{-/-}$ MEFs were allowed for restored expression of distinct Nrf1 isoforms followed by real-time qPCR analysis of Nrf1-target gene expression. The results showed no significant difference between the mRNA levels of Nrf1 expressed in distinct transfected cells.

Acknowledgments

We gratefully acknowledge the help of Dr. Akira Kobayashi (Doshisha University, Japan) for providing both wild-type and Nrf1$^{-/-}$ mouse embryonic fibroblasts (MEFs). We are greatly indebted to Prof. John D. Hayes (University of Dundee) for providing an experimental space to obtain some preliminary data for this study.

Author Contributions

Conceived and designed the experiments: YZ. Performed the experiments: YZ LQ SL YX. Analyzed the data: YZ. Contributed to the writing of the manuscript: YZ. Real-time qPCR: LQ JC. Live-cell imaging: SL. Some immunoblotting: YX. Confocal microscopy: YR.

References

1. Sykiotis GP, Bohmann D (2010) Stress-activated cap'n'collar transcription factors in aging and human disease. Sci. Signal. 3, re3.
2. Koch A, Steffen J, Kruger E (2011) TCF11 at the crossroads of oxidative stress and the ubiquitin proteasome system. Cell Cycle 10, 1200–1207.
3. Steffen J, Seeger M, Koch A, Kruger E (2010) Proteasomal degradation is transcriptionally controlled by TCF11 *via* an ERAD-dependent feedback loop. Mol. Cell 40, 147–158.
4. Radhakrishnan SK, Lee CS, Young P, Beskow A, Chan JY, et al. (2010) Transcription factor Nrf1 mediates the proteasome recovery pathway after proteasome inhibition in mammalian cells. Mol. Cell 38, 17–28.
5. Grimberg KB, Beskow A, Lundin D, Davis MM, Young P (2011) Basic leucine zipper protein Cnc-C is a substrate and transcriptional regulator of the Drosophila 26S proteasome. Mol. Cell Biol. 31, 897–909.
6. Li X, Matilainen O, Jin C, Glover-Cutter KM, Holmberg CI, et al. (2011) Specific SKN-1/Nrf stress responses to perturbations in translation elongation and proteasome activity. PLoS Genet. 7, e1002119.
7. Rushmore TH, Morton MR, Pickett CB (1991) The antioxidant responsive element. Activation by oxidative stress and identification of the DNA consensus sequence required for functional activity. J. Biol. Chem. 266, 11632–11639.
8. Bean TL, Ney PA (1997) Multiple regions of p45 NF-E2 are required for β-globin gene expression in erythroid cells. Nucleic Acids Res. 25, 2509–2515.
9. Johnsen O, Murphy P, Prydz H, Kolsto AB (1998) Interaction of the CNC-bZIP factor TCF11/LCR-F1/Nrf1 with MafG: binding-site selection and regulation of transcription. Nucleic Acids Res. 26, 512–520.
10. Kwak MK, Kensler TW (2011) Targeting Nrf2 signaling for cancer chemoprevention. Toxicol. Appl. Pharmacol. 244, 66–76.
11. Zhang Y, Kobayashi A, Yamamoto M, Hayes JD (2009) The Nrf3 transcription factor is a membrane-bound glycoprotein targeted to the endoplasmic reticulum through its N-terminal homology box 1 sequence. J. Biol. Chem. 284, 3195–3210.
12. Zhang Y, Lucocq JM, Hayes JD (2009) The Nrf1 CNC/bZIP protein is a nuclear envelope-bound transcription factor that is activated by t-butyl hydroquinone but not by endoplasmic reticulum stressors. Biochem. J. 418, 293–310.
13. Higgins LG, Kelleher MO, Eggleston IM, Itoh K, Yamamoto M, et al. (2009) Transcription factor Nrf2 mediates an adaptive response to sulforaphane that protects fibroblasts in vitro against the cytotoxic effects of electrophiles, peroxides and redox-cycling agents. Toxicol. Appl. Pharmacol. 237, 267–280.
14. Xiao H, Lu F, Stewart D, Zhang Y (2013) Mechanisms underlying chemopreventive effects of flavonoids via multiple signaling nodes within Nrf2-ARE and AhR-XRE gene regulatory networks. Curr. Chem. Biol. 7, 151–176.
15. Chan K, Lu R, Chang JC, Kan YW (1996) Nrf2, a member of the NFE2 family of transcription factors, is not essential for murine erythropoiesis, growth, and development. Proc. Natl. Acad. Sci. USA 93, 13943–13948.
16. Xu C, Huang MT, Shen G, Yuan X, Lin W, et al. (2006) Inhibition of 7,12-dimethylbenz(a) anthracene-induced skin tumorigenesis in C57BL/6 mice by sulforaphane is mediated by nuclear factor E2-related factor 2. Cancer Res. 66, 8293–8296.
17. Leung L, Kwong M, Hou S, Lee C, Chan JY (2003) Deficiency of the Nrf1 and Nrf2 transcription factors results in early embryonic lethality and severe oxidative stress. J. Biol. Chem. 278, 48021–48029.
18. Farmer SC, Sun CW, Winnier GE, Hogan BL, Townes TM (1997) The bZIP transcription factor LCR-F1 is essential for mesoderm formation in mouse development. Genes Dev. 11, 786–798.
19. Chan JY, Kwong M, Lu R, Chang J, Wang B, et al. (1998) Targeted disruption of the ubiquitous CNC-bZIP transcription factor, Nrf-1, results in anemia and embryonic lethality in mice. EMBO J. 17, 1779–1787.
20. Kwong M, Kan YW, Chan JY (1999) The CNC basic leucine zipper factor, Nrf1, is essential for cell survival in response to oxidative stress-inducing agents. Role for Nrf1 in γ -gcs(l) and gss expression in mouse fibroblasts. J. Biol. Chem. 274, 37491–37498.
21. Xu Z, Chen L, Leung L, Yen TS, Lee C, et al. (2005) Liver-specific inactivation of the Nrf1 gene in adult mouse leads to nonalcoholic steatohepatitis and hepatic neoplasia. Proc. Natl. Acad. Sci. USA 102, 4120–4125.
22. Ohtsuji M, Katsuoka F, Kobayashi A, Aburatani H, Hayes JD, et al. (2008) Nrf1 and Nrf2 play distinct roles in activation of antioxidant response element-dependent genes. J. Biol. Chem. 283, 33554–33562.
23. Kobayashi A, Tsukide T, Miyasaka T, Morita T, Mizoroki T, et al. (2011) Central nervous system-specific deletion of transcription factor Nrf1 causes progressive motor neuronal dysfunction. Genes Cells 16, 692–703.
24. Lee CS, Lee C, Hu T, Nguyen JM, Zhang J, et al. (2011) Loss of nuclear factor E2-related factor 1 in the brain leads to dysregulation of proteasome gene expression and neurodegeneration. Proc. Natl. Acad. Sci. USA 108, 8408–8413.
25. Kim J, Xing W, Wergedal J, Chan JY, Mohan S (2010) Targeted disruption of nuclear factor erythroid-derived 2-like 1 in osteoblasts reduces bone size and bone formation in mice. Physiol. Genomics 40, 100–110.

26. Itoh K, Wakabayashi N, Katoh Y, Ishii T, Igarashi K, et al. (1999) Keap1 represses nuclear activation of antioxidant responsive elements by Nrf2 through binding to the amino-terminal Neh2 domain. Genes Dev. 13, 76–86.

27. Wang H, Liu K, Geng M, Gao P, Wu X, et al. (2013) RXRα inhibits the Nrf2-ARE signaling pathway through a direct interaction with the Neh7 domain of NRF2. Cancer Res. 73, 3097–3108.

28. Rupert PB, Daughdrill GW, Bowerman B, Matthews BW (1998) A new DNA-binding motif in the Skn-1 binding domain-DNA complex. Nat. Struct. Biol. 5, 484–491.

29. Kusunoki H, Motohashi H, Katsuoka F, Morohashi A, Yamamoto M, et al. (2002) Solution structure of the DNA-binding domain of MafG. Nat. Struct. Biol. 9, 252–256.

30. McMahon M, Itoh K, Yamamoto M, Hayes JD (2003) Keap1-dependent proteasomal degradation of transcription factor Nrf2 contributes to the negative regulation of antioxidant response element-driven gene expression. J. Biol. Chem. 278, 21592–21600.

31. Kobayashi A, Kang MI, Okawa H, Ohtsuji M, Zenke Y, et al. (2004) Oxidative stress sensor Keap1 functions as an adaptor for Cul3-based E3 ligase to regulate proteasomal degradation of Nrf2. Mol. Cell Biol. 24, 7130–7139.

32. Tong KI, Kobayashi A, Katsuoka F, Yamamoto M (2006) Two-site substrate recognition model for the Keap1-Nrf2 system: a hinge and latch mechanism. Biol. Chem. 387, 1311–1320.

33. Padmanabhan B, Tong KI, Ohta T, Nakamura Y, Scharlock M, et al. (2006) Structural basis for defects of Keap1 activity provoked by its point mutations in lung cancer. Mol. Cell 21, 689–700.

34. McMahon M, Thomas N, Itoh K, Yamamoto M, Hayes JD (2006) Dimerization of substrate adaptors can facilitate cullin-mediated ubiquitylation of proteins by a "tethering" mechanism: a two-site interaction model for the Nrf2-Keap1 complex. J. Biol. Chem. 281, 24756–24768.

35. Lo SC, Li X, Henzl MT, Beamer LJ, Hannink M (2006) Structure of the Keap1:Nrf2 interface provides mechanistic insight into Nrf2 signaling. EMBO J. 25, 3605–3617.

36. Ogura T, Tong KI, Mio K, Maruyama Y, Kurokawa H, et al. (2010) Keap1 is a forked-stem dimer structure with two large spheres enclosing the intervening, double glycine repeat, and C-terminal domains. Proc. Natl. Acad. Sci. USA 107, 2842–2847.

37. Nioi P, Nguyen T, Sherratt PJ, Pickett CB (2005) The carboxy-terminal Neh3 domain of Nrf2 is required for transcriptional activation. Mol. Cell Biol. 25, 10895–10906.

38. Katoh Y, Itoh K, Yoshida E, Miyagishi M, Fukamizu A, et al. (2001) Two domains of Nrf2 cooperatively bind CBP, a CREB binding protein, and synergistically activate transcription. Genes Cells 6, 857–868.

39. Zhang J, Hosoya T, Maruyama A, Nishikawa K, Maher JM, et al. (2007) Nrf2 Neh5 domain is differentially utilized in the transactivation of cytoprotective genes. Biochem. J. 404, 459–466.

40. Kim JH, Yu S, Chen JD, Kong AN (2012) The nuclear cofactor RAC3/AIB1/SRC-3 enhances Nrf2 signaling by interacting with transactivation domains. Oncogene 32, 514–527.

41. McMahon M, Thomas N, Itoh K, Yamamoto M, Hayes JD (2004) Redox-regulated turnover of Nrf2 is determined by at least two separate protein domains, the redox-sensitive Neh2 degron and the redox-insensitive Neh6 degron. J. Biol. Chem. 279, 31556–31567.

42. Chowdhry S, Zhang Y, McMahon M, Sutherland C, Cuadrado A, et al. (2013) Nrf2 is controlled by two distinct β-TrCP recognition motifs in its Neh6 domain, one of which can be modulated by GSK-3 activity. Oncogene 32, 3765–3781.

43. Zhang Y, Crouch DH, Yamamoto M, Hayes JD (2006) Negative regulation of the Nrf1 transcription factor by its N-terminal domain is independent of Keap1: Nrf1, but not Nrf2, is targeted to the endoplasmic reticulum. Biochem. J. 399, 373–385.

44. Zhang Y, Lucocq JM, Yamamoto M, Hayes JD (2007) The NHB1 (N-terminal homology box 1) sequence in transcription factor Nrf1 is required to anchor it to the endoplasmic reticulum and also to enable its asparagine-glycosylation. Biochem. J. 408, 161–172.

45. Zhang Y, Ren Y, Li S, Hayes JD (2014) Transcription factor Nrf1 is topologically repartitioned across membranes to enable target gene transactivation through its acidic glucose-responsive domains. PLoS One 9, 1–17(e93456); doi:93410.93137/journal.pone.0093456.

46. Zhang Y, Hayes JD (2013) The membrane-topogenic vectorial behaviour of Nrf1 controls its post-translational modification and transactivation activity. Sci. Rep. 3: 2006, 1–16, doi:10.1038/srep02006.

47. Wang W, Chan JY (2006) Nrf1 is targeted to the endoplasmic reticulum membrane by an N-terminal transmembrane domain. Inhibition of nuclear translocation and transacting function. J. Biol. Chem. 281, 19676–19687.

48. Zhang Y, Li S, Xiang Y, Zhao H, Hayes JD (2014) The selective post-translational processing of Nrf1 yields distinct isoforms that dictate its activity to differentially regulate target genes. (In press of Sci. Rep. MS-13-05858).

49. Chan JY, Han XL, Kan YW (1993) Cloning of Nrf1, an NF-E2-related transcription factor, by genetic selection in yeast. Proc. Natl. Acad. Sci. USA 90, 11371–11375.

50. Luna L, Johnsen O, Skartlien AH, Pedeutour F, Turc-Carel C, et al. (1994) Molecular cloning of a putative novel human bZIP transcription factor on chromosome 17q22. Genomics 22, 553–562.

51. Tsuchiya Y, Morita T, Kim M, Iemura S, Natsume T, et al. (2011) Dual Regulation of the Transcriptional Activity of Nrf1 by beta-TrCP- and Hrd1-Dependent Degradation Mechanisms. Mol. Cell Biol. 31, 4500–4512.

52. Zhong Y, Fang S (2012) Live cell imaging of protein dislocation from the endoplasmic reticulum. J. Biol. Chem. 287, 28057–28066.

53. Zhang Y, Mattjus P, Schmid PC, Dong ZM, Zhong S, et al. (2001) Involvement of the acid sphingomyelinase pathway in UVA-induced apoptosis. J. Biol. Chem. 276, 11775–11782.

54. Chen L, Kwong M, Lu R, Ginzinger D, Lee C, et al. (2003) Nrf1 is critical for redox balance and survival of liver cells during development. Mol. Cell Biol. 23, 4673–4686.

55. Yang H, Magilnick N, Lee C, Kalmaz D, Ou X, et al. (2005) Nrf1 and Nrf2 Regulate Rat Glutamate-Cysteine Ligase Catalytic Subunit Transcription Indirectly via NF-κB and AP-1. Mol. Cell Biol. 25, 5933–5946.

56. Hirotsu Y, Hataya N, Katsuoka F, Yamamoto M (2012) NF-E2-related factor 1 (Nrf1) serves as a novel regulator of hepatic lipid metabolism through regulation of the Lipin1 and PGC-1β genes. Mol. Cell Biol. 32, 2760–2770.

57. Novotny V, Prieschl EE, Csonga R, Fabjani G, Baumruker T (1998) Nrf1 in a complex with fosB, c-jun, junD and ATF2 forms the AP1 component at the TNFα promoter in stimulated mast cells. Nucleic Acids Res. 26, 5480–5485.

58. Prieschl EE, Novotny V, Csonga R, Jaksche D, Elbe-Burger A, et al. (1998) A novel splice variant of the transcription factor Nrf1 interacts with the TNFα promoter and stimulates transcription. Nucleic Acids Res 26, 2291–2297.

59. Zhang Y (2009) Molecular and cellular control of the Nrf1 transcription factor: An integral membrane glycoprotein. VDM Verlag Dr. Müller Publishing House Germany (May 2009). The first edition, pp1–264.

60. Chepelev NL, Bennitz JD, Huang T, McBride A, Willmore WG (2011) The Nrf1 CNC-bZIP protein is regulated by the proteasome and activated by hypoxia. PLoS One 6, e29167.

61. Caterina JJ, Donze D, Sun CW, Ciavatta DJ, Townes TM (1994) Cloning and functional characterization of LCR-F1: a bZIP transcription factor that activates erythroid-specific, human globin gene expression. Nucleic Acids Res 22, 2383–2391.

62. Wang W, Kwok AM, Chan JY (2007) The p65 isoform of Nrf1 is a dominant negative inhibitor of ARE-mediated transcription. J. Biol. Chem. 282, 24670–24678.

63. Kwong EK, Kim KM, Penalosa PJ, Chan JY (2012) Characterization of Nrf1b, a novel isoform of the nuclear factor-erythroid-2 related transcription factor-1 that activates antioxidant response element-regulated genes. PLoS One 7, e48404.

64. Schultz MA, Hagan SS, Datta A, Zhang Y, Freeman ML, et al. (2014) Nrf1 and nrf2 transcription factors regulate androgen receptor transactivation in prostate cancer cells. PLoS One 9, e87204.

65. Herskowitz I (1987) Functional inactivation of genes by dominant negative mutations. Nature 329, 219–222.

66. Sakurai A, Miyamoto T, Refetoff S, DeGroot LJ (1990) Dominant negative transcriptional regulation by a mutant thyroid hormone receptor-beta in a family with generalized resistance to thyroid hormone. Mol. Endocrinol. 4, 1988–1994.

67. Johnsen O, Skammelsrud N, Luna L, Nishizawa M, Prydz H, et al. (1996) Small Maf proteins interact with the human transcription factor TCF11/Nrf1/LCR-F1. Nucleic Acids Res. 24, 4289–4297.

68. Husberg C, Murphy P, Martin E, Kolsto AB (2001) Two domains of the human bZIP transcription factor TCF11 are necessary for transactivation. J. Biol. Chem. 276, 17641–17652.

69. Huang CF, Wang YC, Tsao DA, Tung SF, Lin YS, et al. (2000) Antagonism between members of the CNC-bZIP family and the immediate-early protein IE2 of human cytomegalovirus. J. Biol. Chem. 275, 12313–12320.

70. Wu JL, Lin YS, Yang CC, Lin YJ, Wu SF, et al. (2009) MCRS2 represses the transactivation activities of Nrf1. BMC Cell Biol. 10, 9.

71. Tsuchiya Y, Taniguchi H, Ito Y, Morita T, Karim MR, et al. (2013) The CK2-Nrf1 axis controls the clearance of ubiquitinated proteins by regulating proteasome gene expression. Mol. Cell Biol. 33, 3461–3472.

4

Molecular Insights into the Interaction between *Plasmodium falciparum* Apical Membrane Antigen 1 and an Invasion-Inhibitory Peptide

Geqing Wang[1], Christopher A. MacRaild[1], Biswaranjan Mohanty[1,2], Mehdi Mobli[3], Nathan P. Cowieson[4], Robin F. Anders[5], Jamie S. Simpson[1], Sheena McGowan[6], Raymond S. Norton[1]*, Martin J. Scanlon[1,2]*

1 Medicinal Chemistry, Monash Institute of Pharmaceutical Sciences, Monash University, Parkville, Victoria, Australia, **2** Australian Research Council Centre of Excellence for Coherent X-ray Science, Monash University, Parkville, Victoria, Australia, **3** Centre for Advanced Imaging, University of Queensland, St Lucia, Queensland, Australia, **4** Australian Synchrotron, Clayton, Victoria, Australia, **5** Department of Biochemistry, La Trobe University, Bundoora, Victoria, Australia, **6** Department of Biochemistry and Molecular Biology, Monash University, Clayton, Victoria, Australia

Abstract

Apical membrane antigen 1 (AMA1) of the human malaria parasite *Plasmodium falciparum* has been implicated in invasion of the host erythrocyte. It interacts with malarial rhoptry neck (RON) proteins in the moving junction that forms between the host cell and the invading parasite. Agents that block this interaction inhibit invasion and may serve as promising leads for anti-malarial drug development. The invasion-inhibitory peptide R1 binds to a hydrophobic cleft on AMA1, which is an attractive target site for small molecules that block parasite invasion. In this work, truncation and mutational analyses show that Phe5-Phe9, Phe12 and Arg15 in R1 are the most important residues for high affinity binding to AMA1. These residues interact with two well-defined binding hot spots on AMA1. Computational solvent mapping reveals that one of these hot spots is suitable for small molecule targeting. We also confirm that R1 in solution binds to AMA1 with 1:1 stoichiometry and adopts a secondary structure consistent with the major form of R1 observed in the crystal structure of the complex. Our results provide a basis for designing high affinity inhibitors of the AMA1-RON2 interaction.

Editor: Olivier Silvie, INSERM, France

Funding: The work was funded in part by a grant from the National Health and Medical Research Council of Australia (1025150). RSN is a research fellow of the National Health and Medical Research Council (1059060) (www.nhmrc.gov.au). The funders had no role in study design, data collection and analysis, decision to publish, or preparation of the manuscript.

Competing Interests: The authors have declared that no competing interests exist.

* Email: martin.scanlon@monash.edu (MJS); ray.norton@monash.edu (RSN)

Introduction

Malaria is a deadly infectious disease caused by protozoan parasites of the genus *Plasmodium*. The recently released World Malaria Report estimated that malarial parasites infected over 200 million people worldwide causing 627,000 deaths in 2012 [1]. The increasing incidence of drug resistance, absence of an effective vaccine and lack of diversity amongst current compounds in development renders this ancient disease an ongoing global health problem [2]. Novel anti-malaria therapeutic approaches are urgently required to confront these challenges.

The blood stage of *Plasmodium* infection is the major cause of the clinical symptoms of malaria and the mechanism of erythrocyte invasion is highly conserved in all apicomplexan parasites [3]. Therefore, proteins involved in this process have been actively pursued as targets for both vaccine and drug development. Apical membrane antigen 1 (AMA1), an integral membrane protein that is highly conserved throughout the phylum Apicomplexa, represents one of these protein targets [2]. The initiation of merozoite invasion is marked by formation of the moving junction (MJ), a ring-like protein structure, between the merozoite and the erythrocyte [4]. In our current understanding of

the structure and function of the MJ, AMA1 presents a conserved hydrophobic cleft that interacts with rhoptry neck protein 2 (RON2) [5]. This interaction is essential to the formation of the junction, which commits the parasite to invade [4,6]. Both AMA1 and RON2 are provided by the parasite to enable an active invasion mechanism [7]. AMA1 is initially stored in the parasite micronemes and subsequently translocated to the merozoite surface before invasion, while RON2 is secreted from the parasite rhoptry and transferred to the erythrocyte surface prior to invasion [8–10]. The essential role of AMA1 in host cell invasion has been questioned recently by genetic studies, which showed AMA1-depleted parasites can still form a functional MJ [11,12]. As such, the specific role of AMA1 in host cell invasion remains a matter of debate [13,14], but it is clear that inhibition of the AMA1-RON2 interaction by various agents effectively disrupts invasion and validates AMA1 as a viable therapeutic target [2,15,16]. Specifically, antibodies raised against AMA1 can inhibit invasion by binding to the hydrophobic cleft [17–19], although the inhibition is usually strain-specific [20]. Consistent with these observations, AMA1 evolves under strong selective pressure from the host immune system [21,22], and loops surrounding the hydrophobic cleft are polymorphic [23]. Nonetheless, the AMA1-RON2 interaction is highly conserved. In addition, the interaction between AMA1 and RON2 can be inhibited by peptides. One such peptide, R1, was identified from a random peptide library using phage-display [24,25]. R1 showed a high binding affinity for 3D7 *Pf*AMA1 ($K_D \sim 0.08$ μM) [26] and spans the full-length of the hydrophobic cleft [27,28]. Comparison with the structure of a complex between AMA1 and a peptide derived from RON2 reveals that the two peptides occupy the same region of AMA1 and exhibit structural mimicry [27]. Consistent with these structural studies, R1 can effectively inhibit erythrocyte invasion by malaria parasites *in vitro* [25,28]. Although the inhibition is strain-specific, it has been demonstrated that *N*-methyl modification of R1 broadened its strain specificity [26].

It is evident from the current data that effective targeting of AMA1 from multiple strains requires inhibitors whose interaction is mediated by conserved residues within the hydrophobic cleft, which bind AMA1 without making extensive contact with polymorphic residues. It is likely that this goal will be more easily realized by using smaller molecules as inhibitors. We and others have recently reported the identification of small molecules that bind to AMA1, with the goal of developing these molecules into therapeutically useful antimalarials [15,29]. A common problem faced in small molecule inhibitor design is difficulty in improving the binding affinity and specificity of screening "hits". Identification of binding "hot spots", *i.e.* the subset of residues at the binding interface that contribute most of the free energy to high affinity binding [30], provides important information to guide the design of high-affinity ligands. This is especially critical for targeting protein-protein interactions (PPIs) [31]. As R1 has high binding affinity and makes extensive interactions with the hydrophobic cleft of AMA1, characterization of the AMA1-R1 interaction provides valuable insights into the key interactions that contribute to binding. Indeed, there are many examples showing that small molecule inhibitors can be designed that mimic the interaction of a peptide with a protein target [32–37]. In the current study we have undertaken a detailed biophysical characterization of the interaction of R1 with AMA1 and used computational solvent mapping to identify hot spots at the binding interface. Collectively our data provide a rational basis for designing high-affinity inhibitors of AMA1-RON2 interaction.

Materials and Methods

Expression and purification of AMA1

Domain I+II of 3D7 *Pf*AMA1 (residue 104–442) was expressed, purified and refolded as described [29]. The folding of the purified protein was assessed by monitoring its binding affinity and stoichiometry to R1 using surface plasmon resonance and recording a 1D ^{1}H spectrum, which is characterized in the correctly-folded material by the presence of several upfield-shifted methyl protons (Figure S1 in File S1). Randomly fractionally deuterated (f-^{2}H) AMA1 was prepared by growth of expression cultures in 100% ^{2}H$_2$O/M9 minimal medium supplemented with ^{14}NH$_4$Cl (1 g/L) and protonated ^{12}C-D-glucose (10 g/L). The high-cell-density method was implemented to achieve high protein yield as described in [38]. The hexahistidine (His$_6$) tag of AMA1 was cleaved by tobacco etch virus (TEV) protease in a ratio of 0.02 mg TEV per mg fusion protein in phosphate buffer, pH 8.0 at 4°C for 24 h [39]. The resultant protein was purified on a linear gradient of 0–500 mM NaCl using HiTrap QFF column chromatography (GE healthcare) and dialyzed against 20 mM ammonium bicarbonate solution at 4°C over 2 days before it was lyophilized.

DNA manipulation, expression and purification of R1

R1 peptide was produced recombinantly as an enterokinase-cleavable fusion to thioredoxin. An insert encoding DDDDKVFAEFLPLFSKFGSRMHILK was ligated into pET32a (Novagen) at KpnI/NcoI and transformed into *Escherichia coli* BL21 (DE3). The f-^{2}H, u-^{13}C, ^{15}N-labelled R1 fusion was expressed in 100% ^{2}H$_2$O/M9 minimal medium supplemented with ^{15}NH$_4$Cl (1 g/L) and protonated ^{13}C$_6$-glucose (4 g/L) using the high-cell-density method as described in [38]. The cells were harvested by centrifugation at 5,000 g for 20 min and resuspended in lysis/wash buffer (20 mM Tris-HCl pH 8, 20 mM imidazole, 200 mM NaCl). The cells were lysed by sonication and the supernatants were recovered by centrifugation at 12,000 g for 30 min at 4°C. The His$_6$-tagged R1 fusion in the soluble fraction was purified on a linear gradient of 45–500 mM imidazole by HisTrap column chromatography (GE healthcare). Fractions were analyzed by SDS-PAGE and those containing a band consistent with the expected size of the R1 fusion (~20 kDa) were pooled and dialyzed against enterokinase cleavage buffer (20 mM Tris-HCl pH 7.4, 50 mM NaCl, 2 mM CaCl$_2$, 1 mM EDTA) overnight at 4°C. The fusion protein was then incubated with recombinant enterokinase (Novagen) in a ratio of 0.5 units enterokinase per mg fusion protein at room temperature for 21 h. The sample was then filtered through a 0.22 μm membrane (Millipore, Merck) and purified using HiTrap QFF column chromatography using a gradient of 0–500 mM NaCl in a buffer of 20 mM Tris-HCl pH 8. R1 peptide was finally purified by prep-RP-HPLC using a Phenomenex Luna 5 u C18 column (100×10 mm). The identity and purity (>95%) were confirmed by liquid chromatography mass spectrometry (LCMS) (Figure S2 in File S1). About 1 mg of f-^{2}H, u-^{15}N, ^{13}C-labelled R1 was produced from 0.7 L of minimal medium. ^{2}H incorporation was ~72%, ^{15}N incorporation was ~90%, and ^{13}C incorporation was ~95%.

Synthetic R1 analogues

Truncated and mutant R1 peptides used in the SPR study were synthesized by Mimotopes (Melbourne, Australia) with purity > 90% and all were *N*-terminally acetylated and *C*-terminally amidated.

NMR sample preparation

NMR samples were prepared in a buffer consisting of 20 mM sodium phosphate pH 7, 1 mM EDTA, 0.01% (w/v) sodium azide, 0.2% (w/v) Complete protease inhibitor cocktail (Roche), 50 mM Arg, 50 mM Glu and 6% (v/v) ^{2}H$_2$O unless noted otherwise. For the NMR study of free R1, two samples of u-^{13}C, ^{15}N-labelled R1 at a concentration of 0.4 mM were prepared at pH 5 and pH 7, respectively. To study the AMA1-R1 complex, lyophilized f-^{2}H-labelled AMA1 was added to a sample of f-^{2}H, u-^{13}C, ^{15}N-labelled R1 to give final concentrations of AMA1 and R1 of 320 and 300 μM, respectively. Based on the measured K_D of R1 for AMA1, >90% of the peptide should be bound to AMA1 under these conditions.

NMR spectroscopy

NMR experiments for free R1 were performed at 5°C or 40°C at a ^{1}H frequency of either 500 MHz or 600 MHz on Bruker Avance spectrometers equipped with a TXI-cryoprobe. Chemical shift assignments were made using the following experiments: 2D ^{1}H-^{15}N-HSQC, ^{1}H-^{13}C-HSQC and 3D triple-resonance experiments including HNCACB, CBCA(CO)NH, HBHA(CO)NH and HCCH-TOCSY. All spectra were processed using NMRPipe [40] and analyzed with CARA [41]. All NMR experiments for the AMA1-R1 complex were performed at 40°C in a 5 mm Shigemi tube. The backbone H^N, C^α, and N resonances of f-^{2}H, u-^{13}C, ^{15}N-R1 bound to f-^{2}H-AMA1$_{104-442}$ were assigned using 2D ^{1}H-^{15}N-TROSY HSQC/conventional ^{1}H-^{15}N-HSQC, 3D TROSY-HNCA and TROSY-HN(CO)CA. The 3D TROSY-HNCA was acquired on a Bruker DRX-900 spectrometer equipped with a cryoprobe. Non-uniform sampling was utilized during acquisition, with sampling points chosen randomly from a probability distribution matching the signal decay, as described previously [42]. The spectra were re-constructed using the maximum entropy method with automated parameter selection using the Rowland NMR toolkit [43]. A ^{13}C(F$_2$)-^{1}H(F$_3$) plane of the 3D TROSY-HN(CO)CA was acquired on a Bruker Avance 600 MHz spectrometer. The data were processed using NMRPipe or Topspin 3.0 (Bruker-Biospin) and analyzed with CARA. Chemical shifts are reported relative to sodium 2,2-dimethyl-2-silapentane-5-sulfonate (DSS).

Surface plasmon resonance binding analysis

A Biacore T200 biosensor instrument was used to measure the affinity of the interaction of peptides with 3D7 PfAMA1$_{104-442}$. AMA1 was immobilized onto a CM5 chip as described [29]. Surface plasmon resonance (SPR) experiments were performed at 25°C using HBS-EP (10 mM HEPES, 150 mM NaCl, 3.4 mM EDTA, and 0.05% surfactant P20, pH 7.4) as the running buffer either with (alanine scanning mutagenesis study) or without (truncation study) 1% DMSO. All peptide samples were prepared in the appropriate running buffer. To generate the peptide binding data, peptide at concentrations ranging from 10 nM to 10 μM was injected over immobilized AMA1 at a constant flow rate of 60 μL/min for 1.5 min; peptide dissociation was monitored by flowing running buffer at 60 μL/min for 5 min. The surface was regenerated after each cycle by injecting glycine/HCl at pH 2.0. Sensorgrams were first zeroed on the y-axis and then x-aligned at the start of the injection. Bulk refractive index changes were eliminated by subtracting the reference flow cell responses. For kinetic analysis, k_a and k_d were determined from the processed data sets by globally fitting to a 1:1 binding model. For rapidly associating/dissociating truncated peptides, K_D was determined by fitting to a steady-state affinity model using a fixed R_{max} that was calculated based on the response of R1$_{5-16}$.

Analytical size exclusion chromatography (SEC)

Analytical SEC was performed on a Superdex 75 HR 10/30 column (column dimension 1.0×30 cm, column volume 23.6 mL) at room temperature. Samples (100 μl) containing AMA1 (200 μM) with or without R1 peptide (250 μM) were injected onto the column, which was pre-equilibrated with 20 mM sodium phosphate pH 7. Samples were prepared in NMR buffer (20 mM sodium phosphate pH 7, 1 mM EDTA, 0.01% (w/v) sodium azide, 0.2% (w/v) Complete protease inhibitor cocktail (Roche), 50 mM Arg and 50 mM Glu). The flow rate was maintained at 0.5 mL/min and the elution was monitored by measurement of UV absorbance at 280 nM (A$_{280}$).

Small angle X-ray scattering (SAXS)

SAXS measurements were made at the SAXS-WAXS beamline of the Australian Synchrotron, Melbourne, Australia. For each SAXS measurement, 10×1 s exposures were measured and averaged together after verifying that there was no evidence of radiation damage (systematic change in the shape of the scattering curves as a function of exposure time). During data collection the sample was flowed through a 1.5 mm quartz capillary at a rate of

4 μl/sec to further control for radiation damage. Measurements were performed on a dilution series of AMA1 alone from 3.3 to 0.14 mg/ml in NMR buffer and AMA1+R1 (ratio of 1:1.15) from 3.0 to 0.19 mg/ml in the same buffer. Some concentration-dependent aggregation was observed at protein concentrations above 1 mg/ml as evidenced by increases in Rg and disproportionate increases in I(0) (data not shown). The SAXS data used in this study were from protein at 0.5 mg/ml for AMA1 and 0.75 mg/ml for AMA1:R1. Dilution of the protein below these concentrations did not result in changes to the shape of the scattering curve and calculated molecular weights at these concentrations were consistent with monomeric protein. The molecular weights of the scattering species were estimated from the total forward scatter of the SAXS measurements that were normalised by comparison to water scatter and with reference to the measured protein concentrations. Partial specific volume and scattering length density were calculated using the program MULCh [44]. The monomeric state of the protein was inferred by comparison of the theoretical molecular weight of the protein sequence with the calculated molecular weight from the SAXS experiment. A 1.6 m camera was used with an X-ray energy of 11 keV giving a Q range from 0.01 to 0.5 Å^{-1}. Data were collected on a Pilatus 1M detector (Dektris) and averaging of images, subtraction of blanks and radial integration was performed using the beamline control software ScatterBrain (Australian Synchrotron). Measurements were made at 25°C. Calculation of scattering intensities from molecular models was done using CRYSOL [45]. Radius of gyration (Rg), total forward scatter (I(0)) and P(r) functions were derived using the automated functions in PRIMUS [46] and without manual intervention.

Computational mapping of binding hot spots

FTMAP was employed to map the binding hot spots of AMA1 (http://ftmap.bu.edu/) [47] using the AMA1 structures with PDB ID 3SRJ and 2Z8V, which were downloaded from the Protein Data Bank [18,27]. All ligands and water molecules were removed before mapping. FTMAP searched the global surface of AMA1 with a library of 16 small organic molecules (ethanol, isopropanol, isobutanol, acetone, acetaldehyde, dimethyl ether, cyclohexane, ethane, acetonitrile, urea, methylamine, phenol, benzaldehyde, benzene, acetamide and N,N-dimethylformamide). The small molecule probes have different hydrophobicity and hydrogen bonding capability. FTMAP employs a fast Fourier transform correlation approach to efficiently sample billions of protein-probe complexes [48]. The 2000 most favourable docked positions of each probe were energy-minimized and clustered. The six clusters with the lowest average free energy were selected for each probe. The clusters of different probes were further clustered into consensus sites (CSs) based on the distance between the cluster centres. The details of the FTMAP algorithm are described in [48].

Accession Numbers

Chemical shift assignments for free R1 (pH 5, 40°C) and AMA1-bound R1 (pH 7, 40°C) have been deposited in BMRB under accession codes 19864 and 25134, respectively.

Results and Discussion

Truncation of the R1 peptide

We sought to identify key residues in the interaction of R1 with AMA1. Firstly, in order to define the minimal R1 construct that retains high binding affinity for 3D7 PfAMA1, a series of truncated R1 analogues (Figure 1A) was synthesized and screened by SPR. Kinetic analysis of data generated for native R1 binding to AMA1 produced a K_D of 0.11 µM by globally fitting to a 1:1 binding model (Figure 2A), which is consistent with the reported value (~0.08 µM) [26]. Our previous mutagenesis studies had shown that Phe5, Pro7, Leu8 and Phe9 of R1 were essential for high affinity binding of R1 to AMA1 [49]. This conclusion was supported by the current data, in which the truncated R1$_{11-20}$ showed no binding to AMA1 up to 10 µM (Table 1). Interestingly, R1$_{1-11}$ containing the Phe5-Phe9 segment also displayed no detectable binding to AMA1 up to 10 µM (Table 1), implying that the residues Phe5-Phe9 are necessary but not sufficient for interaction with AMA1, and that other key residues are required to facilitate high affinity binding.

To test this hypothesis, R1$_{4-17}$ and R1$_{5-16}$ peptides were synthesized and their binding affinities were measured by SPR. Since both of these truncated mutants showed fast association and dissociation kinetics (Figure 2B), a steady-state affinity model was used to fit the data, producing K_D values of 0.88 µM for R1$_{4-17}$ and 0.99 µM for R1$_{5-16}$. Although the binding affinity of the peptides was reduced nearly 10-fold relative to native R1, the fact that both peptides retain $K_D < 1$ µM suggests that Val1-Glu4 and His17-Lys20 do not contribute substantially to high affinity binding with AMA1 (Figure 1B). Further truncation to the 11-residue peptide R1$_{5-15}$ resulted in a further ~5-fold reduction in K_D to 4.6 µM. However truncation of this peptide by deletion of Arg15 to generate the 10-residue R1$_{5-14}$ completely abolished measurable binding (Table 1), indicating that Arg15 is essential for high-affinity binding of R1 to AMA1. This result is consistent with the co-crystal structure of AMA1 bound to R1, in which Arg15 of R1 is bound in a pocket at one end of the hydrophobic cleft of

AMA1 (Figure 1B), where it forms four hydrogen bonds and is the residue that contributes the largest proportion to the buried surface in the interface [27]. Therefore, R1$_{5-16}$ was determined to be the minimal construct that retained relatively high binding affinity (~1 µM) to AMA1. This segment of R1 displays remarkable structural similarity to the Ala2031-Met2042 segment of PfRON2, with an RMSD of 1.2 Å over the twelve C^α positions in their respective structures, implying that the high affinity of R1$_{5-16}$ originates from direct mimicry of the natural ligand RON2 as previously suggested (Figure 1C) [27].

Alanine-scanning mutagenesis of the R1 peptide

To identify the key interacting residues of R1$_{5-16}$, alanine-scanning mutagenesis was performed and the binding affinities of the mutants were determined by SPR (Table 2). It was necessary to include 1% DMSO (v/v) in the running buffer for this SPR study to maintain the solubility of all of the peptides. This resulted in a small drop in the affinity of the interaction with R1$_{5-16}$ (Table 2). Previous ELISA assays on four single-point mutants of R1 had demonstrated that mutation of Pro7 abrogated R1 binding, while mutation of Phe5, Leu8 and Phe9 each resulted in 7.5-, 86- and >140-fold reductions in affinity relative to the full-length peptide, respectively [49]. In the current SPR study, substitution of Pro7 to Ala resulted in a 35-fold reduction in affinity for AMA1 relative to R1$_{5-16}$, indicating that Pro7 is one of the residues that are crucial for high affinity binding. In the crystal structure of the AMA1-R1 complex [27], Pro7 does not make any direct contact with AMA1, suggesting that it may play a structural role to maintain the adjacent residues in an appropriate

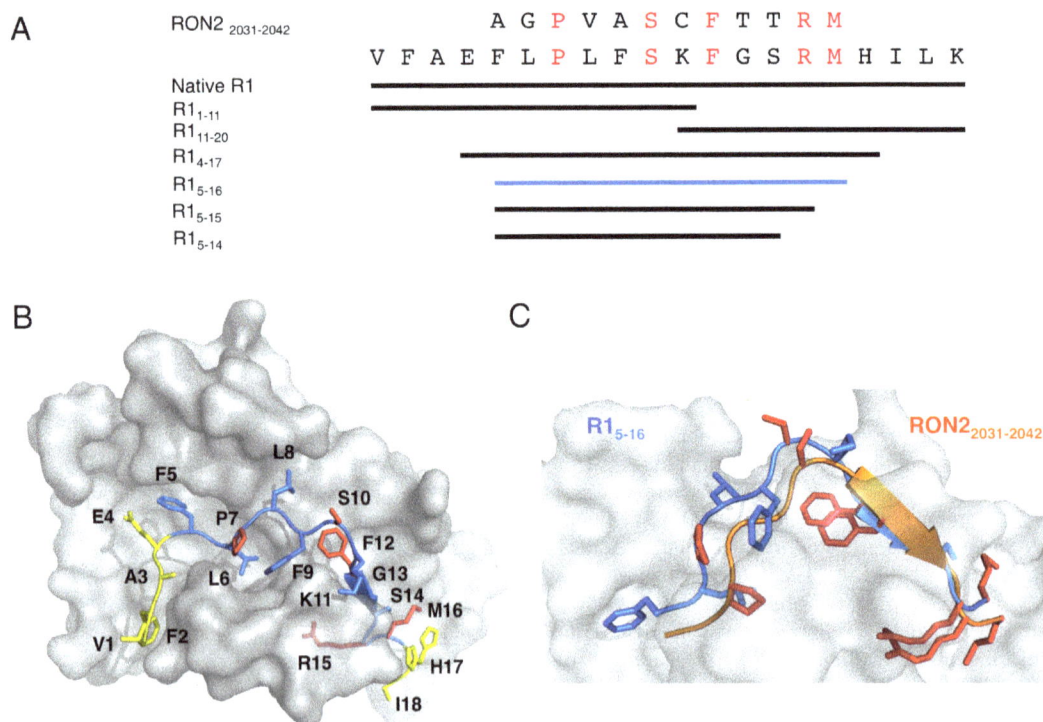

Figure 1. Identification of the minimal binding construct of R1 peptide. A. Amino acid sequences of PfRON2$_{2031-2042}$, native R1 and truncated peptides. Residues that are conserved between R1 and RON2 are highlighted in red. B. Co-crystal structure of PfAMA1 bound to R1 peptide (PDB ID: 3SRJ, [27]). AMA1 is presented as a grey surface; R1 is presented as a cartoon (the minimal binding construct Phe5-Met16 is shown in blue, Val1-Glu4 and His17-Ile18 are in yellow). Side chains of the conserved residues are highlighted in red. The minor form of R1 is omitted in this structure. C. Structural comparison of R1$_{5-16}$ (blue) and PfRON2$_{2031-2042}$ (orange) bound to AMA1. The structure of the R1 peptide bound to PfAMA1 (PDB ID: 3SRJ) superimposed onto the co-crystal structure of PfAMA1-PfRON2 (PDB ID: 3ZWZ, [27]). Only Phe5-Met16 of R1 and Ala2031-Met2042 of PfRON2 are shown for clarity.

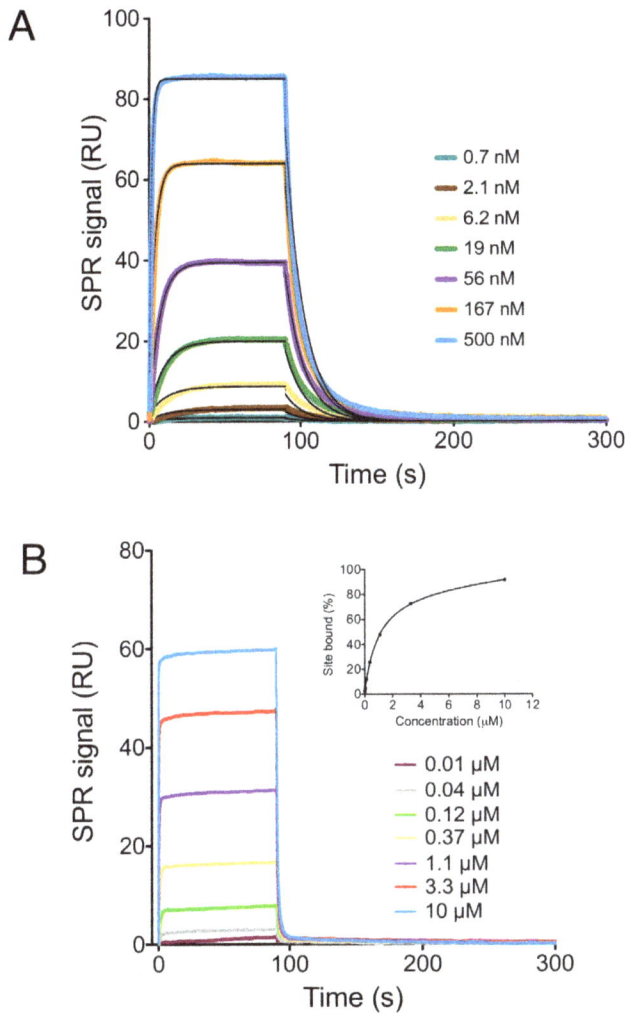

Figure 2. SPR analysis of peptides binding to immobilized 3D7 PfAMA1₁₀₄₋₄₄₂. A series of concentrations, as indicated in sensorgrams, of native R1 (panel A) and truncated R1$_{5-16}$ (panel B) was injected over the AMA1-immobilized surface. k_a and k_d of native R1 were determined by globally fitting to a 1:1 binding model. The apparent equilibrium dissociation constants K_D for other peptides were determined using a steady-state affinity model and are given in Table 1.

conformation for binding (Figure 1B). In addition, substitution of Leu6 to Ala caused a 33-fold reduction in affinity for AMA1 relative to R1$_{5-16}$ (Table 2). Leu6 makes interactions with a cluster of five Tyr residues in AMA1 (Tyr142, Tyr 175, Tyr234, Tyr 236 and Tyr 251). Importantly, Tyr 251 is highly conserved in *Plasmodium* species and has been shown to be essential for AMA1-RON2 interactions [4,50]. Combining current and previous data [49], every residue in the hydrophobic sequence Phe5-Phe9 contributes significantly to AMA1 binding. In the crystal structure of the AMA1-R1 complex, Phe5-Phe9 interacts with a well-defined pocket on one end of the hydrophobic cleft (Figure 1B). All of the above suggest that the pocket is a binding hot spot on AMA1 and potentially an attractive target site.

A substantial drop in affinity (48-fold relative to R1$_{5-16}$) was observed for Ala mutation at Phe12. In the crystal structure, the aromatic ring of Phe12 interacts with two of the key Tyr residues Tyr236 and Tyr251 in the hot spot. In addition, it interacts with Phe183, which was previously identified as a key residue for *Pf*AMA1-*Pf*RON2 interaction [27]. Ala mutation at Phe2038 of RON2 (equivalent to Phe12 of R1, Figure 1A) abolished the binding of RON2 to AMA1 [27]. A 15-fold reduction in affinity relative to R1$_{5-16}$ was observed for the Lys11Ala mutant. This may be caused by disruption of the H-bonds that are observed in the structure between the Lys side chain and Asp227 of AMA1. Mutation of Gly13 resulted in a 21-fold reduction in binding affinity relative to R1$_{5-16}$. Since Gly13 interacts with AMA1 through backbone residues only, this loss in affinity may be the result of conformational changes or steric clashes introduced by the mutation. In contrast to the residues discussed above, individual replacements of Ser10, Ser14 and Met16 with Ala resulted in less than 3-fold reductions in affinity, implying that these residues do not contribute significantly to the binding affinity for AMA1. In the crystal structure of the complex, the side chains of these residues are pointing away from the hydrophobic cleft of AMA1 such that mutation to Ala can be accommodated (Figure 1B) [27].

Consistent with both the truncation studies and the crystal structure, substitution of Arg15 to Ala resulted in largest reduction in affinity (>60-fold relative to R1$_{5-16}$) (Table 2). The importance of the Arg residue at this position is similar to the case with a peptide derived from RON2, where substitution of Arg2041 of RON2 (equivalent to Arg15 of R1, Figure 1A) to Ala abolished the binding to AMA1 [27]. In the structures of their complexes, Arg2041 of *Pf*RON2 interacts with the same pocket of AMA1 as Arg15 of R1 and is the residue that contributes most of the buried

Table 1. Equilibrium dissociation constants (K_D) determined by SPR for the interaction of truncated R1 with 3D7 *Pf*AMA1$_{104-442}$.

Peptide	K_D (µM)[a]
Native R1	0.11
R1$_{1-11}$	No binding[b]
R1$_{11-20}$	No binding[b]
R1$_{4-17}$	0.88
R1$_{5-16}$	0.99
R1$_{5-15}$	4.6
R1$_{5-14}$	No binding[b]

[a] Equilibrium dissociation constants (K_D) were estimated using a kinetic algorithm or a steady-state affinity algorithm available within the Biacore T200 evaluation program.
SPR was performed in HBS-EP running buffer (no DMSO) at 25°C.
[b] No binding event was observed up to a peptide concentration of 10 µM.

Table 2. Equilibrium dissociation constants (K_D) determined by SPR for the interaction of $R1_{5-16}$ mutants with 3D7 PfAMA1$_{104-442}$.

Peptide	Sequence	K_D (μM)[a]
$R1_{5-16}$	Ac-FLPLFSKFGSRM-NH$_2$	1.8±0.04[b]
$R1_{5-16}$ L6A	Ac-F**A**PLFSKFGSRM-NH$_2$	60±12
$R1_{5-16}$ P7A	Ac-FL**A**LFSKFGSRM-NH$_2$	61±25
$R1_{5-16}$ S10A	Ac-FLPLF**A**KFGSRM-NH$_2$	4.8±0.8
$R1_{5-16}$ K11A	Ac-FLPLFS**A**FGSRM-NH$_2$	27±2.1
$R1_{5-16}$ F12A	Ac-FLPLFSK**A**GSRM-NH$_2$	87±5.9
$R1_{5-16}$ G13A	Ac-FLPLFSKF**A**SRM-NH$_2$	39±3.9
$R1_{5-16}$ S14A	Ac-FLPLFSKFG**A**RM-NH$_2$	3.3±0.3
$R1_{5-16}$ R15A	Ac-FLPLFSKFGS**A**M-NH$_2$	>100
$R1_{5-16}$ M16A	Ac-FLPLFSKFGSR**A**-NH$_2$	2.7±0.2

[a] Equilibrium dissociation constants (K_D) were estimated using a steady-state affinity algorithm available within the Biacore T200 evaluation program. The data are expressed as mean ± standard error of the means (SEM). All experiments were conducted on at least three independent occasions.
[b] SPR for $R1_{5-16}$ and its mutants was performed in the presence of 1% DMSO (v/v) in HBS-EP running buffer at 25°C.

surface in the interface (Figure 1C). In addition to R1 and RON2, antibodies IF9 and IgNAR, which bind with high affinity to AMA1, also have either Arg or Lys residues that fit into the same pocket in the hydrophobic cleft in their respective structures [27]. Together, these data confirm that this "Arg pocket" is a binding hot spot on AMA1, which may serve as a pivotal anchor point for RON2 binding and an attractive site for inhibiting the AMA1-RON2 interaction.

Backbone resonance assignments of the AMA1-bound R1 peptide

The crystal structure of the AMA1-R1 complex revealed a somewhat unexpected 2:1 binding stoichiometry, which contrasted with the 1:1 binding observed previously by SPR and ITC studies [27]. To resolve this apparent anomaly, we investigated the AMA1-R1 interaction by solution NMR spectroscopy. A recombinant protein expression system was established to produce uniformly (u-) ^{13}C, ^{15}N-labelled R1 peptide (Figure S2–4 in File S1). Backbone resonance assignments for free u-^{13}C, ^{15}N-labelled R1 were obtained at pH 7 and 5°C using standard triple-resonance experiments. For the free peptide it was necessary to record the spectrum at a lower temperature as several peaks were not observed at 40°C (which was found to be the optimum temperature for recording spectra of the complex), presumably due to their rapid exchange with water (Figure 3). To enable comparison of the free and bound states, amide chemical shifts for free R1 were extrapolated to 40°C by recording a series of ^{1}H-^{15}N-HSQC spectra at increasing temperatures and calculating the temperature dependence of the amide resonances (Table S1 in File S1).

A sample of fractionally deuterated (f-^{2}H), u-^{13}C, ^{15}N-labelled R1 with excess f-^{2}H-labelled AMA1 was prepared for backbone assignment of bound R1. To ensure that all the R1 peptide was in the bound form, samples with different ratios of the R1:AMA1 were also prepared. It was found that when the R1:AMA1 ratio was >1, the ^{1}H-^{15}N-HSQC spectrum contained two sets of peaks corresponding to free R1 and bound R1, respectively (Figure S5 in File S1). This indicates that R1 is in slow exchange with AMA1, which is consistent with its high binding affinity. If two peptides were bound to one AMA1 molecule as shown in the crystal structure, this would either give rise to a second set of bound

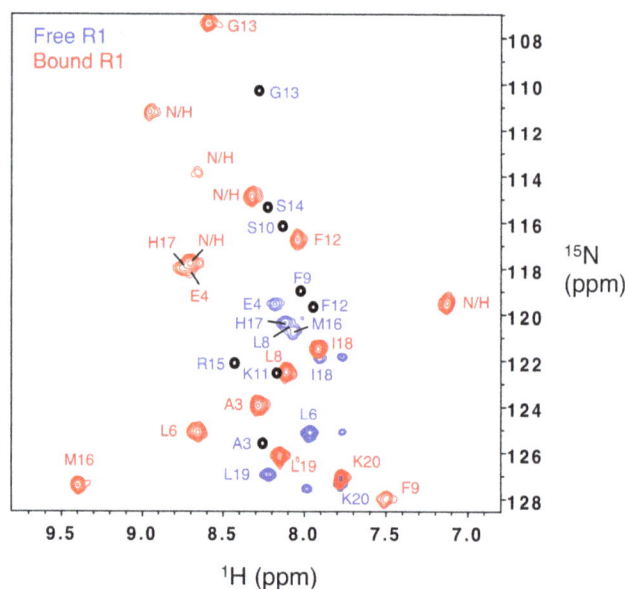

Figure 3. Comparison of the ^{1}H-^{15}N-HSQC spectra of f-^{2}H, u-^{13}C, ^{15}N-labelled R1 in the absence (blue) and presence (red) of a saturating concentration of fractionally deuterated 3D7 PfAMA1$_{104-442}$ at pH 7 and 40°C. Some amide resonances (Ala3, Phe9, Ser10, Lys11, Phe12, Gly13, Ser14 and Arg15) of free R1 were broadened beyond detection at pH 7 and 40°C and their predicted resonances are indicated as black circles in the spectrum (prediction was made as described in the text). N/H indicates unassigned amide resonances of bound R1.

signals in the spectrum or lead to perturbation of the chemical shifts of free R1 in the spectrum recorded with a sub-stoichiometric amount of AMA1; however, no additional peaks or chemical shift perturbations corresponding to a second bound state of R1 were observed. Thus the NMR result supports the 1:1 binding stoichiometry indicated by our SPR data and previous ITC data [27].

R1 is a 20-residue peptide containing a single proline and has a free N-terminal amine. Therefore, a total of 18 peaks were expected in the ^{1}H-^{15}N-HSQC spectrum of bound R1. Of these, 17 were observed for the bound R1 peptide at pH 7 and 40°C (Figure 3), although the peak intensities were non-uniform across the spectrum. Both analytical size-exclusion chromatography (Figure 4) and small-angle X-ray scattering (SAXS) data (Figure 5) indicate that AMA1 interacts with R1 as a monomer, with no evidence for higher order oligomers of protein. The monomeric state is inferred from the SAXS data both from the goodness of fit to the monomeric crystal structures (Figure 5) and from the molecular weight calculated from total forward scattering (37 kDa for apo AMA1 and 41 kDa for AMA1+R1. These values compare to theoretical molecular weights of 41.3 and 43.7 kDa respectively). This suggests that the poor sensitivity of certain residues in the NMR spectra is most likely caused by local conformational exchange in the complex that results in significant broadening for the peaks of affected residues. This effect also resulted in poor sensitivity in 3D experiments and hindered full backbone assignment. Through careful analysis of both TROSY-HNCA and TROSY-HN(CO)CA spectra (Figure S6,7 in File S1), 12 out of 18 expected amide resonances and 15 out of 20 expected C$^\alpha$ resonances were assigned (Table S2 in File S1).

Structural analysis of the AMA1-bound R1 peptide

Free R1 displayed narrow chemical shift dispersion in the proton dimension (7.7 ppm–8.5 ppm) of the ^{1}H-^{15}N-HSQC, consistent with the largely disordered structure in solution that has been observed previously (Figure 3 and Figure S3–4 in File S1) [25]. Upon binding to AMA1 the ^{1}H-^{15}N-HSQC spectrum of the peptide showed broader chemical shift dispersion in the proton dimension (7.0–9.5 ppm), consistent with the peptide assuming a more ordered conformation. The crystal structure of R1 bound to AMA1 identified two R1 peptides bound to AMA1, which were described as the "major" and "minor" states [27]. However, only one set of amide peaks was observed in the ^{1}H-^{15}N-HSQC for bound R1 (Figure 3).

As R1 "minor binder" makes several contacts with R1 "major binder" in the crystal structure, we sought to evaluate the possible structural changes of bound R1 for the 1:1 binding stoichiometry that is observed in solution. Due to the poor sensitivity in 3D experiments, we were not able to solve the solution structure of bound R1 and make direct comparison with the crystal structure. Instead, we probed the secondary structure of bound R1 based on a limited set of assigned C$^\alpha$ chemical shifts and compared that with the secondary structure of bound R1 in the crystal structure. The deviation of the C$^\alpha$ chemical shifts in R1 relative to their random coil values [51] (secondary shifts, $\Delta\delta$) was calculated as these are correlated with the polypeptide backbone torsion angles φ and ψ [52] and can be used to predict the secondary structure of AMA1-bound R1 in solution. The secondary shifts of both free and bound R1 are plotted in Figure 6A. The secondary shifts for free R1 are close to zero. In contrast, bound R1 showed larger secondary shifts. Although the C$^\alpha$ chemical shift of Ser14 remained unassigned, Phe12 and Gly13 showed reasonably strong negative secondary shifts (Phe12 and Gly13 < −1), which is consistent with the presence of extended β-structure in Phe12-Gly13-Ser14 as revealed by the crystal structure of major R1 bound to AMA1 [27]. The C-terminal residues His17-Lys20 of bound R1 showed nearly identical C$^\alpha$ secondary shifts to those of free R1, which is consistent with this region being flexible in solution and suggests that these residues may not be directly involved in the interaction with AMA1.

To further evaluate secondary structure similarity between bound R1 in solution and in the crystal, a comparison was made of C$^\alpha$ chemical shifts, which were determined experimentally for bound R1 in solution and predicted for the major form of R1 in the crystal structure. The predictions for R1 in the crystal structure (Chain C, PDB ID: 3SRJ) were performed using SHIFTX2 [53]. The predicted results are plotted as secondary shifts in Figure 6B.

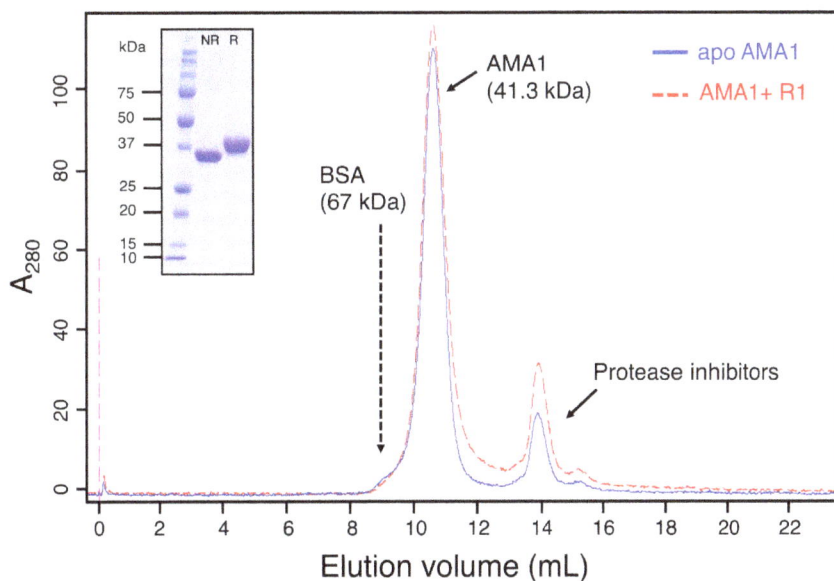

Figure 4. Elution profile of AMA1 in the absence (blue) and presence (red) of R1 peptide on an analytical size exclusion column. The elution time of bovine serum albumin (BSA) was determined for the size exclusion column that was used to elute AMA1 (Superdex 75 HR 10/30, column dimension 1.0×30 cm, column volume 23.6 mL). Both apo AMA1 (200 µM) and R1-bound AMA1 (200 µM AMA1+250 µM R1) showed a similar elution profile. The peak eluting at 10.5 mL is consistent with monomeric AMA1 (MW 41.3 kDa when His tag is not cleaved). SDS-PAGE (inset) confirms only one protein band corresponding to AMA1 is present. NR = non-reducing, R = reducing. The peak at ~14 mL results from the addition of Complete protease inhibitor cocktail (Roche) to the buffer (Figure S8 in File S1), and was also verified by mass spectroscopy.

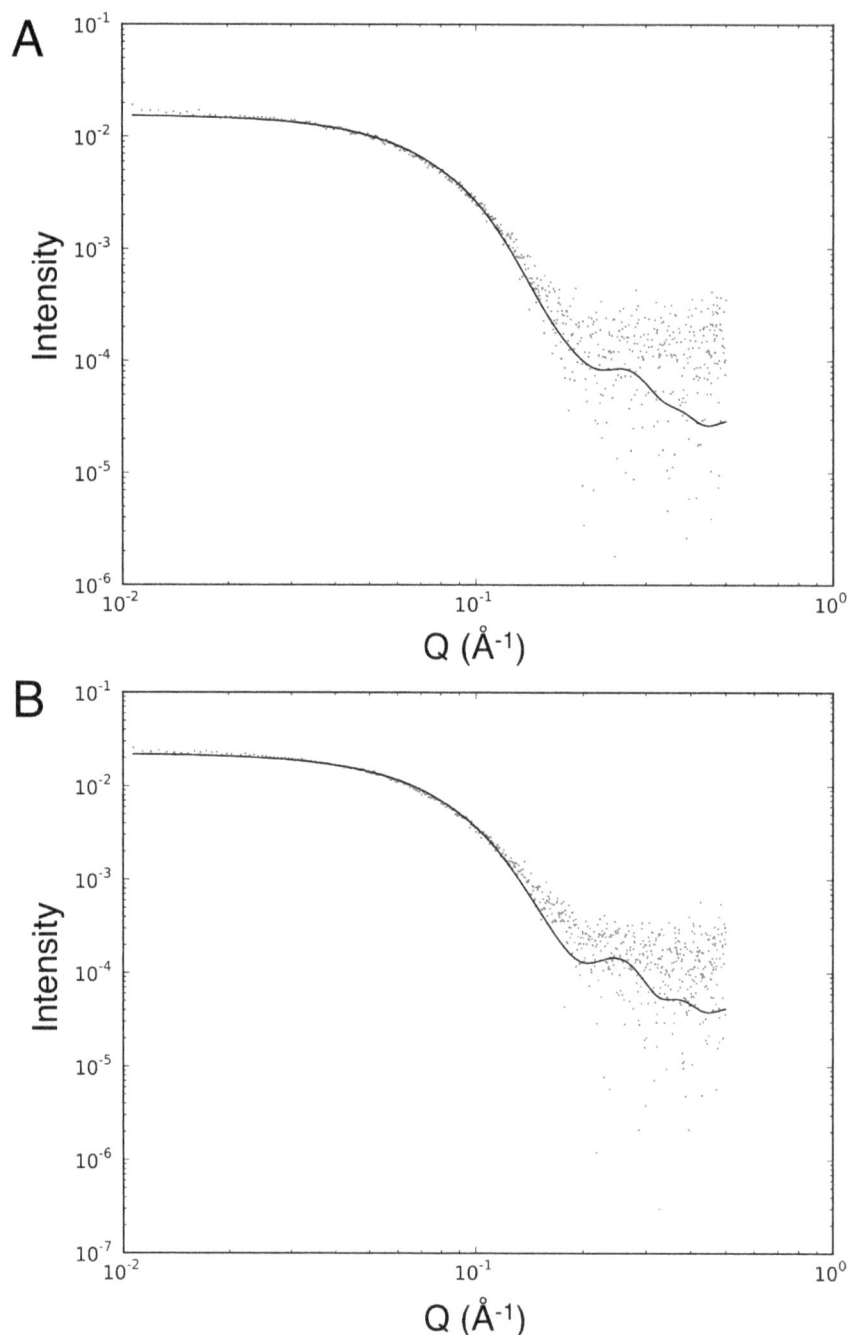

Figure 5. SAXS analysis of AMA1 alone and AMA1 in the presence of R1. (A) AMA1 scattering data fitted to the crystal structure of AMA1 (PDB 1Z40) (Chi-square = 0.61). (B) AMA1+R1 scattering data fitted to the crystal structure of AMA1 bound to R1 (PDB 3SRJ) (Chi-square = 0.72). Q is the momentum transfer vector.

Although the C^{α} chemical shifts of some residues were unassigned or missing, the secondary C^{α} shifts of the experimental data and the crystal structure prediction were strikingly similar. C^{α} chemical shifts were also predicted for the minor R1 in the crystal structure, but the correlation between the predicted data and experimental data was much poorer (correlation coefficient for Glu4-Leu8 of minor R1 = 0.85; correlation coefficient for Glu4-Leu8 of major R1 = 0.99; Figure S9 in File S1). The distinction between the two binding modes was equally unambiguous when chemical shifts were predicted by SPARTA+ instead of SHIFTX2 (Figure S10 in File S1). This suggests that the secondary structure of AMA1-bound R1 in solution is similar to that of the major R1 in the crystal structure. Taken together, the stoichiometry observed in the SPR data we report here, the previous ITC data and comparison of the experimental and predicted NMR data suggest that the minor R1 conformation was most likely an artifact due to the high concentration of R1 peptide used in the crystallographic study.

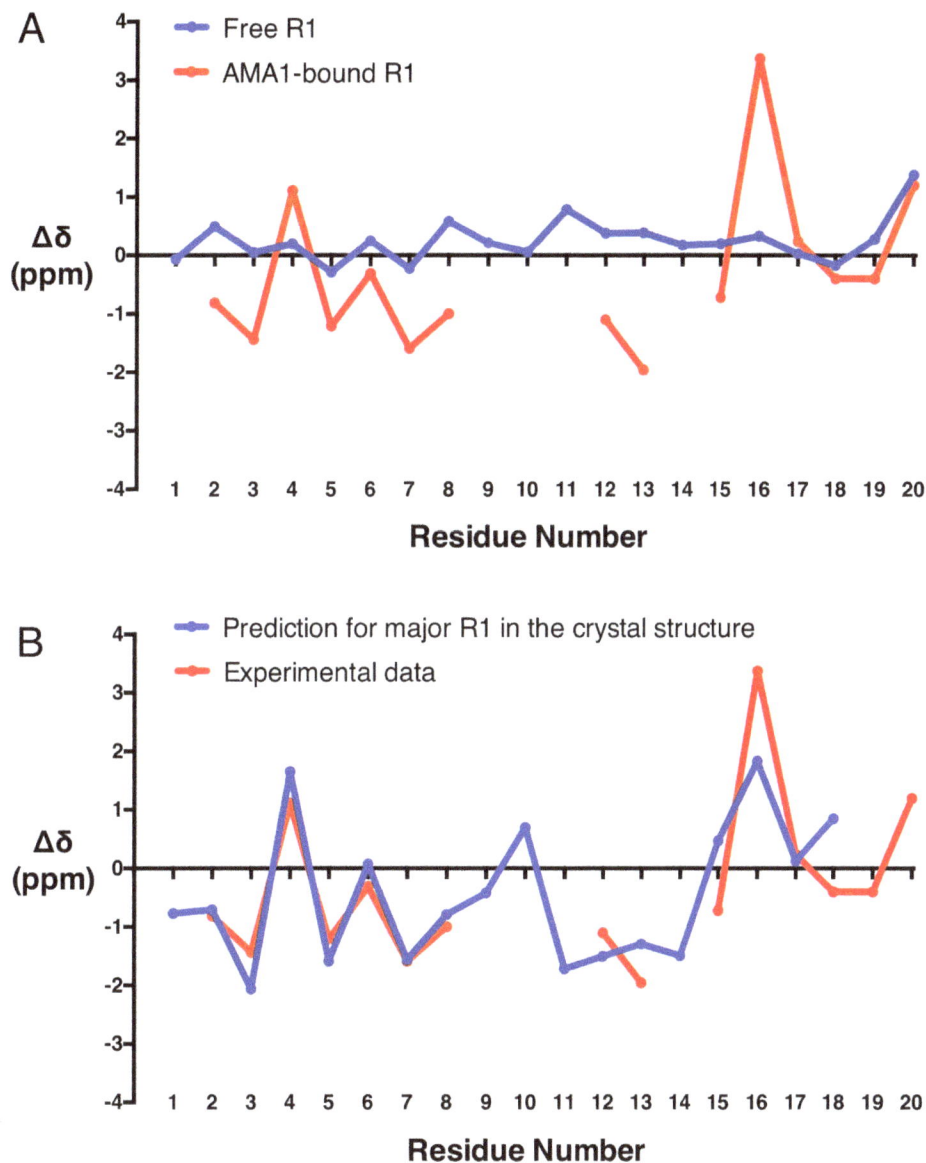

Figure 6. NMR C$^{\alpha}$ secondary chemical shifts. A. The C$^{\alpha}$ secondary shifts for free (blue) and AMA1-bound (red) R1 at pH 7 and 40°C. B. Comparison of the experimentally determined C$^{\alpha}$ secondary shifts of bound R1 (red) and C$^{\alpha}$ secondary shifts predicted for the major form of R1 bound to AMA1 in the crystal structure (Chain C, PDB ID: 3SRJ) using SHIFTX2 (blue) [53].

Computational solvent mapping of AMA1

Using truncation and mutagenesis of R1 peptide, we have identified two binding hot spots on the AMA1 surface that contribute to high affinity of R1 binding. To further assess the capacity of these hot spots to effectively bind small organic molecules, we employed FTMAP, a fragment-based computational solvent mapping algorithm [47]. FTMAP searched the global surface of the AMA1 with a library of 16 small molecule probes that vary in hydrophobicity and hydrogen bonding capability [48]. The regions that bind to probe clusters are designated consensus sites (CS) in FTMAP and the site that binds the highest number of probe clusters is identified as the most druggable. We performed the initial mapping on the structure of 3D7 *Pf*AMA1 co-crystallized with R1 (PDB ID: 3SRJ). Prior to mapping, the bound ligands and water were removed. The five largest consensus sites were located in the same pocket, which was also identified

from SPR analysis as a hot spot for binding Phe5-Phe9 of R1, implicating it as a prospective pocket for small molecule targeting (Figure 7A). The most notable features of the pocket are its hydrophobicity and conservation. The pocket is flanked at one end by a cluster of five Tyr residues (Tyr142, Tyr175, Tyr234, Tyr236 and Tyr251) and at the other end by Leu176, Ala254, Met273 and Phe274 (Figure 7B). Tyr251 is highly conserved across *Plasmodium* species and essential for AMA1-RON2 interactions [4,50]. All the other residues that form the pocket, except Tyr175, are also highly conserved in all known *P. falciparum* sequences [2].

Of the probe clusters identified, CS2 (magenta, 17 probe clusters) overlaps well with Leu6 of R1 and partially with Phe9 of R1; CS5 (grey, 9 probe clusters) overlaps well with Phe5 of R1 (Figure 7A). Probes in CS2 favour hydrogen bonding to phenol hydroxyl groups of Tyr234, Tyr236 and Tyr251. More importantly, the largest consensus site CS1 (cyan, 21 probe clusters) and CS3 (yellow, 13 probe clusters), which are located on the base of

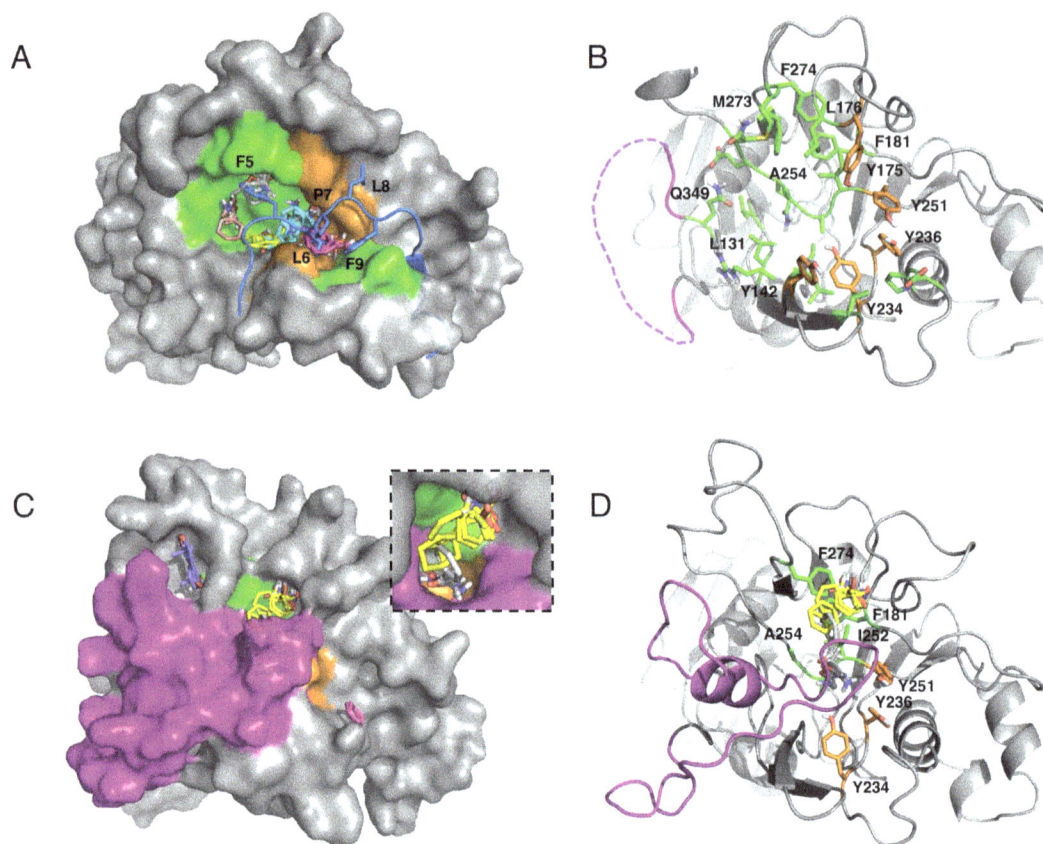

Figure 7. Computational solvent mapping of AMA1 using FTMAP. (A) Mapping results for R1-bound 3D7 *Pf*AMA1 (grey, PDB ID: 3SRJ). R1 peptides and water were removed prior to mapping. The five largest consensus sites, CS1 (cyan, 21 probe clusters), CS2 (magenta, 17 probe clusters), CS3 (yellow, 13 probe clusters), CS4 (salmon, 10 probe clusters) and CS5 (grey, 9 probe clusters) are located in a large hydrophobic pocket that binds to the Phe5-Phe9 segment of R1 peptide. The position of the R1 peptide in the crystal structure is shown for reference (blue). Phe5-Phe9 side chains are displayed as sticks and labelled individually as shown in the figure. The surface of the pocket is coloured according to side chain colours in panel B. (B) The residues that interact with small molecule probe clusters in the pocket are shown as sticks (green/orange). A cluster of five Tyr residues is highlighted in orange. Broken purple line indicates the displaced domain II loop in the R1-bound conformation. (C) Mapping results for IgNAR-bound 3D7 *Pf*AMA1 (grey, PDB ID: 2Z8V). IgNAR and water were removed prior to mapping. Two consensus sites CS3 (yellow, 14 probe clusters) and CS5 (grey, 9 probe clusters) are located in a domain II loop-protected pocket. The surface of the protein is coloured according to colour scheme in panel D. An inset highlights probes that fit into a deep, narrow pocket. (D) Some key interacting residues (highlighted as green/orange sticks), which bind probes when the domain II loop is displaced (panel B), are still solvent accessible to small molecule organic probes when the pocket is partially protected by the domain II loop (purple).

the pocket, revealed the key interactions that are in addition to those formed by the Phe5-Phe9 segment of R1. Probes in CS1 and CS3 make additional interactions with Leu131, Arg143, Leu144, Pro145, Ala253 and Gln255. CS4 (salmon, 10 probe clusters) extends one end of the pocket by interacting with Val129, Gln256 and Gln349 (Figure 7B). Probes in CS4 favour hydrogen bonding to the amide group of Gln349. The mapping results presented here suggest that this hot spot, which interacts with Phe5-Phe9 of R1, is druggable and effectively binds various small organic molecules. Identification of additional key interactions in the hot spot is potentially useful in the development of small molecule inhibitors.

The Phe5-Phe9-interacting hot spot is partially protected by the domain II loop, which is displaced by the binding of R1 and RON2. Part of this hot spot was identified previously as a pocket for small molecule targeting, although it has been suggested that the domain II loop may limit small molecule binding at this site [54]. We have shown that ~420 Å^3 solvent-accessible volume of this pocket is still available for small molecule binding when the domain II loop is not displaced [2]. To further address this issue,

we performed mapping on the structure of 3D7 *Pf*AMA1 co-crystallized with antibody IgNAR (PDB ID: 2Z8V). This is the only AMA1 structure that has a complete description of the domain II loop [18]. IgNAR binds to a region distant from the hot spot and does not induce any significant changes in the structure of AMA1 (C$^\alpha$ RMSD of 0.34 Å between IgNAR-bound and apo AMA1). Our mapping for 2Z8V results showed that two consensus sites CS3 (yellow, 14 probe clusters) and CS5 (grey, 9 probe clusters) are located in the domain II loop-protected pocket (Figure 7C). Importantly, several of the key residues, which bind Phe5-Phe9 of R1 or small molecule probes when the domain II loop is displaced, are still involved in the formation of the loop-protected pocket and remain accessible to small organic molecules (Phe181, Tyr234, Tyr236, Tyr251, Ile252, Ala253, Ala254 and Phe274; Figure 7D). All these residues are highly conserved in *P. falciparum* and their side chain conformations remain almost unchanged when the domain II loop is displaced.

Notably, no consensus sites were found in the Arg pocket in either AMA1 structure. One possible explanation could be that the Arg pocket is relatively small and has a polar surface area.

Although the interactions mediated in the Arg pocket are absolutely crucial for R1/RON2 binding to AMA1, it may be more difficult to identify suitable small molecules to access and interact with this site in isolation.

Conclusions

Using SPR and NMR spectroscopy we have validated that R1 binds to AMA1 in solution with 1:1 stoichiometry, as suggested by previous ITC data [27], and adopts a secondary structure consistent with the major form of R1 observed in the crystal structure of the complex. The minor form of R1 in the crystal structure was not observed in solution and is likely to be a crystallographic artifact. The truncation and mutational studies for R1 presented here have identified several key AMA1-interacting residues scattered along the peptide. Amongst these key residues, the hydrophobic segment Phe5-Leu6-Pro7-Leu8-Phe9, residues Phe12 and Arg15 are those that contribute most to the AMA1 binding affinity. They interact with two distinct binding hot spots, which are located at the two ends of the hydrophobic cleft of AMA1. Both of the pockets are highly conserved across the *P. falciparum* strains and likely to be suitable for designing broad-spectrum AMA1 inhibitors. The "Arg pocket" at one end of the cleft mediates key interactions of several known inhibitory agents, although fragment-based computational solvent mapping on AMA1 suggests that it may be a difficult site to target with small organic molecules because of its small surface area and polar nature. Mimicking the Arg side chain using peptidomimetics based on R1 or RON2 might be a more productive approach to target this important pocket. In contrast, mapping results showed that the Phe5-Phe9-interacting hot spot is druggable and identified key AMA1 residues for small molecule targeting. Our results provide a basis for designing novel high affinity inhibitors of AMA1-RON2 interaction that are effective against the majority of *Pf*AMA1 genotypes.

Supporting Information

File S1 Supporting information. Result S1, Backbone resonance assignments for AMA1-bound R1. Figure S1, 1D ^{1}H spectrum of 3D7 *Pf*AMA1$_{104-442}$ in 20 mM sodium phosphate pH 7, acquired at 600 MHz and 40°C. Figure S2, The purity (> 95%) (A) and mass (B) of f-^{2}H, u-^{13}C, ^{15}N-labelled R1 peptide were verified using LCMS. Figure S3, ^{1}H-^{15}N-HSQC spectrum of 0.4 mM u-^{13}C,^{15}N-labelled R1 at pH 5 and 40°C. Figure S4, Comparison of deviation of ^{1}H chemical shifts (H^N, top panel; H$^\alpha$, bottom panel) from random coil values for previously reported synthetic R1 at pH 4.5, 5°C and ^{13}C, ^{15}N-labelled R1 at pH 7, 40°C (this study). Figure S5, ^{1}H-^{15}N-HSQC spectra of 0.3 mM f-^{2}H, u-^{13}C, ^{15}N-labelled R1 in the presence of increasing concentration of f-^{2}H-AMA1. Figure S6, Strip plot of the 3D TROSY-HNCA spectrum of the f-^{2}H, u-^{13}C, ^{15}N-labelled R1-f-^{2}H-labelled AMA1 complex. Figure S7, The ^{13}C(F$_2$)-^{1}H(F$_3$) plane of the 3D TROSY-HN(CO)CA spectrum of the ^{2}H,^{13}C, ^{15}N-labelled R1-^{2}H-labelled AMA1 complex. Figure S8, Elution profile of NMR buffer on an analytical size exclusion column. Figure S9, C$^\alpha$ secondary shifts predicted for the minor form of R1 in the crystal structure using SHIFTX2. Figure S10, C$^\alpha$ secondary shifts predicted for R1 in the crystal structure using SPARTA+. Table S1, Chemical shifts of free R1 at pH 7 and 40°C. Table S2, Chemical shifts of AMA1-bound R1 at pH 7 and 40°C.

Acknowledgments

The authors thank San Sui Lim, Dr. Martin L. Williams and Dr. Mark D. Mulcair for their kind assistance in this project. The Queensland NMR Network (QNN) is acknowledged for providing access to the 900 MHz spectrometer at University of Queensland.

Author Contributions

Conceived and designed the experiments: GW CAM RFA JSS SM RSN MJS. Performed the experiments: GW CAM BM MM NPC SM. Analyzed the data: GW CAM BM MM NPC SM. Contributed reagents/materials/analysis tools: MM NPC. Wrote the paper: GW CAM BM MM NPC RFA JSS SM RSN MJS.

References

1. World Health Organisation (2013) World Malaria Report 2013.
2. MacRaild CA, Anders RF, Foley M, Norton RS (2011) Apical membrane antigen 1 as an anti-malarial drug target. Curr Top Med Chem 11: 2039–2047.
3. Miller LH, Baruch DI, Marsh K, Doumbo OK (2002) The pathogenic basis of malaria. Nature 415: 673–679.
4. Srinivasan P, Beatty WL, Diouf A, Herrera R, Ambroggio X, et al. (2011) Binding of *Plasmodium* merozoite proteins RON2 and AMA1 triggers commitment to invasion. Proc Natl Acad Sci U S A 108: 13275–13280.
5. Tonkin ML, Roques M, Lamarque MH, Pugniere M, Douguet D, et al. (2011) Host cell invasion by apicomplexan parasites: insights from the co-structure of AMA1 with a RON2 peptide. Science 333: 463–467.
6. Lamarque M, Besteiro S, Papoin J, Roques M, Vulliez-Le Normand B, et al. (2011) The RON2-AMA1 interaction is a critical step in moving junction-dependent invasion by apicomplexan parasites. PLoS Pathog 7: e1001276.
7. Riglar DT, Richard D, Wilson DW, Boyle MJ, Dekiwadia C, et al. (2011) Super-resolution dissection of coordinated events during malaria parasite invasion of the human erythrocyte. Cell Host Microbe 9: 9–20.
8. Triglia T, Healer J, Caruana SR, Hodder AN, Anders RF, et al. (2000) Apical membrane antigen 1 plays a critical role in erythrocyte invasion by *Plasmodium* species. Mol Microbiol 38: 706–718.
9. Mital J, Meissner M, Soldati D, Ward GE (2005) Conditional expression of *Toxoplasma gondii* apical membrane antigen-1 (*Tg*AMA1) demonstrates that *Tg*AMA1 plays a critical role in host cell invasion. Mol Biol Cell 16: 4341–4349.
10. Silvie O, Franetich JF, Charrin S, Mueller MS, Siau A, et al. (2004) A role for apical membrane antigen 1 during invasion of hepatocytes by *Plasmodium falciparum* sporozoites. J Biol Chem 279: 9490–9496.
11. Bargieri DY, Andenmatten N, Lagal V, Thiberge S, Whitelaw JA, et al. (2013) Apical membrane antigen 1 mediates apicomplexan parasite attachment but is dispensable for host cell invasion. Nat Commun 4: 2552.
12. Giovannini D, Spath S, Lacroix C, Perazzi A, Bargieri D, et al. (2011) Independent roles of apical membrane antigen 1 and rhoptry neck proteins during host cell invasion by apicomplexa. Cell Host Microbe 10: 591–602.
13. Yap A, Azevedo MF, Gilson PR, Weiss GE, O'Neill MT, et al. (2014) Conditional expression of apical membrane antigen 1 in *Plasmodium falciparum* shows it is required for erythrocyte invasion by merozoites. Cell Microbiol DOI: 10. 1111/cmi.122287
14. Lamarque MH, Roques M, Kong-Hap M, Tonkin ML, Rugarabamu G, et al. (2014) Plasticity and redundancy among AMA-RON pairs ensure host cell entry of Toxoplasma parasites. Nat Commun. 5:4098.
15. Srinivasan P, Yasgar A, Luci DK, Beatty WL, Hu X, et al. (2013) Disrupting malaria parasite AMA1-RON2 interaction with a small molecule prevents erythrocyte invasion. Nat Commun 4: 2261.
16. Miller LH, Ackerman HC, Su XZ, Wellems TE (2013) Malaria biology and disease pathogenesis: insights for new treatments. Nat Med 19: 156–167.
17. Collins CR, Withers-Martinez C, Bentley GA, Batchelor AH, Thomas AW, et al. (2007) Fine mapping of an epitope recognized by an invasion-inhibitory monoclonal antibody on the malaria vaccine candidate apical membrane antigen 1. J Biol Chem 282: 7431–7441.
18. Henderson KA, Streltsov VA, Coley AM, Dolezal O, Hudson PJ, et al. (2007) Structure of an IgNAR-AMA1 complex: targeting a conserved hydrophobic cleft broadens malarial strain recognition. Structure 15: 1452–1466.
19. Coley AM, Gupta A, Murphy VJ, Bai T, Kim H, et al. (2007) Structure of the malaria antigen AMA1 in complex with a growth-inhibitory antibody. PLoS Pathog 3: 1308–1319.
20. Coley AM, Parisi K, Masciantonio R, Hoeck J, Casey JL, et al. (2006) The most polymorphic residue on *Plasmodium falciparum* apical membrane antigen 1 determines binding of an invasion-inhibitory antibody. Infect Immun 74: 2628–2636.

21. Cortes A, Mellombo M, Mueller I, Benet A, Reeder JC, et al. (2003) Geographical structure of diversity and differences between symptomatic and asymptomatic infections for *Plasmodium falciparum* vaccine candidate AMA1. Infect Immun 71: 1416–1426.

22. Polley SD, Chokejindachai W, Conway DJ (2003) Allele frequency-based analyses robustly map sequence sites under balancing selection in a malaria vaccine candidate antigen. Genetics 165: 555–561.

23. Bai T, Becker M, Gupta A, Strike P, Murphy VJ, et al. (2005) Structure of AMA1 from *Plasmodium falciparum* reveals a clustering of polymorphisms that surround a conserved hydrophobic pocket. Proc Natl Acad Sci U S A 102: 12736–12741.

24. Li F, Dluzewski A, Coley AM, Thomas A, Tilley L, et al. (2002) Phage-displayed peptides bind to the malarial protein apical membrane antigen-1 and inhibit the merozoite invasion of host erythrocytes. J Biol Chem 277: 50303–50310.

25. Harris KS, Casey JL, Coley AM, Masciantonio R, Sabo JK, et al. (2005) Binding hot spot for invasion inhibitory molecules on *Plasmodium falciparum* apical membrane antigen 1. Infect Immun 73: 6981–6989.

26. Harris KS, Casey JL, Coley AM, Karas JA, Sabo JK, et al. (2009) Rapid optimization of a peptide inhibitor of malaria parasite invasion by comprehensive *N*-methyl scanning. J Biol Chem 284: 9361–9371.

27. Vulliez-Le Normand B, Tonkin ML, Lamarque MH, Langer S, Hoos S, et al. (2012) Structural and functional insights into the malaria parasite moving junction complex. PLoS Pathog 8: e1002755.

28. Richard D, MacRaild CA, Riglar DT, Chan JA, Foley M, et al. (2010) Interaction between *Plasmodium falciparum* apical membrane antigen 1 and the rhoptry neck protein complex defines a key step in the erythrocyte invasion process of malaria parasites. J Biol Chem 285: 14815–14822.

29. Lim SS, Debono CO, MacRaild CA, Chandrashekaran IR, Dolezal O, et al. (2013) Development of inhibitors of *Plasmodium falciparum* apical membrane antigen 1 based on fragment screening. Aust J Chem 66: 1530–1536

30. Clackson T, Wells JA (1995) A hot spot of binding energy in a hormone-receptor interface. Science 267: 383–386.

31. Arkin MR, Wells JA (2004) Small-molecule inhibitors of protein-protein interactions: progressing towards the dream. Nat Rev Drug Discov 3: 301–317.

32. James GL, Goldstein JL, Brown MS, Rawson TE, Somers TC, et al. (1993) Benzodiazepine peptidomimetics: potent inhibitors of Ras farnesylation in animal cells. Science 260: 1937–1942.

33. Hirschmann R, Nicolaou KC, Pietranico S, Leahy EM, Salvino J, et al. (1993) De novo design and synthesis of somatostatin non-peptide peptidomimetics utilizing beta-D-glucose as a novel scaffolding. J Am Chem Soc 115: 12550–12568.

34. McDowell RS, Gadek TR, Barker PL, Burdick DJ, Chan KS, et al. (1994) From peptide to non-peptide. 1. The elucidation of a bioactive conformation of the arginine-glycine-aspartic acid recognition sequence. J Am Chem Soc 116: 5069–5076.

35. Damour D, Barreau M, Blanchard JC, Burgevin MC, Doble A, et al. (1996) Design, synthesis and binding affinities of novel non-peptide mimics of somatostatin/sandostatin. Bioorg Med Chem Lett 6: 1667–1672.

36. Blackburn BK, Lee A, Baier M, Kohl B, Olivero AG, et al. (1997) From peptide to non-peptide. 3. Atropisomeric GPIIbIIIa antagonists containing the 3,4-dihydro-1*H*-1,4-benzodiazepine-2,5-dione nucleus. J Med Chem 40: 717–729.

37. Ono K, Takeuchi K, Ueda H, Morita Y, Tanimura R, et al. (2014) Structure-based approach to improve a small-molecule inhibitor by the use of a competitive peptide ligand. Angew Chem Int Ed DOI: 10.1002/anie.201310749

38. Sivashanmugam A, Murray V, Cui C, Zhang Y, Wang J, et al. (2009) Practical protocols for production of very high yields of recombinant proteins using *Escherichia coli*. Protein Sci 18: 936–948.

39. Cabrita LD, Gilis D, Robertson AL, Dehouck Y, Rooman M, et al. (2007) Enhancing the stability and solubility of TEV protease using *in silico* design. Protein Sci 16: 2360–2367.

40. Delaglio F, Grzesiek S, Vuister GW, Zhu G, Pfeifer J, et al. (1995) NMRPipe: a multidimensional spectral processing system based on UNIX pipes. J Biomol NMR 6: 277–293.

41. Keller RLJ (2004) The computer aided resonance assignment tutorial. Zürich, Switzerland: CANTINA Verlag.

42. Mobli M, Stern AS, Bermel W, King GF, Hoch JC (2010) A non-uniformly sampled 4D HCC(CO)NH-TOCSY experiment processed using maximum entropy for rapid protein sidechain assignment. J Magn Reson 204: 160–164.

43. Mobli M, Maciejewski MW, Gryk MR, Hoch JC (2007) An automated tool for maximum entropy reconstruction of biomolecular NMR spectra. Nat Methods 4: 467–468.

44. Whitten AE, Cai S, Trewhella J (2008) MULCh: modules for the analysis of small-angle neutron contrast variation data from biomolecular assemblies. J Appl Cryst 41: 222–226.

45. Svergun D, Barberato C, Koch MHJ (1995) CRYSOL-A program to evaluate x-ray solution scattering of biological macromolecules from atomic coordinates. J Appl Cryst 28: 768–773.

46. Konarev PV, Volkov VV, Sokolova AV, Koch MHJ, Svergun DI (2003) PRIMUS: a Windows PC-based system for small-angle scattering data analysis. J Appl Cryst 36: 1277–1282.

47. Brenke R, Kozakov D, Chuang GY, Beglov D, Hall D, et al. (2009) Fragment-based identification of druggable 'hot spots' of proteins using Fourier domain correlation techniques. Bioinformatics 25: 621–627.

48. Kozakov D, Hall DR, Chuang GY, Cencic R, Brenke R, et al. (2011) Structural conservation of druggable hot spots in protein-protein interfaces. Proc Natl Acad Sci U S A 108: 13528–13533.

49. Lee EF, Yao S, Sabo JK, Fairlie WD, Stevenson RA, et al. (2011) Peptide inhibitors of the malaria surface protein, apical membrane antigen 1: identification of key binding residues. Biopolymers 95: 354–364.

50. Collins CR, Withers-Martinez C, Hackett F, Blackman MJ (2009) An inhibitory antibody blocks interactions between components of the malarial invasion machinery. PLoS Pathog 5: e1000273.

51. Wishart DS, Bigam CG, Holm A, Hodges RS, Sykes BD (1994) ^{1}H, ^{13}C and ^{15}N random coil NMR chemical shifts of the common amino acids. I. investigation of nearest-neighbor effects. J Biomol NMR 5:67–81.

52. Spera S, Bax A (1991) Empirical correlation between protein backbone conformation and C_α and C_β ^{13}C nuclear magnetic resonance chemical shifts. J Am Chem Soc 113: 5490–5492.

53. Han B, Liu Y, Ginzinger SW, Wishart DS (2011) SHIFTX2: significantly improved protein chemical shift prediction. J Biomol NMR 50: 43–57.

54. Tonkin ML, Crawford J, Lebrun ML, Boulanger MJ (2013) *Babesia divergens* and *Neospora caninum* apical membrane antigen 1 structures reveal selectivity and plasticity in apicomplexan parasite host cell invasion. Protein Sci 22: 114–127.

5

Rapid Isolation of Extracellular Vesicles from Cell Culture and Biological Fluids Using a Synthetic Peptide with Specific Affinity for Heat Shock Proteins

Anirban Ghosh[1,2]*, Michelle Davey[1], Ian C. Chute[1], Steven G. Griffiths[1], Scott Lewis[3], Simi Chacko[1], David Barnett[1,2,4], Nicolas Crapoulet[1], Sébastien Fournier[1], Andrew Joy[1], Michelle C. Caissie[1], Amanda D. Ferguson[1], Melissa Daigle[1], M. Vicki Meli[4], Stephen M. Lewis[1,2,5,6], Rodney J. Ouellette[1,2]*

1 Atlantic Cancer Research Institute, Moncton, New Brunswick, Canada, 2 Department of Chemistry and Biochemistry, Université de Moncton, Moncton, New Brunswick, Canada, 3 New England Peptide Inc., Gardner, Massachusetts, United States of America, 4 Department of Chemistry and Biochemistry, Mount Allison University, Sackville, New Brunswick, Canada, 5 Department of Microbiology and Immunology, Dalhousie University, Halifax, Nova Scotia, Canada, 6 Department of Biology, University of New Brunswick, Saint John, New Brunswick, Canada

Abstract

Recent studies indicate that extracellular vesicles are an important source material for many clinical applications, including minimally-invasive disease diagnosis. However, challenges for rapid and simple extracellular vesicle collection have hindered their application. We have developed and validated a novel class of peptides (which we named venceremin, or Vn) that exhibit nucleotide-independent specific affinity for canonical heat shock proteins. The Vn peptides were validated to specifically and efficiently capture HSP-containing extracellular vesicles from cell culture growth media, plasma, and urine by electron microscopy, atomic force microscopy, sequencing of nucleic acid cargo, proteomic profiling, immunoblotting, and nanoparticle tracking analysis. All of these analyses confirmed the material captured by the Vn peptides was comparable to those purified by the standard ultracentrifugation method. We show that the Vn peptides are a useful tool for the rapid isolation of extracellular vesicles using standard laboratory equipment. Moreover, the Vn peptides are adaptable to diverse platforms and therefore represent an excellent solution to the challenge of extracellular vesicle isolation for research and clinical applications.

Editor: Guo-Chang Fan, University of Cincinnati, College of Medicine, United States of America

Funding: The authors thank the Atlantic Innovation Fund from the Atlantic Canada Opportunities Agency, the New Brunswick Innovation Foundation, and Ride for Dad for funding this research. Stephen M. Lewis is supported by a New Investigator Salary Award from the Canadian Institutes of Health Research – Regional Partnerships Program. The funders had no role in study design, data collection and analysis, decision to publish, or preparation of the manuscript.

Competing Interests: Scott Lewis is a coinventor of the Vn peptides and is a salaried employee of New England Peptide, Inc. (NEP), a privately-owned company. He does not own stocks or shares of the company, and as part of his employment contract does not have personal ownership claim to intellectual property developed while at NEP. No other financial competing interests exist. The intellectual property concerning the Vn peptides is protected by patent (Patent # WO 2012/126118 A1).

* Email: anirbang@canceratl.ca (AG); rodneyo@canceratl.ca (RJO)

Introduction

Heat shock proteins (HSPs) are one of the most ancient molecular defense systems. In non-stressed and non-transformed cells, HSPs are ubiquitously expressed in low amounts as intracellular proteins that exhibit various cytoprotective functions, including buffering the cell from stressful conditions, monitoring proper protein folding (chaperones), cellular housekeeping (proteasomes), and presenting antigens to immune cells [1,2]. However, the cytoprotective effects of HSPs are also exploited by transformed cells to promote their own survival. In stressed and cancer cells, intracellular HSP-peptide complexes induce anti-apoptotic effects and act as cytoprotectants by directing damaged proteins for degradation, whereas extracellular HSPs elicit immune responses by carrying a variety of immunogenic peptides [3,4].

Although intracellular chaperones/HSPs have been studied for the last five decades, studies of extracellular HSPs have only begun in recent years. The release of HSPs into the extracellular milieu is emerging as a characteristic of many pathological conditions, including infection and cancer. Recent studies have shown that a broad range of HSP paralogues that are normally restricted to discrete intracellular compartments are relocated to the surface of cancer and infected cells [5–7]. Importantly, the presence of HSPs on the surface of cancer and infected cells is a trait that is not shared by their normal counterparts. Hsp70 is an integral component of the cancer cell membrane *via* its affinity for phosphatidyl serine in the external membrane layer and the glycosphingolipid Gb3 in signaling platforms known as lipid rafts, despite the absence of an externalizing sequence [8]. In addition, exosome/extracellular vesicle-associated extracellular transport of HSPs is evident in many pathological conditions, including cancer [9–15].

Extracellular vesicles (EVs) are a heterogeneous population, both in size and in content, of nano-sized organelles released by most cell types. EVs contain an active cargo of molecules that represent the state of their cell of origin. The release of EVs is a conserved physiological process observed both *in vitro* and *in vivo*. EVs are found in a wide range of biological fluids, including blood, urine, saliva, amniotic fluid, and pleural fluid [16–22]. There are two main groups of extracellular vesicles: exosomes of endosomal origin (40–100 nm in diameter) and shed vesicles (or ectosomes) pinched off from the plasma membrane (50–1000 nm in diameter). We will refer to the collective group as EVs [23]. Pathological conditions, such as cancer, affect the amount and localization of EV protein content. Along with the HSPs, exosomal and EV protein markers include Alix, TSG101, the tetraspanins CD63, CD81, and CD9, HSPs, metalloproteinases, integrins, some glycoproteins, and selectins [24].

We set out to design synthetic peptides that specifically bind to HSPs. The peptide (substrate) binding domain of HSPs is well characterized, especially for Hsp70. In the Hsp70 protein family the substrate binding domain-β (SBD-β) in the C-terminal region forms a hydrophobic binding pocket to bind to substrate peptides or their partner co-chaperones. The well-characterized signature domain of substrate peptides to which the Hsp70 SBD-β binds is called the J-domain. J-domain-containing proteins constitute a conserved family of co-chaperones found in *E.coli* (DnaJ) and humans (Hsp40 and Hsj1) that bind with their partner chaperone, known as a DnaK homologue or Hsc70 respectively [25–27]. The J-domain consists of a four-bundle α-helix, where helices I and IV form the base and helices II and III form a finger-like projection of the structure. A conserved amino acid sequence, HPD (His-Pro-Asp), is located at the tip of the projection [28]. Many structural studies have indicated that the positively charged and hydrophobic amino acid residues of helix II and the HPD sequences of J-domains interact with the hydrophobic peptide binding domain (SBD-β) of the C-terminal parts of HSP70s [17,28–33]. Based on these structural studies of the peptide binding pockets of Hsp70 [25–27,34] we rationalized that: (1) an ideal HSP-binding peptide would be strongly cationic with hydrophobic side chains, consistent with properties conducive to stable association with the peptide binding cleft of Hsp70 isoforms and paralogues and (2) the avidity of those peptides with HSP-binding properties could be screened by counter migration during isoelectric focusing (IEF).

Accordingly, we designed and synthesized a series of peptides (that we collectively named venceremins, or Vn peptides), which were screened for their HSP-binding properties using IEF. Many tested peptides bound HSPs, but during the course of our experiments we discovered that at least one Vn peptide (Vn96) also precipitated small subcellular structures that resemble membrane structures of ER-Golgi origin at low centrifugal speed (10,000×g). These results prompted us to examine the potential of Vn96 as an exosome/EV capture tool from cell culture growth media and biological fluids.

Materials and Methods

Peptides

All the peptides were synthesized at New England peptide (Gardner, US). The Vn96, Vn20 peptides and their use to isolate EVs are patent pending (US 13/824,829. PCT number, PCT/CA2012/050175).

Cell culture and cell lines

Breast cancer cell lines (MCF-10A, MCF-7 and MDB-MB-231) were purchased from the American Tissue and Culture Collection (ATCC) and grown in tissue culture according to the supplier's recommended protocols. The cells were grown to 80–90% confluency, washed four times with serum-free media, and then incubated with a minimal volume of serum-free media required to cover the cells. After four hours of incubation the 'conditioned' cell culture media was collected, followed by removal of cellular material by a two-step centrifugation process (1,000×g and 17,000×g) and/or by filtering with 0.22 μm filters to remove large protein aggregates and other cellular debris. We then precipitated EVs from the collected conditioned cell culture media using either Vn96 peptides or a scrambled version of the Vn96 peptide as a negative control. The above cell lines were also adapted for continuous long-term conditioned cell culture media harvest in compartmentalized flasks (CELLine, AD 1000 bioreactor) designed with a cell-growth chamber that is separated from the bulk cell culture media compartment with a 10 kDa cutoff dialysis membrane. The cell culture media added to the cell-growth chamber were prepared with exosome free (Exo-Free) Fetal Bovine Serum (FBS). FBS was purchased from Wisent Bioproducts (Quebec, Canada, Cat# 080–350). The Exo-free FBS was prepared by centrifugation of FBS at 100,000×g for two hours at 4°C followed by aspiration of the supernatant without disturbing the exosome pellet. The conditioned media were harvested once a week from the cell-growth chamber only. The harvested cell culture media were immediately centrifuged at 1,800×g for five minutes to remove the floating cells, followed by 17,000×g for 15 minutes to remove cellular debris; the prepared material was then stored at 4°C with 5 μl of protease inhibitor cocktail-III (EDM-Millipore) and 0.1% (v/v) ProClin300 (Sigma) as a preservative.

Human sample collection and preparation

This study was reviewed and approved by the Vitalité Health Network Research Ethics Board (New Brunswick, Canada) prior to the beginning of sample collection. Written informed consent was obtained by a Clinical Research Associate from each patient before any blood or urine samples were collected. Whole blood (+ EDTA) was collected from consenting healthy women and breast cancer patients. The plasma layer was collected after centrifugation of the whole-blood (EDTA) at 1,500×g for 15 minutes at room temperature (RT), followed by pre-clearing the plasma by centrifugation at 17,000×g at 4°C for 15 minutes. 5 μl of protease inhibitor cocktail-III and 0.1% (v/v) ProClin300 (preservative) were added to each millilitre of the pre-cleared plasma before archiving at 4°C (short-term) or −80°C (long-term) for storage. Urine samples were collected from consenting male patients scheduled for prostate biopsy subjected to both pre- and post-digital rectal examination (DRE) with prostate massage. The urine samples were centrifuged at 650×g for 10 minutes at RT; supernatants were collected and centrifuged again at 10,000×g for 15 minutes at RT, followed by a final centrifugation at 17,000×g for 15 minutes at RT. Aliquots of 7.5 ml were likewise archived at 4°C or −80°C with 7.5 μl each of protease inhibitor cocktail-III and ProClin300.

EV isolation using Vn peptides

The archived conditioned cell culture media and corresponding 'control' media (unused) were cleared once again by centrifugation at 17,000×g following removal from the archive, and were then incubated with either: 1) biotinylated-Vn96 (b-Vn96) or biotinylated scrambled sequence of Vn96 (b-Scr-Vn96), or, 2) Vn96 or scrambled sequence of Vn96 (Scr-Vn96) overnight at 4°C (long incubation) or 15 minutes at RT (short incubation) with rotation. The peptides were used at either 100 μg/ml or 50 μg/ml of media. The incubated samples were centrifuged at 17,000×g at 4°C for 15 minutes or at 10,000×g for seven minutes at RT using a bench-top microcentrifuge for the long or short incubations, respectively. Semi-translucent precipitates were visible only in case of Vn96 and b-Vn96 incubated samples. All samples were washed three times with phosphate buffered saline (PBS).

The archived plasma samples were thawed and diluted 5 to 10 times with PBS, while the archived urine samples were thawed and used without dilution. The samples were subjected to clearing by centrifugation (17,000×g for 15 min at 4°C) and/or filtration though 0.2 μm pore-size filters. The cleared samples were incubated with 50 μg/ml Vn96 or Scr-Vn96 peptide overnight at 4°C with rotation, followed by precipitation by centrifugation at 17,000×g at 4°C for 15 minutes and three washes with PBS. The precipitated Vn96-EV complexes were processed for either electron microscopy, atomic force microscopy, RNA isolation, or proteomic analysis as described below.

EV and exosome isolation using ultracentrifugation (UCF) and a commercially-available kit

We followed the protocol for EV and/or exosome preparation on a 30% sucrose cushion as described in the 'Current Protocols in Cell Biology' [41] with minor modifications. Briefly, approximately 10 ml of pre-cleared samples were transferred to UCF tubes (SW-40Ti rotors), followed by very careful insertion of a Pasteur pipette into the bottom of the sample in order to layer 500 to 750 μl of 30% sucrose solution in PBS at the bottom of the tube. The samples were centrifuged at 100,000×g for two hours. The exosome-containing sucrose cushions were aspirated carefully using a Pasteur pipette into a new ultracentrifuge tube, diluted to 10 ml with PBS and re-centrifuged at 100,000×g for 90 minutes. The supernatants were discarded and the exosome pellets were carefully resuspended in 50–100 μl of PBS with 5 μl of protease inhibitor. We used ExoQuick for the preparation of EVs from conditioned cell culture media following supplier's instructions.

Electron microscopy

The precipitated Vn96-EV complexes were incubated with 2 μg/ml proteinase K in PBS at 37°C for four hours to disperse the membrane-encapsulated EVs into solution, followed by centrifugation at 17,000×g for 15 minutes during which no visible pellet was observed. The dispersed EVs from the supernatants (5–10 μl) were deposited onto formvar/carbon-coated 200 mesh copper grids for 2–3 minutes, followed by floating on a 100 μl drop of water (on para-film) in a sample-side down orientation for one minute. Fixation was achieved with 3.7% formalin followed by two washes with water. The samples were contrasted with 2% uranyl acetate (w/v) to visualize membranes. The water, 3.7% formalin and 2% uranyl acetate were filtered through 10 kDa cut off filters before use on the EM-grids to remove any particulate contaminants. The dried grids were viewed using a JEOL 6400 electron microscope at the Microscopy and Microanalysis Facility, University of New Brunswick. Minimum three samples and technical repeats were performed to obtain the optimal concentration for visibility.

Atomic force microscopy

Vn96-precipitated EVs were dispersed with proteinase K digestion in 50 μl PBS. The preparation was diluted 1:100 in de-ionized water and adsorbed to freshly cleaved mica sheets that were rinsed with de-ionized water and dried under a gentle stream of nitrogen. Two to four biological repeats were used for each sample type. The samples were scanned in non-contact mode

using a Park Systems XE-100 atomic force microscope equipped with a silicon cantilever (f0~300 kHz, Park Systems). Topographic and phase images were recorded simultaneously at a resolution of 512×512 pixels, at a scan rate of 1 Hz. Image processing was performed using the Park Systems XEI software.

Nanoparticle Tracking Analysis (NTA)

NTA is a method of size-distribution and concentration analysis of nano-particles in liquid, based on their sizes and Brownian motion using the Stokes-Einstein equation. We used NanoSight LM10 with NTA software (V2.3). The Vn96-EV complexes were dispersed by digestion with proteinase K in PBS as described above. UCF-prepared exosomes and Vn96-prepared, proteinase K-digested EVs were subjected to different PBS dilutions (0.1 μm filtered) to find the best windows for NTA video capture. The experiments were repeated at least four times to obtain representative results.

Proteomic analysis

The EV-Vn96 complexes or UCF-purified exosomes were dissolved and heated for five minutes at 85oC in buffer (125 mM Tris pH 6.8 with 2% SDS) to harvest proteins for subsequent analysis. The protein samples were separated on SDS-PAGE and visualized with Coomassie EZBlue stain. Each entire lane was excised into several 2–3 mm long slices and distributed into different microcentrifuge tubes. Each band was treated with 10 mM dithiothreitol and 25 mM iodoacetic acid to reduce internal disulfide bonds and alkylate free cysteine resdues. Fifty microliters of a 20 ng/μL solution of trypsin was added to each band for overnight enzymatic cleavage.

Protein tryptic digest extracts were analyzed by gradient nanoLC-MS/MS using a Quadrupole Orbitrap (Q-Exactive, Thermo-Fisher Scientific) mass spectrometer interfaced to a Proxeon Easy Nano-LC II. Samples were adjusted to 1% aqueous acetic acid and injected (5 μL) onto a narrow bore (20 mm long×100 μm inner diameter) C18 pre-column packed with 5 μm ReproSil-Pur resin (Thermo-Fisher Scientific). High resolution chromatographic separation was then achieved on a Thermo-Scientific Easy C18 analytical column with dimensions of 100 mm by 75 μm i.d. using 3 μm diameter ReproSil-Pur particles. Peptide elution was achieved using an acetonitrile/water gradient system. LC-MS grade water and acetonitrile (EMD Millipore) were both obtained from VWR Canada (Mississauga, ON). Solvent A consisted of 0.1% formic acid in water and solvent B was made up of 90/9.9/0.1 acetonitrile/water/formic acid. Formic acid was purchased from Sigma-Aldrich Canada (Oakville, ON). A linear acetonitrile gradient was applied to the C18 column from 5–30% solvent B in 120 minutes followed by 100% B for 10 minutes at a flow rate of 300 nL/min.

The outlet of the nano-flow emitter on the Q-Exactive (15 μm diameter) was biased to +1.9 kV and positioned approximately 2 mm from the heated (250oC) transfer capillary. The S-lens of the mass spectrometer was maintained at 100 Volts. The Q-Exactive mass spectrometer was calibrated in positive ion mode with mass standards (caffeine, MRFA peptide and Ultramark) every three days as recommended by the instrument manufacturer. Mass spectrometric data was acquired in data dependent mode (DDA, data dependent acquisition) whereby a full mass scan from 350–1500 Th was followed by the acquisition of fragmentation spectra for the five most abundant precursor ions with intensities above a threshold of 20,000. Precursor ion spectra were collected at a resolution setting of 70,000 and an AGC (automatic gain control) value of 1×106. Peptide fragmentation was performed using high energy collision induced dissociation in the HCD and MS/MS spectra were collected in the Orbitrap at a resolution of 17,500 and an AGC setting of 1×105. Peptide precursors were selected using a repeat count of two and a dynamic exclusion period of 20 seconds.

Mass spectrometric protein identification data was analyzed using Proteome Discoverer version 1.3 (Thermo-Fisher Scientific) employing the Sequest scoring algorhithm. A human FASTA database was obtained from UniProt. Searches were performed with the following settings: (a) enzyme specificity of trypsin with two allowed missed cleavages, (b) precursor and fragment tolerances were 10 ppm and 0.8 Da, respectively, (c) a variable modification of methionine oxidation (+15.99 Da), and (d) a fixed modification of cysteine carboxymethylation (+58.00 Da). Proteome Discoverer 1.4 calculated a strict false discovery rate (FDR) of 0.1% based on the results of a decoy (reverse) database search. Proteins were assigned a positive identification if at least two peptides were identified with high confidence. Two biological samples were prepared for each sample type, and one representative dataset for each sample is presented here.

Next-generation RNA sequencing

The conditioned cell culture media (from MCF-7 and MDA-MB-231 cells) were used to isolate EVs using the Vn96 peptide, ExoQuick and ultracentrifugation methods described above. RNA from the isolated EVs was harvested with TRIZOL reagent (Life Technologies) using a protocol adapted for small RNAs. Barcoded cDNA libraries were prepared using RNA-Seq Version 2 kit from Life Technologies following their recommended protocol. Library preparations were assayed for both quality control and quantity using Experion DNA 1K chip (Life Technologies) and diluted to 16 pM concentration. Samples were sequenced using a PGM Sequencer from Life Technologies on a 318 chip following the manufacturer recommended protocol. Each chip was loaded with three samples.

Western-blot analysis

The purified Vn96-EV complexes and UCF-prepared exosomes were dissolved in 4x SDS-loading dye (with or without reducing agents). The proteins were resolved on either 10% or 4–12% gradient SDS-PAGE. The resolved proteins were transferred to either nitrocellulose or PVDF membranes followed by blocking and immunoblotting with indicated antibodies using chemiluminescence and other standard procedures. Each Western-blot experiment was performed at least four times. All the antibodies used were purchased from Santa Cruz Biotechnology.

Results

Selection and validation of HSP-binding peptides

Based on previous knowledge, we reasoned that an ideal HSP-binding peptide would be a 20–30 amino acid cationic peptide with hydrophobic side chains that favor strong interactions with the peptide binding cleft of Hsp70 isoforms [27,28,34,35,36] and other HSP paralogues. We designed a series of peptides that address the flexibility of basic and hydrophobic amino acids with sterically non-bulky residues. Among the candidate sequences screened, we identified peptides that yield complexes with HSPs from different organisms upon counter migration isoelectric focusing (IEF). Fine-tuning of the peptide sequences was carried out by synthesizing analogues of the most promising sequences, followed by their analysis using counter migration IEF. Here we show the results of these counter migration IEF experiments for two peptides (Vn20 and Vn96; depicted in Figure 1B). The peptides were placed at the anode of the IEF gel and recombinant

HSPs placed at the cathode for counter migration. In the absence of counter migrating peptides, the recombinant HSP paralogues moved towards the anode of the IEF gel (Figure 1A). Upon counter migration with Vn peptides, recombinant HSPs were observed closer to the neutral spectrum of the pH gradient, representing complexes formed between the HSPs and the Vn peptides (Figure 1A). Unbound cationic Vn peptides migrated to the cathode end of the gel. A higher affinity of Vn96 over Vn20 for HSPs was observed as a higher proportion of HSP-Vn96 complexes formed compared to HSP-Vn20 complexes formed when similar quantities of both the peptides and the HSPs were loaded on the IEF. Based on these results, Vn96 was selected as a lead peptide for further experiments.

The Vn96 peptide captures HSP complexes and enriches membrane-bound structures from total cell lysates

To further validate the specificity of the Vn96 peptide for HSPs, an affinity pull-down experiment equivalent to immunoprecipitation was designed using cell lysates from the breast cancer cell line MCF-7 prepared in the presence of 1% NP-40 detergent. Streptavidin-coupled Dynabeads were saturated with either biotinylated-Vn96 (b-Vn96) or a biotinylated scrambled sequence of Vn96 (the same amino acids, but arranged in a different order; b-Scr-Vn96), which were used to capture proteins from the cell lysate as described in the methods section. The bound complexes were washed extensively with cell lysis buffer and the captured proteins analyzed by immunoblotting for the indicated HSPs. As shown in Figure 1C, Vn96-coated beads were able to capture different members of the HSP family from the cell lysate, as indicated. In contrast, the scrambled sequence of Vn96 failed to capture the same HSP family members (Figure 1C). These results validated our design strategy and demonstrate that our Vn96 peptide specifically and efficiently binds to HSPs.

While performing the above-described pull-down experiments, we observed visible aggregation of b-Vn96 coated beads in the cell lysate. Both b-Vn96- and b-Scr-Vn96-coated magnetic beads had similar free-flowing suspension properties in lysis buffer, but this property changed for the b-Vn96 beads' post-cell-lysate incubation. To investigate whether this aggregation was due to protein-protein interactions, aliquots of the samples were digested with Proteinase K (2 μg/ml final concentration). The Proteinase K digestion resulted in the beads becoming dispersed in suspension without any visible aggregation (data not shown). Because these aggregates were not observed in the b-Scr-Vn96-coated beads, this confirms Vn96-specific protein interactions. As HSPs are known to be associated with membrane domains on the surface, as well as inside cells [37], we analyzed the Proteinase K-digested supernatants by Transmission Electron Microscopy (TEM) for membrane structures. As shown in Text S1, the Proteinase K-digested supernatant from b-Vn96 samples showed a dense mass of vesicular structures, whereas no such structures were visible in the supernatants from the control sample (b-Scr-Vn96). These membrane structures resembled small vesicles of cytoplasmic origin [38,39]. These data indicate that the Vn96 peptide can capture membrane-bound structures that are associated with HSPs.

The Vn96 peptide precipitates HSP-associated membrane-bound structures from conditioned cell culture growth media

Given the observation that Vn96 could capture HSPs that are associated with membranes, we chose to examine whether Vn96 could capture membrane-bound structures associated with extra-

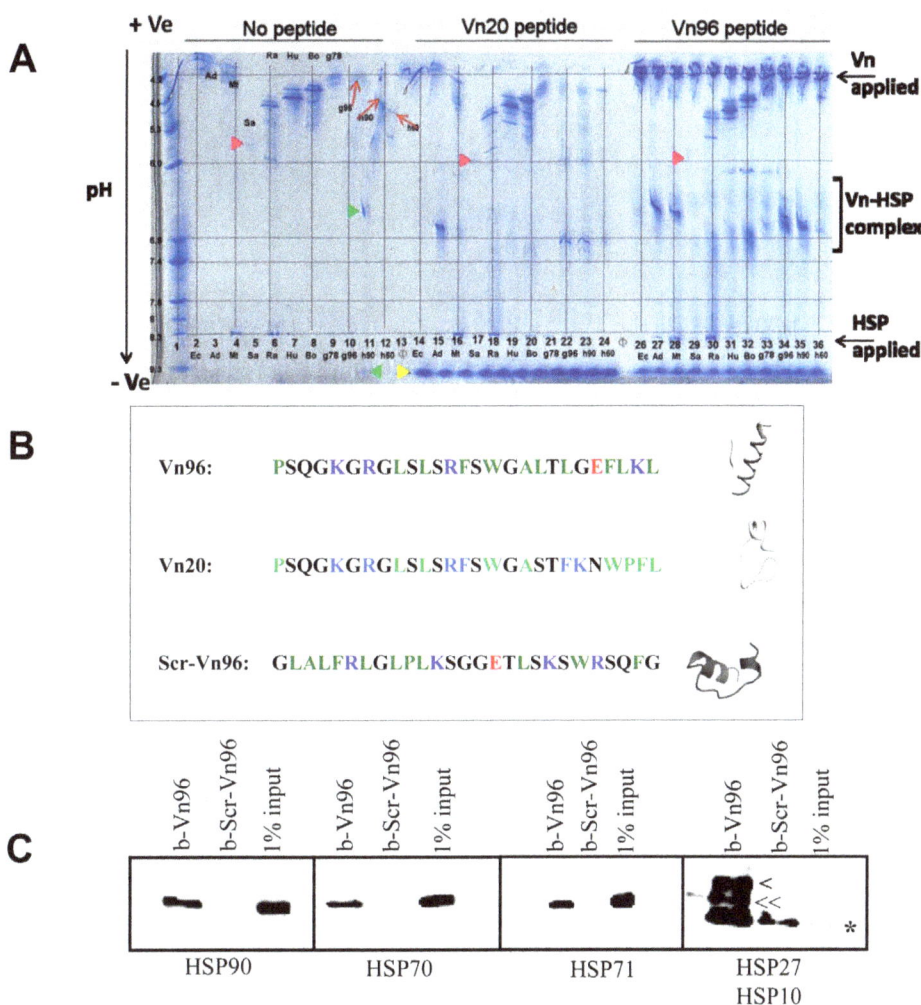

Figure 1. Selection and validation of peptides with HSP-binding properties. A. Peptide selection. Representative demonstration of peptide screening with recombinant HSPs using broad-range pH (3–10) isoelectric focusing (IEF) gels. Samples of 20 μg of the indicated peptides were applied at the anode and 2 μg of purified recombinant HSPs were applied at the cathode. The gradient of pH and electrophoretic directions are indicated on the left side of the gel. Abbreviations of recombinant HSP sources and horizontal lines are used to assist in sample identification in the distorted counter migrations affecting lanes 10–14. The complexes formed during counter migration are indicated at the right of the gel as "Vn-HSP complexes". Estimates of the isoelectric focusing points of unbound gp96, HSP90 and HSP60 in the area of counter migrant distortion are indicated by red arrows (lane 10, 11 and 12). The yellow arrowhead at the bottom indicates unbound Vn peptide isoelectric focusing following counter migration against HSPs. Red arrowheads (lanes 5, 17 and 29) indicate the weakly staining salmon HSP70 (despite standardized dilution). The green arrowhead (middle of lane 11) indicates a complex with HSP90 resulting from Vn20 that has extended binding influence across preceding adjacent lanes. The green arrowhead at the base of the gel (lane 11) indicates the final focusing point of the errantly migrating Vn20. Abbreviations are as follows: Ec, *E.coli* dnaK (lanes 1, 14,26); Ad, *A.davidanieli* HSP70 (lanes 3, 15, 27); Mt, *M.tuberculosis* HSP70 (lanes 4, 16, 28); Sa, Chinook salmon HSP70 (lanes 5, 17, 29); Ra, rat HSP70 (lanes 6, 18, 30); Hu, human HSP70-1 (lanes 7, 19, 31); Bo, bovine HSP70-8 (8, 20, 32); gp78, hamster HSP70-5 (lane 9, 21, 33); gp96, canine GRP96 (lanes 10, 22, 34); h90, human HSP90 (lanes 11, 23, 35); h60, human HSP60 (lanes 12, 24, 36) and as blank lanes (lanes 13 and 25). B. Sequences of Vn96, Vn20 as well as Scrambled-Vn96 (Scr-Vn96) and their predicted 3D structures in aqueous solution using PEP-FOLD server [55]. Red, blue, green, and black amino acid residues are acidic, basic, hydrophobic uncharged and other amino acid residues, respectively. Note that the Vn96 peptide favors a helical conformation. C. Validation of HSP binding by the Vn96 peptide *via* affinity pull-down of HSPs from total cell lysate. MCF-7 breast cancer cells were lysed and processed as described in experimental procedures. Streptavidin-coupled magnetic beads saturated with either biotinylated-Vn96 (b-Vn96) or biotinylated-scrambled sequence of Vn96 (b-Vn96-Scr) peptides were used to perform the pull-down assays. In the immunoblot, 1% volumes of total cell lysate were run as input proteins to compare with proteins bound by the Vn96 peptides. The heat shock proteins tested are indicated. The right lower panel shows both HSP27 (indicated as '>'), HSP10 (indicated as '<<') and a non-specific band (indicated as '*').

cellular HSPs from cell culture conditioned media. To generate conditioned media, the breast cancer cell line MDB-MB-231 was grown in EV-free standard cell culture media as described in the experimental procedures section and subsequently collected for downstream experiments. The conditioned growth media, as well as unused control growth media, was incubated overnight with rotation at 4°C with 100 μg/ml each of b-Vn96 or b-Scr-Vn96

peptide. Translucent precipitates were observed only in the b-Vn96 samples following centrifugation at 17,000×g. The 17,000×g pellets were washed with 1 ml of PBS three times (17,000×g) and the final pellets resuspended in 50 μl of PBS and treated with Proteinase K. TEM studies of Proteinase K-treated b-Vn96 samples revealed populations of vesicles, whereas the b-Scr-Vn96 samples showed no such structures (Figure 2A). The

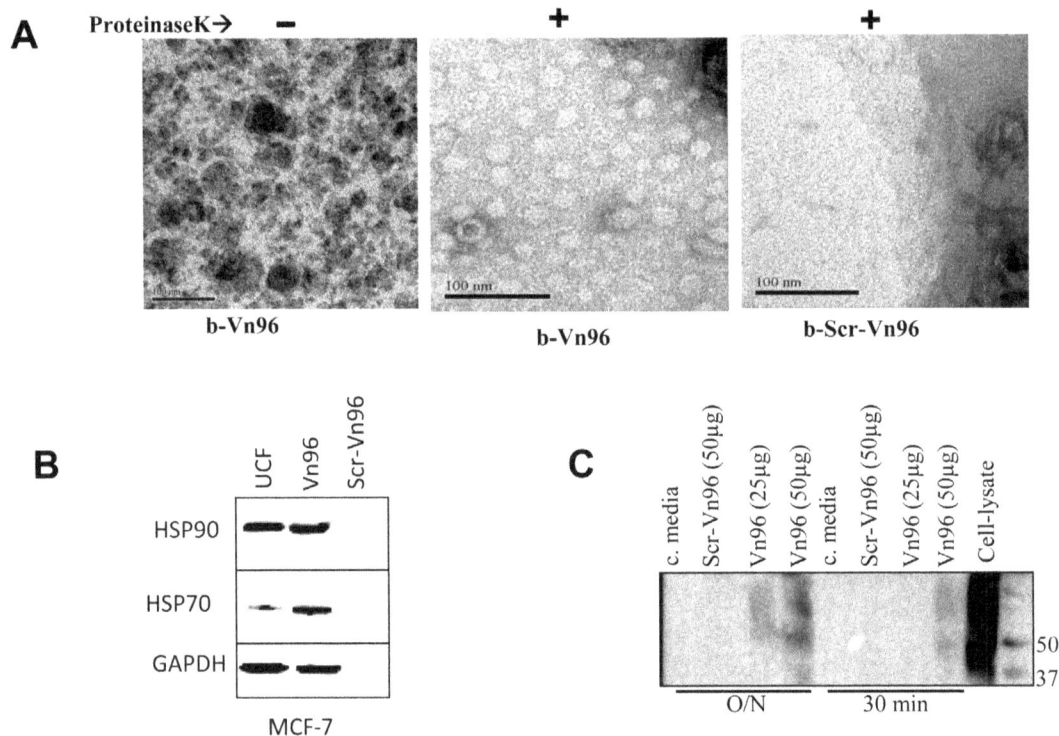

Figure 2. Characterization of extracellular materials precipitated by Vn96. A. Vn96 peptides precipitate vesicular structures from conditioned cell culture media. The biotinylated-Vn96 (b-Vn96) precipitated materials from conditioned cell culture media previously incubated with the MDA-MB-231 breast cancer cell line were subjected to Proteinase K digestion. The transmission electron microscopy analysis were performed on the precipitated material from the b-Vn96 sample (left panel), proteinase K-digested b-Vn96 sample (middle panel), and the proteinase K-digested sample from b-Scr-Vn96 (right panel). The scale bars are 100 nm. B. Identification of exosome markers in the Vn96-purified EVs from conditioned cell culture media. 50 µg each of Vn96 peptide and Scr-Vn96 were incubated with 1 ml of conditioned cell culture media previously incubated with the breast cancer cell line MCF-7 at 4°C for overnight. Exosomes were also isolated from the same conditioned cell culture media by ultracentrifugation (UCF). The presence of HSP70, HSP90 and GAPDH were assessed by immunoblotting. C. CD63 immunoblot. 1 ml of pre-cleared MCF-7 conditioned cell culture media was incubated to precipitate EVs with indicated amount of peptides either overnight (O/N) at 4°C or 30 minutes at room temperature. Total cell lysate of MCF-7 (equivalent to 0.2×10^6 cells) was used as a positive control and conditioned cell culture media alone (c. media) was used as negative control. SDS-PAGE was performed in non-reducing conditions for CD63 immunoblots as recommended by the supplier.

vesicular structures isolated using the b-Vn96 are reminiscent of previously described exosomes and EVs [40]. These results indicate the Vn96 peptide captures membrane-bound structures from cell culture growth media that are potentially EVs.

Identification of canonical EV markers in Vn96-captured membrane structures from conditioned cell growth media

To determine if the membrane-bound structures isolated with b-Vn96 from conditioned cell culture media were indeed EVs, we examined material precipitated with the Vn96 peptide for protein markers of exosomes/EVs by Western-blot analysis. During the course of our experiments, we found that we could also precipitate membrane structures with the Vn96 peptide in the absence of biotinylation and linkage to magnetic beads, but using a similar centrifugation protocol (see methods); we therefore performed downstream experiments using this method.

Vn96 peptide or Scr-Vn96 peptide were added to pre-cleared conditioned cell culture growth media previously incubated with the breast cancer cell line MCF-7; materials were precipitated and harvested as described in the experimental procedures section. Exosomes were purified from the same conditioned cell culture growth media using ultracentrifugation (UCF) on a sucrose cushion as previously described [41]. Western-blot analysis of the material precipitated with Vn96 showed the presence of

HSP70, HSP90, GAPDH (Figure 2B, lane 2), which were also present in the UCF-purified exosomes (Figure 2B, lane 1). Importantly, the amount of EV markers present in Vn96-precipitated material and UCF-purified material were comparable. No signal for EV markers was detected in material precipitated with the Vn96-Scr control peptide (Figure 2B, lane 3).

Similarly, the pre-cleared conditioned cell culture media from MCF-7 cells was incubated with the indicated amount of Vn peptides per ml either overnight (O/N) at 4°C or for 30 minutes at room temperature (Figure 2C). The precipitated materials were subjected to non-reducing SDS-PAGE, followed by anti-CD63 immunoblotting. Our results show that both the overnight and 30 minute incubation protocols precipitate EVs, but at different ratios of Vn96 peptide; specifically, less Vn96 peptide is required when the incubation time is prolonged at 4°C (Figure 2C). Together, these results show that we can precipitate EVs from cell culture growth media using the Vn96 peptide with efficiency comparable to UCF-mediated purification.

The Vn96 peptide precipitates EVs from biological fluids

We wished to further explore whether Vn96 could capture EVs from sources other than cell culture growth media, such as biological fluids. We therefore chose to determine whether Vn96 could capture EVs from urine and plasma. Urine samples were collected from patients (consenting male patients scheduled for prostate biopsy) both pre- and post-digital rectal examination (DRE) with prostate massage. Plasma was collected from consenting healthy women and breast cancer patients.

We first examined whether we could isolate membrane-bound structures from these materials with the Vn96 peptide using TEM and atomic force microscopy (AFM). The plasma samples were diluted ten-fold in PBS before being subjected to Vn96 peptide-mediated precipitation, whereas urine was left undiluted. All samples were subjected to pre-clearing by centrifugation at 17,000×g followed by filtration though 0.22 μm pore size filters. The pre-cleared samples were incubated with 50 μg/ml Vn96 or Scr-Vn96 peptide, followed by precipitation and washes with PBS as described in the methods section. The precipitates were subjected to Proteinase K digestion to obtain a homogenous dispersion of precipitated material, followed by TEM or AFM analyses. As shown in the TEM images (Figure 3A), the size distribution of the membrane structures was similar to the reported sizes of EVs (30 nm to 100 nm). Similarly, AFM analysis in tapping mode was performed for material precipitated from urine by Vn96 and the size distributions are shown in (Figure 3B). Nanoparticle tracking analysis (NTA) of all the samples prepared for Figure 3 was performed (Text S2). It is worth noting that the size distributions of the samples did not match with TEM or AFM measurements (Figures 3A and 3B and Text S2). This discrepancy may be due to the fact that NTA measures the sphere-equivalent hydrodynamic radius from scattered light, and biological samples such as EVs may have a significant hydration shell when dispersed in aqueous solutions, whereas TEM and AFM measure the dry physical structures only. Nonetheless, our results show that Vn96 is able to capture membrane-bound nanoparticles from biological fluids such as plasma and urine.

To determine if the material captured by Vn96 from biological fluids is indeed EVs, we performed Western-blot analysis for canonical EV protein markers. We first isolated material from equal volumes of urine in parallel using the Vn96 peptide and the UCF purification method and assessed for canonical protein markers of EVs by Western-blot analysis (Figure 3C). Urinary EVs precipitated with Vn96 contained canonical EV protein markers (CD9, CD63, CD24, Hsp70, Alix) of comparable or greater abundance than the corresponding UCF-purified exosome sample. Prostate-specific marker, FOLH1 (PSMA) was also detected in post-DRE Vn96-precipitated EVs from the urine of prostate cancer subjects. These data indicate that the Vn96 peptide can precipitate EVs from biological fluids, such as urine.

Comparative proteomic profiling of Vn96-captured EVs from conditioned cell culture growth media and human plasma

To determine if Vn96-mediated capture of EVs results in the isolation of a similar population of EVs as UCF-mediated exosome purification we performed comparative proteomic profiling studies on material isolated from conditioned cell culture growth media and plasma using these methods. For the comparative proteomic studies we used conditioned cell culture growth media used to propagate MCF-10A, MCF-7 and MDA-MB-231 mammary cell lines, which were divided into aliquots that were subjected to each preparation method.

EVs and exosomes were harvested using Vn96 or UCF as described in previous sections. The collected EVs were processed as described in the experimental procedures section. Q-Exactive quadrupole-orbitrap mass spectrometer (Thermo-Fisher Scientific, San Jose, CA) generated spectra were used to search a UniProt protein database with the SEQUEST algorithm (Proteome Discoverer 1.3). Search results were further submitted to Scaffold

4 (Proteome Software, Portland, OR) to generate a minimal list of non-redundant proteins. We extracted the proteome from each sample with 100% probable candidates for Gene Ontology (GO) analysis. As shown in Table 1, GO analysis for cellular components with the proteomes from each sample showed that they originate from extracellular membrane-bound vesicles. More importantly, the proteomes of Vn96-extracted EVs from conditioned cell culture growth media and plasma samples showed highly significant p-values for both the GO terms 'extracellular vesicular exosome' (GO:0070062) and 'Extracellular membrane-bounded organelle' (GO:0065010). Moreover, the proteomes of EVs captured with Vn96 showed a good comparison to the UCF-purified exosome proteomes from the conditioned cell culture growth media or plasma samples as shown in Table 1. These results demonstrate that Vn96 captures a population of EVs that are very similar to exosomes that are purified using the classical UCF method.

Comparative miRNA and other long RNA profiling of Vn96-captured EVs from conditioned cell culture growth media

We wished to further validate that Vn96 isolates a similar population of EVs as other methods. Therefore, we chose to compare the miRNA as well as total RNA cargo of EVs/exosomes purified by different methods (Vn96, UCF, and a commercially-available reagent) from conditioned cell culture growth media used to propagate two breast cancer cell lines, MCF-7 and MDA-MB-231. RNA libraries prepared from isolated EVs were sequenced on the Ion Proton platform (Life Technologies) according to the manufacturer's recommendations, with slight modifications as described in the methods section. Normalization of long RNAs and small RNAs was performed using Reads per Kilobase per Million mapped reads (RPKM) and Trimmed Mean of M-values (TMM) or Lowess methods, respectively. The steps followed for data processing and analysis for profiling the expression of all RNAs and microRNAs is presented as a flowchart in Text S3. The RNA sequence data is archived at Gene Expression Omnibus data repository [GSE58464].

Comparative assessments of miRNA extracted from EVs isolated using Vn96 and the UCF method for one cell line-type revealed very similar profiles with high Pearson correlations, minimal expression variation and less than 5% population variability (Figure 4A). On the other hand, higher dispersion, differential expression and high population variability were observed when miRNA cargos of UCF-purified EVs were profiled for two different cell lines (Figure 4B). Similar wide variations were also observed in the miRNA profiles of EVs precipitated from the same two cell lines using the Vn96 peptide. Furthermore, populations of differentially-expressed miRNAs identified in the EVs of the two cell lines were highly similar irrespective of the isolation method used (UCF or Vn96) as shown in the normalized heat map in Text S4.

Identified miRNAs extracted from MCF-7 cell-line EVs obtained using different methods of EV isolation (including a commercially available EV isolation kit) showed minimal diversifications, as shown in the Venn diagram in Figure 4C (left panel). Greater diversification was observed when the populations of identified miRNA cargos were compared between different cell lines (Figure 4C right panel and Text S5). The highly similar miRNA profiles observed between Vn96 and UCF methods of EV purification from conditioned cell culture growth media further validate Vn96 as a highly specific tool to enrich EVs. RNA profiles of EVs typically show a characteristic enrichment of different species of RNAs (miRNA, miscRNA and lincRNA etc) [42] that

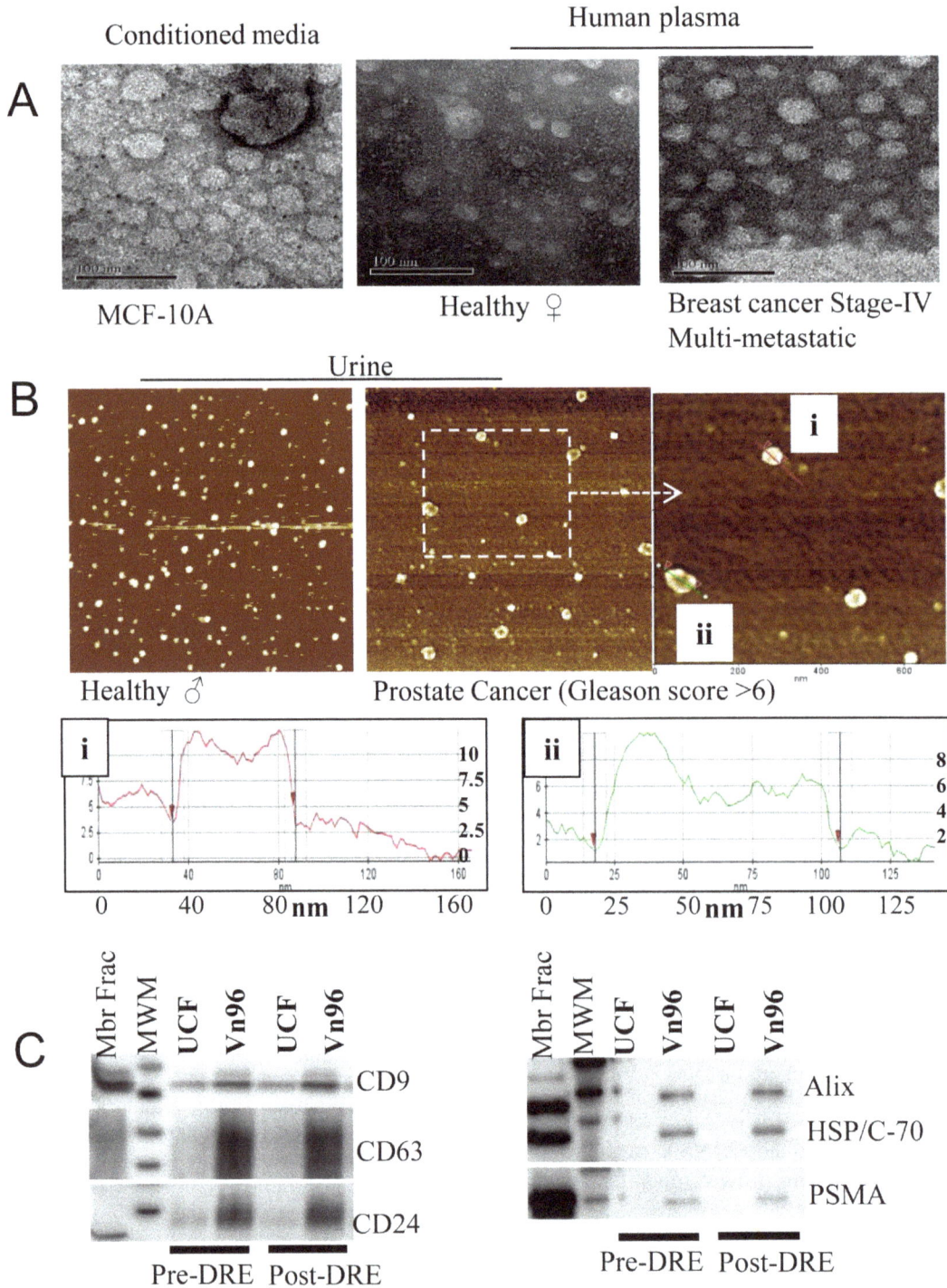

Figure 3. Visualization of Vn96 peptide-precipitated extracellular vesicles (EVs) from biological fluids. Pre-cleared biological fluids (conditioned cell culture media, human plasma and human urine) were used as described in experimental procedures. The Vn96 peptide-precipitated materials were dispersed into solution using Proteinase K digestion prior to microscopic analysis. A. Transmission electron microscopy images of Vn96-precipitated material from the indicated samples (conditioned cell culture media and diluted human plasma). The scale bars are 100 nm. B. Atomic force microscopy (phase) image of the Vn96 peptide-precipitated (Proteinase K digested) EVs from human urine. A differential size distribution pattern is observed between EVs from urine of normal and prostate cancer subjects (equal scale). The enlarged area from the prostate cancer image was used to measure width and thickness of two individual EVs (right panel, i and ii) in nanometers (nm) are shown in the bottom panel. C. Pre- and post-digital rectal exam with prostatic massage (DRE) urine samples were collected from consenting donors. EVs were isolated in parallel from equal volumes of urine using Vn96 and ultracentrifugation (UCF) methods; immunoblot analyses were performed using antibodies against the proteins indicated. Representative results for pre- and post-DRE urine samples from a donor are shown. Mbr Frac = Membrane fraction of prostate adenocarcinoma cell line LNCap and MWM = Protein molecular weight markers.

Table 1. Gene list enrichment analysis for Cellular Component Ontology using ToppGene.

Sample	EV isolation	% GO term	p-value
GO:0070062: Extracellular vesicular Exosome			
Conditioned media: MCF-10A	Vn96	41.27	1.83E-30
Conditioned media: MCF-10A	UC	7.94	4.07E-06
Conditioned media: MCF-7	Vn96	23.81	6.07E-11
Conditioned media: MCF-7	UC	7.94	1.20E-04
Conditioned media: MDA-MB-231	Vn96	20.63	6.53E-19
Conditioned media: MDA-MB-231	UC	20.36	4.07E-19
Human plasma-1	Vn96	14.29	1.99E-10
Human plasma-2	Vn96	12.7	2.59E-08
Human plasma-3	Vn96	15.87	7.92E-11
Human plasma-4	Vn96	11.11	6.70E-19
GO:0065010: Extracellular membrane-bounded organelle			
Conditioned media: MCF-10A	Vn96	41	5.03E-30
Conditioned media: MCF-10A	UC	7.94	5.61E-08
Conditioned media: MCF-7	Vn96	23.08	9.98E-11
Conditioned media: MCF-7	UC	7.69	1.41E-04
Conditioned media: MDA-MB-231	Vn96	20.63	6.53E-19
Conditioned media: MDA-MB-231	UC	20.36	4.07E-19
Human plasma-1	Vn96	13.85	2.68E-10
Human plasma-2	Vn96	12.31	3.36E-08
Human plasma-3	Vn96	15.38	1.10E-10
Human plasma-4	Vn96	10.77	8.38E-07

The list of 100% probable proteins from each sample's proteome was derived and gene list enrichment analysis was carried out using ToppFun (https://toppgene.cchmc.org/) for Cellular Component ontology. ToppGene Suite is being developed at Division of Biomedical Informatics, Cincinnati Children's Hospital Medical Center (BMI CCHMC), Cincinnati, OH 45229. For comparison we also analysed results from two proteomic data-sets [Vesiclepedia ID_44 and Vesiclepedia ID_353] derived from exosomes purified from human plasma using Size exclusion filtration followed by Sucrose density gradient ultracentrifugation (UC), as posted on Vesiclepedia (http://microvesicles.org/index.html). Cellular component ontology analysis using ToppFun (GO:0070062: Extracellular vesicular Exosome) for Vesiclepedia ID_44 and Vesiclepedia ID_353 derived exosomal proteome revealed p-values of 1.15E-09 and 1.92E-11 respectively. Similar analysis for GO:0065010 (Extracellular membrane-bounded organelle) from Vesiclepedia ID_44 and Vesiclepedia ID_353 derived exosomal proteome revealed p-values of 1.54E-09 and 2.66E-11 respectively. The %GO term means the percentage ratio of 'list of proteins as input' over the assigned list of genes for a specific annotation.

differ from total cellular RNA species profiles. For example, the proportion of rRNA is usually decreased by several-fold in EVs in comparison to its proportion in total cellular RNA. Our RNA sequence data reveal similar characteristic patterns [42] of different species of RNAs when compared to UCF and Vn96 methods of EV purification (Text S6).

Together, our data show that Vn96 captures EVs that contain a RNA cargo content that is similar to the established UCF purification method and a commercially-available EV isolation kit.

Discussion

We initially set out to develop HSP-binding peptides that could be used to capture extracellular HSP complexes for further investigation. Our observations during the validation of the peptides led us to discover their potential as exosome or EV capture tools. We found that the Vn96 peptide could capture EVs from conditioned cell culture growth media and biological fluids, such as urine and plasma. Our recent unpublished results also show that Vn96 can capture EVs from mouse and canine plasma, as well as from bovine milk (data not shown). Importantly, we demonstrate that Vn96-mediated EV capture permits the collection of EVs that are both physically and cargo-content similar to EVs/exosomes isolated by the standard UCF-purification method

and a commercially-available EV isolation kit. Unlike other methods, Vn96 permits the collection of EVs from multiple fluid sources using standard laboratory equipment in a minimal amount of time (<40 minutes).

While characterizing Vn96's ability to capture extracellular HSP complexes we observed visibly distinct aggregation patterns in conditioned cell culture growth media and biological fluids when Vn96 was added. We observed no visible aggregation in stock solutions of the peptides (Vn96 and Scrambled-Vn96 in PBS) or the samples to which Scrambled-Vn96 was added. This observation prompted us to investigate the constituents and nature of the aggregates induced by the Vn96 peptide in pre-cleared conditioned cell culture growth media, urine and plasma. We found that Vn96 acts like a 'nano-probe', which enriches vesicular structures that have the properties of exosomes and/or microvesicles (collectively, EVs). We compared Vn96-captured material to exosomes purified by ultracentrifugation using NTA, TEM, AFM, immunoblotting, next-generation sequencing of miRNA cargo, and proteome-based cellular component ontology analysis, and found that they are indeed EVs. Moreover, because the Vn96 peptide can bind to HSPs from multiple species (see Figure 1A), its ability to capture EVs may not be limited to human biological fluids and cell culture samples. Vn96-mediated EV capture may

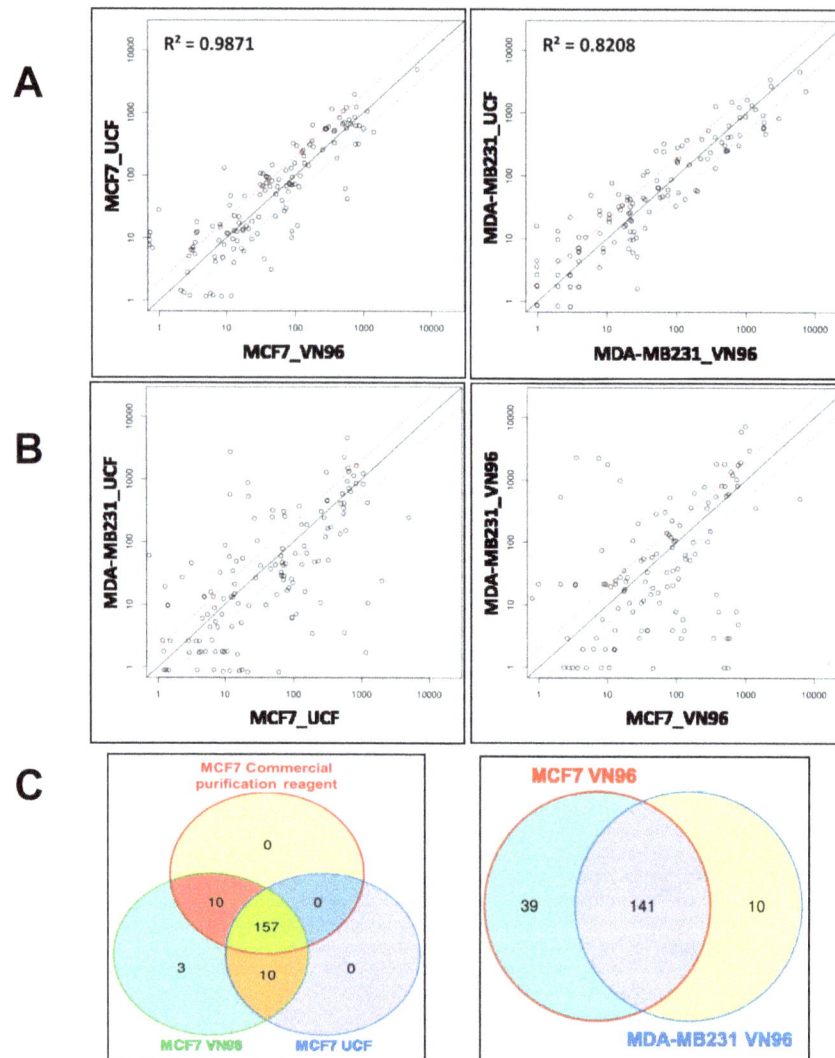

Figure 4. Comparative miRNA-seq data for Vn96- and UCF-purified EVs from conditioned cell culture media. A. Scatter plot comparing normalized expression profiles of miRNAs contained in EVs isolated from the indicated conditioned cell culture media using either ultracentrifugation or the Vn96 peptide. For example, MCF7_UCF and MCF7_VN96 indicate that EVs were purified from conditioned cell culture media previously incubated with MCF-7 cells by ultracentrifugation and the Vn96 peptide, respectively. High Pearson correlations between ultracentrifugation and Vn96 peptide methods of EV purification from the same sample validate Vn96 as an EV purification tool. B. Scatter plot comparing normalized expression profiles of miRNAs contained in EVs isolated from MCF-7 versus MDA-MB-231 conditioned cell culture media using the same purification method. C. Venn diagram of miRNAs contained in EVs isolated from MCF-7 conditioned cell culture media using different methods (ultracentrifugation, Vn96 peptide and a commercially-available exosome purification kit). Less than 10% differences were observed in the miRNA populations between the ultracentrifugation and Vn96 peptide methods, and the commercial kit and Vn96 peptide methods (left panel), but a wider variation in miRNA populations was observed in EVs from different cell lines (right panel).

therefore be applicable to basic research using animal models, as well as diagnostic methods for animal health.

We believe that Vn96 is able to capture EVs due to its interaction with HSPs on their surface, since EV-mediated extracellular transport of HSPs occurs in many pathological conditions [5–15]. However, by virtue of its design the Vn96 peptide forms a cationic alpha helix at physiologically relevant salt and buffer conditions, which may allow Vn96 to gain overall avidity towards ultra-small subcellular structures and other molecules from intracellular and extracellular origin. It is known that alpha-helical cationic peptides can aggregate small multilayered lipid vesicles based on the peptide's ability to form a helical coiled-coil [43] that interacts with and/or inserts into membranes [44–46]; therefore, we cannot rule out the possibility that the

cationic nature of the Vn96 peptide may allow it to directly interact with the membranes of EVs to facilitate their capture. Nonetheless, all of our results confirm that the Vn96 peptide is a useful tool for the collection of EVs from wide variety of sample types, and captures EVs that have characteristics that are equivalent to those obtained by the standard ultracentrifugation isolation method.

The release of EVs is a conserved and essential process of diverse prokaryotic and eukaryotic cells. But this essential process is co-opted during cancer, in which EVs play critical roles in the establishment of cell transformation, cancer progression, metastasis, distal niche formation, stemness, and many aspects of tumor cross-talk with surrounding cells [47]. There is ample evidence that cancer cells produce EVs with cancer-specific signatures, which can be found in body fluids, a finding that opens up new frontiers for cancer diagnostics research. A method that allows the simple and rapid capture of EVs, such as the Vn96 peptide, will permit significant advancement of this field. However, the release of EVs that contain disease signatures is not limited to cancer. Neurons with infectious prion proteins were found to produce EVs that contain the same prions [48]. Similarly, virally-infected host cells release EVs that contain viral factors [49–53], which influence host response. Therefore, the capture of EVs from body fluids represents a possible new approach to minimally-invasive broad-based disease diagnostics.

Vn96-based EV purification provides a simple, efficient, and rapid method of EV enrichment and capture. There are potential benefits of EV enrichment with the Vn96 peptide for both established diagnostics and for new biomarker discovery. Current obstacles to the application of EVs in the clinical setting include difficulties with isolation methods and most prominently enrichment of disease-specific EVs from complex mixtures of vesicular material originating from various cell/tissue types. The current methods available for the isolation of EVs are based on physical characteristics, which can be efficient but are time-consuming, require specialized equipment, and may lack specificity. Similarly, affinity-based methods such as the use of antibody capture are still based on EV 'markers', which appear to vary amongst EV populations [54] and may therefore not be present on all EV species. We have demonstrated that the Vn96 peptide isolates EVs that have clinical value and that the Vn96 peptide compares favourably to current isolation methods in terms of efficiency, cost, and platform versatility as an EV capture tool for discovery research, animal health, and clinical applications.

Supporting Information

Text S1   The Vn96 peptide enriches membrane-bound structures from total cell lysates. A portion of pull-down material shown in Figure 1B was washed with PBS and subjected to Proteinase K digestion. The beads were removed and the suspension was subjected to transmission electron microscopy analysis. A dense vesicular aggregated material, resembling different subcellular vesicles, was observed in the samples from the b-Vn96 pull-down. No such structures were observed in b-Scr-Vn96 samples. The scale bars are 100 nm.

Text S2   Particle size distribution: nanoparticle tracking analysis. The size distribution and relative abundances of the EVs from the samples shown in Figure 3 were measured using nanoparticle tracking analysis as described in the experimental procedures.

Text S3   Flowchart for the analysis of the next-generation RNA-sequencing. Flowchart for the analysis of the next-generation sequencing data for profiling RNA and microRNA expression. RNA libraries prepared from EVs isolated with different methods were sequenced on the Proton platform (Life Technologies). Normalization of long RNA was realized with Reads Per Kilobase per Million mapped reads (RPKM) and small-RNA were normalized with Trimmed Mean of M-values (TMM) or Lowess methods.

Text S4   Heatmap showing the abundance of miRNA. Heatmap showing the abundance of miRNA contained in EVs produced by MCF-7 and MDA-MB-231 cell lines (abundance values normalized with Lowess method). Different methods to isolate EVs were compared (Ultra for ultracentrifugation, VN96 for Vn peptide method, and C.K for commercially-available kit). Missing values are indicated by the grey color. Only miRNAs with zero reads are treated as missing values, whereas miRNAs with 1 or 2 reads are shown in the heatmap.

Text S5   Venn diagram of comparative miRNA expression. miRNA-seq on EVs isolated ultracentrifugation or the Vn96 peptide from cell culture media previously incubated with two different breast cancer cell lines (MCF-7 and MDA-MB-231). Venn diagram comparing miRNA expression between EVs isolated from MCF7 and MDA-MB-231, and between ultracentrifugation and Vn96 peptide methods.

Text S6   Comparative distribution of RNA species contained in EVs. RNA species contained in EVs produced by breast cancer cell lines MCF-7 and MDA-MB-231 that were isolated by ultracentrifugation or the Vn96 peptide method. The right figure is an enlargement of the left figure in order to facilitate the visualization of less abundant RNA species. Proportions of RNA species are similar between isolation methods used. We also observed an enrichment of some RNA species in EVs compared to RNA species contained in the cell. (rRNA represent around 1–10% of all RNA in EVs, while in a cell more than 90% of RNA are rRNA).

Acknowledgments

We thank the consenting patients who provided samples for this study. We thank Susan Belfry of the University of New Brunswick for critical review of electron-microscopy data.

Author Contributions

Conceived and designed the experiments: AG RJO SGG. Performed the experiments: AG M. Davey SL ICC SC SF MCC ADF M. Daigle MVM NC AJ DB SGG. Analyzed the data: AG M. Davey ICC SC SL SF MCC ADF M. Daigle MVM NC AJ DB SGG SML RJO. Contributed reagents/materials/analysis tools: AG M. Davey ICC SC SL SF MCC ADF M. Daigle MVM NC AJ DB SGG SML RJO. Contributed to the writing of the manuscript: AG M. Davey ICC SC SL SF MCC ADF M. Daigle MVM NC AJ DB SGG SML RJO. Validation, application and overall method development to publication: AG. Originated the study: RJO. Constructed and screened the peptide: SGG SL. Protocol development and clinical applications: M. Davey.

References

1. Tamura Y, Torigoe T, Kukita K, Saito K, Okuya K, et al. (2012) Heat-shock proteins as endogenous ligands building a bridge between innate and adaptive immunity. Immunotherapy 8: 841–52.
2. Udono H (2012) Heat shock protein magic in antigen trafficking within dendritic cells: implications in antigen cross-presentation in immunity. Acta Med Okayama 66(1): 1–6.
3. Joly AL, Wettstein G, Mignot G, Ghiringhelli F, Garrido C (2010) Dual role of heat shock proteins as regulators of apoptosis and innate immunity. J Innate Immun. 2(3): 238–47.
4. Kettern N, Rogon C, Limmer A, Schild H, Höhfeld J (2011) The Hsc/Hsp70 co-chaperone network controls antigen aggregation and presentation during maturation of professional antigen presenting cells. PLoS One 20: e16398.
5. Multhoff G (2007) Heat shock protein 70 (Hsp70): membrane location, export and immunological relevance. Methods 43(3): 229–37.
6. Graner MW, Raynes DA, Bigner DD, Guerriero V (2009) Heat shock protein 70-binding protein 1 is highly expressed in high-grade gliomas, interacts with multiple heat shock protein 70 family members, and specifically binds brain tumor cell surfaces. Cancer Sci 100(10): 1870–9.

7. Kotsiopriftis M1, Tanner JE, Alfieri C (2005) Heat shock protein 90 expression in Epstein-Barr virus-infected B cells promotes gammadelta T-cell proliferation in vitro. J Virol 79(11): 7255–61.

8. Multhoff G, Hightower LE (2011) Distinguishing integral and receptor-bound heat shock protein 70 (Hsp70) on the cell surface by Hsp70-specific antibodies. Cell Stress Chaperones 16(3): 251–5.

9. Graner MW, Cumming RI, Bigner DD (2007) The heat shock response and chaperones/heat shock proteins in brain tumors: surface expression, release, and possible immune consequences. J Neurosci 27(42): 11214–27.

10. Khan S, Jutzy JM, Aspe JR, McGregor DW, Neidigh JW, et al. (2011) Survivin is released from cancer cells via exosomes. Apoptosis 16(1): 1–12.

11. Lv LH, Wan YL, Lin Y, Zhang W, Yang M, et al. (2012) Anticancer drugs cause release of exosomes with heat shock proteins from human hepatocellular carcinoma cells that elicit effective natural killer cell antitumor responses in vitro. J Biol Chem 287(19): 15874–85.

12. Graner MW, Alzate O, Dechkovskaia AM, Keene JD, Sampson JH, et al. (2009) Proteomic and immunologic analyses of brain tumor exosomes. FASEB J 23(5): 1541–57.

13. Liu Y, Xiang X, Zhuang X, Zhang S, Liu C, et al. (2010) Contribution of MyD88 to the tumor exosome-mediated induction of myeloid derived suppressor cells. Am J Pathol 176(5): 2490–9.

14. Lancaster GI, Febbraio MA (2005) Exosome-dependent trafficking of HSP70: a novel secretory pathway for cellular stress proteins. J Biol Chem 280(24): 23349–55.

15. Antonio DM (2011) Extracellular heat shock proteins, cellular export vesicles, and the Stress Observation System: A form of communication during injury, infection, and cell damage. Cell Stress Chaperones 16(3): 235–249.

16. Bard MP, Hegmans JP, Hemmes A, Luider TM, Willemsen R, et al. (2004) Proteomic analysis of exosomes isolated from human malignant pleural effusions. Am J Respir Cell Mol Biol 31: 114–121.

17. Pisitkun T, Shen RF, Knepper MA (2004) Identification and proteomic profiling of exosomes in human urine. Proc Natl Acad Sci USA 101: 13368–13373.

18. Palanisamy V, Sharma S, Deshpande A, Zhou H, Gimzewski J, et al Nanostructural and transcriptomic analyses of human saliva derived exosomes. (2010) PLoS One 5:e8577.

19. Michael A, Bajracharya SD, Yuen PS, Zhou H, Star RA, et al. (2010) Exosomes from human saliva as a source of microRNA biomarkers. Oral Dis 16: 34–38.

20. Runz S, Keller S, Rupp C, Stoeck A, Issa Y, et al. (2007) Malignant ascites-derived exosomes of ovarian carcinoma patients contain CD24 and EpCAM. Gynecol Oncol 107: 563–571.

21. Keller S, Ridinger J, Rupp AK, Janssen JW, Altevogt P (2011) Body fluid derived exosomes as a novel template for clinical diagnostics. J Transl Med 9: 86.

22. Street JM, Barran PE, Mackay CL, Weidt S, Balmforth C, et al. (2012) Identification and proteomic profiling of exosomes in human cerebrospinal fluid. J Transl Med 5: 10: 5.

23. Simpson RJ, Kalra H, Mathivanan S (2012) ExoCarta as a resource for exosomal research. J Extracell Vesicle 1: 1–6.

24. Atay S, Godwin AK (2014) Tumor-derived exosomes: A message delivery system for tumor progression. Commun Integr Biol 1: 7(1): e28231.

25. Jiang J, Maes EG, Taylor AB, Wang L, Hinck AP, et al. (2007) Structural basis of J cochaperone binding and regulation of Hsp70. Mol Cell 28(3): 422–33.

26. Nicolaï A, Senet P, Delarue P, Ripoll DR (2010) Human Inducible Hsp70: Structures, Dynamics, and Interdomain Communication from All-Atom Molecular Dynamics Simulations. J. Chem. Theory Comput 6(8): 2501–2519.

27. Gao XC, Zhou CJ, Zhou ZR, Wu M, Cao CY, et al. (2012) The C-terminal helices of heat shock protein 70 are essential for J-domain binding and ATPase activation. J Biol Chem 287(8): 6044–52.

28. Greene MK, Maskos K, Landry SJ (1998) Role of the J-domain in the cooperation of Hsp40 with Hsp70. Proc Natl Acad Sci U S A 95(11): 6108–13.

29. Gässler CS, Buchberger A, Laufen T, Mayer MP, Schröder H, et al. (1998) Mutations in the DnaK chaperone affecting interaction with the DnaJ cochaperone. Proc Natl Acad Sci USA 95(26): 15229–34.

30. Szyperski T, Pellecchia M, Wall D, Georgopoulos C, Wüthrich K (1994) NMR structure determination of the Escherichia coli DnaJ molecular chaperone: secondary structure and backbone fold of the N-terminal region (residues 2–108) containing the highly conserved J domain. Proc. Natl. Acad. Sci. USA 91: 11343–11347.

31. Pellecchia M, Szyperski T, Wall D, Georgopoulos C, Wüthrich K (1996) NMR structure of the J-domain and the Gly/Phe-rich region of the Escherichia coli DnaJ chaperone. J. Mol. Biol 260: 236–250.

32. Qian YQ, Patel D, Hartl FU, McColl D J (1996) Nuclear magnetic resonance solution structure of the human Hsp40 (HDJ-1) J-domain. J. Mol. Biol 260: 224–235.

33. Tsai J, Douglas MG (1996) A conserved HPD sequence of the J-domain is necessary for YDJ1 stimulation of Hsp70 ATPase activity at a site distinct from substrate binding. J. Biol. Chem 271: 9347–9354.

34. Maeda H, Sahara H, Mori Y, Torigo T, Kamiguchi K, et al. (2007) Biological heterogeneity of the peptide-binding motif of the 70-kDa heat shock protein by surface plasmon resonance analysis. J Biol Chem 282(37): 26956–62.

35. Hu J, Wu Y, Li J, Qian X, Fu Z, et al. (2008) The crystal structure of the putative peptide-binding fragment from the human Hsp40 protein Hdj1. BMC Struct Biol 8: 3.

36. Rérole AL, Gobbo J, De Thonel A, Schmitt E, Pais BJP, et al. (2011) Peptides and aptamers targeting HSP70: a novel approach for anticancer chemotherapy. Cancer Res 71(2): 484–95.

37. Triantafilou M, Triantafilou K (2004) Heat-shock protein 70 and heat-shock protein 90 associate with Toll-like receptor 4 in response to bacterial lipopolysaccharide. Biochemical Society Transactions 32: 636–639.

38. Sakashita N, Miyazaki A, Takeya M, Horiuchi S, Chang CC, et al. (2000) Localization of human acyl-coenzyme A: cholesterol acyltransferase-1 (ACAT-1) in macrophages and in various tissues. Am J Pathol 156(1): 227–36.

39. Simon JP, Morimoto T, Bankaitis VA, Gottlieb TA, Ivanov IE, et al. (1998) An essential role for the phosphatidylinositol transfer protein in the scission of coatomer-coated vesicles from the trans-Golgi network. Proc Natl Acad Sci USA 95(19): 11181–6.

40. Salomon C, Sobrevia L, Ashman K, Illanes SE, Mitchell MD, et al. (2013) The Role of Placental Exosomes in Gestational Diabetes Mellitus, Gestational Diabetes - Causes, Diagnosis and Treatment, Dr. Luis Sobrevia (Ed.), ISBN: 978-953-51-1077-4, InTech, DOI: 10.5772/55298. Available: http://www.intechopen.com/books/gestational-diabetes-causes-diagnosis-and-treatment/the-role-of-placental-exosomes-in-gestational-diabetes-mellitus.

41. Th'ery C, Clayton A, Amigorena S, Raposo G (2006) Isolation and Characterization of Exosomes from Cell Culture Supernatants and Biological Fluids. Current Protocols in Cell Biology 3.22.1–3.22.29.

42. Huang X, Yuan T, Tschannen M, Sun Z, Jacob H, et al. (2013) acterization of human plasma-derived exosomal RNAs by deep sequencing. BMC Genomics 14: 319.

43. Vagt T, Zschörnig O, Huster D, Koksch B (2006) Membrane binding and structure of de novo designed alpha-helical cationic coiled-coil-forming peptides. Chemphyschem 7(6): 1361–71.

44. Mihajlovic M, Lazaridis T (2010) Antimicrobial peptides bind more strongly to membrane pores. Biochim Biophys Acta 1798(8): 1494–1502.

45. Hong M, Su Y (2011) Structure and dynamics of cationic membrane peptides and proteins: Insights from solid-state NMR. Protein Sci. 20(4): 641–655.

46. Su Y, Li S, Hong M (2013) Cationic membrane peptides: atomic-level insight of structure-activity relationships from solid-state NMR. Amino Acids 44: 821–833.

47. Azmi AS, Bao B, Sarkar FH (2013) Exosomes in cancer development, metastasis, and drug resistance: a comprehensive review. Cancer Metastasis Rev 32: 623–42.

48. Fevrier B, Vilette D, Archer F, Loew D, Faigle W, et al. (2004) Cells release prions in association with exosomes. Proc Natl Acad Sci USA 101(26): 9683–8.

49. Lenassi M, Cagney G, Liao M, Vaupotic T, Bartholomeeusen K, et al. (2010) HIV Nef is secreted in exosomes and triggers apoptosis in bystander CD4+ T cells. Traffic 11(1): 110–22.

50. Giri PK, Kruh NA, Dobos KM, Schorey JS (2010) Proteomic analysis identifies highly antigenic proteins in exosomes from M. tuberculosis-infected and culture filtrate protein-treated macrophages. Proteomics.10(17): 3190–202.

51. Bhatnagar S, Schorey JS (2007) Exosomes released from infected macrophages contain Mycobacterium avium glycopeptidolipids and are proinflammatory. J Biol Chem 282(35): 25779–89.

52. Bhatnagar S, Shinagawa K, Castellino FJ, Schorey JS (2007) Exosomes released from macrophages infected with intracellular pathogens stimulate a proinflammatory response in vitro and in vivo. Blood 110(9): 3234–44.

53. Pegtel DM, Cosmopoulos K, Thorley-Lawson DA, van Eijndhoven MA, Hopmans ES, et al. (2010) Functional delivery of viral miRNAs via exosomes. Proc Natl Acad Sci USA 107(14): 6328–33.

54. Jia S, Zocco D, Samuels ML, Chou MF, Chammas R, et al. (2014) Emerging technologies in extracellular vesicle-based molecular diagnostics. Expert Rev Mol Diagn 14(3): 307–21.

55. Thévenet P, Shen Y, Maupetit J, Guyon F, Derreumaux P, et al. (2012) PEP-FOLD: an updated de novo structure prediction server for both linear and disulfide bonded cyclic peptides.Nucleic Acids Res. 40(Web Server issue):W288–93. doi: 10.1093/nar/gks419.

6

Involvement of Tumor Macrophage HIFs in Chemotherapy Effectiveness: Mathematical Modeling of Oxygen, pH, and Glutathione

Duan Chen[1], Andrey A. Bobko[2], Amy C. Gross[2], Randall Evans[2], Clay B. Marsh[2], Valery V. Khramtsov[2], Timothy D. Eubank[2]*, Avner Friedman[3,4]*

1 Department of Mathematics and Statistics, University of North Carolina at Charlotte, Charlotte, North Carolina, United States of America, 2 Division of Pulmonary, Allergy, Critical Care and Sleep Medicine, College of Medicine, The Ohio State University, Columbus, Ohio, United States of America, 3 Mathematical Biosciences Institute, The Ohio State University, Columbus, Ohio, United States of America, 4 Department of Mathematics, The Ohio State University, Columbus, Ohio, United States of America

Abstract

The four variables, hypoxia, acidity, high glutathione (GSH) concentration and fast reducing rate (redox) are distinct and varied characteristics of solid tumors compared to normal tissue. These parameters are among the most significant factors underlying the metabolism and physiology of solid tumors, regardless of their type or origin. Low oxygen tension contributes to both inhibition of cancer cell proliferation and therapeutic resistance of tumors; low extracellular pH, the reverse of normal cells, mainly enhances tumor invasion; and dysregulated GSH and redox potential within cancer cells favor their proliferation. In fact, cancer cells under these microenvironmental conditions appreciably alter tumor response to cytotoxic anti-cancer treatments. Recent experiments measured the *in vivo* longitudinal data of these four parameters with tumor development and the corresponding presence and absence of tumor macrophage HIF-1α or HIF-2α in a mouse model of breast cancer. In the current paper, we present a mathematical model-based system of (ordinary and partial) differential equations to monitor tumor growth and susceptibility to standard chemotherapy with oxygen level, pH, and intracellular GSH concentration. We first show that our model simulations agree with the corresponding experiments, and then we use our model to suggest treatments of tumors by altering these four parameters in tumor microenvironment. For example, the model qualitatively predicts that GSH depletion can raise the level of reactive oxygen species (ROS) above a toxic threshold and result in inhibition of tumor growth.

Editor: Daolin Tang, University of Pittsburgh, United States of America

Funding: This work was supported in part by National Science Foundation Award 0635561, NCI 5R00CA131552 (TDE) and R01 HL067167 (CBM). The funders had no role in study design, data collection and analysis, decision to publish, or preparation of the manuscript.

Competing Interests: The authors have declared that no competing interests exist.

* Email: Tim.Eubank@osumc.edu (TDE); afriedman@math.osu.edu (AF)

Introduction

Tumors have distinguishing features from normal tissue. Among the most significant factors in tumor metabolism and physiology are the tissue oxygen concentration, acidity, intracellular glutathione (GSH) concentration and redox status [1–5]; in the sequel we focus on the first three features.

(i) Tissue oxygen level

Clinical investigation has shown that hypoxic regions develop in a wide range of malignancies including cancers of the breast, uterine cervix, and prostate. Inefficient tumor vasculature induces hypoxia which decreases extracellular pH and increases interstitial fluid pressure. Hypoxia-induced transcription factors like HIF-1α regulate VEGF and other glucose-regulating genes like GLUT-1 which augments glucose uptake from the surroundings. This process favors tumor cell proliferation as tumor cells generate 50% of their ATP from glycolysis while normal cells generate only 10%, giving tumor cells an adaptive survival advantage over adjacent normal cells. Further, tumor hypoxia is associated with poor patient prognosis because low oxygen reduces the effectiveness of therapies that require the generation of ROS for cell killing.

(ii) Tumor acidity

Tumor acidity is due to increased lactic acid secretion from the anaerobic metabolism of cancer cells via their expression of tumor M2-PK, a dimeric isoenzyme of pyruvate kinase up-regulated in cancer cells. M2-PK drives pyruvate to lactate, a major energy source in tumors [6]. In turn, tumors have a lower extracellular pH (pH$_e \sim$ 6.7-7.1) [2,7] maintained by increased carbonic anhydrase IX(CAIX) activity compared to normal tissue (pH$_e$ = 7.4) [8,9]. Extracellular acidity results in increased tumor invasion, proliferation, evasion of apoptosis, and cell migration as well as ion trapping of weak base drugs [2]. A sequence of interdisciplinary studies, involving mathematical models and experimental evidences, have been conducted in [10–12], for the tumor-stromal interactions and acid-mediated tumor invasion. More recently, the anti-cancer effects of pH buffer therapy was investigated in [13] and variety of foods were suggested that can contribute to manage cancers.

(iii) Intracellular glutathione

GSH plays a crucial role in balancing redox status in tumor microenvironment [14,15]. Indeed, accumulated evidence indicates that increased level of hydrogen peroxide (H_2O_2) and other reactive oxygen species (ROS) occur in many types of cancer cells compared to their normal counterparts through far greater rates of mitochondrial reduction of superoxide [16]. Within a certain range, increases in ROS promote tumor cell proliferation by activating glucose-regulating genes and production of angiogenesis signaling factors like VEGF, whereas ROS leads to oxidative damage (ROS stress) at levels above a toxicity threshold. As a major intracellular redox buffer and antioxidant for redox adaption [17] and in response to ROS stress, high levels of GSH have been found in various tumor types, being up to several-fold greater than that in surrounding tissues [18].

In the present paper we develop a mathematical model for tumor growth with dynamics of GSH concentration, pH and oxygen tension in the tumor microenvironment. This is a two-scale model: at the tissue level, the interactions between tumor, immune, and endothelial cells, along with corresponding cytokines, are modeled by a set of partial differential equations (PDEs) in a moving domain, in which a velocity field is included to describe the movement of cells, chemicals, and the tumor boundary; at the cellular level, a dynamical system of intracellular chemical interactions between ROS, GSH, and other intermediate molecules is proposed within individual cells. We validate the model by comparing simulations to experimental data, and then use the model to predict tumor growth with intracellular GSH depletion as a possible therapeutic strategy. The model can also be used to monitor the change of pH, GSH and oxygen in tumor as a result of the absence or presence of macrophage HIFαs (HIF-1α or HIF-2α) and corresponding effectiveness of chemotherapeutic drugs. The footprint of these quantities could relate to the efficiency of a drug in terms of the tumor microenvironment. We illustrate this approach by simulation of the course of tumors treated by docetaxel (DTX).

Mathematical Model

In this section we describe a mathematical model representing tumor growth along with dynamics of GSH concentration, pH and oxygen level, by a system of ordinary and partial differential equations. At the tissue level, we have cancer cells interacting with immune cells and the vascular system during angiogenesis, while at the cellular level we have GSH, pH and oxygen concentrations interacting within each cancer cell.

Variables and relations

For tumor growth at the *macroscopic* scale, we have as variables the densities of live and dead tumor cells, macrophages, and endothelial cells (ECs), and the concentrations of cytokines interacting among the cells: monocyte chemoattractant protein-1 (MCP-1/CCL2), vascular endothelial growth factor (VEGF), and soluble VEGF receptor-1 (sVEGFR-1). Two other macroscopic variables are oxygen tension and concentration of hydrogen ions, which can be measured experimentally. At the *microscopic* level, we consider the intracellular concentrations of ROS, GSH, and reduced/oxidized forms of GSH peroxidase (GPx_r/GPx_o). A list of all these variables is given in Table 1.

Relations between macroscopic and microscopic variables are described schematically in Figure 1. ROS (primarily H_2O_2) is an important by-product of aerobic metabolism and plays the role of a double-edged sword [15] in cells: when below a certain toxicity threshold V_{toxic}, a moderate increase in ROS level could promote cell proliferation, but when it is increased above the threshold, the elevated ROS concentration will trigger cell death. GSH is the most abundant antioxidant produced by cancer cells to protect themselves from oxidative stress with the help of the enzyme glutathione peroxidase. On the other hand, large amount of hydrogen ions are produced from glucose or anaerobic metabolism. Low intracellular pH (pH_i) can mediate apoptosis of cancer cells, but the access protons are pumped out by over-expressed proton transporters [19], and this leads to an acidic extracellular environment (low pH_e). Indeed, the experimental measurements in the current work are about pH_e and thus only the extracellular acidosis-induced release of VEGF mentioned in [10] is considered. It was illustrated in [20] that VEGF promoter activity is inversely correlated with tumor extracellular pH *in vivo* in the human glioma xenografts. Additionally, it was concluded in [21,22] that below the toxic threshold, ROS also contributes to upregulation of HIF-1α protein expression, which further enhances VEGF expression. Therefore, the levels of pH and ROS are linked to angiogenesis through VEGF production.

Macroscopic tumor growth model

For model simplicity, the tumor is assumed to be a sphere with radius $r = R(t)$ evolving in time (see Fig. 2), which is embedded in a larger sphere with a fixed radius $r = L$, whose boundary lies in a normal healthy tissue. The proliferation of tumor cells generates an internal pressure and, as a result, a velocity field with radial velocity $v(r,t)$ outward from the center. We assume that all cells and molecules are moving with this velocity; the velocity is zero in the normal tissue. The equations for live and dead tumor cells are defined in the moving domain $[0,R(t)]$ whereas equations for all other cells and chemicals take place in the fixed domain $[0,L]$.

Equations of the macroscopic variables are based on the framework in [23] with some changes due to intracellular reactions. The equation of live cancer cells is:

$$\frac{\partial c}{\partial t} + \frac{1}{r^2}\frac{\partial}{\partial r}(r^2 cv) = \underbrace{\lambda_c(w,C_{ROS})c\left(1 - \frac{c}{c^*}\right)}_{\text{oxygen--and ROS--dependent proliferation}} -$$

$$\underbrace{\mu_{c1}(w)c}_{\text{necrosis}} - \underbrace{\mu_{c2}c}_{\text{apoptosis}}, \quad 0 < r < R(t). \tag{1}$$

Here, the proliferation rate $\lambda_c(w,C_{ROS})$ depends on the oxygen level w and intracellular ROS concentration, C_{ROS}; we assume that $\lambda_c(w,C_{ROS})$ has the form

$$\lambda_c(w,C_{ROS}) = \lambda_c \xi_1(w)\xi_2(C_{ROS}).$$

We also assume that the necrosis rate $\mu_{c1}(w)$ is only oxygen-dependent and the apoptosis rate μ_{c2} is a constant. The forms of the functions of $\xi_1(w)$, $\xi_2(C_{ROS})$, and $\mu_{c1}(w)$ should have the profiles shown in Fig. 3, but for numerical simulation we approximate them by piecewise linear functions.

Table 1. Variables and units of the model.

$c(r,t)$	live tumor cell density (cell/cm^3)
$b(r,t)$	dead tumor cell density (cell/cm^3)
$m(r,t)$	macrophage density (cell/cm^3)
$e(r,t)$	endothelial cell density (cell/cm^3)
$q(r,t)$	M-CSF concentration (g/cm^3)
$p(r,t)$	MCP-1/CCL2 concentration (g/cm^3)
$h(r,t)$	VEGF concentration (g/cm^3)
$s(r,t)$	sVEGFR-1 concentration (g/cm^3)
$w(r,t)$	Oxygen concentration (g/cm^3)
$H(r,t)$	Concentration of H$^+$ (μM)
$C_{ROS}(r,t)$	Concentration of ROS (mostly H$_2$O$_2$) (μM)
$C_{GSH}(r,t)$	Concentration of GSH (μM)
$C_{GPx_r}(r,t)$	Concentration of GPx$_r$ (μM)
$C_{GPx_o}(r,t)$	Concentration of GPx$_o$ (μM)

The functions $\xi_1(w)$ and $\mu_{c1}(w)$ are taken the same as in [23],

$$
\xi_1(w) = \begin{cases} 0 & \text{if } w < w_h, \\[2mm] (w - w_h)/(w_0 - w_h) & \text{if } w_h \leq w \leq w_0, \\[2mm] 1 & \text{if } w > w_0, \end{cases}
$$

$$
\mu_{c1}(w) = \begin{cases} \mu_{c1} & \text{if } w < w_n, \\[2mm] \mu_{c1}(w_h - w)/(w_h - w_n) & \text{if } w_n \leq w \leq w_h, \\[2mm] 0 & \text{if } w > w_h, \end{cases}
$$

where w_n, w_h and w_0 represent thresholds of necroxia, hypoxia and normoxia, respectively.

In the experiments in ovarian cancer in [24], tumor volume was almost doubled when the intracellular ROS level was elevated by 70%, so we estimate by $\dfrac{2-1}{1.7-1} \approx 1.5$, the fold at which ROS level increases proliferation of cancer cells. Accordingly, we take the function $\xi_2(C_{ROS})$ as follows:

$$
\xi_2(C_{ROS}) = \begin{cases} 1 & \text{if } 0 < C_{ROS} < C^0_{ROS} \\[3mm] 1 + \dfrac{0.5}{V_{toxic} - C^0_{ROS}}(C_{ROS} - C^0_{ROS}) & \text{if } C^0_{ROS} \leq C_{ROS} < V_{toxic} \\[3mm] \dfrac{-1.499}{0.2 V_{toxic}}(C_{ROS} - V_{toxic}) + 1.5 & \text{if } V_{toxic} < C_{ROS} < 1.2 V_{toxic}, \\[3mm] 0.001 & \text{if } C_{ROS} > 1.2 V_{toxic}, \end{cases}
$$

where C^0_{ROS} is the typical ROS concentration in cancer cells. The VEGF density satisfies the equation

$$
\frac{\partial h}{\partial t} + \frac{1}{r^2}\frac{\partial}{\partial r}(r^2 h v) = \underbrace{D_h \frac{1}{r^2}\frac{\partial}{\partial r}\left(r^2 \frac{\partial h}{\partial r}\right)}_{\text{diffusion}} + \underbrace{\lambda_h(w, C_{ROS}, H) c \chi_{\{r \leq R(t)\}}}_{\text{produced by cancer cells}}
$$

$$
+ \underbrace{\theta_1 \bar{\lambda}_h(w, H)\frac{q}{q + q_0} m}_{\text{produced by macrophages}} - \underbrace{\bar{\mu}_s s h}_{\text{sVEGFR}-1 \text{ inhibition}} - \underbrace{\mu_h h}_{\text{degradation}} ; \ 0 < r < L.
$$

(2)

In this equation, the diffusion coefficient D_h, binding rate $\bar{\mu}_s$ to sVEGFR-1, and degradation rate μ_h are assumed to be constant. The parameter θ_1 is set as one for normal macrophages but zero for HIF-1α-deficient macrophages [23]. Here, we assume that the VEGF production rate has the form: $\lambda_h(w, C_{ROS}, H) = \lambda_h \xi_3(w) \xi_4(C_{ROS}) \xi_5(H)$. We take

$$
\xi_3(w) = \begin{cases} 0 & \text{if } w < w_n, \\[2mm] (w - w_n)/(w^* - w_n) & \text{if } w_n \leq w < w^* \\[2mm] 1 - 0.7(w - w^*)/(w_0 - w^*) & \text{if } w^* < w \leq w_0, \\[2mm] 0.3 & \text{if } w > w_0, \end{cases}
$$

where $w^* \in (w_h, w_0)$ represents the threshold at which the hypoxic effect is maximal for VEGF production [23].

Up to five-fold increase of the maximum HIF-1α expression was suggested by a cancer model in [25] when the ROS level was elevated. Hence we take the function $\xi_4(C_{ROS})$ to be similar to $\xi_2(C_{ROS})$:

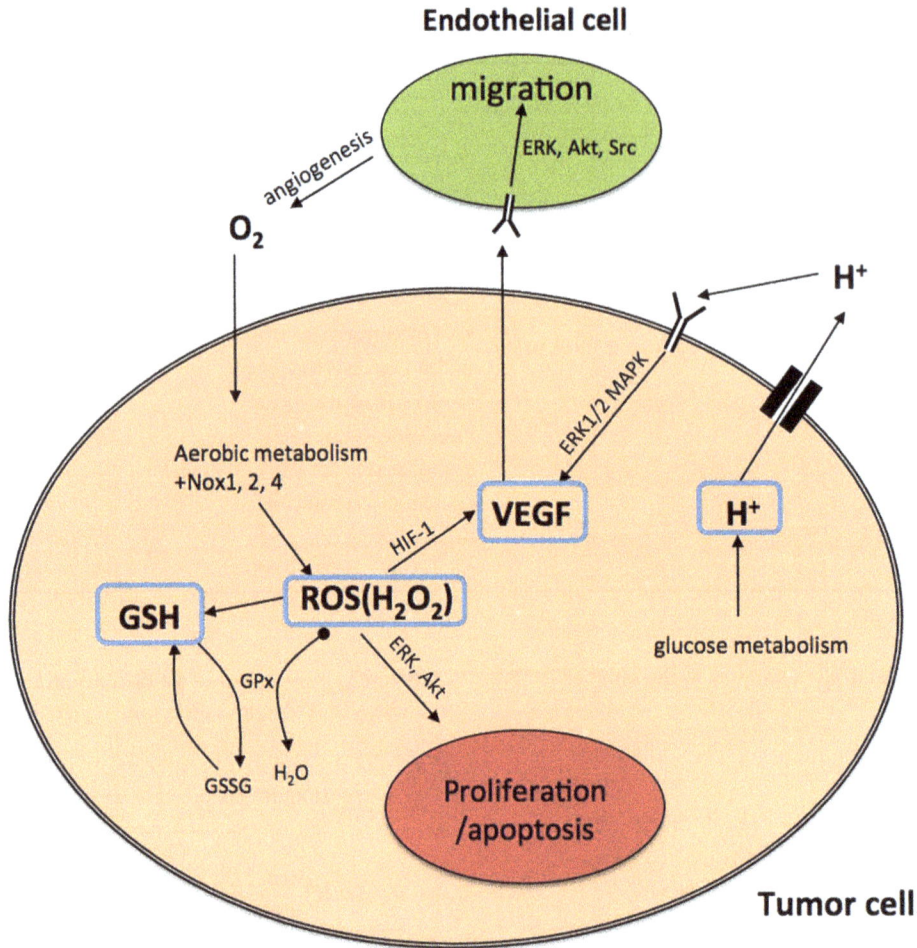

Figure 1. Schematic diagram of the roles of ROS, GSH, and hydrogen ions in cancer cell growth and tumor angiogenesis. (i) ROS is a major by-product of aerobic metabolism and plays a dual role in cancer cell life-cycle: below a certain threshold, increasing amounts of ROS promotes cell proliferation through pathways of extracellular-signal-regulated kinases (ERKs) and cell survival factors such as Akt. However, ROS leads to cell apoptosis when its concentration is over the toxic threshold. Additionally, ROS may play a function in up-regulating HIF-1 expression, which in turn results in increasing the production of angiogenesis factor VEGF. (ii) GSH (glutathione) is the most abundant antioxidant produced by cancer cells to protect themselves from oxidative stress; it can remove ROS (mostly H_2O_2) with the help of enzyme GP_x. (iii) Large amount of hydrogen ions are produced as a consequence of glucose metabolism, and are pumped out by abnormally expressed proton transporters. There is evidence indicating that acidic extracellular environment induces VEGF production through the ERK/MAPK signaling pathway.

$$\xi_4(C_{ROS}) =$$

$$\begin{cases} 1 & \text{if } 0 < C_{ROS} < C_{ROS}^0 \\[2mm] 1 + \dfrac{4}{V_{toxic} - C_{ROS}^0}(C_{ROS} - C_{ROS}^0) & \text{if } C_{ROS}^0 \leq C_{ROS} < V_{toxic} \\[2mm] \dfrac{-4.999}{0.2\,V_{toxic}}(C_{ROS} - V_{toxic}) + 5 & \text{if } V_{toxic} < w < 1.2\,V_{toxic}, \\[2mm] 0.001 & \text{if } C_{ROS} > 1.2\,V_{toxic}. \end{cases}$$

The VEGF promoter activity at extracellular pH = 6.6 is three-fold higher than that at pH = 7.3 [7]. In [26], the pH values are 7 and 6.55 for mammary gland and non-treated MET-1 tumor, which correspond to 1×10^{-7}M and 3×10^{-7}M of hydrogen ion concentrations, respectively. Accordingly we set $H_{low} = 0.1\mu$M, $H_{high} = 0.3\mu$M, and take

$$\xi_5(H) = \begin{cases} 1 & \text{if } H < H_{low}, \\[2mm] \dfrac{2}{H_{high} - H_{low}}(H - H_{low}) + 1 & \text{if } H_{low} \leq H \leq H_{high}, \\[2mm] 3 & \text{if } H > H_{high}. \end{cases}$$

Finally, for simplicity, we set $\bar{\lambda}_h(w,H) = \bar{\lambda}_h \xi_3(w)\xi_5(H)$ [23]. The equation for the concentration of hydrogen ions is given by

$$\frac{\partial H}{\partial t} + \frac{1}{r^2}\frac{\partial}{\partial r}(r^2 H\upsilon) = \underbrace{D_H \frac{1}{r^2}\frac{\partial}{\partial r}\left(r^2 \frac{\partial H}{\partial r}\right)}_{\text{diffusion}} \tag{3}$$

$$+ \underbrace{\lambda_H c \chi_{\{r \leq R(t)\}}}_{\text{produced by cancer cells}} - \underbrace{\mu_H H}_{\text{evacuation}},$$

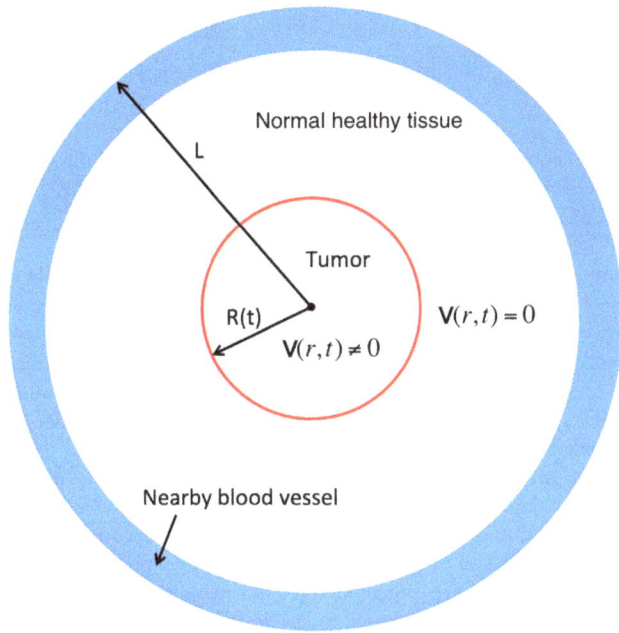

Figure 2. Macroscopic tumor growth: tumor is assumed to be in radially symmetric with radius $R(t)$ evolving in time t; the moving boundary is indicated in red. A healthy normal tissue surrounds the tumor, and the entire simulation domain is a sphere with a fixed radius L. Initially, blood vessel (in blue) is placed in the healthy normal tissue away from the tumor region. Due to abnormal proliferation of cancer cells, there is a radial velocity, $v(r,t)$, within the tumor region, but $v=0$ in the healthy normal tissue.

where $\chi_{\{r \leq R(t)\}} = 1$ if $r \leq R(t)$ and $\chi_{\{r \leq R(t)\}} = 0$ if $r > R(t)$, and where D_H, λ_H and μ_H are the diffusion coefficient, production and evacuation rate of hydrogen ions, respectively; for simplicity, they are assumed to be constant.

The equations for oxygen level w and density of macrophages m appearing in Eqs. (1) – (2), and of the other variables in Table 1 are the same as in [23], except for additional terms involving C_{ROS} and H.

Collecting the equations for all the macroscopic variables listed in Table 1, we have the following system:

$$\frac{\partial c}{\partial t} + \frac{1}{r^2}\frac{\partial}{\partial r}(r^2 cv) = \lambda_c(w, C_{ROS})c\left(1 - \frac{c}{c^*}\right)$$
$$- \mu_{c1}(w)c - \mu_{c2}c, \quad 0 < r < R(t); \tag{4}$$

$$\frac{\partial b}{\partial t} + \frac{1}{r^2}\frac{\partial}{\partial r}(r^2 bv) = \mu_{c1}(w)c + \mu_{c2}c - \mu_b\frac{w}{w_0}mb, \quad 0 < r < R(t); \tag{5}$$

$$\frac{\partial m}{\partial t} + \frac{1}{r^2}\frac{\partial}{\partial r}(r^2 mv) = -\frac{1}{r^2}\frac{\partial}{\partial r}\left(r^2 k_p m \frac{\partial p}{\partial r}\right); \quad 0 < r < L; \tag{6}$$

$$\frac{\partial e}{\partial t} + \frac{1}{r^2}\frac{\partial}{\partial r}(r^2 ev) = -\frac{1}{r^2}\frac{\partial}{\partial r}\left(r^2 k_h e \frac{\partial h}{\partial r}\right), \quad 0 < r < L; \tag{7}$$

$$\frac{\partial q}{\partial t} + \frac{1}{r^2}\frac{\partial}{\partial r}(r^2 qv) = D_q\frac{1}{r^2}\frac{\partial}{\partial r}\left(r^2\frac{\partial q}{\partial r}\right)$$
$$+ \lambda_q c\chi_{\{r \leq R(t)\}} - \mu_q q, \quad 0 < r < L; \tag{8}$$

$$\frac{\partial p}{\partial t} + \frac{1}{r^2}\frac{\partial}{\partial r}(r^2 pv) = D_p\frac{1}{r^2}\frac{\partial}{\partial r}\left(r^2\frac{\partial p}{\partial r}\right)$$
$$+ \lambda_p(w)\frac{q}{q+q_0}m - \mu_p p, \quad 0 < r < L; \tag{9}$$

$$\frac{\partial h}{\partial t} + \frac{1}{r^2}\frac{\partial}{\partial r}(r^2 hv) = D_h\frac{1}{r^2}\frac{\partial}{\partial r}\left(r^2\frac{\partial h}{\partial r}\right)$$
$$+ \lambda_h(w, C_{ROS}, H)c\chi_{\{r \leq R(t)\}} + \theta_1\bar{\lambda}_h(w, H)\frac{q}{q+q_0}m$$
$$- \bar{\mu}_s sh - \mu_h h; \quad 0 < r < L; \tag{10}$$

$$\frac{\partial s}{\partial t} + \frac{1}{r^2}\frac{\partial}{\partial r}(r^2 sv) = D_s\frac{1}{r^2}\frac{\partial}{\partial r}\left(r^2\frac{\partial s}{\partial r}\right)$$
$$+ \theta_2\lambda_s vm - \bar{\mu}_h sh - \mu_s s, \quad 0 < r < L; \tag{11}$$

$$\frac{\partial w}{\partial t} + \frac{1}{r^2}\frac{\partial}{\partial r}(r^2 wv) = D_w\frac{1}{r^2}\frac{\partial}{\partial r}\left(r^2\frac{\partial w}{\partial r}\right)$$
$$+ \lambda_e e - \lambda_m mw - \lambda_w cw, \quad 0 < r < L; \tag{12}$$

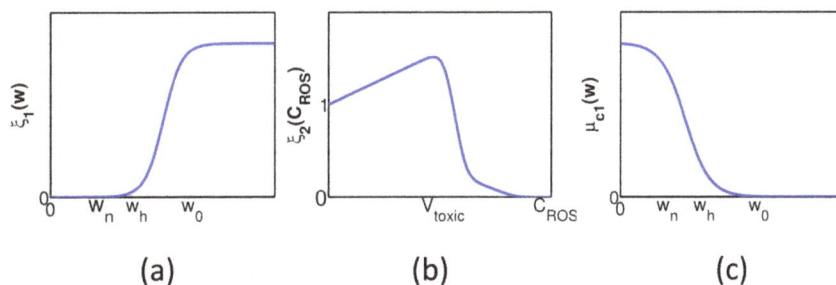

Figure 3. Profiles of the functions ξ_1, ξ_2, and μ_{c1}. Thresholds of necroxia, hypoxia and normoxia are marked as w_n, w_h and w_0, respectively.

$$\frac{\partial H}{\partial t} + \frac{1}{r^2}\frac{\partial}{\partial r}(r^2 H v) = D_H \frac{1}{r^2}\frac{\partial}{\partial r}\left(r^2 \frac{\partial H}{\partial r}\right)$$

$$+ \lambda_H c \chi_{\{r \le R(t)\}} - \mu_H H, \quad 0 < r < L. \tag{13}$$

The radial velocity field $v(r,t)$ and the moving boundary $R(t)$ of the tumor are given by

$$v(r,t) = \frac{1}{\theta c^* r^2}\int_0^r \zeta^2 \left[\lambda_c(w, C_{ROS})c\left(1 - \frac{c}{c^*}\right) - \mu_b \frac{w}{w_0} mb\right]d\zeta$$

$$- \frac{1}{\theta c^*}\left(k_p m \frac{\partial p}{\partial \zeta} + k_h e \frac{\partial h}{\partial \zeta}\right)|_{\zeta = r}, \tag{14}$$

where θ is the fraction of the volume occupied by cells, and

$$\dot{R}(t) = v(R(t),t). \tag{15}$$

Detailed discussions about the corresponding terms and coefficients of Eqs. (5) – (9), (11) – (12) and derivations of (14) – (15) can be found in [23], but for convenience, all the parameters are listed in Table 2.

Intracellular chemical dynamics

As mentioned earlier, H_2O_2 is the major source of ROS. Recent experimental data indicate that an increase of H_2O_2 can explain many hallmarks of cancer, such as cell proliferation, apoptosis resistance, increased angiogenesis, and metastasis [27]. We assume that ROS concentration (which is primarily H_2O_2) is mainly regulated by GSH, although there exist other reducing agents [1]. Removal of H_2O_2 by GSH is associated with a key enzyme, glutathione peroxidase (GPx). In fact, the reactions of intracellular H_2O_2, GPx and GSH are [28]:

$$GPx_r + H_2O_2 + H^+ \xrightarrow{\tilde{k}} GPx_0 + H_2O \tag{16}$$

$$GPx_0 + 2GSH \xrightarrow{\bar{k}} GPx_r + GSSG + H_2O + H^+, \tag{17}$$

where $\tilde{k}$ and $\bar{k}$ are reaction constants, and GPx_r and GPx_0 are the reduced and oxidative forms of GPx, respectively.

Based on (16) and (17), the dynamical system for C_{ROS}, C_{GSH}, C_{GPx_r} and C_{GPx_0} are modeled as the follows:

$$\frac{dC_{ROS}}{dt} = \left(\frac{k_{ROS}w}{w + w_0} - \tilde{k}C_{ROS}C_{GPx_r}\right)\frac{c}{c_0}; \tag{18}$$

$$\frac{dC_{GSH}}{dt} =$$
$$\left[k_{GSH}\left(1 + M_0\frac{\max(C_{ROS} - C_{ROS}^0, 0)}{C_{ROS} + C_{ROS}^0}\right) - \bar{k}C_{GSH}^2 C_{GPx_0} - k_d C_{GSH}\right]\frac{c}{c_0}; \tag{19}$$

$$\frac{dC_{GPx_r}}{dt} = \left(\bar{k}C_{GPx_0}C_{GSH}^2 - \tilde{k}C_{ROS}C_{GPx_r}\right)\frac{c}{c_0}; \tag{20}$$

$$\frac{dC_{GPx_0}}{dt} = \left(\tilde{k}C_{GPx_r}C_{ROS} - \bar{k}C_{GSH}^2 C_{GPx_0}\right)\frac{c}{c_0}. \tag{21}$$

The equations of C_{ROS}, C_{GSH} and concentrations of the other associated molecules take place inside cancer cells. In Eq. (18), ROS is produced at the oxygen level-dependent rate $\frac{k_{ROS}w}{w + w_0}$ and is removed by GPx_r as indicated by Eq. (16). For Eq. (19), GSH is generated inside cancer cells, at a constant rate k_{GSH}. When cells are under oxidative stress (ROS concentration is above normal level C_{ROS}^0, or $C_{ROS} > C_{ROS}^0$), they acquire adaptive mechanisms to counteract the toxicity of increased ROS level by upregulating GSH synthesis. Hence, a C_{ROS}-dependent GSH production is included with $M_0 > 0$. A phenomenological value $M_0 = 2$ is taken in this model since there is no experimental evidence, to the authors' knowledge, about how much GSH production is enhanced due to oxidative stress. Further, GSH is consumed by GPx_0 as indicated by Eq. (17), and degrades at a constant rate k_d. Eqs. (20) and (21) are directly derived from reactions (16) and (17). Since these reactions take place inside cancer cells, in order to couple the ODE system to the macroscopic tumor growth model, all the right-hand sides of Eqs. (18) – (21) are multiplied by $\frac{c}{c_0}$, where c_0 is the reference density of cancer cells.

Initial and boundary conditions

Initial and boundary conditions corresponding to Eqs. (4) – (12) follow those in [23]:

$$c(r,0) = c_0, \quad b(r,0) = 0, \quad r \in [0, R_0], \tag{22}$$

$$m(r,0) = m_0, \quad e(r,0) = \frac{e_0}{1 + e^{(5R_0 - r)/\varepsilon}}, \quad r \in [0, L], \tag{23}$$

$$q(r,0) = p(r,0) = s(r,0) = h(r,0) = g(r,0) = 0, \quad r \in [0, L], \tag{24}$$

$$w(r,0) = w^* + \frac{r^2}{L^2}(w_0 - w^*), \quad r \in [0, L]. \tag{25}$$

At $r = 0$, we have zero flux boundary conditions for w,q,p,s,h,g, and at $r = L$, we impose zero flux boundary conditions for q,p,s,h,g, except for w, and $w(L,t) = w_0$.

Next, for H, we take the initial condition $H(r,0) = H_0$, and the boundary condition $\frac{\partial H}{\partial r} = 0$ at $r = 0$ and $r = L$. For Eqs. (18) – (21), initial conditions are taken as

$$C_{ROS}(0) = C_{ROS}^0, \quad C_{GSH}(0) = C_{GSH}^0,$$
$$C_{GPx_r}(0) = C_{GPx_r}^0, \quad C_{GPx_0}(0) = C_{GPx_0}^0. \tag{26}$$

Table 2. Values and reference of parameters in the macroscopic equations (4) – (15).

Parameter	Dimensional	Reference
μ_{c1}	9.63×10^{-6} s^{-1}	[57,58] and estimated
μ_{c2}	4.80×10^{-6} s^{-1}	[59]
μ_b	4.80×10^{-14} $cm^3cell^{-1}s^{-1}$	[59] and estimated
μ_q	4.80×10^{-5} s^{-1}	[60]
μ_p	2.00×10^{-5} s^{-1}	[61]
μ_s	1.98×10^{-5} s^{-1}	[62]
$\bar{\mu}_s$	1.19×10^{5} $cm^3g^{-1}s^{-1}$	[63] and estinated
μ_h	1.26×10^{-4} s^{-1}	[64]
$\bar{\mu}_h$	3.57×10^{5} $cm^3g^{-1}s^{-1}$	[63] and estimated
μ_H	1.1×10^{-4} s^{-1}	[29]
λ_c	1.6×10^{-5} s^{-1}	[57,58]
λ_q	3.20×10^{-22} $gs^{-1}cell^{-1}$	[65,66] and estimated
λ_p	1.92×10^{-20} $gs^{-1}cell^{-1}$	[67] and estimated
λ_h	1.51×10^{-21} $gs^{-1}cell^{-1}$	[57,61,68,69] and estimated
$\bar{\lambda}_h$	1.83×10^{-20} $gs^{-1}cell^{-1}$	[70,71] and estimated
λ_s	1.86×10^{-20} $gs^{-1}cell^{-1}$	[72] and estimated
λ_2	2.22×10^{-14} $gs^{-1}cell^{-1}$	estimated
λ_m	1.6×10^{-13} $cm^3s^{-1}cell^{-1}$	[73] and estimated
λ_w	2×10^{-13} $cm^3s^{-1}cell^{-1}$	[74,75] and estimated
λ_H	2.2×10^{-11} $\mu M \cdot cm^3 s^{-1} cell^{-1}$	[29]
k_p	6.00 $cm^5g^{-1}s^{-1}$	[57]
k_h	24.00 $cm^5g^{-1}s^{-1}$	[76]
D_p, D_q	2.00×10^{-6} cm^2s^{-1}	[57,61,77] and estimated
D_h, D_s	1.00×10^{-6} cm^2s^{-1}	[76,78] and estimated
D_w	2.00×10^{-5} cm^2s^{-1}	[79] and estimated
D_H	8.0×10^{-5} cm^2s^{-1}	[30,31]
w_0	4.65×10^{-4} gcm^{-3}	[80]
w_n	3.57×10^{-5} gcm^{-3}	[80] and estimated
w^*	1.69×10^{-4} gcm^{-3}	estimated
w_h	1.00×10^{-4} gcm^{-3}	[80] and estimated
θ_1	0 for HIF-1α KO, otherwise 1	estimated
θ_2	0 for HIF-2α KO, otherwise 1	estimated
θ	0.9	[23], estimated
c^*	1.00×10^{9} $cellcm^{-3}$	[81]
c_0	7.20×10^{8} $cellcm^{-3}$	[81,82]
m_0	2.00×10^{8} $cellcm^{-3}$	[81]
e_0	2.50×10^{6} $cellcm^{-3}$	estimated
q_0	$1.00 \times 10^{-9} gcm^{-3}$	Scaling factor
H_0	1.0×10^{-1} μM	[26]

Parameters

Parameters in tumor growth

Values of parameters c_0, m_0, e_0, w_0, w^*, and R_0 are taken the same as in [23] and are listed in Table 2. Additionally, in [29], tumor interstitial pH profiles in normal and neoplastic tissue were measured *in vivo* by a fluorescence ratio imaging microscopy technique. Based on the experimental data, it was concluded in [29] that the production rate of H^+ ranges from 6.17×10^{-12} to 1.78×10^{-10} $\mu M \cdot cm^3 /(s \cdot cell)$, and evacuation rate ranges from 4.47×10^{-5} to $4.60 \times 10^{-4}/s$. We take the geometric means of these observations and set $\lambda_H = 2.20 \times 10^{-11} \mu M \cdot cm^3/(s \cdot cell)$ and

$\mu_H = 1.09 \times 10^{-4}/\text{s}$. The diffusion coefficient of protons is generally much larger than those of ions and chemicals; we take $D_H = 8.0 \times 10^{-5}\text{cm}^2\text{s}^{-1}$, as in [30,31]. Neutral environment is assumed initially, so $H_0 = 1 \times 10^{-7}\text{M} = 0.1\mu\text{M}$ [26].

Parameters in the intracellular dynamics

It was reported in [32] that for seven adherent human tumor cell lines, including colon and breast cancers, the production of hydrogen peroxide is in the range from 0.1 to 1.4 nmol/10^4cells/hour. In a more recent work [33], the superoxide production was measured as 3.71 pmol/2×10^4cells/min for mouse colon carcinoma and 1.21 pmol/2×10^4cells/min for liver hepatoma. We follow the result in [33] and take the ROS production to be in the range of 1×10^{-6} to 3.8×10^{-6}pmol/s for a single tumor cell, after unit conversion. Assuming the typical volume of a cell to be 10^{-9}cm^3, we derive the intracellular ROS production rate k_{ROS} to be

$$k_{ROS} = 1 \text{ to } 3.8 \times 10^{-6} \times \frac{10^{-12}\text{mol}}{10^{-9}\text{cm}^3/\text{s}}$$

$$= 1 \text{ to } 3.8 \times \frac{10^{-18}\text{mol}}{10^{-12}\text{L/s}} = 1 \text{ to } 3.8\mu\text{M/s}.$$

In studies of rat liver mitochondria [28,34], it was reported that the reaction constants of Eqs. (16) – (17) are $2.1 \times 10^7\text{M}^{-1}\text{s}^{-1}$ and $4 \times 10^4\text{M}^{-1}\text{s}^{-1}$, respectively. Thus, after unit conversion we have $\tilde{k} = 21\mu\text{M}^{-1}\text{s}^{-1}$ and $\bar{k} = 4 \times 10^{-2}\mu\text{M}^{-1}\text{s}^{-1}$.

Glutathione synthesis in red blood cells has been measured in [35] and the production rates are 0.5 mmol/L/day and 1.6 mmol/L/day for young and elderly people, respectively. So we take the constant k_{GSH} to be

$$0.5 \text{ to } 1.6 \times \frac{\text{mmol/L}}{\text{Day}} = 0.5 \text{ to } 1.6 \times \frac{10^3\mu\text{M}}{24 \times 3600s}$$

$$= 5.78 \text{ to } 18.5 \times 10^{-3}\mu\text{M/s}.$$

Finally, the degradation rate k_d of GSH is in the range from 3.2 to $9.6 \times 10^{-5}\text{s}^{-1}$ since the half-life of the GSH is between 2 to 6 hours [36,37]. All the parameter values of Eq. (18) – (21) are summarized in Table 3.

Initial conditions

In [28,38], concentrations of H_2O_2 in rat liver cells were found to range from 10^{-9}M to 10^{-7}M, and a base value of $0.2\mu\text{M}$ in tumor cells is estimated from the experiments in [25]. Thus, we take $C_{ROS}^0 = 1.0 \times 10^{-7}\text{M} = 1.0 \times 10^{-1}\mu\text{M}$; this value is also used as the initial condition of C_{ROS}. An upper limit of 700 nM for intracellular levels of H_2O_2 in Jurkat T-cells was suggested in [39,40], beyond which apoptosis was introduced; hence we take the toxicity threshold of ROS in our model to be $V_{toxic} = 700\text{nM} = 700 \times 10^{-3}\mu\text{M} = 7.0 \times 10^{-1}\mu\text{M}$. In the experiments of [26], the average intracellular GSH concentration in mammary gland was 3.3 mM, while in tumor it was 10.7 mM; accordingly we take the initial condition of C_{GSH} to be $C_{GSH}^0 = 3.5 \times 10^3\mu\text{M}$. Cellular concentration of GPx varies from $0.2\mu\text{M}$ to $6.7\mu\text{M}$ in red blood cells and in other cells [34,41], and over 99% of it is in reduced form [41], so we take the

corresponding initial conditions to be $C_{GPx_r}^0 = 1.0\mu\text{M}$ and $C_{GPx_0}^0 = 1.0 \times 10^{-2}\mu\text{M}$.

Results and Discussion

In this section we present model simulations and compare our results with experimental data. All the simulations were carried out with MATLAB (version R2011a Mathworks). The PDEs of parabolic type were numerically solved using package pdepe (MATLAB function for initial-boundary value problems for parabolic-elliptic PDEs in 1D), and the equations of hyperbolic type were solved by the Semi-Lagrangian scheme. The intracellular dynamics were solved by the ODE solver ode15s.

Experimental details

Mice. 6–8 week old C57Bl/6 female mice expressing lysozyme M from the promoter of cre recombinase (LysMcre) were used as wild type control mice. 6–8 week old C57Bl/6 female LysMcre mice also containing homozygous loxP restriction sites surrounding the HIF-1α (LysMcre/HIF-1α$^{fl/fl}$) or HIF-2α (LysMcre/HIF-2α$^{fl/fl}$) genes were used as the experimental groups lacking either HIF-1α or HIF-2α in the myeloid cells.

Tumor model. Met-1 tumor cells isolated from the stage IV tumors of C57Bl/6 PyMT transgenic mice were cultured to 80% confluence then trypsinized, washed, resuspended in RPMI-1640 at 1×10^6 cells per 100 μls and orthotopically implanted into the number four mammary gland of 6–8 week old C57Bl/6 female LysMcre, LysMcre/HIF-1α$^{fl/fl}$, or LysMcre/HIF-2α$^{fl/fl}$ mice. The tumors became palpable approximately 1 week after implantation. Tumor measures were performed 3× per week using calipers and tumor volumes were calculated using the formula volume $= 0.5 \times$ [(large diameter) × (small diameter)^2].

Treatment. Upon tumor palpation, the mice were treated intraperitoneally with 100 μL isotonic saline or Docetaxel (NDC 0409-0201-02, Hospira) 30 mg/kg body weight in 100 μL one time per week. All protocols were approved by The Ohio State University Animal Care and Use Committee, and mice were treated in accordance with institutional guidelines for animal care.

HIF-1α-regulated tumor microenvironment change

Figure 4 shows the comparison between experiments and simulations for tumor volume (in unit of cm^3) changing with time (days). Fig. 4 (a) lists the experimental data in colored columns with error bars. Unless otherwise specified, the red, blue, and green colors represent tumors with wild-type, HIF-1α-, and HIF-2α-deficient macrophages (WT, HIF-1α KO, and HIF-2α KO), respectively. For each type in the longitudinal data, fifteen tumor volumes were measured on each day. The statistical mean of these tumor volumes are calculated and plotted as the heights of the columns, with the error bars as standard deviations. We see that tumors with HIF-1α KO macrophages have volumes as low as one half of those with WT macrophages. By contrast, tumor growth is not inhibited if HIF-2α in macrophages is knocked out. This agrees with our earlier work about the opposing roles of HIF-1α and HIF-2α in mediating tumor angiogenesis [23,42].

In Figs. 4(b) – (d) we compare the model simulations of tumor volume with experiments. In these figures, experimental data are the same as in Fig. 4(a), and are displayed as dots with error bars. For comparison, simulations are plotted in the corresponding colored dash curves. Based on the parameter sensitivity analysis in [23], the parameters $\lambda_c, \lambda_h, \lambda_m$ and λ_w are adjusted to obtain the curves in Fig. 4(b) to fit the experiments; specifically, $\lambda_c = 1.6 \times 10^{-5}\text{s}^{-1}$, $\lambda_h = 1.5 \times 10^{-21}\text{gs}^{-1}\text{cell}^{-1}$,

$\lambda_m = 1.6 \times 10^{-13} \text{cm}^3\text{s}^{-1}\text{cell}^{-1}$, and $\lambda_w = 2.0 \times 10^{-13}\text{cm}^3\text{s}^{-1}$ cell^{-1}. Fixing these parameters but setting $\theta_1 = 0$ or $\theta_2 = 0$ in Eqs. (10) − (11), we obtained the simulations for tumor growth with HIF-1α or HIF-2α-deficient macrophages and displayed them in Figs. 4(c) − (d), respectively. The agreement of the numerical simulations with experiments is fairly good. The R squared score [43,44] is used to quantify the goodness of fit; the values for the cases of WT, HIF-1α KO, and HIF-2α KO are 0.9630, 0.9184, and 0.8917, respectively. Based on the comparison, we proceed to use the model with the same parameters to calculate other quantities in the tumor microenvironment.

Intracellular GSH concentration in normal tissues, non-treated tumors, and GM-CSF treated tumors were explored in [26]. It was concluded that GSH concentration in cancer cells is significantly higher compared with that in normal tissues, and it is lowered when tumor growth is suppressed by GM-CSF treatment. Therefore, based on the previous conclusion that HIF-1α KO inhibits tumor growth, we hypothesized that the GSH concentration in tumors with HIF-1α KO macrophages is lower than that in tumors with WT or HIF-2α KO macrophages. This hypothesis was verified by both experiments and simulations. Figure 5 (a) displays the experiments of GSH concentration (in unit of Molar) against time (days). Similarly, the column heights represent the mean values of the GSH concentration in tumors and the error bars are standard deviations. Note that only one of the quantities (GSH, oxygen and pH) can be measured on each tumor, so that the total number of data point is five per day. From the figure we can see that GSH concentration in tumors with HIF-1α KO macrophages (blue bars) is significantly lower, whereas tumors with WT (red) and HIF-2α KO macrophages (green) have similar and higher level of GSH concentration in general, except for the measurement on the last day.

Figs. 5(b) − (d) show the simulations corresponding to the three groups of experiments in Fig. 5(a). Since the total sample size is relatively small, the R squared is not calculated. In these simulations, initial average GSH concentration is 0.0125 M. In tumors with WT macrophages, there is no significant change in GSH concentration and after 30 days it is 0.0118 M. A similar pattern is observed in the tumor with HIF-2α KO macrophages. By contrast, the GSH concentration in tumors with HIF-1α KO macrophages eventually decays to 0.0048 M in a linear fashion

over the same period of time. In Figs. 5(c) − (d), we notice that the model did not reproduce the sudden increases of GSH concentration occurring between day 20 and 27 as indicated in the experiments. This suggests that there is an additional latent mechanism for the GSH concentration growth.

Tumors usually have a more acidic environment (a lower pH_e) than normal tissue and the pH_e is elevated in the GM-CSF treated tumors [26]. Accordingly, we hypothesize that acidosis will be relieved in tumors with HIF-1α KO macrophages, although there could be other factors contributing to the pH when tumor microenvironment is altered. Figure 6 (a) shows the experimental results regarding the level of pH_e: the level is 6.8 in tumors with HIF-1α KO macrophages, compared with of 6.6 in tumors with WT macrophages. Surprisingly, as indicated in the figure, the pH_e in tumors with HIF-2α KO macrophages is also raised up to a similar level as in tumors with HIF-1α KO macrophages.

Part of these features are captured in the model simulations: the simulated pH_e in tumors with WT macrophages is generally below 6.8 (Fig. 6 (b)) and it is elevated above this number in tumors with HIF-1α KO macrophages (Fig. 6 (c)). However, the simulations underestimate the pH_e of tumors with HIF-2α KO macrophages (or over-estimate the H^+ concentration), as seen in Fig. 6(d). The reason could be that we have only taken into account the impact of HIFs on cancer cells while other cells could also contribute to the concentration of hydrogen ions. It is interesting to notice that, in Fig. 6(a), the experimental data of the pH_e level for tumors with the three types of macrophages, all peak on day 13. This feature is also observed in our corresponding simulations in Figure 6 (b) − (d), although the peak values shift to around day 10.

Figure 7 displays the experiments and model simulations of oxygen tension (in units of mmHg). The experimental data of averaged oxygen level taken at several time points are shown in Fig. 7 (a). Since there are relatively large variations among the individual mice, it is difficult to draw conclusions about the impact of HIF-1α or HIF-2α KO on oxygen tension that is independent of the tumor volume. We therefore proceed from another perspective, to represent the experimental data for the individual mice instead of taking the average. In Fig. 7 (b) the oxygen level is plotted against tumor volume. For better comparison, weighted nonlinear squares fitting was applied (with the reciprocal of experimental variance as weights) to obtain the colored curves fitting to the corresponding dots for each group. Fig. 7(b) suggests

Table 3. Values and reference of parameters in the intracellular dynamics of Eqs. (18) − (21).

Parameter	value and unit	Reference
k_{ROS}	2.0 μMs^{-1}	[32]
k_{GSH}	2.27 μMs^{-1}	[35], estimated
k_d	$8.0 \times 10^{-5}\text{s}^{-1}$	[36,37]
$\tilde{k}$	20 $\mu\text{M}^{-1}\text{s}^{-1}$	[34]
$\bar{k}$	4×10^{-2} $\mu\text{M}^{-1}\text{s}^{-1}$	[34]
C_{ROS}^0	1.0×10^{-1} μM	[25,28,38]
C_{GSH}^0	3.5×10^3 μM	[26]
$C_{GPx_r}^0$	1.0 μM	[34,41], estimated
C_{GPx0}^0	1.5×10^{-2} μM	[34,41], estimated
V_{toxic}	7.0×10^{-1} μM	[39,40]

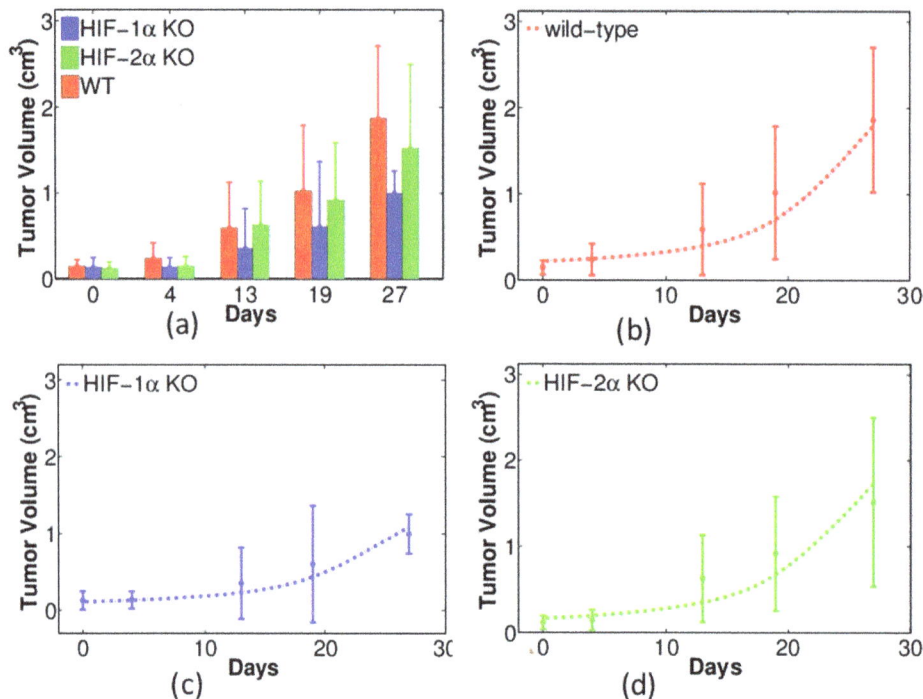

Figure 4. Experiments and simulations of tumor volume with wild-type, HIF-1α- and HIF-2α-deficient macrophages (WT, HIF-1α KO, and HIF-2α KO). Horizontal axis represents time (in days) and vertical axis scales tumor volume (in units of cm³). (a): Experimental data of tumor volumes with error bars (standard deviations). Red: WT; Blue: HIF-1α KO; Green: HIF-2α KO. (b)-(d): Comparison of experiments (dots with error bars) and numerical simulations (dash curves) for tumor volumes with WT, HIF1-α, and HIF-2α KO macrophages, respectively.

that tumors with HIF-1α KO macrophages generally have lower oxygen levels than in WT and in HIF-2α KO macrophages. By contrast, tumors with HIF-2α KO macrophages have higher oxygen levels; this is consistent with the conclusions in [42], and the model simulations in Figs. 7(c) – (d). qualitatively agree with this conclusion.

The GSH-ROS axis

Intracellular dynamics between ROS and GSH have significant impact on cell's life-cycle, signaling processes, and tumor angiogenesis. Thus, ROS-mediated mechanisms could be used to devise strategies to interfere with the life-cycle of cancer cells in order to inhibit tumor growth. ROS level can be regulated by GSH concentration. In [45], L-Buthionine (BSO) treatment was utilized in a human B lymphoma cell line to achieve intracellular GSH depletion. As a consequence, ROS level was increased and a variety of apoptotic signals of cancer cells were induced even when there were no external apoptotic stimuli. In the current work, we use our model to perform simulations on the effects of GSH depletion in tumor growth.

Figure 8 displays the results of regulating intracellular GSH concentration in tumors with WT macrophages 8(a), 8 (c) and HIF-1α-deficient macrophages 8(b), 8(d). GSH depletion is simulated by augmenting the GSH degradation coefficient k_d in Eq. (19) to different extents. The red, green, and blue curves are results with no depletion (k_d), moderate depletion ($10 \times k_d$), and severe depletion ($20 \times k_d$), respectively. Fig. 8(a) and 8(b) show the intracellular ROS concentrations in case of WT- and HIF-1α-deficient macrophages, respectively. In both cases, when k_d is increased 10 fold, the ROS levels are elevated but still remain below the assumed toxic threshold (0.7 μM), as indicated by the green curves in Figs. 8(a) and 8 (b). Consequently, the corre-

sponding tumor growth, shown by the green curves in Fig. 8(c) and 8(d) are actually promoted, because ROS at this level helps cancer proliferation. By contrast, as shown by the blue curves in the figure, when k_d is increased by 20 fold, the ROS levels are elevated above the toxic threshold, and then they damages cancer cells. As a consequence, the tumor growth is suppressed.

By carefully comparing the simulation results in Fig 8 (a) and 8 (b), we notice that the ROS level in tumors with HIF-1α-deficient macrophages is slightly less than that in tumors with WT macrophages. This seems to be contradictory to our previous simulations that with HIF-1α KO macrophages, GSH concentration in cancer cells is reduced and hence the ROS level is supposed to increase. This apparent contradiction can be explained by the assumption made in the model that ROS production is oxygen level dependent (first term of the right hand side of Eq. (18)): since there is less oxygen in tumors with HIF-1α KO macrophages, ROS production is actually reduced in cancer cells.

The therapeutic strategy of GSH depletion is to selectively raise ROS level above the toxic threshold in cancer cells; however, the model indicates that HIF-1α knockout in macrophages could reduce intracellular ROS production in tumor cells. Thus, by GSH depletion, tumor volume reduction with HIF-1α KO macrophages may be less significant than in tumors with WT macrophages. As shown in Fig. 8 (c), severe depletion of GSH reduces tumor volume from 1.875 cm³ to 1.183 cm³ on day 27, or a 37% reduction; on the other hand, in tumors with HIF-1α KO macrophages, as indicated by Fig. 8 (d), the same amount of GSH depletion reduces the tumor volume from 1.260 cm³ to 1.043 cm³, or a 17% reduction.

In the above simulations, the treatment of GSH depletion was assumed to start at the beginning of tumor growth. But we also simulated the effects of GSH depletion ($20 \times k_d$) starting at

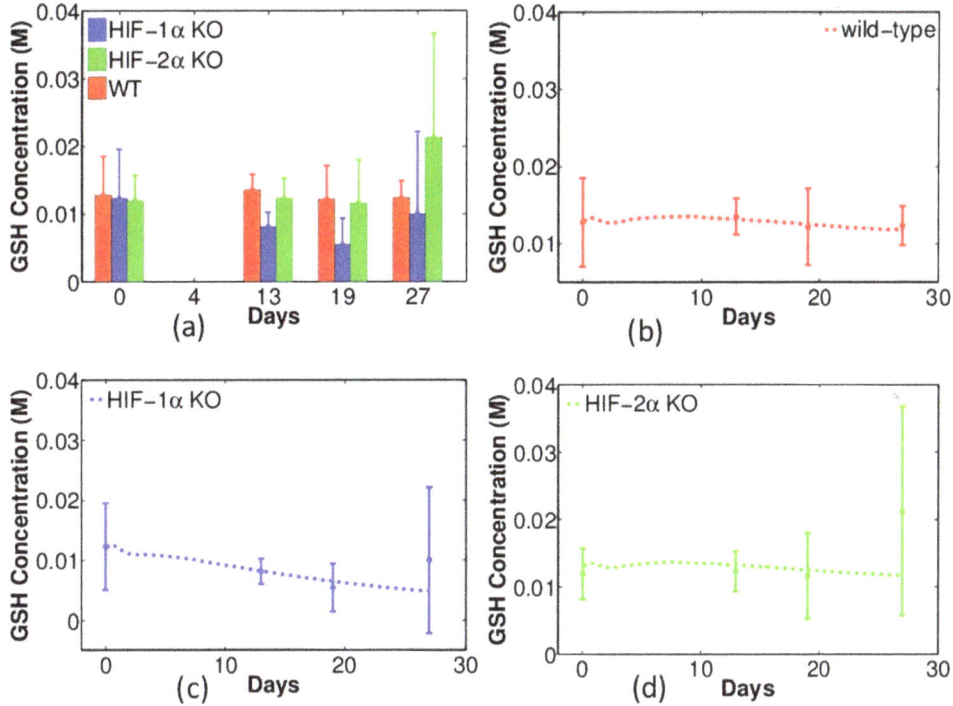

Figure 5. Experiments and simulations of intracellular GSH concentration ([GSH]) in tumors with wild-type, HIF-1α- and HIF-2α-deficient macrophages (WT, HIF-1α KO, and HIF-2α KO). Horizontal axis represents time (in days) and vertical axis scales [GSH] in units of Molar. (a): Experimental data of [GSH] with error bars. Red: WT; Blue: HIF-1α KO; Green: HIF-2α KO. (b) – (d): Comparison of experiments (dots with error bars) and numerical simulations (dash curves) of [GSH] for tumors with WT, HIF-1α, and HIF-2α KO macrophages, respectively.

different times of tumor growth. In Figure 9 (a), the ROS levels with GSH depletion starting on the first, the ninth, and the fourteenth day of tumor growth are presented in red, green and blue curves, respectively. Fig. 9 (b) shows the corresponding tumor volumes with these treatments. We see that earlier treatment of GSH depletion will maintain the ROS level above the toxicity threshold for a longer time, and thus has a better effect in suppressing tumor growth.

Effectiveness of docetacxel treatment

HIFs can regulate tumor microenvironment including GSH concentration, pH, and oxygen tension. Since changes in the tumor microenvironment can have significant impact on both tumor growth and efficacy of chemotherapies, another set of experiments was performed to determine the effectiveness of docetaxel (DTX) chemotherapy for tumors with HIF-1α- and HIF-2α-deficient macrophages.

Figure 10 shows the experiments of non-treated (black bars) and DTX-treated tumor growth (white bars), with WT, HIF-1α KO and HIF-2α KO macrophages in 10(a)- 10(c), respectively; the black columns of day 13 is normalized by one, and the white columns correspond to tumor volume relative to non-treated tumor. Comparing the black and white bars, we conclude that tumor environment with HIF-1α KO macrophages are responding better to the DTX-treatment: tumor volume is reduced to less than 40% of the non-treated tumor, as seen in Fig. 10(b). By contrast, Fig. 10(a) shows that the DTX-treatment has very limited effects (tumor volume is reduced by less than 10%) for tumors with WT macrophages. DTX seems to have no effect on tumors with HIF-2α KO macrophages, as shown in Fig. 10(c).

Our model can be used to simulate tumor growth with DTX treatment and predict the corresponding characteristics of tumor

microenvironment which were not monitored in the above experiments. But before we perform the simulations we need to modify the model in order to incorporate the effect of DTX-treatment. It is known that DTX increases the apoptotic rate of cancer cells by binding to microtubules during mitosis. It is also known [42,46] that the efficacy of the drug depends on the level of oxygen. Accordingly, we take in Eq. (1) a modified apoptotic rate:

$$\tilde{\mu}_{c2} = \begin{cases} \mu_{c2}, & \text{for non-treated tumors,} \\ \\ \theta_4 \mu_{c2} \eta(w), & \text{for DTX-treated tumors,} \end{cases} \quad (27)$$

where

$$\eta(w) = \frac{w_0}{w}. \quad (28)$$

and $\theta_4 > 1$. Figure 11 shows that with the choice of $\theta_4 = 3$ the model simulations are in good fit with the experimental results in Fig. 10. Note that a different set of mice were used in the experiments recorded in Fig. 10 from those in the previous experiments. Hence our simulations in the non-treated case correspond to the mice in Fig. 10, not in Fig. 4.

We can now use the model to predict the change of tumor microenvironment associated with the DTX treatment. Figure 12 shows the model simulations of GSH concentration, pH, and oxygen tension in (a) – (c), respectively. Each panel displays the effect of the combination of DTX treatment and HIF-1α knockout. The red and blue solid curves are for non-treated tumor with WT and HIF-1α KO macrophages, respectively; the green and magenta dashed curves are for the corresponding tumor

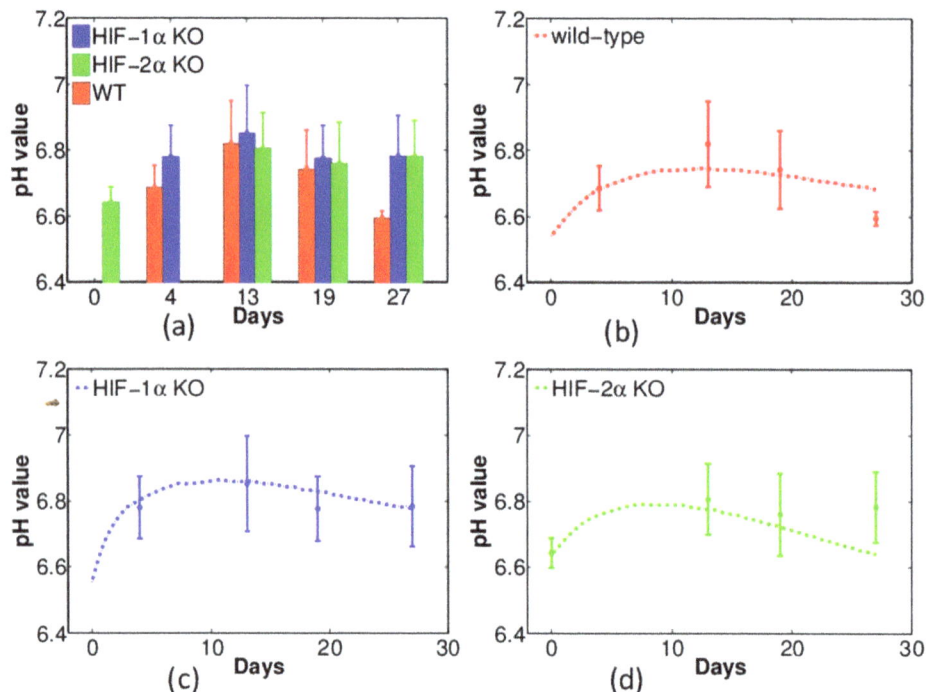

Figure 6. Experiments and simulations of pH in tumors with wild-type, HIF-1α-, and HIF-2α-deficient macrophages (WT, HIF-1α KO, and HIF-2α KO). Horizontal axis represents time (in days) and vertical axis shows the pH value. (a): Experimental data of pH against time with error bars. Red: WT; Blue: HIF-1α KO; Green: HIF-2α KO. (b) – (d): Comparison of experiments (dots with error bars) and numerical simulations (dash curves) of pH in tumor with WT, HIF1-α, and HIF-2α KO macrophages, respectively.

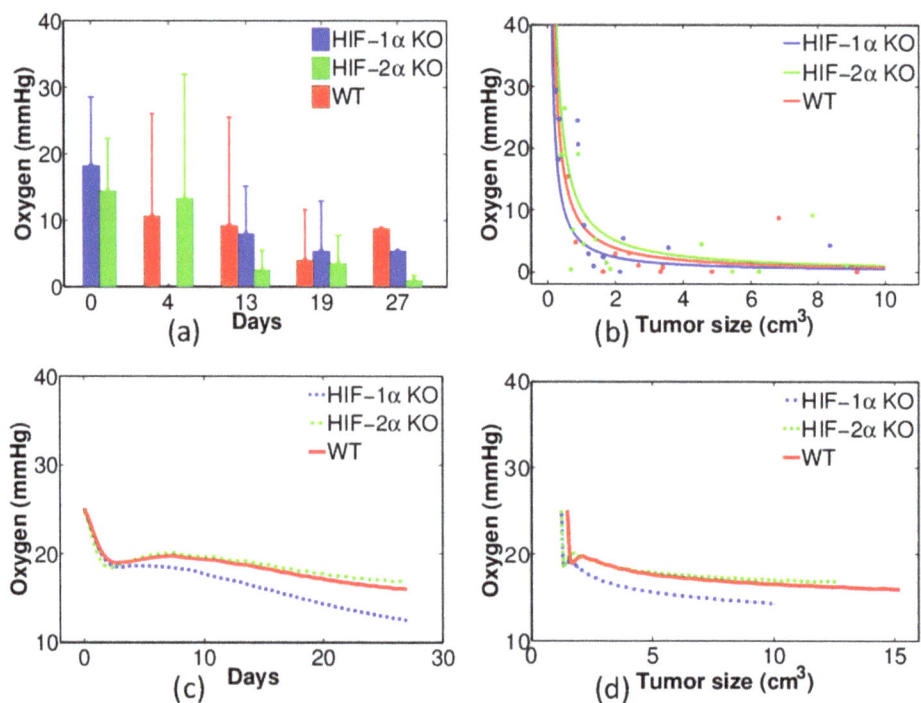

Figure 7. Experiments and simulations of oxygen tension of tumors with wild-type, HIF-1α- and HIF-2α-deficient macrophages (WT, HIF-1α KO and HIF-2α KO). (a): Experimental data of oxygen tension (mmHg) against time (days). Red: WT; Blue: HIF-1α KO; Green: HIF-2α KO; (b): Same experiments aligned with tumor volumes (dots) and the correspondingly fitted curves; (c): Numerical simulations of oxygen tension against time; (d): Numerical simulations of oxygen tension aligned with tumor volumes.

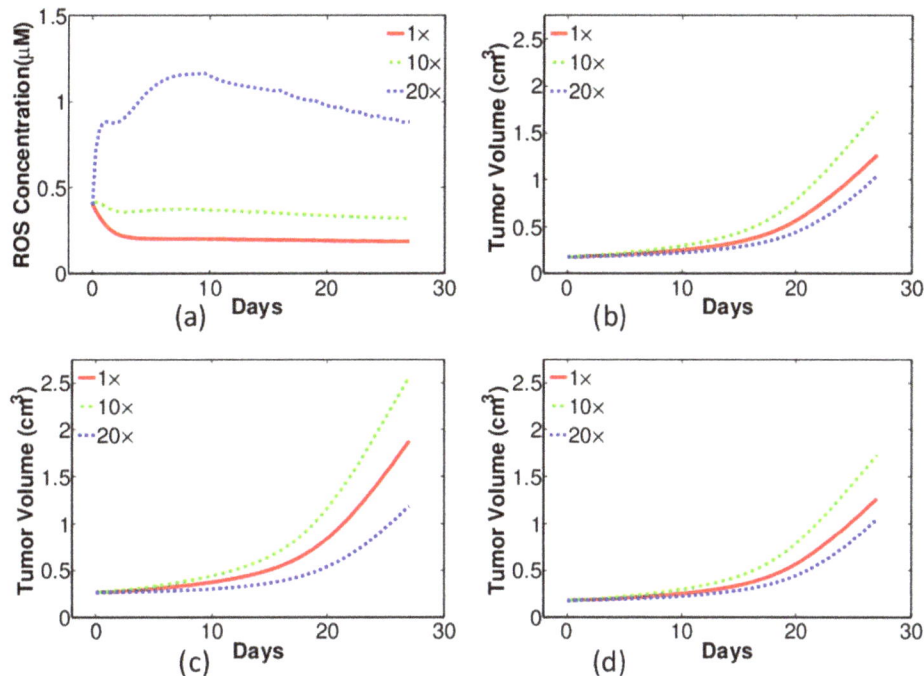

Figure 8. Simulations of intracellular ROS concentration (first row) and tumor growth (second row) with different levels of GSH depletion in tumors with wild-type macrophages (left column) and HIF-1α KO macrophages (right column). Red: no depletion (k_d); Green: moderate depletion ($10 \times k_d$); Blue: severe depletion ($20 \times k_d$). (a–b): ROS levels (μM) against time (days); (c–d): the corresponding tumor volume (cm^3).

with the DTX treatment. Comparing the blue and green curves, we conclude that HIF-1α KO in macrophages significantly lowers GSH concentration and reduces oxygen tension in tumor microenvironment than DTX treatment does. Recalling Fig. 10 (b) or Fig. 11 (b), we see that there is a correlation between the effectiveness of DTX and reduced levels of GSH concentration, increased pH, and reduced oxygen tension.

Figure 13 shows the simulated change of tumor growth with DTX treatment and the parameter variations. For clear comparison, the simulation with the same parameters as in Figs. 11 and 12 are in red curves, and the tumor volume on the last day is normalized by one. In these simulations, the parameter λ_H in Eq. (13) is increased by three times ($3\lambda_H$) to approximate the "proton addition" and the resulting tumor growth curves are in green, while the parameter μ_H is increased to $3\mu_H$ to simulate "proton depletion' and the corresponding tumor growth is in blue. Fig. 13 (a) and (b) are for pH variations with WT and HIF-1α macrophages, respectively. We conclude from the simulations that proton addition (or pH lowering) will reduce the DTX efficacy while proton deletion (or pH enhancing) will increase the efficacy of DTX. These phenomena are enhanced in tumors with WT macrophages than in tumors with HIF-1α-deficient macrophages.

Figure. 13 (c) and (d) are for oxygen variations with WT and HIF-1α macrophages, respectively. As before, the result with the same parameters as in Figs. 11 and 12 are shown in red curves and the volume on the last day is normalized by one. In these simulations, the parameter λ_e in Eq. (12) is increased to $1.5\lambda_e$ and reduced to $0.5\lambda_e$ for the "oxygen addition" and "oxygen depletion", respectively. We conclude that DTX is more effective with lower oxygen tension, while the efficacy of DTX shows no obvious differences in tumors with WT and HIF-1α-deficient macrophages.

Modeling enhanced therapeutic effectiveness

The power of mathematical modeling lies in the ability to alter variables that can be difficult or impossible to manipulate through experimentation and predict changes in outcome to the system. Such predictions are increasingly more valuable when the model system has been validated and correspond to data collected from *in vitro* or *in vivo* experimentation. Using modeling predictions generated from experiments performed on PyMT breast tumors in mice with wild type macrophages or mice with macrophages deficient in either HIF-1α or HIF-2α, we set out to predict enhanced therapeutic effectiveness to inhibit breast tumor growth based on changes in tumor intracellular glutathione, tumor pH, and tumor oxygen tension in the presence of the chemotherapy agent, docetaxel.

Summary of model validation by experimental data

- 1) Tumors with macrophages deficient in HIF-1α grow slower than tumors with wild type macrophages (Fig. 4).
- 2) Tumors with macrophages deficient in HIF-1α have reduced levels of intracellular GSH while tumors with wild type macrophages maintain higher intracellular GSH levels (Fig. 5).
- 3) Tumors with wild type macrophages have a reduced pH compared to tumors with HIF-1α- or HIF-2α-deficiency (Fig. 6).
- 4) Tumors with HIF-1α-deficient macrophages have less average oxygen than tumors with wild type macrophages (Fig. 7).
- 5) Docetaxel is markedly more effective in reducing tumor growth rates in tumors with HIF-1α-deficient macrophages than tumors from either wild type or HIF-2α-deficient macrophages (Fig. 11).

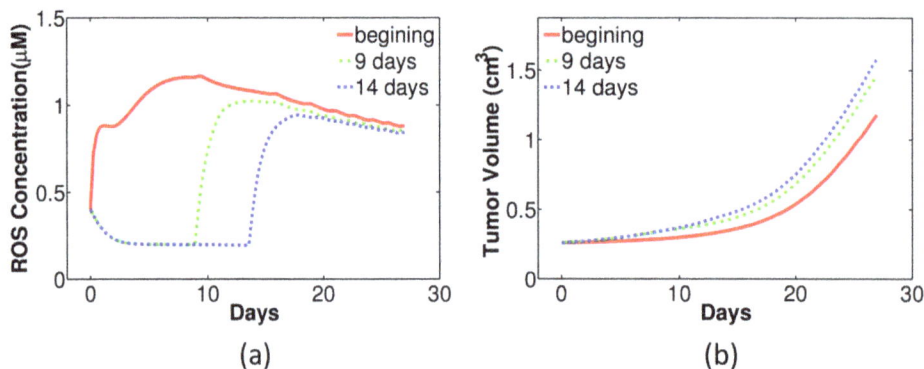

Figure 9. Simulations of intracellular ROS concentration and tumor growth with severe GSH depletion ($20 \times k_d$) at different time in tumors with wild-type macrophages. Red: GSH depletion at the beginning; Green: GSH depletion starts from the ninth day; Blue: GSH depletion from the fourteenth day. (a) ROS levels (μM) against time (days); (b) the corresponding tumor volume (cm^3).

Summary of model predictions

- 1) Depleting tumor intracellular GSH by $10 \times$ enhances tumor growth in tumors containing either wild type macrophages or HIF-1α-deficient macrophages. To the contrary, depleting GSH $20 \times$ inhibits tumor growth rates in tumors with wild type macrophages but has little or no effect on tumors with HIF-1α-deficient macrophages (Fig. 8).

- 2) Depleting tumor intracellular GSH starting at treatment day 1 maximally enhances free ROS leading to slower tumor growth rates in tumors with wild type macrophages, but does not have such an effect on macrophages deficient in HIF-1α, most likely because GSH levels in tumors with HIF-1α-deficient macrophages are already depleted (Fig. 9).

- 3) Changing tumor pH with DTX treatment alters tumor growth rates more in tumors with wild type macrophages than in tumors with HIF-1α-deficient macrophages (Fig. 13) while adding or reducing oxygen with DTX treatment had no differential effect on tumors with wild type macrophages or those tumors with macrophage HIF-1α-deficiency (Fig. B(c)(d)).

Our modeling alleges a major contributor to docetaxel effectiveness in inhibiting tumor growth is linked to HIF-1α-deficient macrophage regulation of intracellular tumor GSH

levels. Studies are underway in our laboratory demonstrating that tumor cells co-cultured with HIF-1α-deficient macrophages regulate the expression of tumor cell GSH-building enzymes. Indeed, studies have reported that increased tumor cell GSH levels and overexpression of GSH-synthesizing enzymes both predict a poor prognosis [47] and lead to reduced sensitivity to chemotherapy [48–54]. Glutathione is not translated as most other proteins; it is a tripeptide synthesized from the amino acids L-cysteine, L-glutamic acid, and glycine and made in two ATP-dependent steps: First, γ-glutamylcysteine is synthesized from L-glutamate and cysteine by the enzyme γ-glutamylcysteine synthetase. Second, glycine is added to γ-glutamylcysteine by the enzyme glutathione synthetase. Downregulation of these key GSH-building enzymes, along with membrane transporters like γ-glutamyl transferase in tumor cells, restrict their ability to compensate for ROS build-up, thus making them more susceptible to high ROS as well as limiting their ability to neutralize chemotherapy drugs like docetaxel by GSH. Our study suggests that therapies directed at promoting tumor cell apoptosis, as do most standard chemotherapy compounds, would be greatly enhanced in combination with a small molecule inhibitor specific for macrophage HIF-1α. Unexpectedly, because tumors with macrophages deficient in HIF-1α display reduced average oxygen tension, our modeling predicts that a similar treatment strategy would be ineffective for ROS-

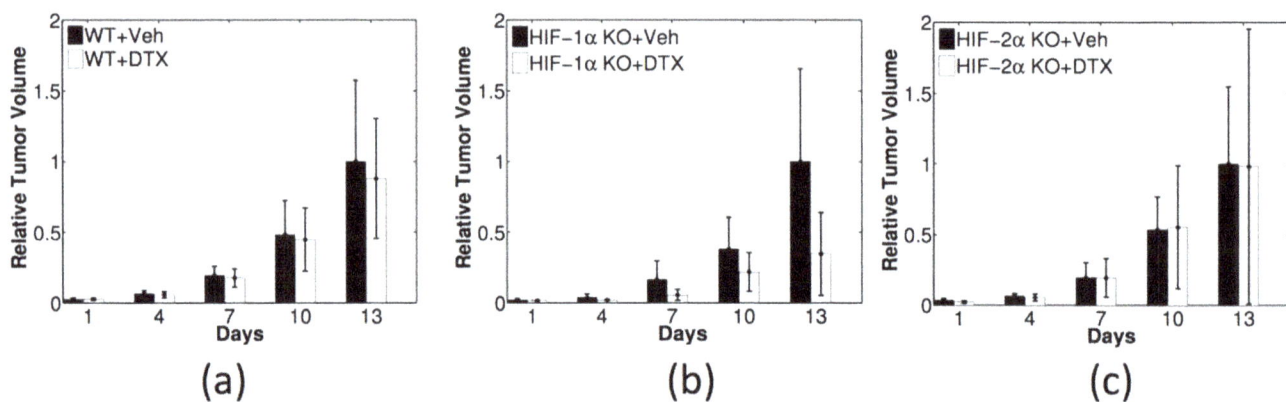

Figure 10. Experiments of testing DTX efficacy in tumors with wild-type, HIF-1α-, and HIF-2α-deficient macrophages (WT, HIF-1α KO and HIF-2α KO), in (a)-(c), respectively. Black: non-treated tumors (Veh); white: DTX-treated tumors. Relative tumor volume is obtained by dividing the volume of the treated tumor by the volume of non-treated tumor at the last day of each case.

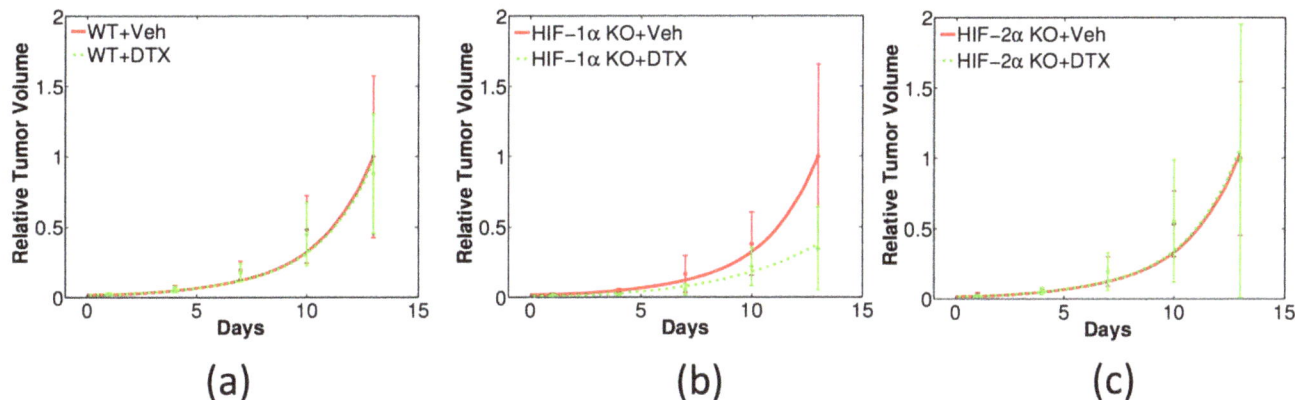

Figure 11. Comparison of simulations (colored curves) with the experiments from Figure 10 (dots with error bars) for DTX effectiveness in tumors with wild-type, HIF-1α-, and HIF-2α-deficient macrophages (WT, HIF-1α KO and HIF-2α KO) in (a) – (c), respectively. Tumor volumes are normalized in the same way as in Figure 10.

generated killing treatments such as radiation therapy which requires oxygen.

Conclusions

Tumor growth and effectiveness of chemotherapies greatly depend on the chemical tumor microenvironment. Thus, development of approaches, experimentally and numerically, to study dynamical changes in the tumor microenvironment may provide a key tool for anti-cancer drugs screening and optimization of anticancer therapies. In this work, we focused on several parameters which determine the chemical tumor microenvironment including GSH concentration, pH level and oxygen tension. The use of L-Band electron paramagnetic resonance (EPR) technology and probes developed specifically for each parameter allow for *in vivo*, real-time longitudinal analysis of mouse models of breast cancer. In this model, compared to normal mammary gland tissue, solid tumors generally have lower oxygen tension, lower extracellular pH, and higher intracellular GSH concentration, emulating the environmental parameters of human cancers. Interestingly, we found that this tumor microenvironment can also be altered by the absence or presence of macrophage HIF-1α or HIF-2α. Experiments had been performed to measure changes in GSH concentration, pH level and oxygen tension as their

associated tumors progressed. Concomitantly, experiments were carried out to investigate the effectiveness of docetaxel treatment on tumors with wild-type, HIF-1α- and HIF-2α-deficient macrophages. In this paper we developed a mathematical model that simulates tumor growth along with the dynamics of GSH concentration, pH, and oxygen tension and how these parameters are altered by the macrophage HIF subunits. The model is multiscale: interactions among cancer cells, immune system, endothelial cells, oxygen level, hydrogen ions, and corresponding cytokines were described at the tissue level by a coupled system of partial differential equations with a moving boundary, while chemical dynamics among GSH, ROS and other molecules are modeled by a set of ordinary differential equations at the cellular level. The model was validated by the comparison of simulations with experimental data from the prospective of intracellular GSH, pH, and oxygen tension in tumors grown in wild-type (LysMcre), HIF-1α-deficient (LysMcre/HIF-1α$^{flox/flox}$) and HIF-2α-deficient (LysMcre/HIF-2α$^{flox/flox}$) mice. Next the model was extended to include treatment with docetaxel (DTX), a chemotherapeutic drug that inhibits disassembly of microtubules during mitotic cell division thus initiating apoptosis. The model for the case of DTX treatment was validated by comparing the simulation with experimental results for tumor growth under DTX treatment, with or without macrophage HIF-1α or HIF-2α. Clinical trials

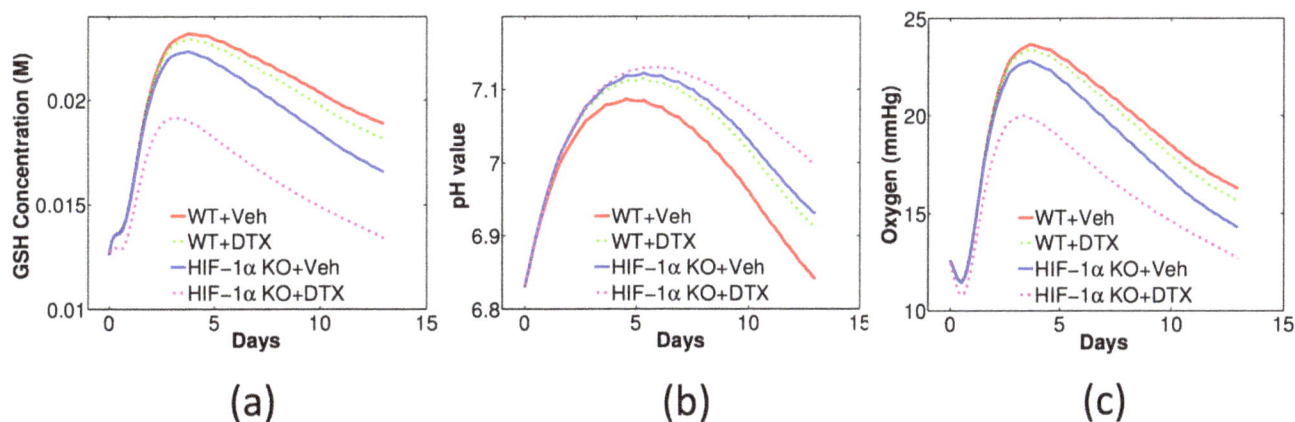

Figure 12. Model simulations of intracellular GSH concentration (a), pH (b), and oxygen tension (c) changing with time in DTX-treated and non-treated tumors, combined with WT or HIF-1α KO macrophages.

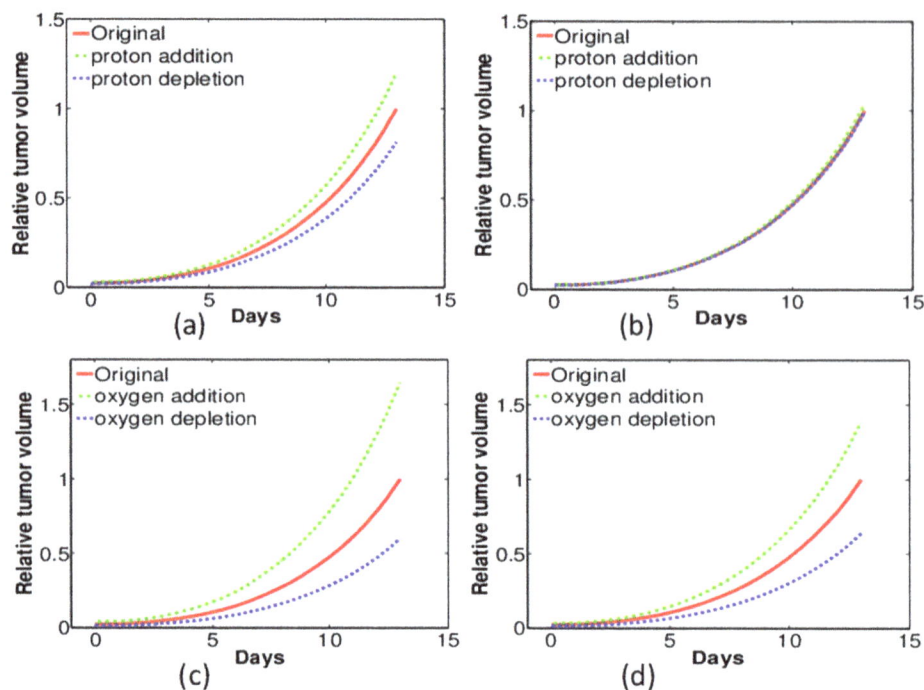

Figure 13. Model simulations of tumor growth with DTX treatment and parameter variations. (a) Proton variation with WT macrophages; (b) Proton variation with HIF-1α KO macrophages; (c) oxygen variation with WT macrophages; (d) oxygen variation with HIF-1α KO macrophages. The tumor volume without pH or oxygen variation on the last day is normalized to one.

involving therapeutic manipulation of tumor cell GSH, GSH-building enzymes, and targeting of transcription factors inhibiting these mechanisms are abundant (reviewed extensively in [55]). But our experimental and modeling data demonstrates that contribution of the tumor microenvironment, specifically from tumor macrophages, in the regulation of tumor cell GSH should be considered. Our model suggests an intriguing possibility that tumor-associated macrophages, specifically through HIF-1α activity, can augment tumor intracellular GSH to help tumor cells develop a resistance to therapy. Our experimental data and modeling predictions were obtained using the PyMT orthotopic breast tumor implantation model to understand the role of HIF transcription factors in regulating the chemical tumor microenvironment and a consequence on chemotherapy effectiveness. It

would be interesting to perform similar longitudinal experiments tracking tumor GSH, pH, and oxygen in transgenic PyMT mice with wild type macrophages which spontaneously form mammary tumors starting at 4 weeks of age and progress through all four stages similar to human breast cancer [56] to understand the changes in these parameters as the tumor progresses to malignancy.

Author Contributions

Conceived and designed the experiments: AAB CBM TDE. Performed the experiments: ACG RE VVK. Analyzed the data: DC AF. Contributed reagents/materials/analysis tools: DC AF. Wrote the paper: DC TDE AF.

References

1. Cook J, Gius D, Wink D, Krishna MC, Russo A, et al. (2004) Oxidative Stress, Redox, and the Tumor Microenvironment. Seminars in Radiation Oncology 14: 259–266.
2. Gillies R, Raghunand N, Garcia-Martin M, Gatenby R (2004) pH imaging. A review of pH measurement methods and applications in cancers. IEEE Eng Med Biol Mag 23: 57–64.
3. Khramtsov VV, Gillies RJ (2014) Janus-Faced Tumor Microenvironment and Redox. Antioxid Redox Signal 21: 723–729.
4. Ortega AL, Mena S, Estrela J (2011) Glutathione in Cancer Cell Death. Cancers 3: 1285–1310.
5. Vaupel P, Mayer A (2007) Hypoxia in cancer: significance and impact on clinical outcome. Cancer Metastasis Rev 26: 225–239.
6. Sonveaux P, Vegran F, Schroeder T, Wergin MC, Verraxet J, et al. (2008) Targeting lactate-fueled respiration selectively kills hypoxic tumor cells in mice. J Clin Invest 118: 3930–42.
7. Xu L, Fukumura D, Jain RK (2002) Acidic extracellular pH induces Vascular Endothelial Growth Factor (VEGF) in human Glioblastoma cells via ERK1/2 MARK signaling pathway. The Journal of Biological Chemistry 277: 11368–11374.
8. Raghunand N, Altbach M, van Sluis R, Baggett B, Taylor C, et al. (1999) Plasmalemmal pH-gradients in drug-sensitive and drug-resistant MCF-1 human breast carcinoma xenografts measured by 31P magnetic resonance spectrpscopy. Biochem Pharmacology 57: 309–312.
9. Stubbs M, Bhujwalla Z, Tozer G, Rodrigues L, Maxwell R, et al. (1992) An assessment of 31P MRS as a method of measuring pH in rat tumours. NMR Biomed, 5: 351–359.
10. Gatenby RA, Gawlinski ET, Gmitro AF, Kaylor B, Gillies R (2006) Acid-Mediated Tumor Invasion: a Multidisciplinary Study. Cancer Research 66: 5216–5223.
11. Smallbone K, Gatenby RA, Maini P (2008) Mathematical modeling of tumour acidity. Journal of Theoretical Biology 255: 106–112.
12. Martin NK, Gaffney E, Gatenby R, Maini P (2010) Tumour-stromal interactions in acid-mediated invasion: A mathematical model. Journal of Theoretical Biology 267: 461–470.
13. Ribeiro MdLC, Silva AS, Bailey KM, Kumar NB, Sellers TA, et al. (2012) Buffer Therapy for Cancer. J Nutr Food Sci 2: 6.
14. Li S, Yan T, Yang JQ, Oberley T, Oberley L (2000) The Role of Cellular Glutathione Peroxidase Redox Regulation in the Suppression of Tumor Cell Growth by Manganese Superoxide Dismutase. Cancer Research 60: 3927–3939.
15. Trachootham D, Alexandre J, Peng H (2009) Targeting cancer cells by ROS-mediated mechanisms: a radical therapeutic approach? Nature Reviews 8: 579–591.

16. Kawanishi S, Hiraku Y, Ponlaor S, Ma N (2006) Oxidative and nitrative DNA damage in animals and patients with inflammatory diseases in relation to inflammation-related carcinogenesis. Biol Chem 387: 365–372.

17. Schafer FQ, Buettner GR (2001) Redox environment of the cell as viewed through the redox state of the glutathione disulfide/glutathione couple. Free Radic Biol Med 30: 1191–212.

18. Roshchupkina GI, Bobko AA, Bratasz A, Reznikov VA, Kuppusamy P, et al. (2008) In vivo EPR measurement of glutathione in tumor-bearing mice using improved disulfide biradical probe. Free Radic Biol Med 45: 312–320.

19. Harguindey S, Arranz JL, Wahl ML, Orive G, Reshkin SJ (2009) Proton Transport Inhibitors as Potentially Selective Anticancer Drugs. Anticancer Research 29: 2127–2136.

20. Dai L, Jain RK (2002) Acidic extracellular pH induces Vascular Endothelial Growth Factor (VEGF) in human Glioblastoma cells via ERK1/2 MAPK signaling pathway. The Journal of Biological Chemistry 277: 11368–11374.

21. Coso S, Harrison I, Harrison CB, Vinh A, Sobey CG, et al. (2012) NADPH Oxidase as Regulators of Tumor Angiogenesis: Current and Emerging Concepts. Antioxidants and Redox Signaling 16: 1229–1247.

22. Manda G, Nechifor MT, Neagu TM (2009) Reactive Oxygen Species, Cancer and Anti-Cancer Therapies. Current Chemical Biology 3: 342–366.

23. Chen D, Roda JM, Marsh CB, Eubank TD, Friedman A (2012) Hypoxia Inducible Factors-mediated inhibition of cancer by GM-CSF: A mathematical model. Bulletin of Mathematical Biology 74: 275–77.

24. Hu Y, Rosen DG, Zhou Y, Feng L, Yang G, et al. (2005) Mitochondrial Manganese-Superoxide Dismutase Expression in Ovarian Cancer: Role in cell proliferation and response to oxidative stress. The Journal of Biological Chemistry 280: 39485–39492.

25. Qutub AA, Popel AS (2008) Reactive Oxygen Species Regulated Hypoxia-Inducible Factor 1α Differentially in Cancer and Ishemia. Molecular and Cellular Biology 28: 5106–5119.

26. Bobko AA, Eubank TD, Voorhees JL, Efimova OV, Kirilyuk IA, et al. (2012) In Vivo Monitoring of pH, Redox Status, and Glutathione Using L-Band EPR for Assessment of Therapeutic Effectiveness in Solid Tumors. Magnetic Resonance in Medicine 67: 1827–1836.

27. Lopez-Lazaro M (2007) Dual role of hydrogen peroxide in cancer: Possible relevance to cancer chemoprevention and therapy. Cancer Letters 252: 1–8.

28. Ng CF, Schafer FQ, Buettner GR, Rodgers V (2007) The rate of cellular hydrogen peroxide removal shows dependency on GSH: Mathematical insight in *in vivo* H_2O_2 and GPx concentrations. Free Radical Research 41: 1201–1211.

29. Martin G, Jain R (1994) Noninvasive measurement of interstitial pH profiles in normal and neoplastic tissue using fluorescence ratio imaging microscopy. Cancer Research 54: 5670–4.

30. Wraight C (2006) Chance and design–Proton transfer in water, channels and bioenergetic proteins. Biochimica et Biophysica Acta 1757: 886–912.

31. Luz Z, Meiboom S (1964) The activation energies of proton transfer reactions in water. J Am Chem Soc 86: 4768.

32. Szatrowski TP, Nathan CF (1991) Production of Large Amounts of Hydrogen Peroxide by Human Tumor Cells. Cancer Research 51: 794–798.

33. Laurent A, Nicco C, Chereau C, Goulvestre C, Alexande J, et al. (2005) Controlling Tumor Growth by Modulating Endogenous Production of Reactive Oxygen Species. Cancer Research 65: 948–956.

34. Antunes F, Salvador A, Pinto R (1995) PHGPx and phospholipase A_2/GPx: comparative importance on the reduction of hydroperoxides in rat liver mitochondria. Free Radic Bio Med 19: 669–677.

35. Sekhar R, Patel S, Guthikonda A, Reid M, Balasubramanyam A, et al. (2011) Deficient synthesis of glutathione underlies oxidative stress in aging and can be corrected by dietary cysteine and glycine supplementation. Am J Clin Nutr 94: 847–53.

36. Ookhten M, Hobdy K, Corvasce M, Aw T, Kaplowitz M (1985) Sinusoidal efflux of glutathione in the perfused rat liver. J Clin Invest 75: 258–265.

37. Lauterburg B, Adams J, Mitchell J (1984) Hepatic glutathione homeostasis in the rat: efflux accounts for glutathione turnover. Hepatology 4: 586–590.

38. Oshino N, Chance B, Sies H, Bucher T (1973) The role of H_2O_2 generation in perfused rat liver and the reaction of catalase compound I and hydrogen donors. Arch Biochem Biophys 154: 117–131.

39. Antunes F, Cadenas E (2001) Cellular titration of apoptosis with steady state concentrations of H_2O_2: submicromolar levels of H_2O_2 induce apoptosis through Fenton chemistry independent of the cellular thiol state. Free Radic Biol Med 30: 1008–1018.

40. Stone J (2004) An assessment of proposed mechanisms for sensing hydrogen peroxide in mammalian systems. Atch Biochem Biophys 422: 119–124.

41. Flohe L (1978) Glutathione peroxidase: fact and fiction. Ciba Foundation Symp 65: 95–122.

42. Eubank T, Roda J, Liu H, O'Neill T, Marsh C (2011) Opposing Roles for HIF-1α and HIF-2α in the Regulation of Angiogenesis by Mononuclear Phagocytes. Blood 117: 323–32.

43. Steel R, Torrie J (1960) Principles and Procedures of Statistics with Special Reference to the Biological Sciences. McGraw Hill.

44. Colin Cameron A, Windmeijer FA, Gramajo H, Cane D, Khosla C (1997) An R-squared measure of goodness of fit for some common nonlinear regression models. Journal of Econometrics 77: 1790–2.

45. Armstrong J, Steinauer K, Hornung B, Irish J, Lecane P, et al. (2002) Role of glutathione depletion and reactive oxygen species generation in apoptotic

46. Roda JM, Summer LA, Evans R, Philips GS, Marsh CB, et al. (2011) Hypoxia-Inducible Factor-2α regulates GM-CSF-Derived soluble Vascular Endothelial Growth Factor Receptor 1 production from macrophages and inhibits tumor growth and angiogenesis. The Journal of Immunology 187: 1970–1976.

47. Bard S, Noel P, Chauvin F, Quash G (1986) gamma-Glutamyltranspeptidase activity in human breast lesions: an unfavourable prognostic sign. Br J Cancer 53: 637–42.

48. O'Brien M, KD T (1996) Glutathione and related enzymes in multidrug resistance. Eur J Cancer 32: 967–78.

49. Wang J, Yi J (2008) Cancer cell killing via ROS: to increase or decrease, that is the question. Cancer Biol Ther 7: 1875–84.

50. Calvert P, KS Y, Hamilton T, O'Dwyer P (1998) Clinical studies of reversal of drug resistance based on glutathione. Chem Biol Interact 111–112: 213–24.

51. Godwin A, Meister A, O'Dwyer P, Huang C, Hamilton T, et al. (1992) High resistance to cisplatin in human ovarian cancer cell lines is associated with marked increase of glutathione synthesis. Proc Natl Acad Sci USA 89: 3070–4.

52. Mulcahy R, Untawale S, Gipp J (1994) Transcriptional up-regulation of gamma-glutamylcysteine synthetase gene expression in melphalan-resistant human prostate carcinoma cells. Mol Pharmacol 46: 909–14.

53. Hochwald S, Rose D, Brennan M, Burt M (1997) Elevation of glutathione and related enzyme activities in high-grade and metastatic extremity soft tissue sarcoma. Ann Surg Oncol 4: 303–9.

54. Lewis A, Hayes J, Wolf C (1988) Glutathione and glutathione-dependent enzymes in ovarian adenocarcinoma cell lines derived from a patient before and after the onset of drug resistance: intrinsic differences and cell cycle effects. Carcinogenesis 9: 1283–7.

55. Traverso N, Ricciarelli R, Nitti M, Marengo B, Furfaro A, et al. (2013) Role of glutathione in cancer progression and chemoresistance. Oxid Med Cell Longev: 972913.

56. Lin E, Jones J, Li P, Zhu L, Whitney K, et al. (2003) Progression to malignancy in the polyoma middle T oncoprotein mouse breast cancer model provides a reliable model for human diseases. Am J Pathol 163: 2113–23.

57. Owen MR, Byrne HM, Lewis CE (2004) Mathematical modeling of the use of macrophages as vehicles for drug delivery to hypoxic tumour sites. Journal of Theoretical Biology 226: 377–391.

58. Qian B, Deng Y, Hong Im J, Muschel RJ, Zou Y, et al. (2009) A distinct macrophage population mediates metastatic breast cancer cell extravasation, establishment and growth. PLoS One 4: e6562.

59. Breward CJW, Byrne HM, Lewis CE (2001) Modeling the interactions between tumour cells and a blood vessel in a microenvironment within a vascular tumour. European Journal of Applied Mathematics 12: 529–556.

60. Tang S, Liu H, Rao Q, Geng Y, Zheng G, et al. (2000) Internalization and half-life of membrane-bound macrophage colony-simulating factor. Chinese Sc Bull 45: 1697–1703.

61. Owen MR, Sherratt JA (1997) Pattern formation and spatiotemporal irregularity in a model for macrophage tumour interactions. Journal of Theoretical Biology 189: 63–80.

62. Wathen K, Sarvela J, Stenman F, Stenman U, Vuorela P (2011) Changes in serum concentrations of soluble vascular endothelial growth factor receptor-1 after pregnancy. Human Reproduction 26: 221–226.

63. Wu FTH, Stefanini MO, Gabhann FM, Popel AS (2009) A compartment model of VEGF distribution in humans in the presence of soluble VEGF receptor-1 acting as a ligand trap. PLoS One 4: e5108.

64. Plank M, Sleeman B, PF J (2004) A mathematical model of tumour angiogenesis, regulated by vascular endothelial growth factor and the angiopoietins. Journal of Theoretical Biology 229: 435–454.

65. Utting JC, Flanagan AM, Brandao-Burch A, Orriss IR, Arnett TR (2010) Hypoxia stimulates osteoclast formation from human peripheral blood. Cell Biochemistry and Function 28: 374–380.

66. Oren H, Duman N, Abacioglu H, Ozkan H, Irken G (2001) Association between serum Macrophage Colony-Stimulating Factor levels and Monocyte and Thrombocyte Counts in healthy, hypoxic, and septic term neonates. Pediatrics 108: 329–32.

67. Bosco MC, Puppo M, Pastorino S, Mi Z, Melillo G, et al. (2004) Hypoxia selectively inhibits Monocyte Chemoattractant Protein-1 production by macrophages. The Journal of Immunology 172: 1681–1690.

68. Braunstein S, Karpisheva K, Pola C, Goldberg J, Hochman T, et al. (2007) A hypoxia-controlled cap-dependent to cap-independent translation switch in breast cancer. Molecular Cell 28: 501–512.

69. Pyaskovskaya ON, Kolesnik DL, Kolobov AV, Vovyanko SI, Solyanik GI (2008) Analysis of growth kinetics and proliferative heterogeneity of lewis lung carcinoma cells growing as unfed culture. Experimental Oncology 30: 269–275.

70. Me llo G, Sausville E, Cloud K, Lahusen T, Varesio L, et al. (1999) Flavopiridol, a protein kinase inhibitor, down-regulates hypoxic induction of vascular endothelial growth factor expression in human monocytes. Cancer Research 59: 5433–5437.

71. Vicioso L, Gonzalez F, Alvarez M, Ribelles N, Molina M, et al. (2006) Elevated serum levels of vascular endothelial growth factor are associated with tumor-associated macrophages in primary breast cancer. American Journal of Clinical Pathology 125: 111–118.

72. Wu FTH, Stefanini MO, Gabhann FM, Kontos CD, Annex BH, et al. (2010) VEGF and soluble VEGF receptor-1 (sFlt-1) distributions in peripheral arterial

disease: an in silico model. Am J Physiol Heart Circ Physiol 298: H2174–H2191.

73. Girgis-Gabardo A, Hassell J (2008) Scale-up of breast cancer stem cell aggregate cultures to suspension bioreactors. Biotechnol Prog 22: 801–810.

74. Butterworth E A, Cater D (1967) Effect of lysolecithin on oxygen uptake of tumour cells polymorphonuclear leucocytes lymphocytes and macrophages in vitro. British Journal of Cancer 21: 373389.

75. Chen Y, Cairns R, Papandreou I, Koong A, Denko NC (2009) Oxygen consumption can regulate the growth of tumors, a new perspective on the warburg effect. PLoS One 4: 27033.

76. Schugart RC, Friedman A, Zhao R, Sen CK (2008) Wound angiogenesis as a function of tissue oxygen tension. PNAS 105: 2628–2633.

77. Casciari JJ, Sotirchos SV, Sutherland RM (1988) Glucose diffusivity in multicellular tumor spheroids. Cancer Research 48: 3905–3909.

78. Gabhann FM, Popel AS (2004) Model of competitive binding of vascular endothelial growth factor and placental growth factor to VEGF receptors on endothelial cells. Am J Physiol Heart Circ Physiol 286: H153–H164.

79. Macdougall JDB, Mccabe M (1967) Diffusion coefficient of oxygen through tissues. Nature 215: 1173–1174.

80. Vaupel P, Mayer A, Briest S, Hockel M (2003) Oxygenation gain factor: A novel parameter characterizing the association between hemoglobin level and the oxygenation status of breast cancers. Cancer Research 63: 7634–7637.

81. Eubank T, Robert RD, Khan M, Curry J, Nuovo GJ, et al. (2009) Granulocyte Macrophage Colony-Stimulating Factor inhibits breast cancer growth and metastasis by invoking an anti-angiogenic program in tumor-educated macrophages. Cancer Research 69: 2133–2140.

82. Vincensini D, Dedieu V, Eliat PA, Vincent C, Bailly C, et al. (2007) Magnetic resonance imaging measurements of vascular permeability and extracellular volume fraction of breast tumors by dynamic Gd-DTPA-enhanced relaxometry. Magnetic Resonance Imaging 25: 293302.

7

Expression Profiling of Selected Glutathione Transferase Genes in *Zea mays* (L.) Seedlings Infested with Cereal Aphids

Hubert Sytykiewicz*, Grzegorz Chrzanowski, Paweł Czerniewicz, Iwona Sprawka, Iwona Łukasik, Sylwia Goławska, Cezary Sempruch

Siedlce University of Natural Sciences and Humanities, Department of Biochemistry and Molecular Biology, Siedlce, Poland

Abstract

The purpose of this report was to evaluate the expression patterns of selected glutathione transferase genes (*gst1*, *gst18*, *gst23* and *gst24*) in the tissues of two maize (*Zea mays* L.) varieties (relatively resistant Ambrozja and susceptible Tasty Sweet) that were colonized with oligophagous bird cherry-oat aphid (*Rhopalosiphum padi* L.) or monophagous grain aphid (*Sitobion avenae* L.). Simultaneously, insect-triggered generation of superoxide anion radicals ($O_2^{\cdot-}$) in infested *Z. mays* plants was monitored. Quantified parameters were measured at 1, 2, 4, 8, 24, 48 and 72 h post-initial aphid infestation (hpi) in relation to the non-infested control seedlings. Significant increases in *gst* transcript amounts were recorded in aphid-stressed plants in comparison to the control seedlings. Maximal enhancement in the expression of the *gst* genes in aphid-attacked maize plants was found at 8 hpi (*gst23*) or 24 hpi (*gst1*, *gst18* and *gst24*) compared to the control. Investigated *Z. mays* cultivars formed excessive superoxide anion radicals in response to insect treatments, and the highest overproduction of $O_2^{\cdot-}$ was noted 4 or 8 h after infestation, depending on the aphid treatment and maize genotype. Importantly, the Ambrozja variety could be characterized as having more profound increments in the levels of *gst* transcript abundance and $O_2^{\cdot-}$ generation in comparison with the Tasty Sweet genotype.

Editor: Keqiang Wu, National Taiwan University, Taiwan

Funding: This research was financially supported by the National Science Centre (NCS, Poland) under the grant no. N N310 733940. The funders had no role in study design, data collection and analysis, decision to publish, or preparation of the manuscript.

Competing Interests: The authors have declared that no competing interests exist.

* Email: huberts@uph.edu.pl

Introduction

Maize (*Zea mays* L.) has increasingly emerged as a pivotal model plant species (Poaceae family, Panicoideae subfamily) that is widely used in a variety of genetic and ecotoxicological experiments [1–3]. During the last decade, its world production and utilization in many sectors of industrial production was substantially increased; therefore, it is important to get better insight into the complex mechanisms underlying maize tolerance towards a vast array of biotic and abiotic stressors [4–5]. Among the numerous insects attacking *Z. mays* plants, destructive influence of cereal aphids (Hemiptera, Aphidoidea) colonization should be underlined [6–8]. These phloem feeding parasites are involved in severe exploitation of the host systems, resulting in a broad range of detrimental effects, such as mechanical injuries of the stylet-penetrated tissues, local chlorosis or necrosis, deformations of organs, biomass reduction, significant disturbances of cellular homeostasis and transmission of pathogenic viruses. The harmfulness of the aphid attack is linked to the suppression of photosynthesis, diminution in chlorophyll content, intensive removal of water and photosynthates from the sieve elements [9–12]. Recently, there has been evidence showing that the severity of aphid-induced damages is largely associated with the composition of species-specific elicitors present in the salivary secretions injected into the host tissues [13–14]. Importantly, an aphid-triggered oxidative burst in tissues of host systems colonized by these hemipterans has scarcely been reported [6,15]. On the other hand, it has been documented that cereal aphids evoked a significant decrease in ascorbate content in triticale and deterioration of the antioxidative capacity toward DPPH (1,1-diphenyl-2-picrylhydrazyl) radicals in maize plants [6,16]. It should be noted that cellular redox imbalance in plant cells due to a chronic overproduction of various reactive oxygen species (ROS) may result in profound oxidative damages of lipids, polysaccharides, proteins and nucleic acids [15–16].

Cytosolic glutathione transferases (GSTs, E.C.2.5.1.18) embrace a multifunctional superfamily of enzymes participating in many physiological processes involved in plant growth and development, shoot regeneration and adaptability to adverse environmental stimuli [17]. Plant GSTs catalyze the nucleophilic substitution or addition reactions of endogenous substrates and xenobiotics with glutathione molecules, leading to the synthesis of less toxic compounds with greater solubility in water, which secondarily improves their vacuolar sequestration [18–19]. Additionally, glutathione transferases are involved in scavenging of excessive amounts of ROS generated in plant tissues under oxidative stress conditions, and they participate in the signal transduction pathways, cellular responses to auxins and cytokinins, as well as metabolic turnover of cinnamic acid and anthocyanins [20–21]. According to Dixon et al. [22], AtGSTZ1-1 from *Arabidopsis thaliana* L. possesses maleylacetone isomerase activity and participates in tyrosine degradation. Furthermore, GSTs display glutathione-peroxidase activity associated with the reduction of hydroperoxides [23]. Some authors have proposed that the activation of glutathione transferases in plants exposed to different stressors is associated with an increased ability to neutralize the lipid hydroperoxides synthesised in oxidatively damaged membranes [24–25]. It has been previously reported that GST isoforms overexpressed in transgenic plants markedly augment tolerance levels to herbicide treatment and oxidative stress [26–27]. Consistent with these observations, tau-GST from *Lycopersicon esculentum* Mill., elevated resistance to hydrogen peroxide-stimulated stress and repressed *Bax*-stimulated apoptosis in transformed yeast cells [28]. Likewise, upregulation of several plant glutathione transferases in catalase-deficient mutants were reported [29].

There are numerous studies indicating a rapid and substantial increase in the activity of various plant GST isozymes and differential regulation of *gst* genes influenced by multifarious external factors (e.g. heavy metals, herbicides, drought, low and high temperatures, UV radiation, exogenous application of chemical inducers of oxidative stress, insect infestation and fungal or viral infection) [30–32]. However, there is a lack of published data concerning expression profiling of the *gst* genes and superoxide anion radical ($O_2^{\bullet-}$) production in the seedlings of maize varieties exposed to cereal aphid colonization. It may be assumed that mono- and oligophagous aphids differentially affect the transcriptional activity of *gst* genes and $O_2^{\bullet-}$ generation in tissues of maize genotypes, exhibiting diverse resistance levels to the aphid infestation. To verify this hypothesis, the relative quantification of four *gst* genes (*gst1*, *gst18*, *gst23* and *gst24*) was performed and the amount of $O_2^{\bullet-}$ was measured in the seedlings of *Z. mays* Ambrozja (susceptible) and Tasty Sweet (relatively resistant) varieties infested by monophagous grain aphid (*Sitobion avenae* F.) or oligophagous bird cherry-oat aphid (*Rhopalosiphum padi* L.). The study was also aimed at assessing whether the scale of aphid-triggered changes in the levels of the analysed parameters may be dependent on the insect density.

Methods

Plant material

The seeds of two investigated *Z. mays* varieties (Ambrozja and Tasty Sweet) were acquired from local commercial grain suppliers: Reheza (Moszna, Poland) and PNOS S.A. (Ożarów Mazowiecki, Poland). Before performing the bioassays, intact maize seeds without any visible damages were surface sterilized as described previously [32]. Subsequently, portions of plant material (5 seeds of each cultivar per plate; four replicates) were subjected to potato dextrose agar (PDA) plate screening in order to confirm the absence of mycoflora, according to the method of Adejumo et al. [33]. Ambrozja genotype has previously been classified as relatively resistant, whereas Tasty Sweet is susceptible to the cereal aphids' infestation [6]. Maize seeds were sown in round plastic pots (10×9 cm; diameter × height) filled with general-purpose horticultural substrate and no additional fertilization was applied. Seedlings were grown in a climate chamber at 22±2°C/ 16±2°C (day/night) with a light intensity of 100 μM m^{-2} s^{-1}, a long-day photoperiod (L16: D8) and a relative humidity of 65±5%. It is important to note that only health maize seedlings of similar height were included during the experiments.

Aphids

Wingless parthenogenetic females of *R. padi* and *S. avenae* aphids were collected from the field crops within the Siedlce district, Poland (52°09′54″N, 22°16′17″E). The authors state that no specific permissions were required for the sampling of aphids in this location, and confirm that the field studies did not involve endangered or protected species. The collected females were transferred to the seedlings of common wheat (*Triticum aestivum* L.) cv. Tonacja in the Department of Biochemistry and Molecular Biology, University of Natural Sciences and Humanities (Siedlce, Poland). New wheat seedlings were provided every week, and the aphids were reared for a year in the climate chamber under the conditions described above. Adult apterous females of the cereal aphids used in the leaf infestation experiments originated from the mother stock cultures of parthenogenetic individuals.

Infestation experiments

Leaves of 14-day-old maize seedlings (Ambrozja and Tasty sweet cultivars) were colonized with 10, 20, 40, or 60 adult wingless females of the relevant cereal aphids (*R. padi* or *S. avenae*) per plant. The control groups of seedlings were not infested with hemipterans. The levels of relative expression of the selected *gst* genes (*gst1*, *gst18*, *gst23* and *gst24*) and $O_2^{\bullet-}$ generation in *Z. mays* seedling leaves were determined 1, 2, 4, 8, 24, 48, and 72 h after initial insect infestation (hpi). Maize plants infested with aphids and the non-infested (control) plants were isolated in gauze-covered plastic cylinders (20×50 cm; diameter × height). At the end of each variant of biotests, the aphids were removed from the plants and, subsequently, the seedling leaves were excised and used immediately for further analytic procedures.

Determination of superoxide anion radical generation in the maize seedlings

The formation of $O_2^{\bullet-}$ was measured by the reduction of nitroblue tetrazolium (NBT), according to the method of Chaitanya and Naithani [34] with necessary modifications. Freshly collected *Z. mays* seedling leaves were cut into small pieces, and 0.5 g of the plant material was homogenized in 5 cm^3 of ice-cold phosphate buffer (100 mM, pH 7.2) with 1 mM diethyldithiocarbamate (superoxide dismutase inhibitor). The homogenate was filtered through four layers of nylon mesh and centrifuged at 19 000×g for 20 min at 4°C. A portion of the supernatant (0.2 cm^3) was combined with 0.8 cm^3 of the phosphate buffer and 0.1 cm^3 of 25 mM NBT (Sigma-Aldrich, Poland), and then, the mixture was incubated at 25°C for 5 min. Absorbance values of the sample before the incubation (A_0) and after the incubation period (A_S) were determined at 540 nm using an Epoch UV-Vis microplate spectrophotometer (BioTek, USA). The amount of $O_2^{\bullet-}$ in *Z. mays* seedling leaves was calculated using the following formula: $\Delta A_{540} = A_S - A_0$, and it was expressed as ΔA_{540} (min^{-1} g^{-1}) fresh weight.

Isolation of total RNA and cDNA synthesis

The insect-infested and non-infested seedling leaves of both investigated *Z. mays* genotypes were collected and homogenized immediately in liquid nitrogen by employing a sterile ceramic mortar and pestle. Total RNA was extracted with the application of *Spectrum Plant Total RNA Kit* (Sigma Aldrich, Poland) and, subsequently, trace amounts of genomic DNA were degraded using the *On-Column DNase I Digestion Set* (Sigma Aldrich, Poland). The quantitative-qualitative evaluation of the RNA samples was conducted with the use of an Epoch UV-Vis microplate spectrophotometer (BioTek, USA). High-quality RNA preparates ($A_{260/280}$ >2.0; $A_{260/230}$ >1.8) were exclusively accepted for the reverse-transcription reaction. Synthesis of complementary DNA (cDNA) was performed with the use of *RevertAid Premium First Strand cDNA Synthesis Kit* (Fermentas, Poland). It should be noted that the protocol scheme with oligo(dT)$_{18}$ primers was applied. Additionally, two negative controls (NTC – no template control, and NRT – no reverse transcriptase) were included.

Gene expression quantification

The relative expression of the target *gst* genes in foliar tissues of the aphid-infested and non-infested (control) *Z. mays* seedlings was estimated using the quantitative real-time reverse-transcription polymerase chain reaction (qRT-PCR). The glyceraldehyde-3-phosphate dehydrogenase (*gapdh*) gene was used as the internal reference [6]. Transcriptional activity of four *gst* genes (*gst1*, *gst18*, *gst23* and *gst24*) was measured with the application of TaqMan Gene Expression Assays (Life Technologies, Poland). The selection of target genes was based on their regulation under specific stress conditions (*gst1* has widely been described as a molecular marker of oxidative stress in maize tissues and *gst23* has been thought to be associated with multiple disease resistance, whereas expression of *gst18* and *gst24* genes was markedly altered under fungal infections) [19,32]. Reference sequences and unique assay names (IDs) of the quantified *gst* transcripts are listed in Table S1. The reaction mixtures (20 mm^3 final volume) contained 10 mm^3 2× TaqMan Fast Universal PCR Master Mix, 1 mm^3 20× TaqMan gene expression assay solution, 4 mm^3 template (cDNA) and 5 mm^3 RNase-free deionised water. Detection of the fluorescence signals was carried out on the StepOne Plus Real-Time PCR System equipped with StepOnePlus Software v2.3 (Life Technologies, USA). Amplification plots were obtained under the following thermal cycling conditions: initial activation of Ampli-Taq Gold DNA polymerase at 95°C (20 s) and, subsequently, 40 cycles of 95°C (1 s) and 60°C (20 s). Relative gene expression was estimated according to the comparative C_T ($\Delta\Delta C_T$) method [35], and the results are reported as the mean *n*-fold change ± standard deviation (SD) in the specific transcript amount of the aphid-stressed plants compared to the relevant non-infested control plants. Three biological and three technical replicates were included for each tested sample.

Statistical analysis

The data are presented as the mean ± SD of three independent experiments. Each group of aphid-stressed and non-infested maize plants consisted of ten seedlings of a similar height. Factorial analysis of variance (ANOVA) was applied to assess the effects of four experimental indicators (maize cultivar, hemipteran species, insect abundance and aphid exposure period) as well as their interdependence. Afterwards, a post-hoc Tukey's test was performed (*p* values less than 0.05 were considered significant). Statistical analyses were carried out with the implementation of STATISTICA 10 software (StatSoft, Poland).

Results

Effects of cereal aphids colonization on $O_2^{\bullet-}$ generation in *Z. mays* seedlings

Both *R. padi* and *S. avenae* aphids accelerated $O_2^{\bullet-}$ production in the colonized Ambrozja and Tasty Sweet maize cultivars compared with the relevant control plants (Table 1, 2). Bird cherry-oat aphid infestation led to a greater increase in $O_2^{\bullet-}$ amounts than grain aphid attack. For example, at 4 hpi,

colonization of Tasty Sweet or Ambrozja plants with bird cherry-oat aphids at the highest density (60 per seedling) led to 65 and 209% increases in the $O_2^{\bullet-}$ levels relative to the control, respectively, whereas infestation of these cultivars with the same number of *S. avenae* and insect exposure time led to 49 and 117% increases in $O_2^{\bullet-}$, respectively. Ambrozja seedlings that were attacked with either aphid species were characterized by significantly higher production of $O_2^{\bullet-}$ than the insect-stressed Tasty Sweet plants (Table 1, 2). The lowest initial number of both aphid species (10 per seedling) resulted in slight increments in the superoxide anion radicals content in the leaves of both maize varieties in relation to the control. Plants treated with higher numbers of aphids showed proportionally greater levels of $O_2^{\bullet-}$ accumulation. Consequently, the largest differences in aphid-stimulated production of $O_2^{\bullet-}$ between the two maize cultivars were observed at the highest insect density (60 per seedling). For example, *R. padi*–stressed Tasty Sweet plants generated 2–26% more $O_2^{\bullet-}$ (depending on duration of aphid colonization) than seedlings attacked by *S. avenae*, whereas *R. padi*–stressed Ambrozja plants had 4–91% greater rates of $O_2^{\bullet-}$ formation than *S. avenae*–infested seedlings. Additionally, slightly more superoxide anion radicals production was found in the non-infested Ambrozja seedlings than in Tasty Sweet plants (Table 1, 2). Importantly, the duration of aphid infestation had a strong influence on the generation of $O_2^{\bullet-}$ in leaves of both *Z. mays* genotypes. Comparative analysis of all treatments revealed that the lowest level of $O_2^{\bullet-}$ generation occurred at 1 hpi (2–9% increase, depending on the aphid infestation level) relative to the control. Maximal $O_2^{\bullet-}$ formation was observed at 4 hpi in Tasty Sweet seedlings infested with 60 individuals of bird cherry-oat aphid or grain aphid, and in Ambrozja seedlings colonized with 20–60 *R. padi* or 40–60 *S. avenae* aphids per plant. For the other tested bioassay variants, the highest $O_2^{\bullet-}$ generation occurred after 8 hpi compared to the non-stressed seedlings. Prolonged aphid feeding resulted in a progressive decrease in the amount of analysed ROS in comparison to maximal changes observed after 4–8 h of aphid colonization. Furthermore, factorial analysis of variance (ANOVA) revealed significant effects of the experimental indicators and their interactions on levels of $O_2^{\bullet-}$ production in the maize seedlings (Table 3).

Transcriptional activity of *gst1* gene in the aphid-stressed maize seedlings

The conducted biotests demonstrated that short-term feeding of the examined cereal aphids (*R. padi* or *S. avenae*) did not influence the amount of *gst1* mRNA transcript in the seedlings of Ambrozja and Tasty Sweet maize genotypes (Figure 1). Two hours after initial infestation, the low abundance of aphids (10–20 individuals per plant) did not alter the gene expression, but a higher number of insects (40–60 per seedlings) stimulated a slight increment in transcriptional activity of the target gene (from 5% increase in Tasty Sweet plants colonized with 60 *S. avenae* to a 26% increase in Ambrozja plants infested with the same number of *R. padi* aphids). After 4 and 8 hpi, the levels of *gst1* transcript gradually enhanced in both maize varieties colonized with the tested aphid species, with the exception of two aphid treatments (10 and 20 insects per plant) at 4 hpi when there were no changes in the relative expression of the analysed gene in Tasty Sweet genotype. The highest accumulation of the *gst1* transcript amount in the aphid-infested maize seedlings of both *Z. mays* genotypes occurred at 24 hpi and 60 aphids per plant (4.3–5.5-fold elevations in Ambrozja, and 2.4–3.1-fold increases in Tasty Sweet seedlings, depending on the aphid species). However, extended insect colonization (48–72 hpi) resulted in a gradually lower gene

Table 1. Levels of $O_2^{\bullet-}$ generation (ΔA_{540} min^{-1} g^{-1} fresh weight) in leaves of the maize seedlings colonized with *R. padi*.

Time intervals of aphid infestation (hpi)	Aphid abundance (per plant)				
	0	**10**	**20**	**40**	**60**
Ambrozja genotype					
0	0.47±0.03a	0.47±0.03a	0.47±0.03a	0.47±0.03a	0.47±0.03a
1	0.47±0.02b	0.47±0.02b	0.47±0.02b	0.49±0.03ab	0.51±0.04a
2	0.48±0.04b	0.48±0.04b	0.49±0.03b	0.52±0.04ab	0.58±0.03a
4	0.47±0.03d	0.50±0.04d	0.81±0.06c	1.23±0.08b	1.45±0.10a
8	0.49±0.04d	0.65±0.04c	0.70±0.05c	0.91±0.06ab	1.03±0.06a
24	0.48±0.04d	0.57±0.05c	0.62±0.05bc	0.69±0.04b	0.87±0.07a
48	0.48±0.03cd	0.55±0.04c	0.57±0.03bc	0.63±0.03b	0.81±0.06a
72	0.49±0.04c	0.53±0.03c	0.54±0.04bc	0.60±0.04ab	0.76±0.05a
Tasty Sweet genotype					
0	0.42±0.02a	0.42±0.02a	0.42±0.02a	0.42±0.02a	0.42±0.02a
1	0.42±0.02a	0.42±0.02a	0.42±0.02a	0.43±0.02a	0.44±0.02a
2	0.44±0.03a	0.44±0.03a	0.45±0.02a	0.47±0.03a	0.49±0.04a
4	0.43±0.02bc	0.44±0.03bc	0.47±0.03b	0.50±0.05b	0.71±0.08a
8	0.43±0.02d	0.50±0.04cd	0.53±0.05b	0.62±0.06a	0.66±0.06a
24	0.45±0.04bc	0.48±0.03b	0.51±0.04ab	0.54±0.05a	0.61±0.05a
48	0.44±0.03bc	0.45±0.04bc	0.50±0.04b	0.51±0.05ab	0.58±0.05a
72	0.43±0.03b	0.43±0.03b	0.46±0.03ab	0.48±0.03ab	0.55±0.05a

Values are the means ± standard deviation (SD) of three independent experiments (10 plants per repeat); hpi-hours post-initial insect infestation; the different letters in rows denote significant differences according to Tukey's test (P≤0.05).

Table 2. Levels of $O_2^{\bullet-}$ generation (ΔA_{540} min^{-1} g^{-1} fresh weight) in leaves of the maize seedlings colonized with *S. avenae*.

Time intervals of aphid infestation (hpi)	Aphid abundance (per plant)				
	0	**10**	**20**	**40**	**60**
Ambrozja genotype					
0	0.47±0.03a	0.47±0.03a	0.47±0.03a	0.47±0.03a	0.47±0.03a
1	0.47±0.02a	0.47±0.02a	0.47±0.02a	0.48±0.03a	0.49±0.03a
2	0.48±0.04ab	0.48±0.04ab	0.48±0.04ab	0.50±0.04ab	0.54±0.04a
4	0.47±0.03c	0.49±0.03c	0.52±0.05c	0.88±0.07ab	1.02±0.08a
8	0.49±0.04c	0.60±0.04b	0.67±0.05b	0.65±0.04b	0.85±0.06a
24	0.48±0.04b	0.54±0.03b	0.63±0.05ab	0.62±0.05ab	0.76±0.05a
48	0.48±0.03bc	0.52±0.04b	0.59±0.03ab	0.59±0.03ab	0.68±0.04a
72	0.49±0.04bc	0.51±0.03b	0.54±0.04b	0.55±0.03b	0.63±0.05a
Tasty Sweet genotype					
0	0.42±0.02a	0.42±0.02a	0.42±0.02a	0.42±0.02a	0.42±0.02a
1	0.42±0.02a	0.42±0.02a	0.42±0.02a	0.43±0.02a	0.43±0.03a
2	0.44±0.03a	0.44±0.03a	0.44±0.03a	0.45±0.02a	0.48±0.04a
4	0.43±0.02b	0.45±0.03b	0.45±0.02b	0.47±0.03b	0.64±0.05a
8	0.43±0.02b	0.48±0.03ab	0.51±0.05a	0.58±0.04a	0.55±0.04a
24	0.45±0.04ab	0.47±0.02ab	0.49±0.04a	0.51±0.04a	0.52±0.05a
48	0.44±0.03ab	0.44±0.03ab	0.47±0.03a	0.49±0.03a	0.50±0.04a
72	0.43±0.03ab	0.43±0.04ab	0.44±0.03ab	0.46±0.02a	0.48±0.03a

Values are means ± standard deviation (SD) of three independent experiments (10 plants per repeat); hpi-hours post-initial insect infestation; different letters in rows denote significant differences according to Tukey's test (P≤0.05).

Table 3. Factorial ANOVA results for tested indicators (*Z. mays* cultivar, hemipteran species, insect abundance and aphid exposure period) and interdependence between these parameters affecting $O_2^{\bullet-}$ formation in the maize seedlings.

Tested factors and interactions	Df	F	p
Maize cultivar (C)	1	175.2	≤0.001
Hemipteran species (S)	2	87.9	≤0.001
Insect abundance (A)	3	68.2	≤0.001
Aphid exposure period (EP)	7	52.7	≤0.001
S × C	2	19.6	≤0.001
S × A	6	14.5	≤0.001
C × A	3	27.1	≤0.001
S × EP	14	14.9	≤0.001
C × EP	7	18.5	≤0.001
A × EP	21	10.6	≤0.001
S × C × A	6	12.4	≤0.001
S × C × EP	14	9.9	≤0.001
S × A × EP	42	8.7	≤0.001
C × A × EP	21	4.9	≤0.001
S × C × A × EP	42	3.7	≤0.008

Df-degrees of freedom; p-values less than 0.05 were considered significant; F-ratio is defined as the variance between samples/the variance within samples.

expression in relation to the levels recorded at 24 hpi. Generally, *R. padi* infestation led to a more profound increase in the transcriptional activity of the *gst1* gene in comparison with *S. avenae* (e.g. 120% higher increase in Ambrozja and 69% increment in Tasty Sweet plants, at 24 hpi and 60 aphids per plant). The results of factorial ANOVA confirmed a significant influence of the analysed indicators and their interactions on expression of the *gst1* gene in the maize seedlings (Table 4).

Amount of *gst18* transcript in the insect-injured *Z. mays* seedlings

The performed analyses revealed that the transcriptional activity of the *gst18* gene in tissues of both maize cultivars remained unaffected after 1 or 2 h of aphid colonization (Figure 2). The 4 h infestation with a higher density of insects (40–60 per seedling) resulted in slightly enhanced levels of gene expression (16–112% increment), whereas a lower abundance (10–20 aphids per plant) did not evoke any alternations compared to the control. Eight hours after the initial infestation, the transcriptional activity of the *gst18* gene in seedlings of the investigated *Z. mays* cultivars gradually increased in proportion to the number of hemipterans per plant (21–82% increase in Tasty Sweet and 27–440% increase in Ambrozja variety). It is important to note, that the highest stimulation of target gene expression occurred at 24 hpi. Colonization of maize plants with *R. padi* aphids at this time point led to 1.4–4.1-fold and 2.1–6.2-fold elevations in the transcript abundance in Tasty Sweet and Ambrozja seedlings, accordingly, whereas the grain aphid attack resulted in 1.3–3.4 -fold and 1.7–5.5-fold increases in the corresponding maize cultivars. During the next two periods of aphid infestation the scale of upregulation of the *gst18* gene in both maize genotypes was less pronounced (1.2–4.9-fold increments at 48 hpi; 1.1–4.2 -fold elevations at 72 hpi, depending on the aphid treatments). Importantly, *R. padi*-colonized maize plants were characterized with greater amounts of the target transcript (20–97% Tasty Sweet and 61–163% Ambrozja) in relation to *S. avenae*-attacked seedlings. Furthermore, it was evidenced that the aphid-infested

Ambrozja plants responded with much greater increases in the *gst18* gene expression compared to the infested Tasty Sweet genotype (e.g. 50–258% greater increments at 60 insects per plant). Statistical analysis confirmed the considerable impact of tested parameters and their interrelation on expression of the analysed gene in the investigated *Z. mays* plants (Table 4).

Relative expression of *gst23* gene in maize plants colonized with cereal aphids

Results concerning the expression levels of the *gst23* gene in the aphid-infested seedlings of *Z. mays* are depicted in figure 3. It has been found that feeding *S. avenae* or *R. padi* for 1 h did not evoke any disturbances in the transcriptional activity of the target gene in tissues of the investigated maize cultivars. Insect feeding for 2 h did not result in any changes in the *gst1* gene expression in *S. avenae*-infested Tasty Sweet plants, regardless of the number of aphids per plant. Likewise, the tested cereal aphids (10–20 insects per seedling) did not affect the transcriptional activity of the analysed gene in both maize genotypes (Ambrozja or Tasty Sweet). However, higher numbers of aphids (40–60 per plant) led to an elevation in the *gst23* transcript abundance, ranging from 10% in Tasty Sweet plants colonized by 40 *R. padi* aphids to 42% increase in Ambrozja seedlings infested by 60 insects per plant. Further extension of colonization period (4 hpi) resulted in a continuous increase (3–132%) in *gst23* gene expression in the maize tissues compared with the relevant control plants. The maximal induction of the target gene in aphid-attacked maize plants occurred at 8 hpi. At this time point, 10–60 *R. padi* per plant evoked 1.9–2.8 -fold and 2.3–7.2-fold increases in the levels of transcript accumulation in Tasty Sweet and Ambrozja plants, respectively. Infestation with *S. avenae* (10–60 aphids per seedling) caused 1.6–2.2-fold and 2.1–5.3-fold elevations in Tasty Sweet and Ambrozja varieties, respectively. Prolonged exposure to aphids (24–72 hpi) could be linked to a progressively lower upregulation of the *gst23* gene in comparison with the changes observed at 8 hpi. Interestingly, long-term colonization (72 hpi) by the grain aphid did not influence the analysed transcript amount in Tasty Sweet

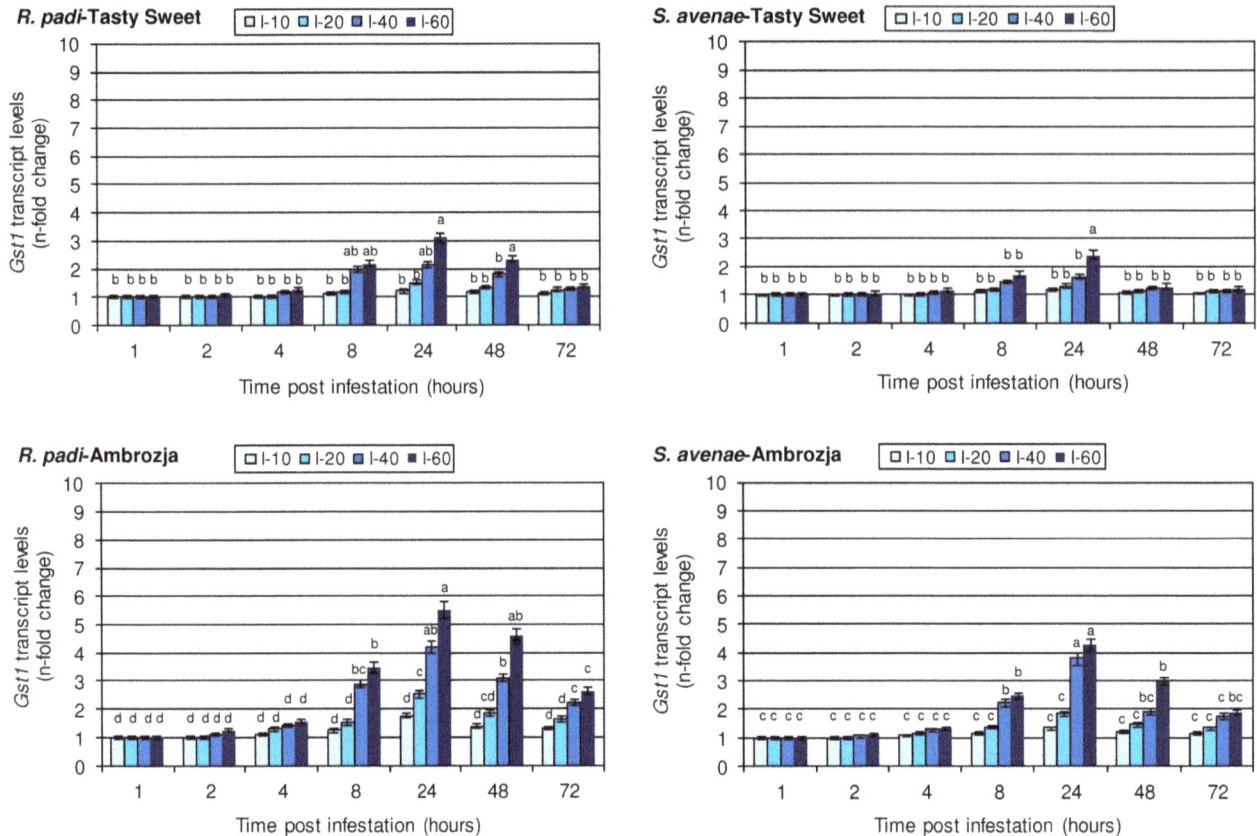

Figure 1. Influence of the tested cereal aphids on *gst1* gene expression in the seedlings of Ambrozja and Tasty Sweet maize cultivars. Values signify the mean *n*-fold changes in the *gst1* transcript abundance in the aphid-stressed *Z. mays* plants in comparison with the non-infested group of seedlings. Error bars represent the standard deviation ($\pm$ SD). For each maize-aphid treatment, three independent biological replicates were accomplished. The obtained gene expression data were normalized to the *gapdh* gene. The different letters above the SD bars designate significant differences among compared plants at P$\leq$0.05 based on the Tukey's test. I-10, I-20, I-40 and I-60 are the levels of aphid infestation (10, 20, 40 and 60 insects per plant, accordingly).

plants in relation to the relative non-infested seedlings. Comparative analyses revealed that the bird cherry-oat aphid caused a more noticeable augmentation of *gst23* gene expression (11–170%, depending on the aphid treatment and maize cultivar) in comparison with *S. avenae* aphids. Moreover, elevation of the transcriptional activity of the *gst23* gene in maize plants occurred in parallel with increasing aphid densities per plant. A markedly higher transcript amount was found in the insect-stressed Ambrozja seedlings compared to Tasty Sweet plants. For example, after 8 h infestation, 60 *R. padi* aphids stimulated 2.8- and 7.2-fold increments in Tasty Sweet and Ambrozja plants, respectively, whereas feeding the same number of *S. avenae* individuals led to 2.2- and 5.3-fold increases in Tasty Sweet and Ambrozja varieties, respectively. The results of factorial ANOVA analysis proved that there was a significant impact of the investigated parameters and their interdependence on the transcriptional activity of the *gst23* gene in the maize seedlings (Table 5).

Abundance of *gst24* transcript in *Z. mays* seedlings infested with the cereal aphids

Relative expression data of the *gst24* gene in aphid-colonized maize seedlings are presented in figure 4. Transcriptional activity of the target gene in tissues of both tested *Z. mays* cultivars infested with *R. padi* or *S. avenae* remained at the same levels after 1 hpi, when compared to the respective control seedlings. In maize plants

exposed to insect infestation for 2 h, only subtle accumulation of the *gst24* transcript was recorded (3–10% increase in Tasty Sweet seedlings, and 6–24% elevation in Ambrozja plants). Prolonged aphid colonization (4–8 hpi) was associated with a steady enhancement in the expression of the analysed gene from 5% elevation in *S. avenae*-infested Tasty Sweet plants to 133% increment in *R. padi*-attacked Ambrozja seedlings, compared to the controls. The highest enhancement in the transcript amount for tissues of the aphid-infested maize plants was found at 24 hpi (e.g. 60 *R. padi* aphids influenced 2.5-fold and 4.5-fold increases in Tasty Sweet and Ambrozja seedlings, respectively, whereas the same abundance of *S. avenae* affected 2.0- and 3.8-fold increments in the relevant maize genotypes). It should be emphasized that insect infestation for 48 and 72 h resulted in a gradually decreasing upregulation of *gst24* gene expression in *Z. mays* seedlings of the investigated cultivars in relation to the changes demonstrated after 24 h. Furthermore, the aphid-attacked Ambrozja plants responded to a higher elevation in the transcriptional activity of the target gene when compared with Tasty Sweet variety (e.g. 12–205% larger increase at the highest level of aphid infestation). It was additionally demonstrated that there was a higher abundance of the target mRNA transcript in *R. padi*-infested maize cultivars in comparison with *S. avenae*-stressed seedlings. It is important to underline that the scale of alternations in the gene expression was proportional to densities of the tested hemipterans on the seedlings

Table 4. Factorial ANOVA results for tested indicators (*Z. mays* cultivar, hemipteran species, insect abundance and aphid exposure period) and interactions between these parameters affecting *gst1* and *gst18* transcript amounts in the maize seedlings.

Tested factors and interactions	Df	F	p	F	p
		gst1 gene		*gst18* gene	
Maize cultivar (C)	1	986.8	≤0.001	916.9	≤0.001
Hemipteran species (S)	2	852.4	≤0.001	1645.2	≤0.001
Insect abundance (A)	3	1447.0	≤0.001	1362.5	≤0.001
Aphid exposure period (EP)	7	1078.3	≤0.001	1573.8	≤0.001
S × C	2	561.4	≤0.001	965.2	≤0.001
S × A	6	374.2	≤0.001	1258.9	≤0.001
C × A	3	309.2	≤0.001	724.5	≤0.001
S × EP	14	407.4	≤0.001	1419.8	≤0.001
C × EP	7	275.1	≤0.001	1160.4	≤0.001
A × EP	21	223.2	≤0.001	583.9	≤0.001
S × C × A	6	79.5	≤0.001	185.3	≤0.001
S × C × EP	14	77.8	≤0.001	306.0	≤0.001
S × A × EP	42	58.0	≤0.001	163.2	≤0.001
C × A × EP	21	45.2	≤0.001	142.5	≤0.001
S × C × A × EP	42	13.8	≤0.001	40.5	≤0.001

Df-degrees of freedom; *p*-values less than 0.05 were considered significant; F-ratio is defined as the variance between samples/the variance within samples.

of the investigated maize varieties. The maximal abundance of bird cherry-oat aphids (60 insects per plant) led to 7–50% and 9–71% higher increments of *gst24* gene expression in the Tasty Sweet and Ambrozja plants, respectively, relative to the number of grain aphids. The statistical analysis evidenced significant effects of the tested variables and their interconnections in terms of *gst24* gene expression in *Z. mays* plants (Table 5).

Discussion

Monophagous *Sitobion avenae* F. (grain aphid) and oligophagous *Rhopalosiphum padi* L. (bird cherry-oat aphid) become serious pest species building up numerous colonies on many maize varieties grown in Poland, especially during warm and moist vegetative seasons [36–38]. Despite many research groups conducting extensive studies on the complex plant-aphid interactions, the participation of these hemipterans in the generation of oxidative stress and the functioning of the antioxidant defence network in the host systems still remain to be unraveled. To the best of our knowledge, this is the first report evaluating the impact of *R. padi* or *S. avenae* infestations on the expression profiles of the four genes encoding glutathione transferase isozymes (GSTF1, GST18, GST23 and GST24), as well as the levels of superoxide anion radical generation in the seedlings of susceptible (Tasty Sweet) and relatively resistant (Ambrozja) maize genotypes.

Aphid salivary glands produce a battery of hydrolytic enzymes that participate in the cleavage of primary and secondary cell walls, plasma membranes, and a variety of intracellular compounds. Additionally, salivary secretions of these hemipterans contain various elicitors, metabolic regulators, and phytotoxic constituents that trigger cascades of local and/or systemic defensive reactions as well as the processes of premature senescing, apoptosis, or necrosis within the colonized plant systems [39–41]. Studies have documented that proteinaceous effectors (Mp10 and Mp42) from *M. persicae* are capable of enhancing the defence systems in *Nicotiana benthamiana* Dom. plants, whereas two elicitors of *Macrosiphum euphorbiae* Thom., Me10 and Me23, possess the ability to suppress the host reactions in order to facilitate prolonged phloem feeding [13-14]. Aphid saliva infiltration and profound ultrastructural damages induced by insect mouthparts in the host tissues may be linked to excessive ROS release in the attacked organs. Superoxide anion radical is one of the major and most deleterious reactive oxygen species generated in plant cells both in the normal physiological state and in response to adverse environmental stimuli. It was found that the seedlings of both maize varieties colonized with *R. padi* or *S. avenae* aphids responded an early overproduction of $O_2^{\cdot-}$ in comparison to the non-stressed control. The maximal enhancement in the superoxide anion radical generation in *Z. mays* seedlings was noted after 4–8 h of aphid feeding. Interestingly, a more marked elevation in $O_2^{\cdot-}$ amounts occurred in the seedlings of Ambrozja (relatively resistant) plants in relation to Tasty Sweet (susceptible) cultivar. These observations are coherent with the results obtained by Mai et al. [15] who ascertained that *Pisum sativum* L. plants infested with the pea aphid (*Acyrthosiphon pisum* Harr.) possessed substantially higher amounts of $O_2^{\cdot-}$ relative to the insect-free control. Furthermore, the most significant increase in excessive $O_2^{\cdot-}$ formation was found at the highest infestation level (30 aphids per seedling). According to these authors, the prolonged aphid feeding resulted in the progressive increase in $O_2^{\cdot-}$ levels within the attacked plants (e.g. 1.46- and 1.81-fold increments in relation to the reference plants at 24 and 96 hpi, accordingly). Moreover, it was reported that Russian wheat aphid (*Diuraphis noxia* Mordv.) markedly augmented the biosynthesis of hydrogen peroxide in resistant wheat plants in relation to the aphid-susceptible line. The oxidative burst in plants is associated with a dramatic increase in superoxide anion radicals' production at early stages of the exposure to various biotic stressing factors [28]. This phenomenon is linked with subsequent oxidative wave passing throughout plant tissues, leading to triggering the defence networks in the hosts, on the one hand, and possible suppression of the growth and development of herbivorous insects, on the

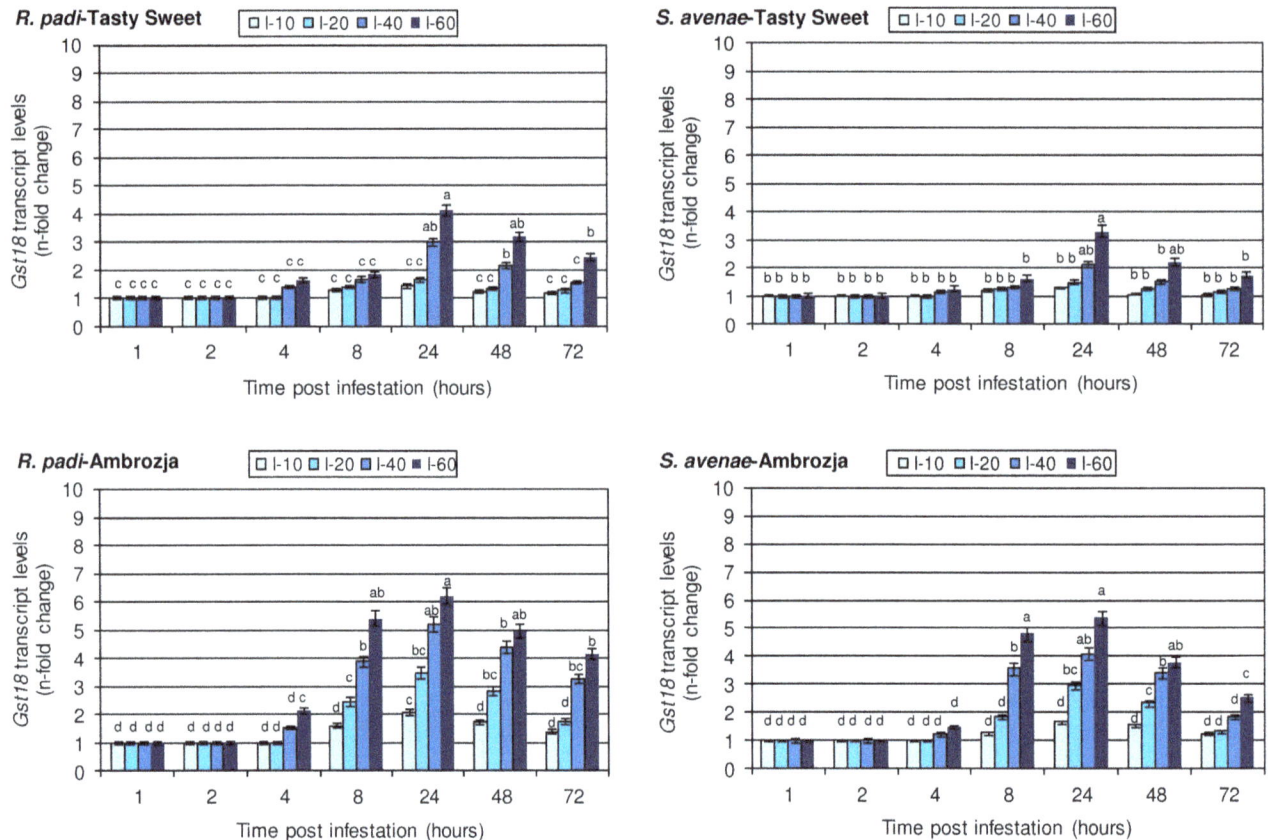

Figure 2. Influence of the tested cereal aphids on *gst18* gene expression in the seedlings of Ambrozja and Tasty Sweet maize cultivars. Values signify the mean *n*-fold changes in the *gst18* transcript abundance in the aphid-stressed *Z. mays* plants in comparison with the non-infested group of seedlings. Error bars represent the standard deviation (± SD). For each maize-aphid treatment, three independent biological replicates were accomplished. The obtained gene expression data were normalized to the *gapdh* gene. The different letters above the SD bars designate significant differences among compared plants at P≤0.05 based on the Tukey's test. I-10, I-20, I-40 and I-60 are the levels of aphid infestation (10, 20, 40 and 60 insects per plant, accordingly).

other hand [42]. In order to overcome the excessive accumulation of this highly reactive and cytotoxic ROS form, the superoxide anion radicals are converted in the dismutation reaction to molecular oxygen (O_2) and less toxic hydrogen peroxide (H_2O_2) [6,15–16,39]. Furthermore, we revealed very slight changes in superoxide anion radicals content in non-infested maize seedlings of both tested cultivars with duration of experimental time, but the recorded differences were not statistically significant. It is probable that isolation of *Z. mays* seedlings with the cover gauze could cause a minor mechanical stress influencing negligible fluctuations in $O_2^{\bullet-}$ amount.

Plants have developed a number of defence mechanisms that are involved with protecting the cells from the detrimental impact of exaggerated ROS formation in response to a variety of abiotic and biotic stresses [43–46]. Until now, it has been identified at least 42 genes encoding diverse isozymes of glutathione transferase in maize [47]. In recent years, an important role of cytosolic GSTs in the alleviation of oxidative stress in plant tissues has been increasingly described [48–52]. The GSTs predominantly occur as homo- or heterodimers, with subunits of 23–30 kDa [53]. It should be underlined that among diverse groups of GST isozymes, only Tau and Phi classes are plant specific [50]. The performed molecular studies revealed that the cereal aphid infestations led to significant increases in the relative expression of analysed *gst* genes (*gst1*, *gst18*, *gst23* and *gst24*) in the seedling leaves of both *Z. mays*

genotypes, exhibiting distinct susceptibility levels to the insect colonization. Time-course analysis revealed that the target genes encoding the relevant GST isoenzymes (GSTF1, GST18, GST23 and GST24) were maximally upregulated at different aphid exposure periods (*gst23* at 8 hpi; *gst1*, *gst18*, and *gst24* genes at 24 hpi). Interestingly, the bird cherry-oat aphid infestation caused more considerable increments in the amounts of all tested *gst* transcripts in the maize plants compared to grain aphid feeding. Additionally, relatively resistant Ambrozja plants that were attacked by the cereal aphids were characterized with a higher stimulation of the transcriptional activity of the *gst* genes in relation to the susceptible Tasty Sweet plants. There have been limited reports published evidencing aphid-stimulated transcriptional reprogramming in the attacked host plants [54–58]. Microarray experiments performed by Kuśnierczyk and co-workers revealed that feeding of *M. persicae* or *Brevicoryne brassicae* L. for 72 h led to significant alternations in the transcriptional activity of 13 *gst* genes in 22–30-day-old plants of three tested *Arabidopsis thaliana* ecotypes (Landsberg *erecta*/L*er*/, Cape Verde Islands/Cvi/, and Wassilewskija/Ws/) [54]. The aphid colonization (8–12 insects per leaf) resulted in the overexpression of most analysed *gst* genes in the plants representing the tested ecotypes when compared to the non-stressed control. The opposite tendency was identified in the expression patterns of GSTU18 and GSTU20 transcripts (0.23–1.60-fold and 0.26–1.89-

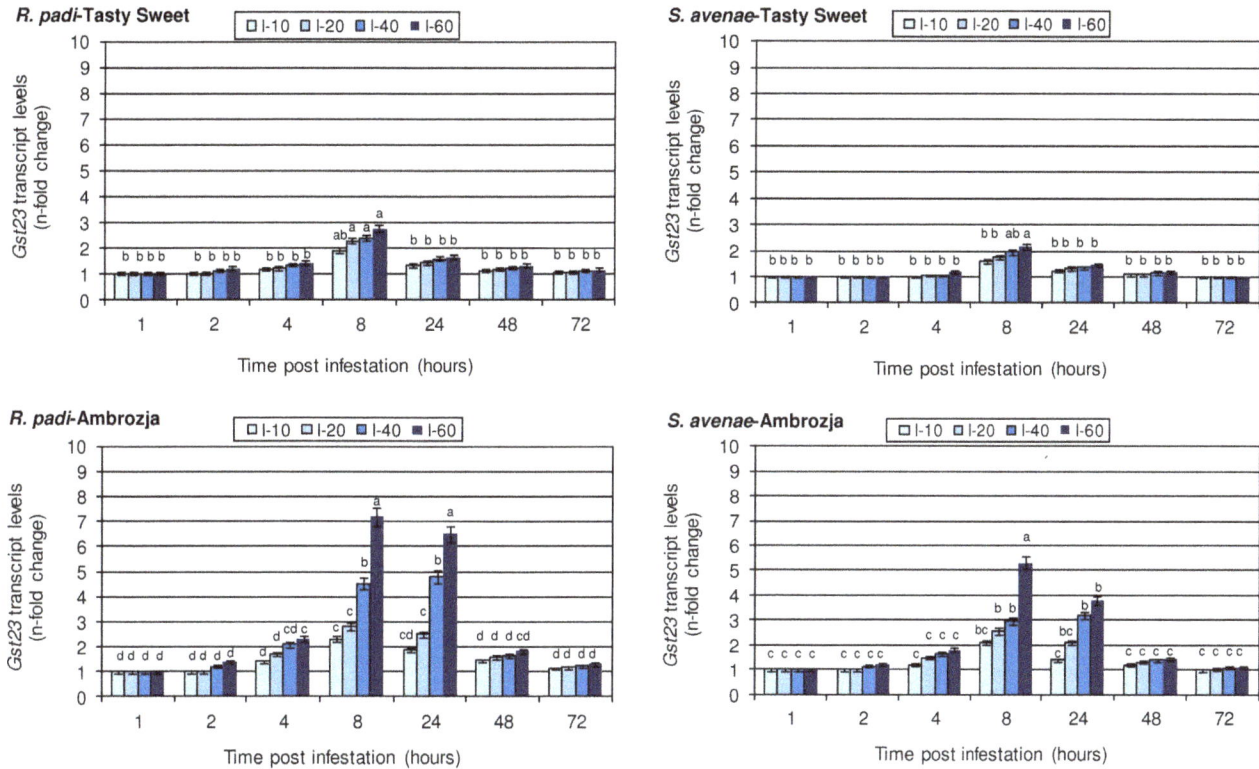

Figure 3. Influence of the tested cereal aphids on *gst23* gene expression in the seedlings of Ambrozja and Tasty Sweet maize cultivars. Values signify the mean *n*-fold changes in the *gst23* transcript abundance in the aphid-stressed *Z. mays* plants in comparison with the non-infested group of seedlings. Error bars represent the standard deviation ($\pm$ SD). For each maize-aphid treatment, three independent biological replicates were accomplished. The obtained gene expression data were normalized to the *gapdh* gene. The different letters above the SD bars designate significant differences among compared plants at $P \leq 0.05$ based on the Tukey's test. I-10, I-20, I-40 and I-60 are the levels of aphid infestation (10, 20, 40 and 60 insects per plant, accordingly).

Table 5. The factorial analysis of variance of tested indicators (*Z. mays* cultivar, hemipteran species, insect abundance and aphid exposure period) and interactions between these parameters affecting *gst23* and *gst24* transcript amounts in the maize seedlings.

Tested factors and interactions	Df	F	p	F	p
		gst23 gene		**gst24 gene**	
Maize cultivar (C)	1	1142.5	≤ 0.001	573.2	≤ 0.001
Hemipteran species (S)	2	748.3	≤ 0.001	1229.2	≤ 0.001
Insect abundance (A)	3	1325.2	≤ 0.001	495.5	≤ 0.001
Aphid exposure period (EP)	7	1409.8	≤ 0.001	1050.9	≤ 0.001
S × C	2	721.4	≤ 0.001	147.6	≤ 0.001
S × A	6	438.9	≤ 0.001	120.8	≤ 0.001
C × A	3	166.3	≤ 0.001	29.5	≤ 0.001
S × EP	14	844.6	≤ 0.001	275.0	≤ 0.001
C × EP	7	275.0	≤ 0.001	50.6	≤ 0.001
A × EP	21	196.5	≤ 0.001	47.3	≤ 0.001
S × C × A	6	45.9	≤ 0.001	16.7	≤ 0.001
S × C × EP	14	69.4	≤ 0.001	7.4	≤ 0.004
S × A × EP	42	53.1	≤ 0.001	12.8	≤ 0.001
C × A × EP	21	27.7	≤ 0.001	3.6	≤ 0.006
S × C × A × EP	42	10.5	≤ 0.001	1.5	≤ 0.017

Df-degrees of freedom; *p*-values less than 0.05 were considered significant; F-ratio is defined as the variance between samples/the variance within samples.

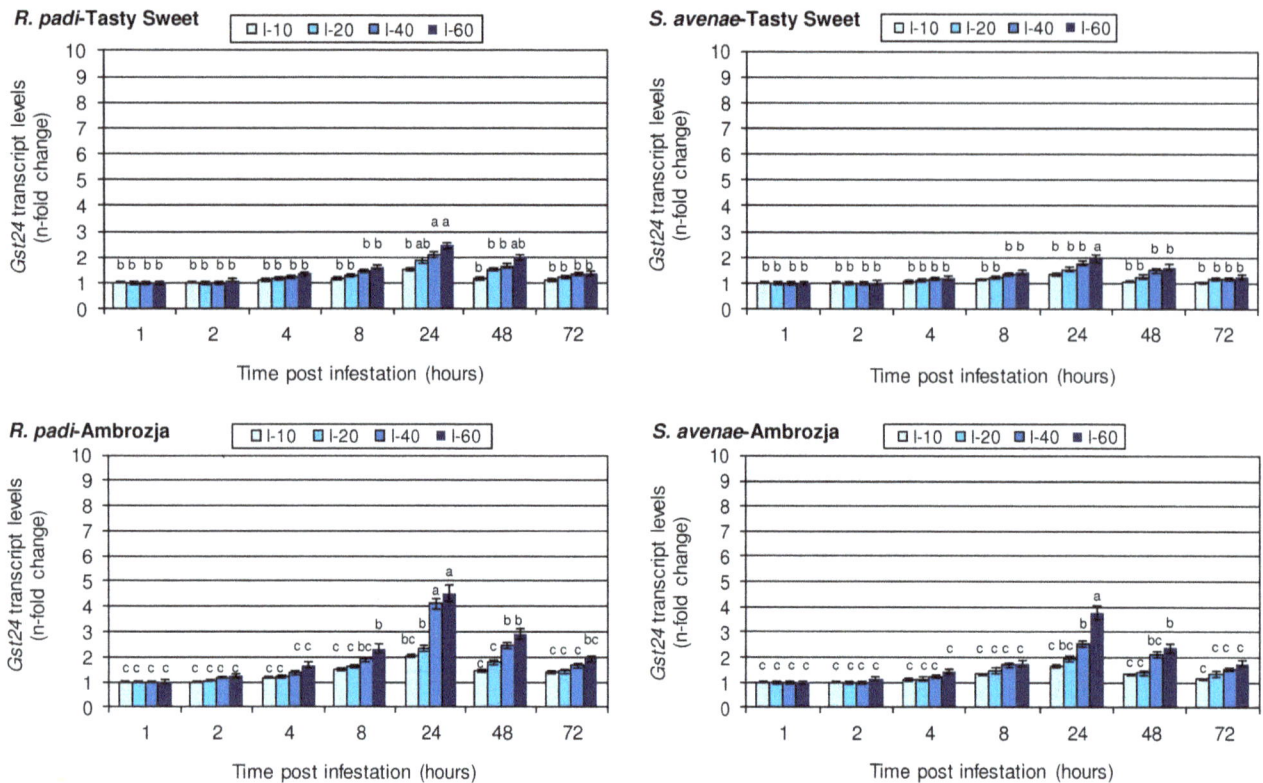

Figure 4. Influence of the tested cereal aphids on *gst24* gene expression in the seedlings of Ambrozja and Tasty Sweet maize cultivars. Values signify the mean *n*-fold changes in the *gst24* transcript abundance in the aphid-stressed *Z. mays* plants in comparison with the non-infested group of seedlings. Error bars represent the standard deviation ($\pm$ SD). For each maize-aphid treatment, three independent biological replicates were accomplished. The obtained gene expression data were normalized to the *gapdh* gene. The different letters above the SD bars designate significant differences among compared plants at P$\leq$0.05 based on the Tukey's test. I-10, I-20, I-40 and I-60 are the levels of aphid infestation (10, 20, 40 and 60 insects per plant, accordingly).

fold down regulation, respectively) in the investigated ecotypes in relation to the insect-free plants. Additionally, these authors demonstrated upregulation of the glutathione-conjugate transporter (MRP4) in the aphid-injured Cape Verde Islands and Wassilewskija plants. Another infestation experiments conducted by Kuśnierczyk et al. revealed that 21–25-day-old *A. thaliana* plants (Landsberg *erecta*/L*er*/ecotype) infested with *B. brassicae* (4 aphids per leaf) were characterized with an early enhancement (at 6 hpi) of the amount of ATGST6, ATGST7 and ATGST10 transcripts relative to the control. Furthermore, prolonged aphid colonization (48 hpi) led to the strong upregulation of four glutathione transferase (ATGSTU3, ATGSTU10, ATGSTU11, ATGSTL1) genes, as well as increases in the transcript amounts of two glutathione S-conjugate transporters (MRP3 and MRP4) [55]. Similarly, Moran et al. elucidated that infestation of *A. thaliana* with *M. persicae* aphids for 72 h resulted in 2.9-fold- and 4.8-fold elevations in the expression of *gst1* and *gst11* genes, respectively, compared to the non-treated control [56]. Stotz et al. also ascertained that the diamondback moth (*Plutella xylostella* L.), feeding on the rosette leaves of wild-type *A. thaliana* plants, influenced profound increments in the expression of *gst2* and *gst6* genes compared to the insect-free control [57]. Likewise, Bandopadhyay and co-workers evidenced that the transcriptional activity of genes encoding various glutathione transferase isoforms may vary significantly depending on the duration of the aphid exposure period [58]. According to the cited authors, *Rorippa indica* L. plants infested with mustard aphids (*Lipaphis erysimi*/L./Kalt.) responded with a 2.5-fold elevation in the transcriptional

activity of the *AT1G78370* gene (glutathione transferase AtTAU20) at 12 hpi, but a dissimilar trend occurred when the insect colonization was extended to 48 hpi. Furthermore, it should be noted that other biotic stressors, such as pathogenic fungi or microorganisms are able to trigger notable modifications in the expression patterns of several *gst* genes within the hosts. For example, it has been elucidated that maize plants infected with *Ustilago maydis* possessed an increased transcriptional activity of seven transferase glutathione genes (*gst15*, *gst18*, *gst20*, *gst24*, *gst25*, *gst30*, and *gst36*) after 12 h post-fungal inoculation [59]. The upregulation levels ranged from a 3.1-fold increase of the *gst18* gene expression to a 108-fold increment in the *gst30* transcript abundance compared to non-treated plants. Microarray data achieved by Luo et al. revealed that the expression of transferase glutathione genes in maize kernels of aflatoxin-resistant (Eyl25) and aflatoxin-susceptible (Eyl31) lines differentially responded 72 h after inoculation with *Aspergillus flavus* [60]. Some authors have suggested that the induction of *gst* genes is involved in limiting the adverse effects of oxidative stress within plant tissues, including the reduction of cell death events occurring as a result of the hypersensitive reactions [17,59–60].

In the present study, *R. padi* infestation contributed to a substantially greater upregulation of the analysed *gst* genes and to the increases in the $O_2^{\bullet-}$ generation in seedlings of both Ambrozja and Tasty Sweet genotypes in comparison to grain aphid feeding. Oligophagous bird cherry-oat aphids alternate the host plants between members of the *Prunus* genus (winter hosts) and a broad set of *Poaceae* species (summer hosts), whereas the life cycle of monophagous *S. avenae* is associated with numerous grasses and cereals [10–11,61]. Greater diversity of plant systems colonized with *R. padi* indicates a higher adaptation of this hemipteran species to the chemical composition of the hosts. Conceivable sources of distinct biochemical and molecular effects in aphid-infested maize plants may be caused by differences in the insect salivary compounds and specific routes of stylet insertion throughout the plant tissues. It has been reported that salivary secretions of the bird cherry-oat aphid contain a wide spectrum of biocatalysts, which are responsible for the hydrolysis of structural macromolecules in the primary and secondary cell walls, and plasmalemma [62–63]. Furthermore, microscopic observations conducted by some researchers have documented additional profound injuries within the mesophyll cells of both winter and summer hosts, whereas *S. avenae* infestation resulted in a much lower range of ultrastructural damages, and they displayed a typical intercellular mode of mouthparts passage within the winter wheat Sakva plants [64–65]. According to Urbańska et al., the bird cherry-oat aphid has evolved an adaptive enzymatic mechanism that allows detoxification of harmful cyanogenic constituents present in the leaves of primary hosts [66]. Łukasik et al. provided valuable findings, indicating that *R. padi* feeding caused greater depletion in the content of ascorbate and greater stimulation of ascorbate peroxidase activity in the triticale seedlings compared to *S. avenae* aphids [16]. Similarly, Sytykiewicz revealed that bird cherry-oat aphid infestation of maize plants evoked a more significant decrease in the total antioxidant capacity towards the DPPH (1,1-diphenyl-2-picrylhydrazyl) radical in relation to grain aphid colonization [6]. It may be assumed that a decreased efficacy of DPPH radical scavenging activity in aphid-infested maize plants might be associated with a continuous pressure of biotic stressing factor (aphid colonization) that triggered the oxidative stress in the host systems. It is particularly evident when massive and/or prolonged aphid infestation occurred. It is likely that the pool of available antioxidants under stressful conditions significantly depressed the total antioxidative capacity of extracts derived from the infested maize seedlings when compared to the control. On the other hand, lower contents of ascorbate and glutathione were evidenced, as well as higher levels of ascorbate peroxidase and glutathione transferase activities in tissues of the bird cherry-oat aphid, in comparison with *S. avenae* individuals, which proves that there are significant differences in the functioning of the antioxidative machinery within these cereal aphids [12,67].

This report provides new insight into the molecular basis of highly complex antioxidative responses of the model maize plants colonized with cereal aphids. It was demonstrated, there is differential regulation of four *gst* genes, encoding various isoforms of glutathione transferase in the insect-challenged seedling leaves of *Z. mays*, representing high and low susceptibility to the aphid colonization. The obtained results revealed insect-triggering oxidative stress and the crucial role of glutathione transferases in constituting complex defence reactions in the attacked host systems. In order to gain a better understanding of the elicitation of the plant defence reactions which occur at the early stages of aphid infestation in maize plants, the extended molecular analyses comprising transcriptome-wide screening of other aphid-regulated genes, as well as identification of low molecular and regulatory RNA molecules (e.g. miRNA) and assessing their gene expression profiles should be performed.

Supporting Information

Table S1 The set of *Z. mays* glutathione transferase genes analysed with the application of *TaqMan Gene Expression Assays*[#]. [#] *TaqMan Gene Expression Assays* used in the performed experiments were developed and supplied by Life Technologies (Poland).

Author Contributions

Conceived and designed the experiments: HS GC PC. Performed the experiments: HS. Analyzed the data: HS IS. Wrote the paper: HS IŁ CS SG.

References

1. Bender RS, Haegele JW, Ruffo ML, Below FE (2013) Nutrient uptake, partitioning, and remobilization in modern, transgenic insect-protected maize hybrids. Agron J 105: 161–170.
2. Dukowic-Schulze S, Harris A, Li J, Sundararajan A, Mudge, et al. (2014) Comparative transcriptomics of early meiosis in *Arabidopsis* and maize. J Genet Genomics 41: 139–152.
3. Zhao Y, Cai M, Zhang X, Li Y, Zhang J, et al. (2014) Genome-wide identification, evolution and expression analysis of mTERF gene family in maize. PLoS One 9: e94126. doi: 10.1371/journal.pone.0094126.
4. Bosak EJ, Seidl-Adams IH, Zhu J, Tumlinson JH (2013) Maize developmental stage affects indirect and direct defense expression. Environ Entomol 42: 1309–1321.
5. Seidl-Adams I, Richter A, Boomer K, Yoshinaga N, Degenhardt J, et al. (2014) Emission of herbivore elicitor-induced sesquiterpenes is regulated by stomatal aperture in maize (*Zea mays*) seedlings. Plant Cell Environ (in press). doi: 10.1111/pce.12347. Article first published online: 2014 May 13.
6. Sytykiewicz H (2014) Differential expression of superoxide dismutase genes in aphid-stressed maize (*Zea mays* L.) seedlings. PLoS One 9: e94847. doi:10.1371/journal.pone.009484715.
7. Czapla A, Kurczak P, Kiełkiewicz M (2011) Elementy bionomii mszycy różano-trawowej (*Metopolophium dirhodum* Walker) na wybranych odmianach kukurydzy. Prog Plant Prot 51: 787–793. (In Polish).
8. Strażyński P (2008) Aphid fauna (Hemiptera, Aphidoidea) on maize crops in Wielkopolska – species composition and increase in number. Aphids and Other Homopterous Insects 14: 123–128.
9. Sprawka I, Goławska S, Czerniewicz P, Sytykiewicz H (2011) Insecticidal action of phytohemagglutinin (PHA) against the grain aphid, *Sitobion avenae*. Pestic Biochem Phys 100: 64–69.
10. Halarewicz A, Gabryś B (2012) Probing behavior of bird cherry-oat aphid *Rhopalosiphum padi* (L.) on native bird cherry *Prunus padus* L. and alien invasive black cherry *Prunus serotina* Ehrh. in Europe and the role of cyanogenic glycosides. Arthropod-Plant Inte 6: 497–505.
11. Zielińska L, Trzmiel K, Jeżewska M (2012) Ultrastructural changes in maize leaf cells infected with maize dwarf mosaic virus and sugarcane mosaic virus. Acta Biol Cracov Bot 54: 97–104.
12. Sempruch C, Horbowicz M, Kosson R, Leszczyński B (2012) Biochemical interactions between triticale (*Triticosecale*; Poaceae) amines and bird cherry-oat aphid (*Rhopalosiphum padi*; Aphididae). Biochem Syst Ecol 40: 162–168.
13. Rodriguez PA, Stam R, Warbroek T, Bos JI (2014) Mp10 and Mp42 from the aphid species *Myzus persicae* trigger plant defenses in *Nicotiana benthamiana* through different activities. Mol Plant Microbe Interact 27: 30–39.
14. Pitino M, Hogenhout SA (2013) Aphid protein effectors promote aphid colonization in a plant species-specific manner. Mol Plant Microbe Interact 26: 130–139.
15. Mai VC, Bednarski W, Borowiak-Sobkowiak B, Wilkaniec B, Samardakiewicz S, et al. (2013) Oxidative stress in pea seedling leaves in response to *Acyrthosiphon pisum* infestation. Phytochemistry 93: 49–62.
16. Łukasik I, Goławska S, Wójcicka A (2012) Effect of cereal aphid infestation on ascorbate content and ascorbate peroxidase activity in triticale. Pol J Environ Stud 21: 1937–1941.
17. Gong H, Jiao Y, Hu WW, Pua EC (2005) Expression of glutathione-S-transferase and its role in plant growth and development *in vivo* and shoot morphogenesis *in vitro*. Plant Mol Biol 57: 53–66.
18. Dixon DP, Hawkins T, Hussey PJ, Edwards R (2009) Enzyme activities and subcellular localization of members of the *Arabidopsis* glutathione transferase superfamily. J Exp Bot 60: 1207–1218.
19. Wisser RJ, Kolkman JM, Patzoldt ME, Holland JB, Yu J, et al. (2011) Multivariate analysis of maize disease resistances suggests a pleiotropic genetic basis and implicates a GST gene. Proc Natl Acad Sci USA 108: 7339–7344.
20. Cummins I, Wortley DJ, Sabbadin F, He Z, Coxon CR, et al. (2013) Key role for a glutathione transferase in multiple-herbicide resistance in grass weeds. Proc Natl Acad Sci USA 110: 5812–5817.

21. Lan T, Yang ZL, Yang X, Liu YJ, Wang XR, et al. (2009) Extensive functional diversification of the *Populus* glutathione S-transferase supergene family. Plant Cell 21: 3749–3766.

22. Dixon DP, Cole DJ, Edwards R (2000) Characterisation of a zeta class glutathione transferase from *Arabidopsis thaliana* with a putative role in tyrosine catabolism. Arch Biochem Biophys 384: 407–412.

23. Wagner U, Edwards R, Dixon DP, Mauch F (2002) Probing the diversity of the *Arabidopsis* glutathione S-transferase gene family. Plant Mol Biol 49: 515–532.

24. Kumar M, Yadav V, Tuteja N, Johri AK (2009) Antioxidant enzyme activities in maize plants colonized with *Piriformospora indica*. Microbiology 155: 780–790.

25. Dean JD, Goodwin PH, Hsiang T (2005) Induction of glutathione S-transferase genes of *Nicotiana benthamiana* following infection by *Colletotrichum destructivum* and *C. orbiculare* and involvement of one in resistance. J Exp Bot 56: 1525–1533.

26. Yu T, Li YS, Chen XF, Hu J, Chang X, et al. (2003) Transgenic tobacco plants overexpressing cotton glutathione S-transferase (GST) show enhanced resistance to methyl viologen. J Plant Physiol 160: 1305–1311.

27. Chronopoulou EG, Labrou NE (2009) Glutathione transferases: Emerging multidisciplinary tools in red and green biotechnology. Recent Pat Biotechnol 3: 211–223.

28. Kampranis SC, Damianova R, Atallah M, Toby G, Kondi G, et al. (2000) A novel plant glutathione S-transferase/peroxidase suppresses *Bax* lethality in yeast. J Biol Chem 275: 29207–29216.

29. Queval G, Issakidis-Bourguet E, Hoeberichts FA, Vandorpe M, Gakière B, et al. (2007) Conditional oxidative stress responses in the *Arabidopsis* photorespiratory mutant *cat2* demonstrate that redox state is a key modulator of day length-dependent gene expression, and define photoperiod as a crucial factor in the regulation of H_2O_2-induced cell death. Plant J 52: 640–657.

30. Jain M, Ghanashyam C, Bhattacharjee A (2010) Comprehensive expression analysis suggests overlapping and specific roles of rice glutathione S-transferase genes during development and stress responses. BMC Genomics 11: 1471–2164.

31. Dall'Asta P, Rossi GB, Arisi ACM (2013) Abscisic acid-induced antioxidant system in leaves of low flavonoid content maize. IV Simposio Brasileiro de Genetica Molecular de Planta (SBGMP), Bento Gonçalves, Brasil, 8-12 April. pp. 26.

32. Sytykiewicz H (2011) Expression patterns of glutathione transferase gene (*Gst1*) in maize seedlings under juglone-induced oxidative stress. Int J Mol Sci 12: 7982–7995.

33. Adejumo TO, Hettwer U, Nutz S, Karlovsky P (2009) Real-time PCR and agar plating method to predict *Fusarium verticillioides* and fumonisin B content in Nigerian maize. J Plant Protect Res 49: 399–404.

34. Chaitanya KSK, Naithani SC (1994) Role of superoxide, lipid peroxidation and superoxide dismutase in membrane perturbation during loss of viability in seeds of *Shorea robusta* Gaertn.f. New Phytol 126: 623–627.

35. Livak KJ, Schmittgen TD (2001) Analysis of relative gene expression data using real-time quantitative PCR and the $2^{-\Delta\Delta Ct}$ method. Methods 25: 402–408.

36. Pieńkosz A, Leszczyński B, Warzecha R (2005) Podatność kukurydzy na mszyce zbożowe. Prog Plant Prot 45: 989–992. (In Polish).

37. Krawczyk A, Miętkiewski R, Hurej M (2006) Owadobójcze grzyby porażające mszyce żerujące na kukurydzy. Prog Plant Prot 46: 378–381. (In Polish).

38. Bereś PK, Pruszyński G (2008) Ochrona kukurydzy przed szkodnikami w produkcji integrowanej. Acta Sci Pol ser Agricultura 7: 19–32. (In Polish).

39. Morkunas I, Mai VC, Gabryś B (2011) Phytohormonal signaling in plant responses to aphid feeding. Acta Physiol Plant 33: 2057–2073.

40. Will T, van Bel AJ (2008) Induction as well as suppression: How aphid saliva may exert opposite effects on plant defense. Plant Signal Behav 3: 427–430.

41. Will T, Furch AC, Zimmermann MR (2013) How phloem-feeding insects face the challenge of phloem-located defenses. Front Plant Sci 4: 336. doi: 10.3389/fpls.2013.00336.

42. War AR, Paulraj MG, Ahmad T, Buhroo AA, Hussain B, et al. (2012) Mechanisms of plant defense against insect herbivores. Plant Signal Behav 7: 1306–1320.

43. Santamaria ME, Martínez M, Cambra I, Grbic V, Diaz I (2013) Understanding plant defence responses against herbivore attacks: an essential first step towards the development of sustainable resistance against pests. Transgenic Res 22: 697–708.

44. Morkunas I, Formela M, Marczak L, Stobiecki M, Bednarski W (2013) The mobilization of defence mechanisms in the early stages of pea seed germination against *Ascochyta pisi*. Protoplasma 250: 63–75.

45. Prince DC, Drurey C, Zipfel C, Hogenhout SA (2014) The leucine-rich repeat receptor-like kinase BRASSINOSTEROID INSENSITIVE1-ASSOCIATED KINASE1 and the cytochrome P450 PHYTOALEXIN DEFICIENT3 contribute to innate immunity to aphids in *Arabidopsis*. Plant Physiol 164: 2207–2219.

46. Basantani M, Srivastava A, Sen S (2011) Elevated antioxidant response and induction of tau-class glutathione S-transferase after glyphosate treatment in *Vigna radiata* (L.) Wilczek. Plant Physiol Biochem 99: 111–117.

47. McGonigle B, Keeler SJ, Lau SMC, Koeppe MK, O'Keefe DP (2000) A genomics approach to the comprehensive analysis of the glutathione S-transferase gene family in soybean and maize. Plant Physiol 124: 1105–1120.

48. Moons A (2005) Regulatory and functional interactions of plant growth regulators and plant glutathione S-transferases (GSTs). Vitam Horm 72: 155–202.

49. Wang Y, Tang Y, Zhang M, Cai F, Qin J, et al. (2012) Molecular cloning and functional characterization of a glutathione S-transferase involved in both anthocyanin and proanthocyanidin accumulation in *Camelina sativa* (Brassicaceae). Genet Mol Res 11: 4711–4719.

50. Kitamura S, Akita Y, Ishizaka H, Narumi I, Tanaka A (2012) Molecular characterization of an anthocyanin-related glutathione S-transferase gene in cyclamen. J Plant Physiol 169: 636–642.

51. Abedini R, Zare S (2013) Glutathione S-transferase (GST) family in barley: identification of members, enzyme activity, and gene expression pattern. J Plant Physiol 170: 1277–1284.

52. Liu YJ, Han XM, Ren LL, Yang HL, Zeng QY (2013) Functional divergence of the glutathione S-transferase supergene family in *Physcomitrella patens* reveals complex patterns of large gene family evolution in land plants. Plant Physiol 161: 773–786.

53. Lo Piero AR, Mercurio V, Puglisi I, Petrone G (2010) Different roles of functional residues in the hydrophobic binding site of two sweet orange tau glutathione S-transferases. FEBS J 277: 255–262.

54. Kuśnierczyk A, Winge P, Midelfart H, Armbruster WS, Rossiter JT, et al. (2007) Transcriptional responses of *Arabidopsis thaliana* ecotypes with different glucosinolate profiles after attack by polyphagous *Myzus persicae* and oligophagous *Brevicoryne brassicae*. J Exp Bot 58: 2537–2552.

55. Kuśnierczyk A, Winge P, Jørstad TS, Troczyńska J, Rossiter JT, et al. (2008) Towards global understanding of plant defence against aphids-timing and dynamics of early *Arabidopsis* defence responses to cabbage aphid (*Brevicoryne brassicae*) attack. Plant Cell Environ 31: 1097–1115.

56. Moran PJ, Cheng Y, Cassell JL, Thompson GA (2002) Gene expression profiling of *Arabidopsis thaliana* in compatible plant-aphid interactions. Arch Insect Biochem Physiol 51: 182–203.

57. Stotz HU, Pittendrigh BR, Kroymann J, Weniger K, Fritsche J, et al. (2000) Induced plant defense responses against chewing insects. Ethylene signaling reduces resistance of *Arabidopsis* against Egyptian cotton worm but not diamondback moth. Plant Physiol 124: 1007–1018.

58. Bandopadhyay L, Basu D, Sikdar SR (2013) Identification of genes involved in wild crucifer *Rorippa indica* resistance response on mustard aphid *Lipaphis erysimi* challenge. PLoS One 8: e73632. doi: 10.1371/journal.pone.0073632.

59. Doehlemann G, Wahl R, Horst RJ, Voll LM, Usadel B, et al. (2008) Reprogramming a maize plant: transcriptional and metabolic changes induced by the fungal biotroph *Ustilago maydis*. Plant J 56: 181–195.

60. Luo M, Brown RL, Chen ZY, Menkir A, Yu J, et al. (2011) Transcriptional profiles uncover *Aspergillus flavus*-induced resistance in maize kernels. Toxins 3: 766–786. doi: 10.3390/toxins3070766.

61. Czerniewicz P, Leszczyński B, Chrzanowski G, Sempruch C, Sytykiewicz H (2011) Effects of host plant phenolics on spring migration of bird cherry-oat aphid (*Rhopalosiphum padi* L.). Allelopathy J 27: 309–316.

62. Rao SA, Carolan JC, Wilkinson TL (2013) Proteomic profiling of cereal aphid saliva reveals both ubiquitous and adaptive secreted proteins. PLoS One 8: e57413. doi: 10.1371/journal.pone.0057413.

63. Urbańska A, Niraz S (1990) Anatomiczne i biochemiczne aspekty żerowania mszyc zbożowych. Zesz Probl Post Nauk Roln 392: 201–213. (In Polish).

64. Urbańska A (2010) Histochemical analysis of aphid saliva in plant tissue. EJPAU ser Biology 13: #26. Available: www.ejpau.media.pl/volume13/issue4/art-26.html. Accessed 2014 Oct 13.

65. Sytykiewicz H, Leszczyński B (2008) Monitoring of anatomical changes within the bird cherry shoots evoked by *Rhopalosiphum padi* L. (Hemiptera, Aphididae). XXIII International Congress of Entomology, Durban, Republic of South Africa, 6–12 July, abstract no. 811.

66. Urbańska A, Leszczyński B, Matok H, Dixon AFG (2002) Cyanide detoxifying enzymes of bird cherry oat aphid. EJPAU ser Biology 5: #01. Available: www.ejpau.media.pl/volume5/issue2/biology/art-01.html. Accessed 2014 Oct 13.

67. Łukasik I (2006) Effect of *o*-dihydroxyphenols on antioxidant defence mechanisms of cereal aphids associated with glutathione. Pesticides 3-4: 67–73.

8

Site-Specific Chemoenzymatic Labeling of Aerolysin Enables the Identification of New Aerolysin Receptors

Irene Wuethrich[9], Janneke G. C. Peeters[9], Annet E. M. Blom, Christopher S. Theile, Zeyang Li, Eric Spooner, Hidde L. Ploegh*, Carla P. Guimaraes

Whitehead Institute for Biomedical Research, Department of Biology, Massachusetts Institute of Technology, Cambridge, Massachusetts, United States of America

Abstract

Aerolysin is a secreted bacterial toxin that perforates the plasma membrane of a target cell with lethal consequences. Previously explored native and epitope-tagged forms of the toxin do not allow site-specific modification of the mature toxin with a probe of choice. We explore sortase-mediated transpeptidation reactions (sortagging) to install fluorophores and biotin at three distinct sites in aerolysin, without impairing binding of the toxin to the cell membrane and with minimal impact on toxicity. Using a version of aerolysin labeled with different fluorophores at two distinct sites we followed the fate of the C-terminal peptide independently from the N-terminal part of the toxin, and show its loss in the course of intoxication. Making use of the biotinylated version of aerolysin, we identify mesothelin, urokinase plasminogen activator surface receptor (uPAR, CD87), glypican-1, and CD59 glycoprotein as aerolysin receptors, all predicted or known to be modified with a glycosylphosphatidylinositol anchor. The sortase-mediated reactions reported here can be readily extended to other pore forming proteins.

Editor: Ludger Johannes, Institut Curie, France

Funding: Funding provided by R01 AI087879, http://grants.nih.gov/grants/funding/r01.htm. The funders had no role in study design, data collection and analysis, decision to publish, or preparation of the manuscript.

Competing Interests: The authors have declared that no competing interests exist.

* Email: ploegh@wi.mit.edu

[9] These authors contributed equally to this work.

Introduction

Pore-forming toxins (PFTs) comprise the largest category of bacterial virulence factors [1]. One of the better studied examples is aerolysin secreted by *Aeromonas hydrophila* [2]. Aerolysin forms a homo-heptameric pore that spans the plasma membrane of the target cell [3] [4], leading to depletion of small ions [5] [6] [7], rapid loss of ATP, and ultimately cell death [8].

Aerolysin is secreted as an inactive monomeric precursor, proaerolysin, comprising a 43-residue C-terminal peptide (CP) [9] (Fig. 1A). The CP has chaperone features and appears to be required in the course of synthesis to properly fold proaerolysin into its soluble form. It not only prevents aggregation but also impedes premature pore formation by controlling the onset of heptamerization [10]. Proaerolysin is known to bind to N-glycosylated glycosylphosphatidylinositol (GPI)-anchored proteins at the target cell surface [11] [12]. Not only is the glycan important for binding but also the polypeptide to which it is attached [13].

Maturation of proaerolysin to aerolysin involves proteolytic cleavage in a flexible loop that precedes the C-terminal peptide. Furin is thought to play a major role in this process, but other proteases at the plasma membrane may participate as well [14] [15]. Following cleavage, monomers oligomerize to form a prepore complex on the cell surface [16], a step that requires release of the C-terminal peptide [17]. Removal of the C-terminal peptide induces the transition from prepore to the pore complex. The aerolysin heptamer undergoes a drastic concerted conforma-tional change of the extramembranous region, accompanied by a vertical collapse of the complex, which ultimately leads to the insertion of a water-filled transmembrane beta-barrel into the lipid bilayer [17]. The CP is not part of the functional pore, as inferred from tryptophan fluorescence and energy transfer measurements [18]. Its fate after separation from the heptamer is unknown.

Insights into the mechanism of aerolysin intoxication have been obtained without the possibility of labeling discrete domains of the toxin at will. Being able to do so might allow a more detailed examination of the role and fate of each of the specific domains. It is still unclear which domains of aerolysin bind to the proteinaceous moieties of its receptors. Chemical labeling of exposed Lys or Cys residues usually results in a heterogeneous population of labeled proteins, making it impossible to accurately assess the identity of the molecular species responsible for activity. To overcome this technical challenge, we explore sortase-based site-specific chemoenzymatic labeling [19–21]. This allows us to investigate the fate of individual N- and C-terminal domains, while preserving toxin activity. Attachment of a single fluorophore at the very C-terminus of the C-terminal peptide makes it possible to directly visualize this chaperone's departure during aerolysin intoxication. Attachment of a single biotin group at the N-terminus of aerolysin enables us to identify novel cell surface receptors.

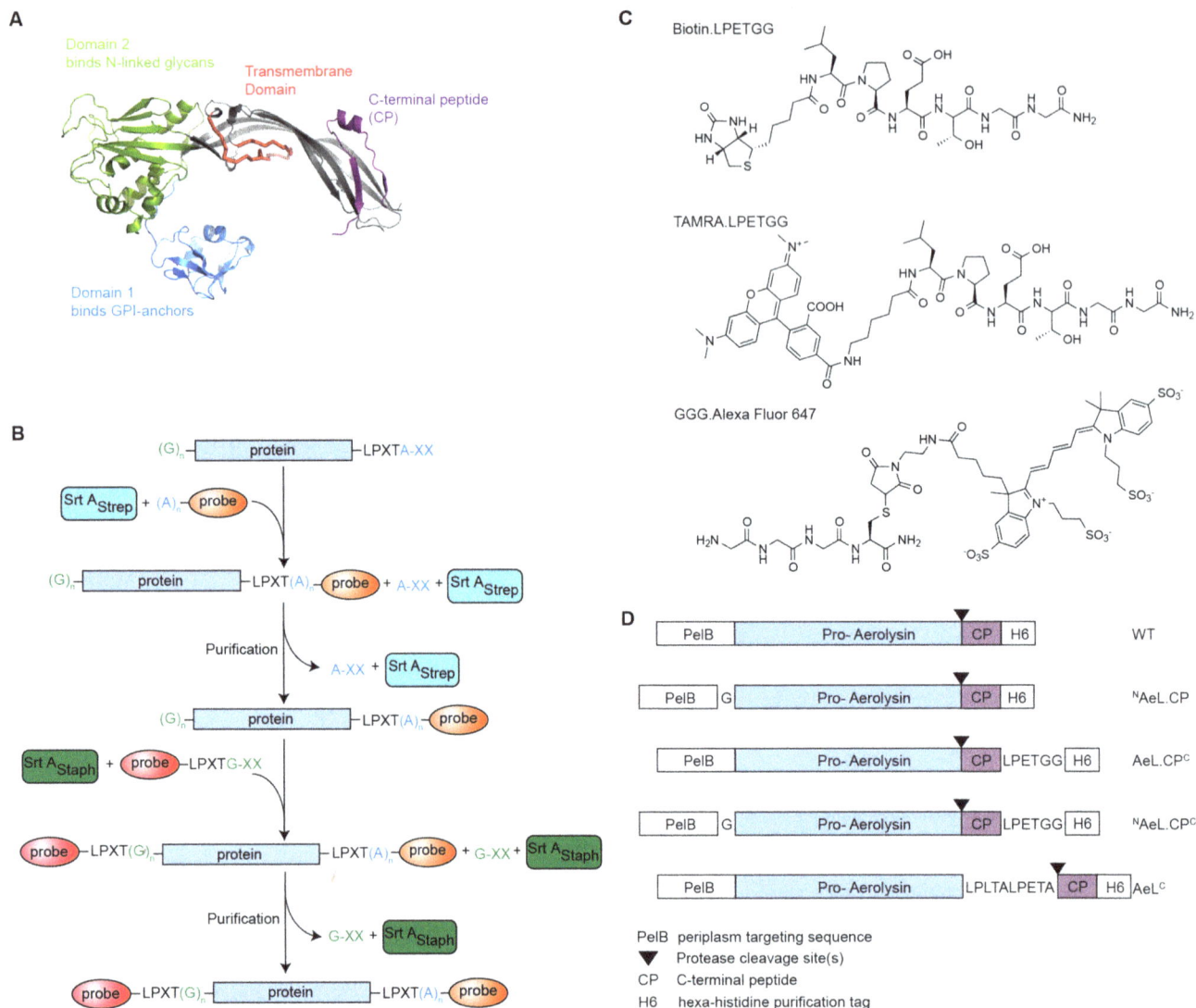

Figure 1. Strategies for site-specific labeling of proaerolysin. A Structure of the proaerolysin monomer (PDB: 1PRE). Proaerolysin consists of several different domains, two of which are responsible for receptor binding (domains 1 and 2), one containing the trans-membrane domain, and the C-terminal peptide (CP), which functions as a chaperone and dissociates from the rest of the complex upon heptamer association and pore formation. **B** Sortase reaction mechanism. C-terminal sortagging: sortase cleaves after threonine in the context of its recognition motif resulting in the formation of a new covalent bond with the N-terminus of an added oligoglycine or oligoalanine nucleophile coupled to a label of choice. N-terminal sortagging: the N-terminal glycine of proaerolysin is recognized as a nucleophile by sortase and conjugated to an LPXTG/A probe bearing a label. **C** Structures of probes used in this study. Not depicted is AAA.Alexa Fluor 647, which is similar to GGG.Alexa Fluor 647, but with alanine replacing glycine. PelB: periplasm targeting sequence, cleaved off by the producer bacteria upon export of proaerolysin to the periplasm. H6: hexahistidine handle for affinity purification. Protease cleavage sites are recognized by target cell surface proteases such as furin. CP: C-terminal peptide, serves as a chaperone for proaerolysin. Upon its loss, proaerolysin is converted to mature aerolysin (AeL). **D** Scheme for wild type (WT) and sortaggable versions of proaerolysin with their designations. The LPXTG/A pentapeptides are sortase recognition motifs.

Materials and Methods

Antibodies, cell lines, constructs

Antibodies against CD59 (sc-28805) and mesothelin (sc-50427) were purchased from Santa Cruz Biotechnology. HRP-coupled secondary anti-rabbit antibody was from BD Biosciences. HeLa cells were purchased from American Type Culture Collection and cultured in Dulbecco's Modified Eagle Medium (DMEM) supplemented with 10% Fetal Bovine Serum (FBS). KBM7 cells were a kind gift from the T. R. Brummelkamp lab, and were described previously [22]. KBM7 cells were maintained in Iscove's Modified Dulbecco's Medium (IMDM) supplemented with 10%

FBS. The wild type proaerolysin construct [23] was a generous gift from F. G. van der Goot. Sortaggable variants were cloned by site-directed mutagenesis using the QuikChange kit (Agilent Technologies) following the manufacturer's instructions and using the following primers: NAeL.CP (introduction of a single glycine at N-terminus),

forward: 5'-AGCCGGCGATGGCCGGTATGGCAGAGC-CCGTC-3',

reverse: 5'-GACGGGCTCTGCCATACCGGCCATCGCC-GGCT-3'; AeL.CPC (introduction of LPETGG at C-terminus), forward: 5′-GCGTGACCCCTGCTGCCAATCAACTAC-CAGAGACCGGTGGACTCGAGCACCACCACCACCACC-ACTGAGATCC-3′,

reverse: 5′-GGATCTCAGTGGTGGTGGTGGTGGTGCT-CGAGTCCACCGGTCTCTGGTAGTTGATTGGCGCAGG-GGTCACGC-3′. ^{N}AeL.CP^C, was built using the forward primer 5′-AGCCGGCGATGGCCGGTATGGCAGAGCCCGTC-3′, and the reverse primer 3′-GGATCTCAGTGGTGGTGG-TGGTGGTGCTCGAGTCCACCGGTCTCTGGTAGTTGA-TTGGCAGCAGGGGTCACGC-5′ with a PCR on WT proaer-olysin template using the Expand High Fidelity PCR system (Roche Diagnostics). AeL.^{C} (introduction of LPLTALPETA motive upstream of the C-terminal peptide) was done in a two-step-manner using QuikChange, according to the manufacturer's instructions:

5′-AGATCGGTGCTCCCCTCCCGCTCACTGCTGACA-GCAAGGGTG-3′,

3′-CACCTTGCTGTCAGCAGTGAGCGGGAGGGGAGC-ACCGATCT-5′;

5′-TCCCCTCCCGCTCACTGCTCTCCCGGAGACTGC-TGACAGCAAGGTGCGTCG-3′,

3′-CGACGCACCTTGCTGTCAGCAGTCTCCGGGAGA-GCAGTGAGCGGGAGGGGA-5′.

Expression and purification proaerolysin

Overnight cultures of *E. coli* BL21 (DE3) pLysS (Promega) transformed with the various aerolysin constructs and grown at 30°C were diluted 1:50 with LB broth supplemented with 200 µg/mL ampicillin plus 35 µg/mL chloramphenicol, and incubated at 37°C, shaking at 220 rpm, to an optical density of 0.5–0.6 at 600 nm. Expression of proaerolysin was induced with 1 mM isopropyl-beta-D-1-thiogalactopyranoside (IPTG) (Sigma), and the temperature was lowered to 26°C. After 4–5 hours, cells were harvested and centrifuged at 6000×g, 4°C for 20 min. Subsequent steps were carried out at 4°C. Cell pellets were resuspended in 10 ml lysis buffer per 1 L expression culture: 50 mM Tris-HCl pH 7.5, 300 mM NaCl, 0.5 mg/ml polymixin B (Sigma) supple-mented with complete protease cocktail inhibitors (Roche) and 50 µg/ml phenylmethylsulfonyl fluoride (PMSF) (Sigma). The suspension was agitated for 45 minutes at 4°C and centrifuged at 6000×g for 30 min at 4°C. The supernatant was incubated at 4°C with 0.25 ml bed volume NiNTA agarose (Qiagen) per 1 L culture, overnight, with gentle rotation. The resin was washed with 20 column volumes of 50 mM Tris-HCl pH 7.5, 300 mM NaCl, 10 mM imidazole. The protein was eluted with 5 column volumes 50 mM Tris-HCl pH 7.5, 300 mM NaCl, 150 mM imidazole. The fractions were subjected to buffer exchange to 50 mM Tris-HCl pH 7.5, 300 mM NaCl, using a PD-10 desalting column (GE Healthcare). 10% (v/v) glycerol was added to the protein preparations, aliquots were snap-frozen, and stored at −80°C. Protein concentration was determined by Bradford assay (Bio-Rad Laboratories).

Toxicity assay

0.5×10^5 KBM7 WT cells were incubated for 1 h at 37°C with different concentrations of each of the aerolysin variants (as indicated in the figures) in a total volume of 100 µL. Cells were washed twice with cold PBS and resuspended in PBS containing 1 µg/mL propidium iodide and analyzed by flow cytometry. The percentage of PI negative controls was set to 100%, and the 50% lethal dosis (LC50) calculated in R. 0.001 was added to all concentration values to avoid taking a log2 of 0.

Flow cytometry

Data acquisition was performed on a FACS Calibur HTS (BD Biosciences) using the CellQuest Pro (BD Biosciences) software. Data were analyzed with FlowJo (Tree Star Inc.).

Sortase expression, purification, immobilization. Sortase expression, purification, immobilization

Sortase A (SortA) from *Staphylococcus aureus* (SrtA$_{Staph}$) and SortA from *Streptococcus pyogenes* (SrtA$_{Strep}$) were expressed and purified as described previously [21] [20]. Additionally we used a heptamutant form of Sortase A from *S. aureus* (SrtA$_{staph7M}$), which combined previously described mutations to give Ca^{2+} independence and increased activity [24] [25]. SrtA was immobilized on cyanogen bromide activated sepharose beads (Sigma) in a ratio of 1 g dry beads per 30 mg SrtA$_{Staph}$ or 40 mg SrtA$_{Staph7M}$. The beads were swelled in 50 mL of 1 mM HCl for five washes of five minutes each at 4°C. After extensive washing with ice-cold water the sortase was coupled to the beads in 100 mM NaHCO$_3$ and 500 mM NaCl for 2 hrs at 25°C or O.N. at 4°C (make sure the storage buffer of the SortA is exchanged as Tris will react with the beads). Finally, the coupled beads were washed and stored as a 50% bead slurry in 50 mM Tris (pH 7.4) and 150 mM NaCl at 4°C. All washes/filtrations were done in a plastic capped fritted column and the buffers were removed between steps by vacuum filtration. For long-term storage more than one week add 20% glycerol and store aliquots at −20°C.

Synthesis of sortase probes and sortase labeling

GGG.TAMRA, AAA.AF647, TAMRA.LPETGG and Bio-tin.LPETGG were synthesized as described in [20] [21]. Soluble sortase labeling reactions with SrtA$_{strep}$ and SrtA$_{staph}$ were performed as described [19] [20] [21] [26]. The SrtA$_{staph7M}$ has increased activity and reactions took place at 4°C and with 20% of sortase in relation to proaerolysin. Additionally, Ca^{2+} is no longer needed in the coupling buffer. Sortase immobilized to cyanogen bromide beads was filtered from the reaction solution. Otherwise reaction conditions are the same as the soluble sortase.

Fluorescence image scan

Fluorescence scans were obtained using a variable mode imager (Typhoon 9200; GE Healthcare).

SDS PAGE, Coomassie staining, and Immunoblot

SDS-PAGE was performed as described [27]. Gels were stained with Coomassie Brilliant Blue R250 (Thermo Scientific) according to the manufacturer's instructions. Proteins were blotted onto polyvinylidene difluoride (PVDF) membranes and probed with the appropriate antibodies, followed by chemoluminescence detection using Western Lightning ECL detection kit (Perkin Elmer Life Sciences) and exposure to XAR-5 films (Kodak).

Fluorescence microscopy

HeLa cells grown on coverslips were washed with ice-cold DMEM media and incubated on ice for 30 minutes with the appropriate concentrations of labeled or unlabeled aerolysin (as indicated in the figures). Cells were washed 3 times with ice-cold PBS, fixed with 4% paraformaldehyde in PBS for 20 minutes at room temperature to prevent activity of plasma membrane-associated proteases that cleave off the C-terminal peptide, washed with PBS, incubated for 1 minute in PBS containing 1 µg/mL Hoechst stain, and mounted with glycerol on coverslips. Alterna-tively, cells were shifted to 37°C after Hoechst staining. All images were collected on a PerkinElmer Ultraview Multispectral Spinning

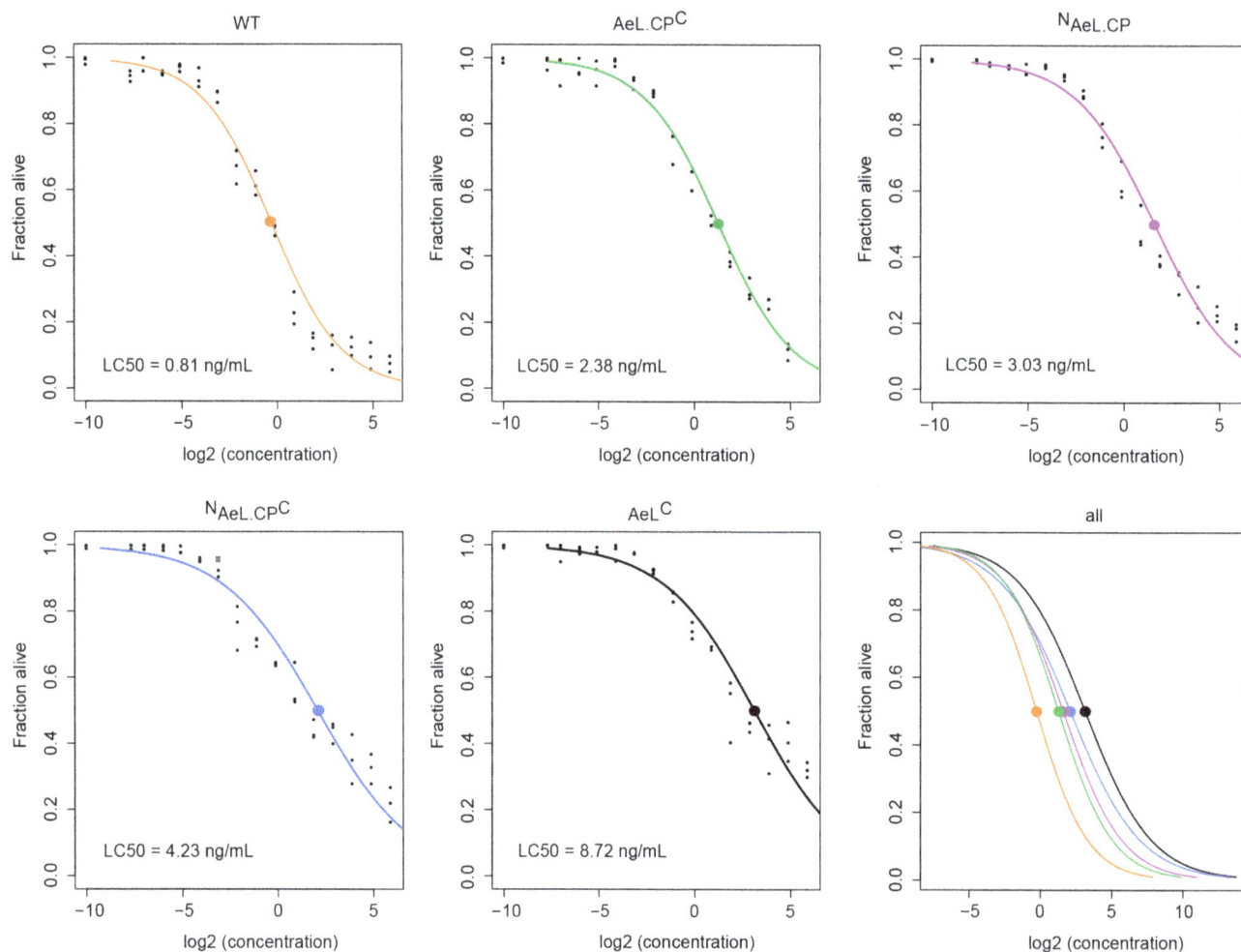

Figure 2. Impact of aerolysin modification on toxic activity. Aerolysin variants were titrated on KBM7 cells. 0.5×10^5 cells per sample were incubated with toxin for 1 hour at 37°C in a total volume of 100 μL, stained with propidium iodide (PI), and the PI negative percentage determined by flow cytometry. The concentration range for the aerolysin variants ranged from 60 ng/mL to 4 pg/μL. Every condition was tested in triplicate. The percentage of PI negative controls was set to 100%, and the 50% lethal dose (LC50) calculated in R. 0.001 was added to all concentration values to avoid taking a log2 of 0.

Disk Confocal Microscope equipped with a Yokogawa CSU-22 spinning disk confocal on a Zeiss Axiovert 200 motorized inverted microscope with Chroma 488/568/647 and 458/515/647 triple dichroic mirrors and Prior emission filter wheel, Perkin Elmer laser launch with 100 mW argon gas laser (488 nm, 514 nm), 100 mW krypton gas laser (568 nm), and 405 nm, 440 nm and 640 nm solid state lasers with AOTF for laser line selection/ attenuation and fiber-optic delivery system, a Zeiss 1.4 NA oil immersion 63x objective lens and a Prior piezo-electric objective focusing device for maintenance of focus. Images were acquired with a Hamamatsu ORCA ER cooled CCD camera controlled with Volocity software. Confocal images were collected using an exposure time of 500 ms and 1×1 binning. For time-lapse microscopy, laser power was set to 77% for the 100 mW krypton gas laser (568 nm), and to 100% for the 640 nm 40 mW solid-state laser. Number of frames: 1 per image. Acquisition frequency: 1 frame per 25 seconds. Brightness was adjusted on displayed images (identically for compared image sets) using Fiji software.

Immunoprecipitation

~ 10^7 HeLa cells per condition were incubated with 120 μg [Biotin]AeL.CP or WT aerolysin for 30 min at 4°C, washed, scraped, and lysed in buffer containing 0.5% (v/v) NP40, 10 mM Tris-HCl pH 7.4, 150 mM NaCl, 5 mM $MgCl_2$, supplemented with complete protease cocktail inhibitors (Roche) and 50 μg/ml phenylmethylsulfonyl fluoride (PMSF) (Sigma). Immunoprecipitations were performed for 3 h at 4°C with rotation using 20 μL neutravidin-sepharose beads (Thermo Scientific) per sample. Samples were eluted by boiling in reducing sample buffer and subjected to SDS-PAGE, followed by immunoblotting or mass spectrometry.

Mass spectrometry

Bands were excised, reduced, alkylated and digested with trypsin at 37C overnight. The resulting peptides were extracted, concentrated and injected onto a Dionex RSLCnano HPLC equipped with a self-packed Jupiter 3 μm C18 analytical column (0.075 mm by 10 cm, Phenomenex). Peptides were eluted using standard reverse-phase gradients. The effluent from the column

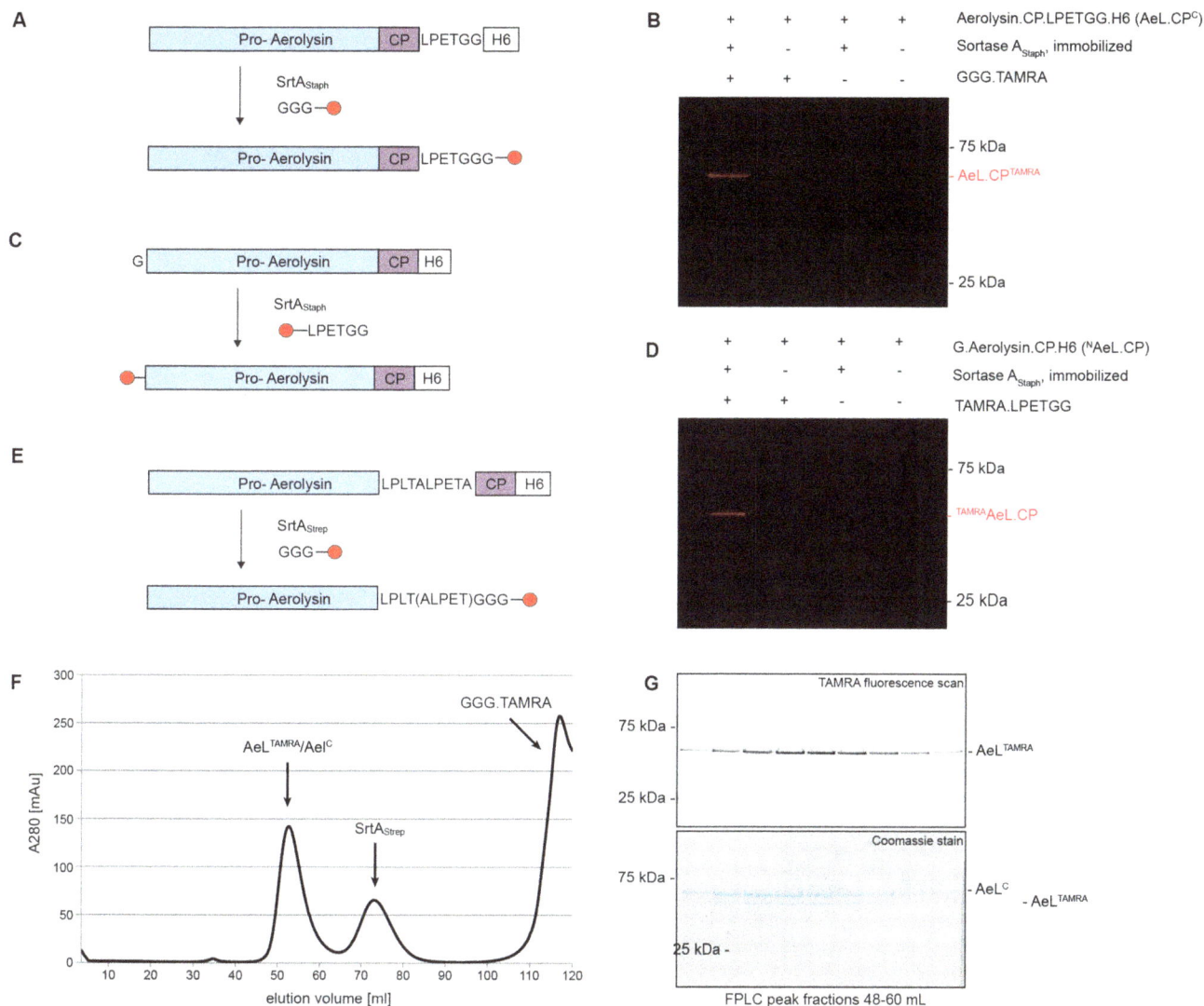

Figure 3. Installation of a single label on proaerolysin. The fluorophore carboxytetramethylrhodamine (TAMRA) was installed at the N-terminus of aerolysin (^{N}AeL.CP), at the C-terminus of aerolysin upstream of the CP (AeL^C) and at the C-terminus of the C-terminal peptide (AeL.CP^C) with sortase. **A, C, E** Schematic representation of the sortagging reactions using of ^{N}AeL.CP, AeL.CP^C, AeL^C respectively. **B, D** Sortagging of ^{N}AeL.CP and AeL.CP^C, respectively, with respective control conditions, resolved by SDS PAGE and imaged with a fluorescence scanner. Product is visible by fluorescent signal. SrtA$_{Strep}$ and SrtA$_{Staph}$ recognize and cleave LPXTA and LPXTG motives, respectively. **F** Purification of labeled AeL^{TAMRA}, gel filtration. The first peak in the A280 elution profile corresponds to aerolysin, the second to sortase, and the third to free nucleophile. **G** Analysis of the first peak of the gel filtration elution profile with SDS PAGE followed by fluorescence image scan and Coomassie stain. A fraction of AeL^C is not converted to fluorescent product.

was analyzed using a Thermo Orbitrap Elite mass spectrometer (nanospray configuration) operated in a data dependent manner. The resulting fragmentation spectra were correlated against the known database using SEQUEST. Scaffold Q+S (Proteome Software) was used to provide consensus reports for the identified proteins.

Results

Strategies for site-specific labeling of proaerolysin

Sortases A (SrtA) recognize a pentapeptide motif specific to an individual bacterial enzyme, e.g., LPXTG for SrtA from *Staphylococcus aureus* (SrtA$_{Staph}$) and LPXTA for SrtA from *Streptococcus pyogenes* (SrtA$_{Strep}$) (where X is any aminoacid). SrtA cleaves the peptide bond between the threonine and glycine or

alanine, respectively, yielding a thioacyl intermediate, which is then resolved by a nucleophilic attack of the N-terminus of an oligoglycine- or oligoalanine-containing nucleophile (Fig. 1*B*). This results in the formation of a new peptide bond [28] [26]. Because SrtA$_{Staph}$ and SrtA$_{Strep}$ enzymes are orthogonal to one another it is possible to introduce two distinct labels into one and the same protein or virus [29] [30] [31].

Hexa-histidine tags have been genetically installed at the C-terminus of proaerolysin [23]. However, site-specific fluorescent labeling of the C-terminus of mature aerolysin has not previously been attempted, an essential requirement for live-cell imaging. Using sortases we installed biotin and fluorophore probes onto different domains of proaerolysin (Fig. 1*C*). Labels were placed at: the N-terminus of proaerolysin (^{N}AeL.CP), the C-terminus of the C-terminal peptide (AeL.CP^C), the C-terminus of aerolysin

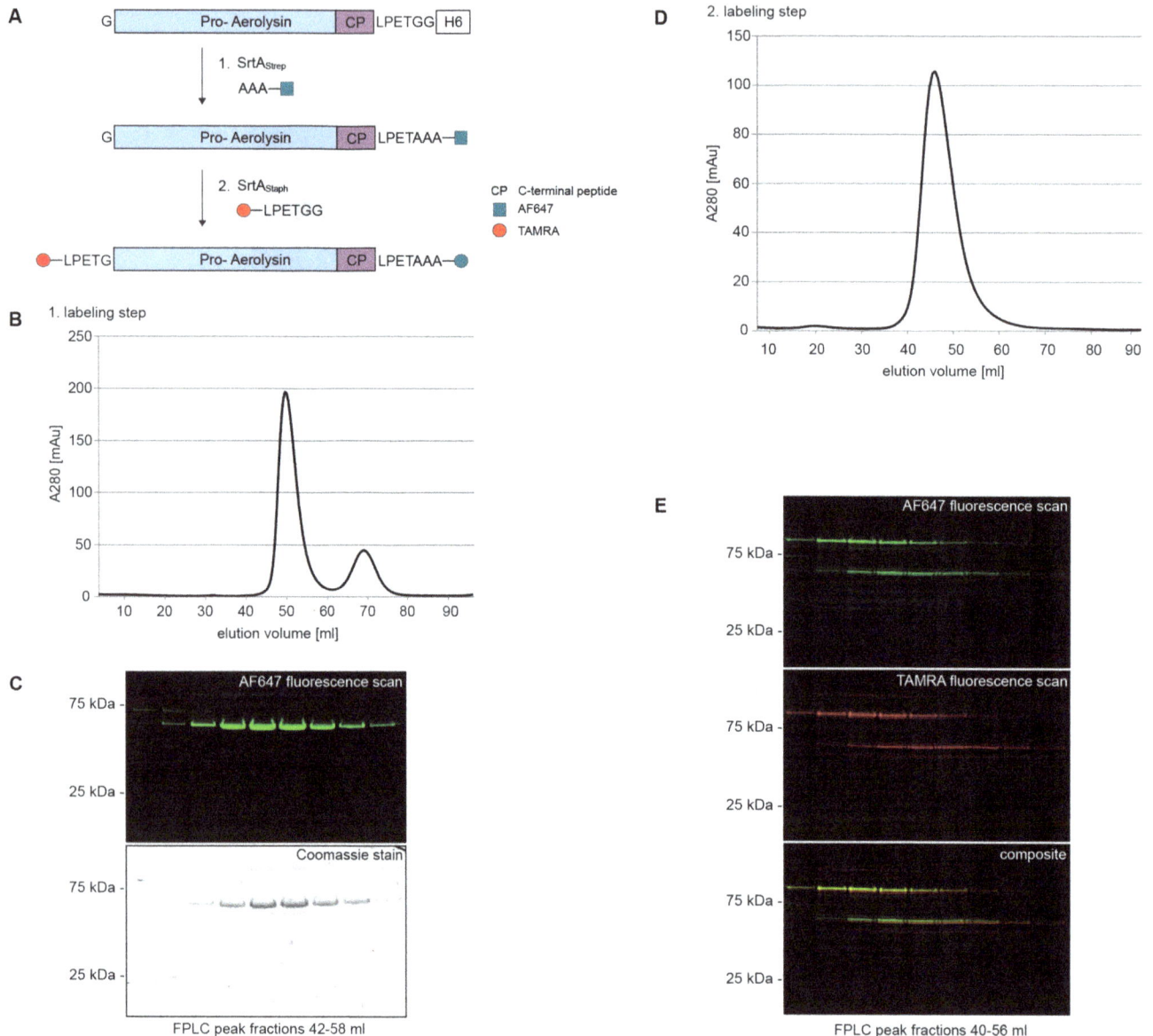

Figure 4. Double-labeling of proaerolysin. Double-labeling was achieved with a two-step approach. **A** Schematic representation of the dual labeling strategy of proaerolysin. **B** We used SrtA$_{Strep}$ to install an oligoalanine coupled to the fluorophore AF647 at the C-terminus of proaerolysin, followed by a gel filtration purification step. **C** Elution profiles were analyzed by SDS-PAGE, fluorescence scan and coomassie stain. **D** The reaction product was subjected to the second round of sortagging with SrtA$_{Staph7M}$ and LPETG-coupled TAMRA fluorophore for N-terminal labeling. SrtA$_{Staph}$ does not recognize or cleave LPXTA, hence the C-terminal label remains intact. A single peak is observed on the elution profile as immobilized sortase was used for the reaction and removed prior to gel filtration. **E** Elution profiles were analyzed by SDS-PAGE followed by fluorescence scan.

preceeding the chaperone (AeL^C), as well as creating a double-label variant (^{N}AeL.CP^C). The different sortaggable proaerolysin versions are schematically diagrammed in Fig. 1*D*.

Aerolysin activity

The different versions of sortaggable aerolysin were titrated on KBM7 cells. Toxin concentrations ranging from 60 ng/mL to 4 pg/mL were assayed in triplicate. The assay was performed for all aerolysin versions and concentrations in a single experiment on aliquots of the same batch of cells. Cells (3.5×10^5 per sample) were intoxicated for 1 hour at 37°C, washed, stained with propidium iodide and analyzed by flow cytometry. The percentage of live cells was determined and the median lethal concentration (LC50)

calculated (Fig. 2). Compared to wild type (WT) aerolysin, all of the modified versions showed a slight decrease in toxicity. The difference was greatest for AeL^C, which was ~10 fold less toxic than the WT. Modifying the N terminus with a single glycine impaired toxicity ~3 fold. This was comparable to the loss of activity observed for the C-terminal modified version. Modification of both the N and C terminus of proaerolysin reduced toxic activity further and revealed the toxicity of ^{N}proAeL^C to be intermediate between the WT and AeL^C.

Installation of a single label on proaerolysin

Proaerolysin was labeled at either its N- or C-terminus with a peptide coupled to carboxy-tetramethylrhodamine (TAMRA)

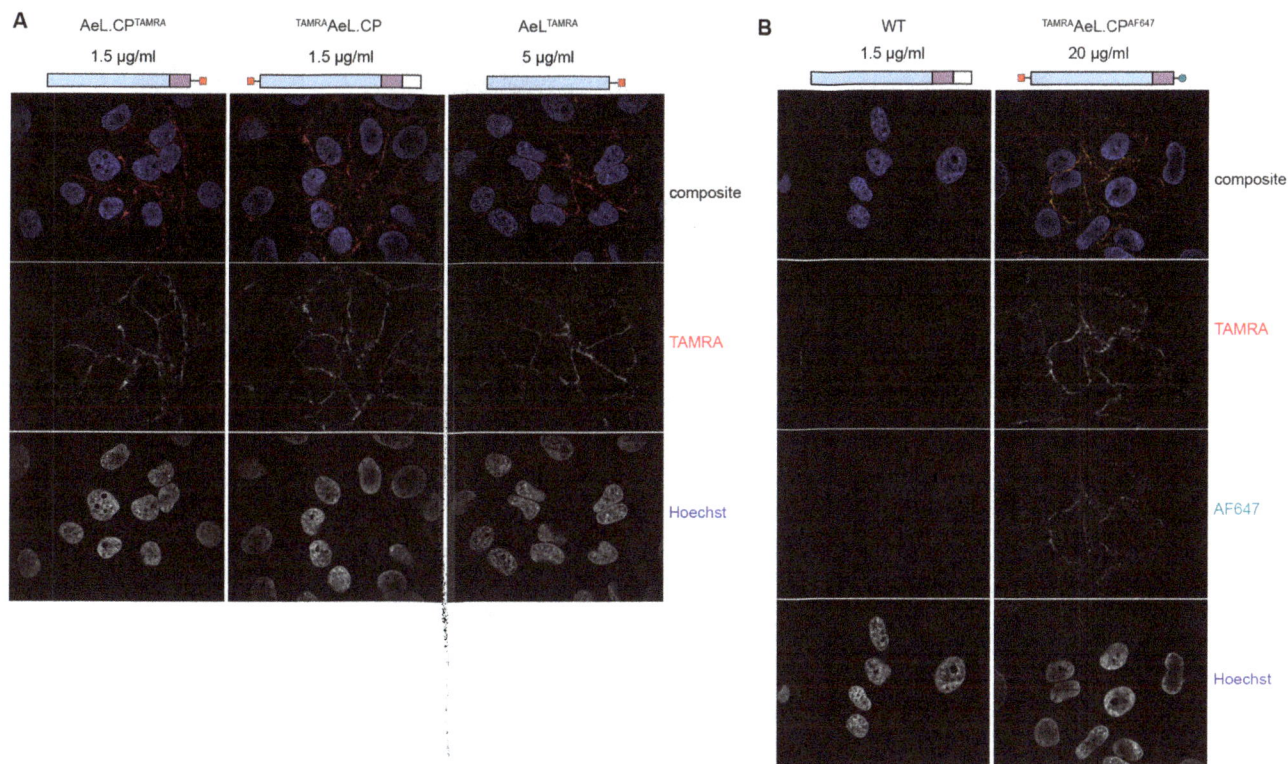

Figure 5. Aerolysin imaging. Aerolysin variants, fluorescently labeled, bind to the cell surface of HeLa cells. Images were acquired by confocal fluorescence microscopy. **A** Single labeled aerolysin versions. For comparable signal intensity, different aerolysin concentrations were required as indicated. **B** Double-labeled aerolysin and unlabeled aerolysin control.

[Figs. 3A and 3C]. Fluorescent product was observed only when all the components of the labeling reaction mixture were co-mixed. No background labeling detected (Fig. 3B and 3D). The labeling efficiency was near-quantitative, as previously demonstrated for cholera toxin [32] and various other proteins [30] [33]. N-terminally labeled proaerolysin (^{Tamra}AeL.CP) migrated slightly faster on SDS-PAGE than the C-terminally labeled AeL.CP^{TAMRA}.

To label AeL^C, we introduced a tandem sortase recognition site, LPLTALPETA, upstream of the protease cleavage site(s) that precede(s) the C-terminal peptide (Fig. 3E). We empirically determined that installation of a single sortase recognition motif, either LPLTA or LPETG, was insufficient to yield a good substrate for sortase and failed to yield a labeled product (data not shown).

Sortagged product was purified by fast protein liquid chromatography (FPLC) to separate the product from free dye-conjugated nucleophile and sortase (Fig. 3F). The fractions of the elution profile containing aerolysin were resolved by reducing SDS-PAGE and analyzed by fluorescence scan followed by coomassie staining. For this construct, labeling was incomplete (yield <50%) (Fig. 3G). Prolonged incubation times, different reaction temperatures and increasing the concentration of nucleophile did not further improve the extent of labeling (data not shown).

Double-labeling of proaerolysin

Labeling with two different probes was achieved by combining sortases with different specificities, SrtA$_{Staph}$ and SrtA$_{Strep}$, such that the product of the first reaction was not recognized as a substrate for the second (Fig. 4A). In the first step, the C-terminus of ^{N}AeL.CP^C was reacted with AAA.Alexa Fluor 647 by SrtA$_{Strep}$

with near-complete labeling efficiency. The product was purified by FPLC and used as a substrate for the second labeling reaction (Fig. 4B). The elution peak containing ^{N}AeL.CP^{AF647} also contained a minor fraction of higher and lower molecular weight species (Fig. 4C), the identity of which is not known.

TAMRA.LPETGG was appended to the N-terminus of ^{N}AeL.CP^{AF647} in a second labeling step. We used immobilized SrtA$_{Staph7M}$ to simplify sortase removal. Free nucleophile was removed by size exclusion chromatography (Fig. 4D). Labeling was monitored by SDS-PAGE, followed by fluorescence imaging (Fig. 4E). Two prominent polypeptides were visible in both channels (AF647: peudo color green; TAMRA: pseudo color red), one around 50 kDa, and a second around 100 kDa. In addition, a third polypeptide with an apparent molecular weight of 150 kDa was detected in the TAMRA channel but not in the AF647 channel. Image overlay showed co-localization of the 50 and 100 kDa species, most probably oligomers that lost the CP.

Aerolysin imaging. Next we checked whether the different labeled proaerolysin versions would still bind to cells. Cell preparation, incubation, and the subsequent washing steps prior to fixation were done at 4°C to prevent activity of cell surface proteases that would otherwise activate proaerolysin. Confocal fluorescence microscopy revealed a rim-staining pattern for single-labeled proaerolysin (Fig. 5A). To acquire images with the same image acquisition settings (laser intensity, exposure time, gain), 3.3 times more (5 µg/mL) AeL^{TAMRA} had to be added to cells compared to both ^{TAMRA}AeL.CP and AeL.CP^{TAMRA} (1.5 µg/mL). 20 µg/mL double-labeled ^{TAMRA}AeL.CP^{AF647} was required for an adequate signal to noise ratio (Fig. 5B). Both fluorophores were visible as rim staining, and co-localized at the plasma membrane. Shifting the intoxicated cells to 37°C for 10 minutes

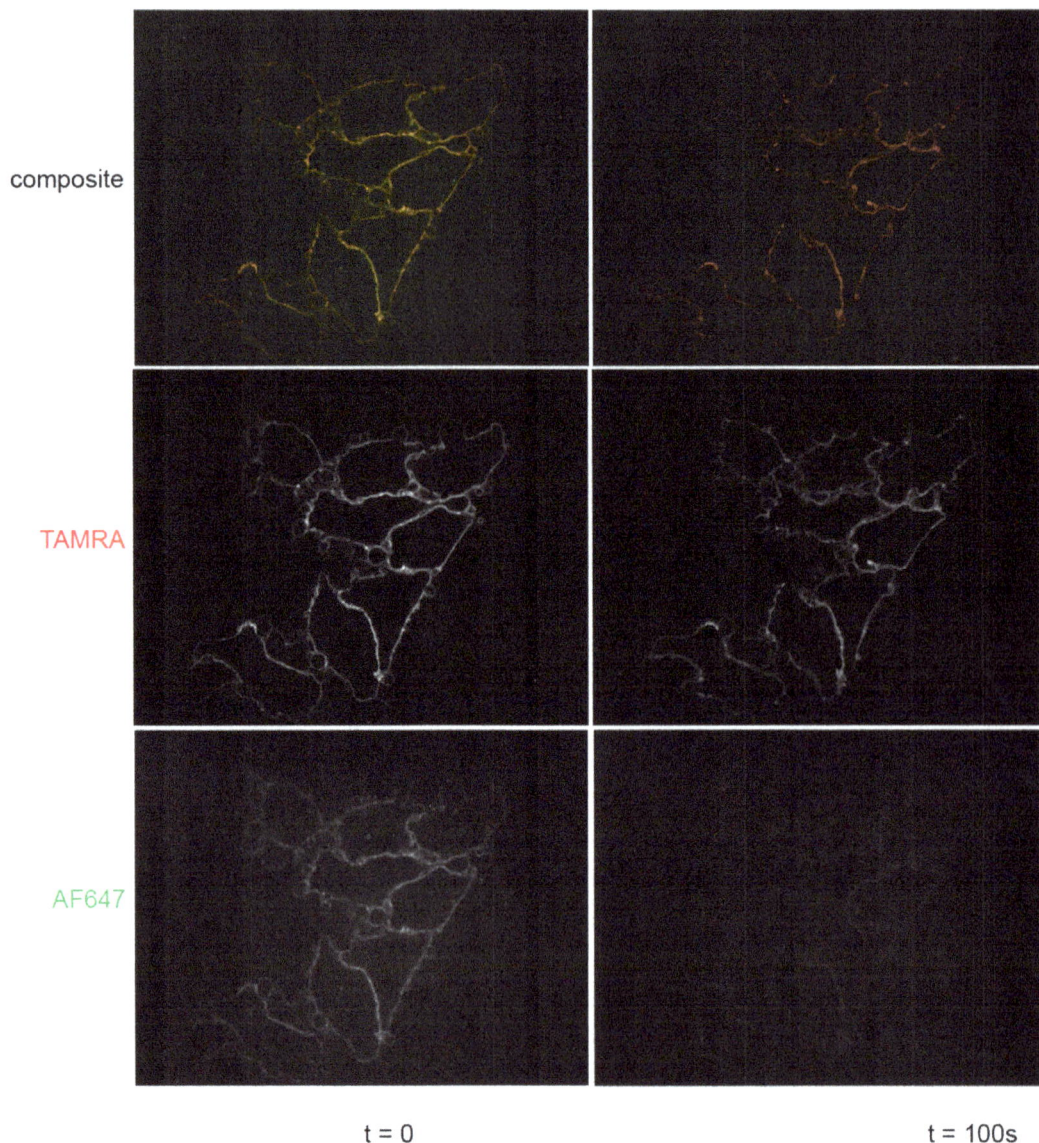

Figure 6. Dissociation of the C-terminal chaperone in the course of intoxication. HeLa cells were incubated with ^TAMRA^AeL.CP^AF647^ for 30 minutes at 4°C, washed, and the temperature shifted to 37°C. Images were acquired by confocal microscopy.

prior to imaging resulted in cell detachment, indicative of intoxication (data not shown).

Dissociation of the C-terminal chaperone in the course of intoxication. We used the double-labeled version of aerolysin to monitor the fate of the C-terminal chaperone during aerolysin intoxication. HeLa cells were incubated with ^TAMRA^AeL.CP^AF647^ for 30 minutes on ice, washed, and then shifted to 37°C. Confocal microscopy showed an initial overlapping surface staining pattern for both fluorophores. The intensity of the AF647 signal decreased over time to almost background level in ~120 seconds, whereas the signal for TAMRA suffered loss of intensity to a much smaller extent and remained well above background (Fig. 6). This is indicative of separation of the two labels, and hence consistent with loss of the C-terminal peptide.

Identification of new aerolysin receptors. Aerolysin was sort°gged with biotin at its N-terminus (Fig. 7A) and incubated with HeLa cells at 4°C. Upon cell lysis, using a mild detergent, biotinylated aerolysin and its bound materials were recovered with neutravidin beads. The eluted proteins were separated on a reducing SDS-PAGE gel, and analyzed by mass spectrometry. Five GPI-anchored proteins were identified: mesothelin, urokinase plasminogen activator surface receptor (uPAR, CD87), glypican-1, complement decay accelerating factor (CD55), and CD59 glycoprotein; each represented by multiple exclusive unique peptide coverage (Fig. 7B). Interaction was confirmed for mesothelin and CD59 by immunoblot in an independent experiment (Fig. 7C).

Discussion

Aerolysin is the first example of a pore-forming toxin to which a site-specific, chemoenzymatic labeling strategy has been applied. Sortagging allows maximal versatility in the choice of functionalities to be installed [28] [33]. Sortase accepts protein substrates in their native tertiary or quaternary structure. This eliminates two common problems of genetic fusion proteins: aggregation and

A

IP: neutravidin beads
IB: anti-biotin

C

IP: neutravidin beads
IB: anti-mesothelin

IP: neutravidin beads
IB: anti-CD59

B Aerolysin interactors, Mass Spectrometry Hits

MSLN_HUMAN UniProt Q13421 Mesothelin 12 exclusive unique peptides
MALPTARPLLGSCGTPALGSLLFLLFSLGWVQPSRTLAGETGQEAAPLDGVLANPPNISSLSPRQLLGFPCAEVS
GLSTERVRELAVALAQKNVKLSTEQLRCLAHRLSEPPEDLDALPLDLLLFLNPDAFSGPQACTRFFSRITKANVDL
LPRGAPERQRLLPAALACWGVRGSLLSEADVRALGGLACDLPGRFVAESAEVLLPRLVSCPGPLDQDQQEAAR
AALQGGGPPYGPPSTWSVSTMDALRGLLPVLGQPIIRSIPQGIVAAWRQRSSRDPSWRQPERTILRPRFRRREVE
KTACPSGKKAREIDESLIFYKKWELEACVDAALLATQMDRVNAIPFTYEQLDVLKHKLDELYPQGYPESVIQHLGY
LFLKMSPEDIRKWNVTSLETLKALLEVNKGHEMSPQAPRRPLPQVATLIDRFVKGRGQLDKDTLDTLTAFYPGYL
CSLSPEELSSVPPSSIWAVRPQDLDTCDPRQLDVLYPKARLAFQNMNGSEYFVKIQSFLGGAPTEDLKALSQQN
VSMDLATFMKLRTDAVLPLTVAEVQKLLGPHVEGLKAEERHRPVRDWILRQRQDDLDTLGLGLQGGIPNGYLVLD
LSMQEALSGTPCLLGPGPVLTVLALLLASTLA

UPAR_HUMAN UniProt Q03405 Urokinase plasminogen activator surface receptor (CD87) 6 exclusive unique
peptides
MGHPPLLPLLLLLHTCVPASWGLRCMQCKTNGDCRVEECALGQDLCRTTIVRLWEEGEELELVEKSCTHSEKTN
RTLSYRTGLKITSLTEVVCGLDLCNQGNSGRAVTYSRSRYLECISCGSSDMSCERGRHQSLQCRSPEEQCLDVV
THWIQEGEEGRPKDDRHLRGCGYLPGCPGSNGFHNNDTFHFLKCCNTTKCNEGPILELENLPQNGRQCYSCK
GNSTHGCSSEETFLIDCRGPMNQCLVATGTHEPKNQSYMVRGCATASMCQHAHLGDAFSMNHIDVSCCTKSGC
NHPDLDVQYRSGAAPQPGPAHLSLTITLLMTARLWGGTLLWT

GPC1_HUMAN UniProt P35052 Glypican-1 14 exclusive unique peptides
MELRARGWWLLCAAAALVACARGDPASKSRSCGEVRQIYGAKGFSLSDVPQAEISGEHLRICPQGYTCCTSEM
EENLANRSHAELETALRDSSRVLQAMLATQLRSFDDHFQHLLNDSERTLQATFPGAFGELYTQNARAFRDLYSEL
RLYYRGANLHLEETLAEFWARLLERLFKQLHPQLLLPDDYLDCLGKQAEALRPFGEAPRELRLRATRAFVAARSF
VQGLGVASDVVRKVAQVPLGPECSRAVMKLVYCAHCLGVPGARPCPDYCRNVLKGCLANQADLDAEWRNLLD
SMVLTDKFWGTSGVESVIGSVHTWLAEAINALQDNRDTLTAKVIQGCGNPKVNPQGPGPEEKRRRGKLAPRER
PPSGTLEKLVSEAKAQLRDVQDFWISLPGTLCSEKMALSTASDDRCWNGMARGRYLPEVMGDGLANQINNPEV
EVDITKPDMTIRQQIMQLKIMTNRLRSAYNGNDVDFQDASDDGSGSGSGDGCLDDLCSRKVSRKSSSSRTPLTH
ALPGLSEQEGQKTSAASCPQPPTFLLPLLLFLALTVARPRWR

DAF_HUMAN UniProt P08174 Complement decay-accelerating factor (CD55) 7 exclusive unique peptides
MTVARPSVPAALPLLGELPRLLLLVLLCLPAVWGDCGLPPDVPNAQPALEGRTSFPEDTVITYKCEEESFVKIPGEK
DSVICLKGSQWSDIEEFCNRSCEVPTRLNSASLKQPYITQNYFPVGTVVEYECRPGYRREPSLSPKLTCLQNLKW
STAVEFCKKKSCPNPGEIRNGGQIDVPGGILFGATISFSCNTGYKLFGSTSSFCLISGSSVQWSDPLPECREIYCPAP
PQIDNGIIQGERDHYGYRQSVTYACNKGFTMIGEHSIYCTVNNDEGEWSGPPPECRGKSLTSKVPPTVQKPTTV
NVPTTEVSPTSQKTTTKTTTPNAQATRSTPVSRTTKHFHETTPNKGSGTTSGTTRLLSGHTCFTLTGLLGTLVTM
GLLT

CD59_HUMAN UniProt P13987 CD59 glycoprotein 2 exclusive unique peptides
MGIQGGSVLFGLLLVLAVFCHSGHSLQCYNCPNPTADCKTAVNCSSDFDACLITKAGLQVYNKCWKFEHCNFND
VTTRLRENELTYYCCKKDLCNFNEQLENGGTSLSEKTVLLLVTPFLAAAWSLHP

signal peptide (removed from mature form)
MS coverage
Lipidation (GPI anchor)
removed from mature form

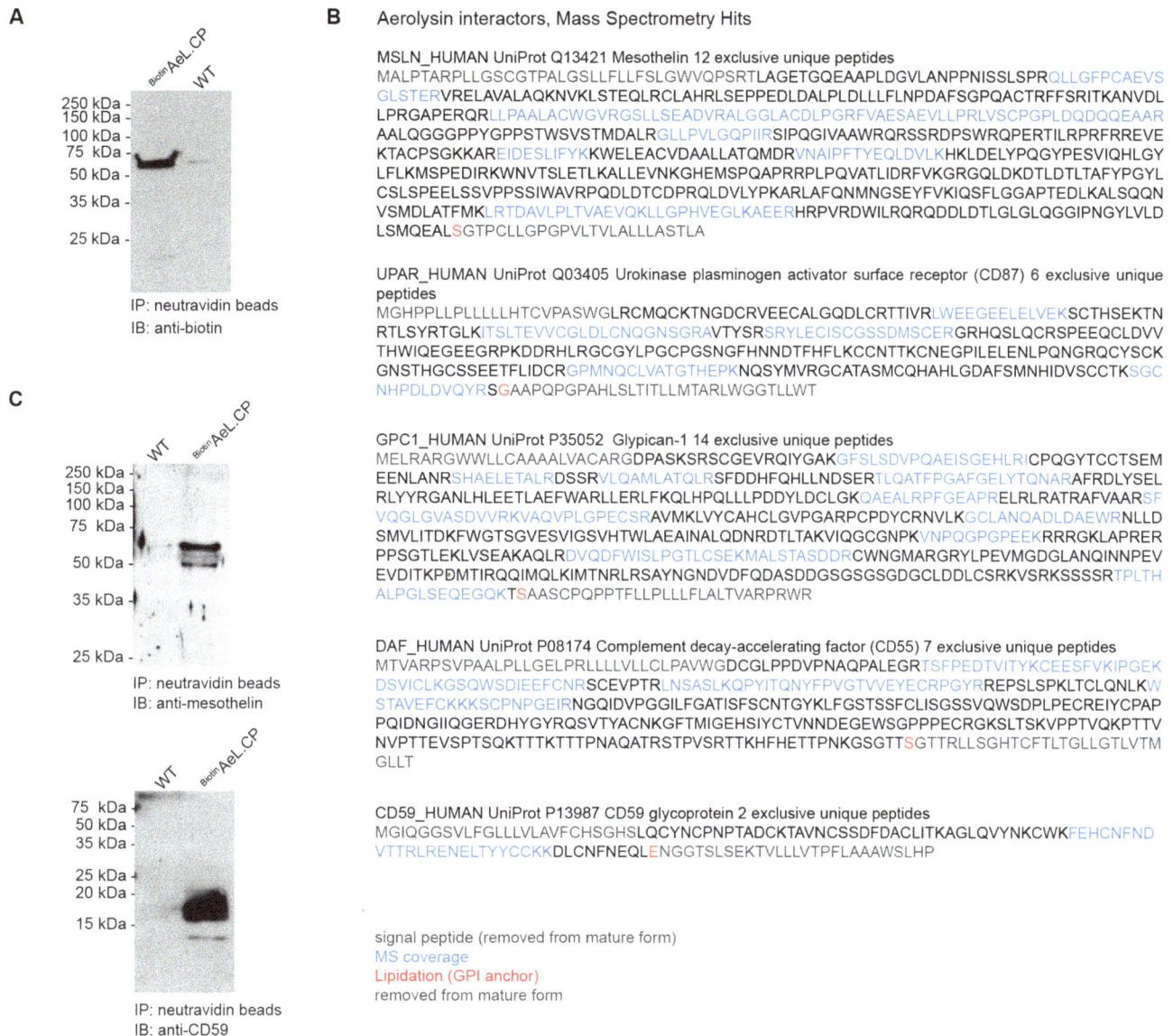

Figure 7. Identification of new aerolysin receptors. BiotinAeL.CP was used to identify new GPI-anchored proteins that bind Aerolysin. **A** Biotin.LPETG was attached to the N-terminus of proaerolysin via sortagging. The purified reaction product was analyzed by immunoblot. **B** HeLa cells were incubated with BiotinAeL.CP for 3 hours at 4°C and subsequently lysed with 0.5% NP-40. After pull-down with neutravidin beads, proteins were eluted, analyzed by SDS-PAGE, and subjected to mass spectrometry. Five GPI-anchored proteins were identified. UniProt accession codes are indicated. Peptides identified by mass spectrometry, lipidated amino acids, signal peptides, as well as peptides cleaved off from the pro-proteins are highlighted. **C** Binding of BiotinAeL.CP to mesothelin and to CD59 was verified by immunoblot.

non-functional folding. Aerolysin tolerates only subtle modifications [34]. A dramatic conformational change must take place for the soluble aerolysin monomer to form the homo-heptameric pore upon binding to a suitable receptor [17]; a point mutation can lock the protein into a particular conformation [35] and/or impede oligomerization [9], preventing toxicity. This leaves only three sites readily amenable to site-specific alteration: the very N-terminus, the very C-terminus, and the slightly more variable sequence that flanks the protease cleavage site(s) preceding the C-terminal peptide [36]. While it is true that addition of a few residues, a single glycine at the N-terminus of aerolysin, or the LPXTG/A motif diminishes toxicity (anywhere from a factor of 3 up to a factor of 10, Fig. 2), we have shown that enzymatic modification of any of these three aerolysin sites nonetheless yields

a functional product fully capable of intoxication. It is not immediately obvious from the aerolysin crystal structure whether the N-terminus is critically engaged in receptor binding or pore-formation [37]. Genetic appendage of an affinity tag at the C-terminus of the C-terminal peptide for purification purposes is standard, but its effect on toxicity has not been systematically investigated. As observed in this study, internal modification of aerolysin has the most detrimental effect. The anomalous mobility on SDS-PAGE of TAMRAproAeL compared to proAeLTAMRA (calculated molecular weights: 55.5 kDa for TAMRAproAeL and 54.8 kDa for AeL.CPTAMRA) we attribute to the relative positioning of the fluorophores, and the incomplete denaturation and/or differences in SDS binding. Alternatively TAMRAAeL.CP may have lost its C-terminal peptide, which has a molecular weight of approximately 3.7 kDa. However, unless proteases were present in the sortagging reaction, or aerolysin could somehow activate autocatalytically, this we consider less likely. AeL.CP^{TAMRA} was clearly not affected in this manner, or it would have lost its fluorescence.

We show that modification of mature aerolysin not only at its N-terminus and C-terminal end of the CP (in the context of the holotoxin), but also at the newly generated C-terminus after cleavage of the CP is readily achievable using sortase. After the CP is cleaved off by the sortase reaction, it remains associated with non-receptor-bound aerolysin and continues to exert its function as a chaperone. By inhibiting aggregation and premature pore-formation, it maintains the molecule's toxic potential [10].

Introduction of five additional amino acids in addition to the sortase recognition site was necessary to achieve successful sortase-mediated modification. Placing the LPXTG/A pentapeptide in a flexible loop generally increases flexibility of the protein backbone and thus accessibility, which in certain cases is required for proper sortase action [28] [32]. We did not determine which of the two motifs within this tandem sequence was recognized and used by sortase. The fact that modification of this site by addition of amino acids only modestly affects toxicity suggests that the new C-terminus has no critical function in the mature pore. Still, even though the five amino acid extension rendered the site accessible to sortase, the sortagging reaction could not be driven to completion. Presumably there is residual steric hindrance that interferes with accessibility for sortase.

Site-specific N- and C-terminal labeling of a single polypeptide using sortases of different specificity has been demonstrated in earlier work [29]. Applying the same strategy on aerolysin, obtained data are entirely consistent with double labeling. SrtA$_{Strep}$ not only accepts oligoalanine, but also oligoglycine as a nucleophile, albeit with different kinetics [29] [38]. The N-terminal glycine of ^{N}AeL.CP^{label} acts as a nucleophile, and can resolve the substrate-sortase thioacyl intermediate. This may result in concatenation of a fraction of aerolysin monomers, and would explain the detection of the 100 kDa and 150 kDa protein bands in the double-labeling reactions. These molecular weights are compatible with dimer and trimer formation, respectively. Dimers of aerolysin appear to contribute to the protein's stability and have been detected in solution [39] [40] [41]. Dissociation in the presence of SDS is dependent on detergent concentration. Van der Goot et al. reported that "the dimer begins to come apart at 0.0125% SDS and is nearly completely dissociated by 0.025% detergent" [39]. The SDS concentration in our system is 0.1% (w/v), which should be sufficient to achieve denaturation. However, we know of several examples where non-covalent oligomers might be formed in the presence of SDS or resist to denaturing conditions, for example, the Cholera toxin B subunit [21]. How the enzymatic modification of aerolysin affects these properties is not known.

Installation of fluorescent tags does not compromise the ability of aerolysin to bind to its receptors. The different toxin amounts required to achieve equivalent binding reflect the differences in LC50 observed for the unlabeled, sequence-modified aerolysin variants. Moreover, the immediate detachment of the adherent HeLa cells shortly after temperature shift is a clear indication of the toxicity of the aerolysin variants. In the case of AeL^{TAMRA},

where the unlabeled fraction constitutes the majority after reaction, it is not possible to infer toxicity of the labeled fraction, although the unlabeled, altered sequence of the input aerolysin preparation used for labeling is of course toxic.

With the double labeled ^{TAMRA}AeL.CP^{AF647} construct in hand, we could not only confirm the toxicity of the labeled fraction itself, but also visualize the loss of aerolyin's C-terminal peptide in the course of intoxication by microscopy. We thus confirm the previous findings of van der Goot et al that the chaperone is not part of the functional pore and separates from the active toxin [18].

A further application of sortagged aerolysin is the identification of new GPI-anchored human cell surface proteins that serve as receptors for the toxin. Previously, it was known that aerolysin binds to a subset of N-glycanated GPI-anchored proteins [11] [12] [42], where not only the GPI anchor, but also the receptor polypeptide moiety plays a role [13]. Plasma membrane micro-domains act as a concentration platform for such GPI-anchored proteins [43]. The earliest identified receptor was Thy-1 from mouse lymphocytes [44]. Others are an unidentified 80 kD protein on baby hamster kidney cells [6], an unidentified 47 kD receptor on rat erythrocytes [45], the variant surface glycoprotein (VSG) of *Trypanosoma brucei* over-expressed in mammalian cells, *Leishmania major* CD63, but only when expressed in Chinese hamster ovary cells [11], and murine contactin [46]. In addition, aerolysin binds to human complement decay accelerating factor (CD55) [47]. In our assay we were able to detect CD55, attesting to the power of our approach, along with mesothelin, urokinase plasminogen activator surface receptor (uPAR, CD87), glypican-1, and CD59 glycoprotein, a novel set of molecularly identified GPI-anchored proteins not previously associated with aerolysin binding. Of note, CD59 was specifically excluded as an aerolysin receptor in previous work [13]. We speculate that the reason for this observed difference might be of a technical nature. The fact that we identify CD59 with two different analysis methods, immunoblot and mass spectrometry, makes us confident that CD59 is a true interaction partner of aerolysin.

Sortagging converts aerolysin into a versatile and valuable tool to study the 'GPI-ome' as a means of further characterizing lipid rafts where most GPI-anchored proteins are clustered [48], The sortagging strategy described here should be applicable also to other members of the bacterial pore-forming toxin family and may facilitate further biophysical studies on membrane interactions and pore formation.

Acknowledgments

We are grateful to Gisou van der Goot for the plasmid encoding proaerolysin; Thijn Brummelkamp for KBM7 cells; Wendy Salmon for assistance with fluorescence microscopy; George Bell for statistics; Lenka Kundrat for help with figures; and all the Ploegh laboratory members for helpful discussions and suggestions.

Author Contributions

Conceived and designed the experiments: IW JGCP CPG. Performed the experiments: IW JGCP AEMB. Analyzed the data: IW JGCP AEMB HLP CPG. Contributed reagents/materials/analysis tools: IW JGCP AEMB CST ZL ES. Contributed to the writing of the manuscript: IW HLP CPG.

References

1. Alouf JE (2005) The Comprehensive Sourcebook of Bacterial Protein Toxins. Third Edition. Academic Press. ISBN: 978-0-12-088445-2.
2. Bernheimer AW, Avigad LS (1974) Partial characterization of aerolysin, a lytic exotoxin from Aeromonas hydrophila. Infection and Immunity 9: 1016–1021.
3. Wilmsen HU, Leonard KR, Tichelaar W, Buckley JT, Pattus F (1992) The aerolysin membrane channel is formed by heptamerization of the monomer. EMBO J 11: 2457–2463.

4. Moniatte M, van der Goot FG, Buckley JT, Pattus F, van Dorsselaer A (1996) Characterisation of the heptameric pore-forming complex of the Aeromonas toxin aerolysin using MALDI-TOF mass spectrometry. FEBS Lett 384: 269–272.

5. Wilmsen HU, Pattus F, Buckley JT (1990) Aerolysin, a hemolysin from Aeromonas hydrophila, forms voltage-gated channels in planar lipid bilayers. J Membr Biol 115: 71–81.

6. Abrami L, Fivaz M, Glauser PE, Parton RG, van der Goot FG (1998) A pore-forming toxin interacts with a GPI-anchored protein and causes vacuolation of the endoplasmic reticulum. The Journal of Cell Biology 140: 525–540.

7. Krause KH, Fivaz M, Monod A, van der Goot FG (1998) Aerolysin induces G-protein activation and Ca2+ release from intracellular stores in human granulocytes. J Biol Chem 273: 18122–18129.

8. Fennessey CM, Ivie SE, McClain MS (2012) Coenzyme depletion by members of the aerolysin family of pore-forming toxins leads to diminished ATP levels and cell death. Mol Biosyst 8: 2097–2105.

9. Pernot L, Schiltz M, van der Goot FG (2010) Preliminary crystallographic analysis of two oligomerization-deficient mutants of the aerolysin toxin, H132D and H132N, in their proteolyzed forms. Acta Crystallogr Sect F Struct Biol Cryst Commun 66: 1626–1630.

10. Iacovache I, Degiacomi MT, Pernot L, Ho S, Schiltz M, et al. (2011) Dual chaperone role of the C-terminal propeptide in folding and oligomerization of the pore-forming toxin aerolysin. PLoS Pathog 7: e1002135.

11. Diep DB, Nelson KL, Raja SM, Pleshak EN, Buckley JT (1998) Glycosylphosphatidylinositol anchors of membrane glycoproteins are binding determinants for the channel-forming toxin aerolysin. J Biol Chem 273: 2355–2360.

12. Hong Y, Ohishi K, Inoue N, Kang JY, Shime H, et al. (2002) Requirement of N-glycan on GPI-anchored proteins for efficient binding of aerolysin but not Clostridium septicum alpha-toxin. EMBO J 21: 5047–5056.

13. Abrami L, Velluz M-C, Hong Y, Ohishi K, Mehlert A, et al. (2002) The glycan core of GPI-anchored proteins modulates aerolysin binding but is not sufficient: the polypeptide moiety is required for the toxin-receptor interaction. FEBS Lett 512: 249–254.

14. Howard SP, Buckley JT (1985) Activation of the hole-forming toxin aerolysin by extracellular processing. J Bacteriol 163: 336–340.

15. Abrami L, Fivaz M, Decroly E, Seidah NG, Jean F, et al. (1998) The pore-forming toxin proaerolysin is activated by furin. J Biol Chem 273: 32656–32661.

16. van der Goot FG, Pattus F, Wong KR, Buckley JT (1993) Oligomerization of the channel-forming toxin aerolysin precedes insertion into lipid bilayers. Biochemistry. 32(10): 2636–42.

17. Degiacomi MT, Iacovache I, Pernot L, Chami M, Kudryashev M, et al. (2013) Molecular assembly of the aerolysin pore reveals a swirling membrane-insertion mechanism. Nat Chem Biol. 9(10): 623–9.

18. van der Goot FG, Hardie KR, Parker MW, Buckley JT (1994) The C-terminal peptide produced upon proteolytic activation of the cytolytic toxin aerolysin is not involved in channel formation. J Biol Chem 269: 30496–30501.

19. Popp MW, Antos JM, Ploegh HL (2009) Site-specific protein labeling via sortase-mediated transpeptidation. Curr Protoc Protein Sci Chapter 15: Unit15 3.

20. Guimaraes CP, Witte MD, Theile CS, Bozkurt G, Kundrat L, et al. (2013) Site-specific C-terminal and internal loop labeling of proteins using sortase-mediated reactions. Nat Protoc 8: 1787–1799.

21. Theile CS, Witte MD, Blom AEM, Kundrat L, Ploegh HL, et al. (2013) Site-specific N-terminal labeling of proteins using sortase-mediated reactions. Nat Protoc 8: 1800–1807.

22. Carette JE, Guimaraes CP, Varadarajan M, Park AS, Wuethrich I, et al. (2009) Haploid genetic screens in human cells identify host factors used by pathogens. Science 326: 1231–1235.

23. Iacovache I, Paumard P, Scheib H, Lesieur C, Sakai N, et al. (2006) A rivet model for channel formation by aerolysin-like pore-forming toxins. EMBO J 25: 457–466.

24. Chen I, Dorr BM, Liu DR (2011) A general strategy for the evolution of bond-forming enzymes using yeast display. Proceedings of the National Academy of Sciences 108: 11399–11404.

25. Hirakawa H, Ishikawa S, Nagamune T (2012) Design of Ca2+-independent Staphylococcus aureus sortase A mutants. Biotechnol Bioeng 109: 2955–2961.

26. Popp MW, Antos JM, Grotenbreg GM, Spooner E, Ploegh HL (2007) Sortagging: a versatile method for protein labeling. Nat Chem Biol 3: 707–708.

27. Laemmli UK (1970) Cleavage of structural proteins during the assembly of the head of bacteriophage T4. Nature 227: 680–685.

28. Popp MW, Ploegh HL (2011) Making and breaking peptide bonds: protein engineering using sortase. Angew Chem Int Ed Engl 50: 5024–5032.

29. Antos JM, Chew, Guimaraes CP, Yoder NC, Grotenbreg GM, et al. (2009) Site-specific N- and C-terminal labeling of a single polypeptide using sortases of different specificity. J Am Chem Soc 131: 10800–10801.

30. Hess GT, Cragnolini JJ, Popp MW, Allen MA, Dougan SK, et al. (2012) M13 bacteriophage display framework that allows sortase-mediated modification of surface-accessible phage proteins. Bioconjug Chem 23: 1478–1487.

31. Hess GT, Guimaraes CP, Spooner E, Ploegh HL, Belcher AM (2013) Orthogonal labeling of M13 minor capsid proteins with DNA to self-assemble end-to-end multiphage structures. ACS Synth Biol 2: 490–496.

32. Guimaraes CP, Carette JE, Varadarajan M, Antos J, Popp MW, et al. (2011) Identification of host cell factors required for intoxication through use of modified cholera toxin. The Journal of Cell Biology 195: 751–764.

33. Witte MD, Theile CS, Wu T, Guimaraes CP, Blom AEM, et al. (2013) Production of unnaturally linked chimeric proteins using a combination of sortase-catalyzed transpeptidation and click chemistry. Nat Protoc 8: 1808–1819.

34. Diep DB, Lawrence TS, Ausió J, Howard SP, Buckley JT (1998) Secretion and properties of the large and small lobes of the channel-forming toxin aerolysin. Mol Microbiol 30: 341–352.

35. Tsitrin Y, Morton CJ, el-Bez C, Paumard P, Velluz M-C, et al. (2002) Conversion of a transmembrane to a water-soluble protein complex by a single point mutation. Nat Struct Biol 9: 729–733.

36. van der Goot FG, Lakey J, Pattus F, Kay CM, Sorokine O, et al. (1992) Spectroscopic study of the activation and oligomerization of the channel-forming toxin aerolysin: identification of the site of proteolytic activation. Biochemistry 31: 8566–8570.

37. Tucker AD, Parker MW, Tsernoglou D, Buckley JT (1990) Crystallization of a proform of aerolysin, a hole-forming toxin from Aeromonas hydrophila. J Mol Biol 212: 561–562.

38. Race PR, Bentley ML, Melvin JA, Crow A, Hughes RK, et al. (2009) Crystal structure of Streptococcus pyogenes sortase A: implications for sortase mechanism. J Biol Chem 284: 6924–6933.

39. van der Goot FG, Ausió J, Wong KR, Pattus F, Buckley JT (1993) Dimerization stabilizes the pore-forming toxin aerolysin in solution. J Biol Chem 268: 18272–18279.

40. Fivaz M, Velluz MC, van der Goot FG (1999) Dimer dissociation of the pore-forming toxin aerolysin precedes receptor binding. J Biol Chem 274: 37705–37708.

41. Barry R, Moore S, Alonso A, Ausió J, Buckley JT (2001) The channel-forming protein proaerolysin remains a dimer at low concentrations in solution. J Biol Chem 276: 551–554.

42. Howard SP, Buckley JT (1982) Membrane glycoprotein receptor and hole-forming properties of a cytolytic toxin. Biochemistry 21: 1662–1667.

43. Abrami L, van der Goot FG (1999) Plasma membrane microdomains act as concentration platforms to facilitate intoxication by aerolysin. The Journal of Cell Biology 147: 175–184.

44. Nelson KL, Raja SM, Buckley JT (1997) The glycosylphosphatidylinositol-anchored surface glycoprotein Thy-1 is a receptor for the channel-forming toxin aerolysin. J Biol Chem 272: 12170–12174.

45. Cowell S, Aschauer W, Gruber HJ, Nelson KL, Buckley JT (1997) The erythrocyte receptor for the channel-forming toxin aerolysin is a novel glycosylphosphatidylinositol-anchored protein - Cowell −2003 - Molecular Microbiology - Wiley Online Library. Mol Microbiol 25: 343–350.

46. MacKenzie CR, Hirama T, Buckley JT (1999) Analysis of receptor binding by the channel-forming toxin aerolysin using surface plasmon resonance. J Biol Chem 274: 22604–22609.

47. Andrew AJ, Kao S, Strebel K (2011) C-terminal Hydrophobic Region in Human Bone Marrow Stromal Cell Antigen 2 (BST-2)/Tetherin Protein Functions as Second Transmembrane Motif. Journal of Biological Chemistry 286: 39967–39981.

48. Simons K, Sampaio JL (2011) Membrane Organization and Lipid Rafts. Cold Spring Harb Perspect Biol. 3(10): a004697 doi: 10.1101/cshperspect.a004697.

9

An Odorant-Binding Protein Is Abundantly Expressed in the Nose and in the Seminal Fluid of the Rabbit

Rosa Mastrogiacomo[1⁹], Chiara D'Ambrosio[2⁹], Alberto Niccolini[3], Andrea Serra[1], Angelo Gazzano[3], Andrea Scaloni[2]*, Paolo Pelosi[1]*

1 Department of Agriculture, Food and Environment, University of Pisa, Pisa, Italy, **2** Proteomics & Mass Spectrometry Laboratory, ISPAAM, National Research Council, Napoli, Italy, **3** Department of Veterinary Sciences, University of Pisa, Pisa, Italy

Abstract

We have purified an abundant lipocalin from the seminal fluid of the rabbit, which shows significant similarity with the sub-class of pheromone carriers "urinary" and "salivary" and presents an N-terminal sequence identical with that of an odorant-binding protein (rabOBP3) expressed in the nasal tissue of the same species. This protein is synthesised in the prostate and found in the seminal fluid, but not in sperm cells. The same protein is also expressed in the nasal epithelium of both sexes, but is completely absent in female reproductive organs. It presents four cysteines, among which two are arranged to form a disulphide bridge, and is glycosylated. This is the first report of an OBP identified at the protein level in the seminal fluid of a vertebrate species. The protein purified from seminal fluid is bound to some organic chemicals whose structure is currently under investigation. We reasonably speculate that, like urinary and salivary proteins reported in other species of mammals, this lipocalin performs a dual role, as carrier of semiochemicals in the seminal fluid and as detector of chemical signals in the nose.

Editor: Sabato D'Auria, CNR, Italy

Funding: The authors have no support or funding to report.

Competing Interests: The authors have declared that no competing interests exist.

* Email: andrea.scaloni@ispaam.cnr.it (A. Scaloni); ppelosi@agr.unipi.it (PP)

⁹ These authors contributed equally to this work.

Introduction

Odorant-binding proteins (OBPs) of vertebrates are a sub-class of lipocalins [1–2], a protein super-family including retinol-binding protein [3], ß-lactoglobulin [4] and many other members that differ for amino acid sequence and physiological function but share the highly conserved structure of the ß-barrel, a sort of cup made of 8 antiparallel ß-sheets enclosing a binding cavity for hydrophobic ligands [5-10]. Vertebrate OBPs are binding proteins of about 150–160 amino acids firstly identified in the nasal epithelium of mammals and classified as carriers for odorants and pheromones [11–17]. Several members of this family have been isolated from different mammals, such as bovine, pig, rabbit and others [18–25], as well as in amphibians [26]. OBPs bind to a large variety of small organic molecules, including odorants and pheromones, with a broad specificity and dissociation constants in the micromolar range [9,27–31].

Despite the detailed structural and functional information available for several OBPs, their physiological role in olfaction is still not clear [15–17,32–33]. A carrier for hydrophobic odorants across the aqueous nasal mucus seems reasonable, but a more specific function in detecting chemical messengers cannot be excluded. This idea is based on the expression of several OBPs in the same species, with different and complementary spectra of binding [30,34]. Moreover, there is clear evidence that insect OBPs, a class of proteins structurally different from those of vertebrates, but probably with similar functions [35], are often required for a correct detection of odors and pheromones [36–37], and are also involved in the discrimination of different semiochemicals [38–39].

Whatever their role and detailed mechanism of action, it is reasonable to hypothesise that OBPs from vertebrates might be involved in the detection of pheromones, rather than general odorants. This idea is suggested by the small number of OBP sub-types reported in mammals, as compared to those from insects, and their expression in the vomeronasal organ (an organ dedicated to pheromone perception) [40–42] or in glands of the nasal respiratory epithelium [43], but not in the olfactory mucosa. The sole exception of the human OBP, which was detected in the mucus of the olfactory cleft, but not in the lower nasal regions [44], might be explained with the fact that the vomeronasal organ is absent or non-functioning in humans. However, strong evidence for the involvement of OBPs in detecting pheromones comes from their expression in organs dedicated to the synthesis and the delivery of pheromones [33]. In fact, OBPs similar or identical to those identified in the nose have also been reported as expressed in non-sensory organs and secreted in biological fluids involved in pheromonal communication. Best studied examples include the "major urinary proteins" (MUPs) of mouse and rat [7,45–48], which are synthesised in the liver and excreted in the urine at concentrations of several mg/mL, the "salivary proteins" (SALs) of the boar, abundantly produced by the submaxillary glands [10,19,34], and the so-called "aphrodisin" identified in the vaginal secretion of the hamster [49–50]. In each species, these proteins are produced in the above-mentioned organs in a sex-specific fashion, while they are expressed in the nose equally in both sexes [51]. When released in the urine, saliva or other secretions, such proteins are loaded with organic compounds known to be the species-specific pheromones, while in the nose they are void. In particular, it has been reported that murine MUPs, when excreted in the urine, are complexed with known animal pheromones, such as 2-sec-butylthiazoline and 3,4-dehydro-*exo*-brevicomin [47,52]. Similarly, pig SALs, when isolated from the saliva, carries the boar-specific pheromones 5α-androst-16-en-3-one and 5α-androst-16-en-3-ol [19].

Although the few cases reported above have been studied in detail, the use of OBPs as carriers of pheromones to be released in the environment might be much more common and widespread. The sweath of horses contains large amounts of an OBP-like protein complexed with putative semiochemicals [25], while the salivary lipocalins of several mammals, often reported as allergens [53–55], might perform similar functions. Chemical communication in the rabbit has not been widely studied. A single pheromone has been so far described, namely the volatile compound 2-methyl-2-butenal, which was isolated from the milk and shown to trigger a very clear and robust response in the puppies [56–57]. Information on rabbit OBPs is limited to our previous work reporting the isolation and partial characterization of three members from the nasal tissue [18,23]. The present study was aimed at further investigating the putative role of rabbit OBPs as carriers of pheromones to be released in the environment and describes an OBP expressed only in the nose of both sexes and in seminal fluid.

Experimental Procedures

Materials

Rabbit bodies were kindly provided by a local abbattoir and dissected within an hour after death or kept at −20°C for a few days. Rabbit seminal fluid was collected using an all-glass artificial vagina equipped with a jacket where warm water was circulated.

Ethics statement

All operations were carried out in strict accordance with the recommendations for handling laboratory animals of the National Research Council (CNR) of Italy. The protocol was approved by the Committee on the Ethics of Animal Experiments of the Italian CNR (Permit Number: 01-2014 of February 18, 2014). All efforts were made to minimize suffering of the animals.

RNA extraction and cDNA synthesis

Total RNA was extracted using TRI Reagent (Sigma), following the manufacturer's protocol. cDNA was prepared from total RNA by reverse transcription, using 200 units of SuperScript™ III Reverse Transcriptase (Invitrogen) and 0.5 mg of an oligo-dT primer in a 50 μL reaction volume. The mixture also contained 0.5 mM of each dNTP (GE-Healthcare), 75 mM KCl, 3 mM MgCl2, 10 mM DTT and 0.1 mg/ml BSA in 50 mM Tris-HCl, pH 8.3. The reaction mixture was incubated at 50°C for 60 min and the product was directly used for PCR amplification or stored at −20°C.

Polymerase chain reaction

Aliquots of 1 μL of crude cDNA were amplified in a Bio-Rad Gene Cycler thermocycler, using 2.5 units of *Thermus aquaticus*

DNA polymerase (GE-Healthcare), 1 mM of each dNTP (GE-Healthcare), 1 μM of each PCR primer, 50 mM KCl, 2.5 mM MgCl$_2$ and 0.1 mg/ml BSA in 10 mM Tris-HCl, pH 8.3, containing 0.1% v/v Triton X-100. At the 5′ end, we used a specific primer (rabOBP3-fw: 5′-CACAGCCACTCGGA-3′) corresponding to the sequence encoding the first five amino acids of the mature protein. At the 3′ end, we used an oligo-dT to first obtain the correct sequence of the gene, then a specific primer (rabOBP3-rv: 5′-TTAGGCGGCTCCGCCGTC-3′) encoding the last five residues and the stop codon, to check the presence of the gene in different tissues. After a first denaturation step at 95°C for 5 min, we performed 35 amplification cycles (1 min, at 95°C; 30 sec, at 50°C; 1 min, at 72°C) followed by a final step of 7 min, at 72°C.

Cloning and sequencing

The crude PCR products were ligated into a pGEM (Promega) vector without further purification, using a 1:5 (plasmid:insert) molar ratio and incubating the mixture overnight, at room temperature. After transformation of *E. coli* XL-1 Blue competent cells with the ligation products, positive colonies were selected by PCR using the plasmid's primers SP6 and T7 and grown in LB/ampicillin medium. DNA was extracted using the Plasmid MiniPrep Kit (Euroclone) and custom sequenced at Eurofins MWG (Martinsried, Germany).

Preparation of the tissue extracts

Crude extracts were prepared by homogenization of the corresponding tissues in 10 mL of 20 mM Tris-HCl pH 7.4 (Tris buffer) per gram of tissue, using a Polytron homogenizer, followed by centrifugation at 20,000×g for 20 min. The clear supernatant was immediately used for SDS-PAGE and Western blotting experiments.

Purification of the seminal protein

Lipocalins from rabbit seminal fluid were purified through a 1×30 cm Superose 12 column in 50 mM ammonium bicarbonate, as previously reported [23]. Selected fractions were then pooled, dialysed against 20 mM Tris-HCl, pH 7.4, and applied to a 1.5×25 cm Whatman DE-52 column. Elution was performed using a linear 0.1–0.4 M NaCl gradient, in 20 mM Tris-HCl, pH 7.4. Each fraction was analysed using 12% SDS-PAGE.

Protein digestion and peptide separation

Rabbit seminal fluid OBP was resolved by SDS-PAGE, excised from the gel, triturated, *in-gel* reduced, S-alkylated and digested with trypsin, as previously reported [56]. Gel particles were extracted with 25 mM NH$_4$HCO$_3$/acetonitrile (1:1 v/v) by sonication, and digests were concentrated. Peptide mixtures were either desalted using μZipTipC$_{18}$ pipette tips (Millipore) before MALDI-TOF-MS analysis, directly analyzed by nanoLC-ESI-LIT-MS/MS (see below) or simply resolved on an Easy C$_{18}$ column (100×0.075 mm, 3 μm) (Proxeon) using a linear gradient of acetonitrile containing 0.1% trifluoroacetic acid in aqueous 0.1% trifluoroacetic acid, at a flow rate of 300 nL/min, for 80 min. In the latter case, collected fractions were concentrated and analyzed by MALDI-TOF-MS.

Protein alkylation under native conditions

Protein samples for disulfide assignment were alkylated with 1.1 M iodoacetamide in 0.25 M Tris-HCl, 1.25 mM EDTA, and 6 M guanidinium chloride, pH 7.0, at 25 °C for 1 min in the dark. Samples were separated from excess salts and reagents by passing the reaction mixture through a PD10 column (Amersham Biosciences), as previously reported [59]. Protein samples were finally digested and resolved by LC as mentioned above.

Glycopeptide enrichment

To isolate glycopeptides, rabbit seminal fluid OBP digest aliquots were solved in 80% acetonitrile, 2% formic acid and loaded on GELoader tips (Eppendorf, Germany), which were plugged with 3M Empore C8 extraction disk material (3M Bioanalytical Technologies, MN) and packed with ZIC-HILIC (200 Å, 10 μm, zwitterionic sulfobetaine functional groups) resin (Sequant, Sweden) [60]. Loaded microcolumns were washed twice with 15 μL of 80% acetonitrile, 2% formic acid. Glycopeptides were first eluted with 10 μL of 2% formic acid and then with 5 μL of 50% acetonitrile, 2% formic acid; pooled fractions were analyzed by MALDI-TOF-MS, as described below.

Peptide deglycosylation and disulfide reduction

Glycopeptides were directly deglycosylated on the MALDI target by treatment with 0.2 U of PNGase F (Roche) in 50 mM NH₄HCO₃, pH 8, at 37 °C, for 1 h. Then, 2 μL of 0.1% trifluoroacetic acid was added to reaction mixtures, which were desalted on μZipTipC18 pipette tips (Millipore) before MALDI-TOF-TOF-MS analysis [61].

Disulfide-containing peptides were directly reduced on the MALDI target by treatment with 10 mM mM DTT in 50 mM NH₄HCO₃, pH 8, at 37 °C, for 1 h. Then, 2 μL of 0.1% trifluoroacetic acid was added to reaction mixtures, which were desalted on μZipTipC18 pipette tips (Millipore) before MALDI-TOF-TOF-MS analysis [61].

MS analysis

Peptide mixtures were analyzed by nLC-ESI-LIT-MS/MS using a LTQ XL mass spectrometer (ThermoFinnigan, USA) equipped with a Proxeon nanospray source connected to an Easy-nLC (Proxeon, Denmark) [58]. They were resolved on an Easy C_{18} column (100×0.075 mm, 3 μm) (Proxeon) using a linear gradient of acetonitrile containing 0.1% formic acid in aqueous 0.1% formic acid, at a flow rate of 300 nL/min, for 25 min. Spectra were acquired in the range m/z 400–1800. Acquisition was controlled by a data-dependent product ion scanning procedure over the 3 most abundant ions, enabling dynamic exclusion (repeat count 1 and exclusion duration 1 min). The mass isolation window and collision energy were set to m/z 3 and 35%, respectively.

During MALDI-TOF-MS analysis, entire protein digests or selected peptide fractions were loaded on the instrument target together with 2,5-dihydroxy-benzoic acid (10 mg/mL in 70% v/v acetonitrile, 0.1% v/v trifluoroacetic acid) or α-cyano-4-hydroxycinnamic acid (saturated solution in 30% v/v acetonitrile, 0.1% v/v trifluoroacetic acid) as matrices, using the dried droplet technique; a 384-spot ground steel plate (Bruker Daltonics) was used to this purpose. Spectra were acquired in the m/z range 500–5000 on a Bruker Ultraflextreme MALDI-TOF-TOF instrument (Bruker Daltonics) operating either in reflectron mode or linear mode. Instrument settings were: pulsed ion extraction = 100 ns, laser frequency = 1000 Hz, number of shots per sample = 2500–5000 (random walk, 500 shots per raster spot). Mass spectra were calibrated externally using nearest neighbour positions loaded with Peptide Calibration Standard II (Bruker Daltonics), with quadratic calibration curves. MS/MS spectra were acquired in LIFT mode. Data were elaborated using the FlexAnalysis software (Bruker Daltonics).

nLC-ESI-LIT-MS/MS data were searched by using MASCOT (version 2.2.06) (Matrix Science, UK) against an updated rabbit EST database containing available protein sequences (NCBI 28/11/2013, 212376 sequences). As searching parameters, we used a mass tolerance value of 2 Da for precursor ion and 0.8 Da for ion fragments, trypsin trypsin and/or slymotrypsin (cleavage at Lys, Arg, Phe, Tyr, Trp and Leu) as proteolytic enzymes, a missed cleavages maximum value of 2, Cys carbamidomethylation and Met oxidation as fixed and variable modification, respectively. Protein candidates with more than 2 assigned unique peptides with an individual Mascot ion score >25 and a significant threshold ($p<0.05$) were further considered for protein identification. In the case of glycopeptides or disulfide-containing peptides, MALDI-TOF mass signals were assigned to peptides, glycopeptides or disulfide-containing peptides using the GPMAW 4.23 software (Lighthouse Data, Denmark). This software generated a mass/fragment database output based on protein sequence, protease selectivity, nature of the amino acids susceptible to eventual glycosylation/oxidation and the molecular mass of the modifying groups. Searching parameters were set as mentioned above; mass values were matched to protein regions using a 0.02% mass tolerance value. MALDI-TOF-TOF searching parameters were set with tolerances of 100 ppm and 0.5 Da for MS and MS/MS data, respectively. Glycosylation or disulfide assignments were always confirmed by additional MS experiments on deglycosylated or reduced peptides, respectively.

Ligand-binding experiments

The affinity of the fluorescent probe N-phenyl-1-naphthylamine (1-NPN) was measured by titrating a 2 μM solution of the protein with aliquots of 1 mM 1-NPN solved in methanol to reach final concentrations of 2–16 μM. The probe was excited at 337 nm and the maximum emission wavelength was 415 nm. Dissociation constant was evaluated using GraphPad Prism software. Affinities of other ligands were measured in competitive binding assays, by titrating a solution containing the protein and 1-NPN both at the concentration of 4 μM with 1 mM solutions of each competitor in methanol to reach final concentrations of 0–16 μM. Dissociation constants of the competitors were calculated from the concentrations of ligand halving the initial fluorescence value of 1-NPN (IC_{50}), using the equation:

$$K_D = IC_{50}/1 + 1 - NPN/K_{1-NPN}$$

1-NPN being the free concentration of 1-NPN and K_{1-NPN} being the dissociation constant of the complex protein/1-NPN.

Results

Identification and purification of an OBP from the rabbit seminal fluid

With the aim of identifying OBPs expressed in rabbit non-sensory organs, we verified the occurrence of a protein in the male semen that showed a cross-reactivity with a polyclonal antiserum raised against the boar salivary lipocalin (pig SAL) [19]. This protein, which migrated in SDS-PAGE as a blurred band at about 23 kDa, was very abundant in the seminal liquid but was not present in the sperm cells. Figure 1 reports the electrophoretic analysis of the supernatant and the pellet obtained by centrifugation of the crude semen. The weaker cross-reactivity of the pellet was due to a contamination with the seminal fluid and disappeared completely after washing the pellet three times with buffer. Protein concentration in the semen was estimated to be about 10–20 mg/mL. This protein was then purified by gel filtration chromatography on a Superose-12 column, followed by anion-exchange

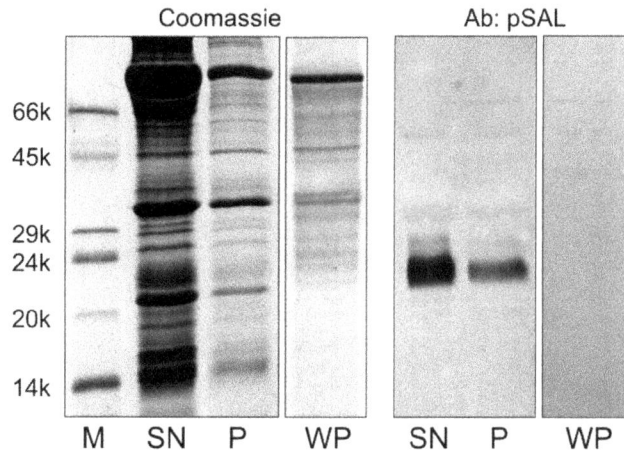

Figure 1. SDS-PAGE analysis of rabbit sperm and corresponding Western blotting. SN, soluble fraction; P, sperm cells; WP, sperm cells after washing three times with buffer. A strong cross-reactivity with a polyclonal antiserum raised against pig SAL [19] was observed for a protein migrating at about 23 kDa. Staining was much stronger in the soluble fraction; the weak reactivity observed for the sperm cells disappeared after washing the cells, thus indicating the absence of the protein in this sample.

chromatography on a DE-52 resin. Figure 2 reports the SDS-PAGE profile of selected fractions from the first purification step, together with the corresponding Western blotting, as well as of the purified protein that was used for further studies.

In order to characterize the nature of this seminal protein, we performed a MALDI-TOF peptide mass fingerprinting analysis on its tryptic digest following reduction with dithiothreitol and alkylation with iodoacetamide (data not shown). MS results matched to a sequence reported in the NCBI EST database (entry EL341998) annotated as UTE-7, which corresponded to a cDNA isolated from rabbit uterus. The sequence at the protein N-terminus of UTE-7 is identical with that of a rabbit OBP (rabOBP3) we had previously isolated from the nasal tissue [23]. Since the identity of some nucleotides in the EST entry mentioned

above was not determined and the sequence was partial, we again cloned the corresponding cDNA and sequenced it; data are reported in Supplementary Figure S1. Our analysis provided a complete nucleotide assignment, together with very few base corrections, finally ascertaining a corresponding protein sequence as made of 161 amino acids. Finally, massive peptide mapping nanoLC-ESI-LI-MS/MS experiments on a tryptic digest ascertained the nature of the protein N- and C-terminus, verifying about 93% of its amino acid sequence (Table S1).

Tissue expression

To detect the site of synthesis for this seminal protein, we performed PCR experiments on samples of cDNA prepared from different parts of male and female reproductive organs. To first identify the full sequence of the gene (Figure S1), we used a specific primer at the 5'-end encoding the first five amino acids of the sequence reported in the database as UTE-7 (acc. no: EL341998) and an oligo-dT at the 3'-end. Then, we used the same primer at the 5'-end and a second specific primer at the 3'-end encoding the last five residues and the stop codon, to check for the presence of this gene in different organs. In particular, olfactory and respiratory epithelium from both sexes, prostate, epididymis, testis, uterus, uterine tubes, ovaries, vagina and vaginal vestibule were evaluated. Amplification bands were obtained only for the prostate as well as for the respiratory epithelium of both sexes. Parallel cloning and sequencing of samples from these tissues always yielded the same sequence (Figure S1), excluding the occurrence of various protein isoforms. The specificity of protein expression in these tissues was confirmed at the protein level by Western-blotting experiments (Figure 3). On this basis, we can conclude that the protein previously named as UTE-7 is not produced in the uterus, nor in any part of the female reproductive system, but was probably found in such organ as result of a sample contamination. On the other hand, the sequence we report here very likely corresponds to the protein (rabOBP3) we had

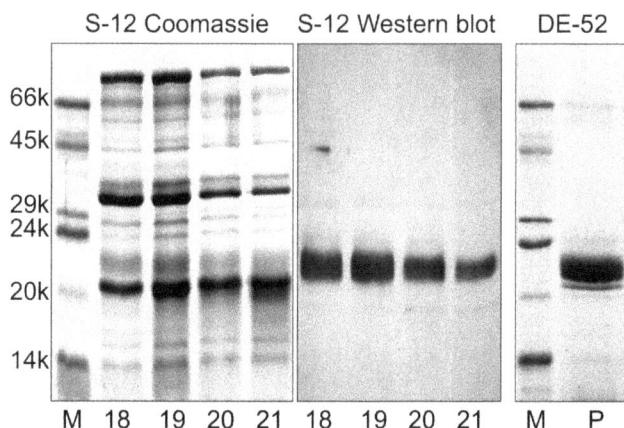

Figure 2. Purification of the rabbit seminal fluid OBP. A sample of crude seminal fluid, as obtained after sperm centrifugation, was resolved at first by gel filtration chromatography on a Superose-12 column and then by anion-exchange chromatography on a DE-52 column (see Materials and Methods section for details). The protein was eluted as a pure component, as verified by SDS-PAGE.

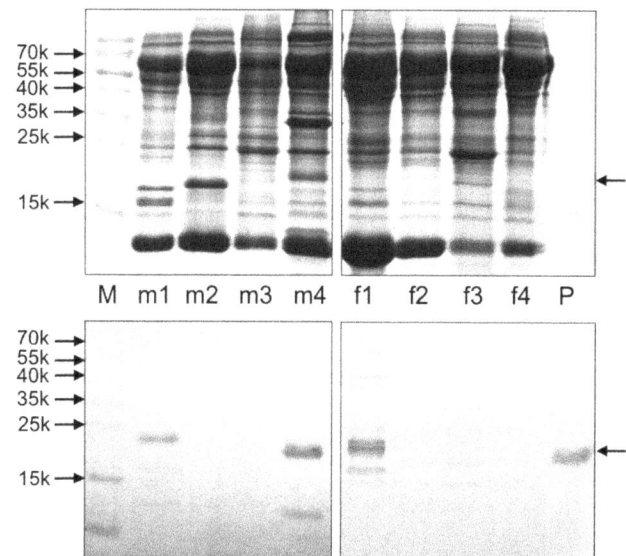

Figure 3. Expression of rabOBP3 in different tissues of male (m) and female (f) rabbit individuals. SDS-PAGE analysis of different rabbit tissues and corresponding Western blotting are shown. M: molecular weight markers; m1: nasal respiratory tissue; m2: epididymis; m3: testis; m4: prostate; f1: nasal respiratory tissue; f2: uterine tubes; f3: ovaries; f4: uterus; P: purified rabOBP3.

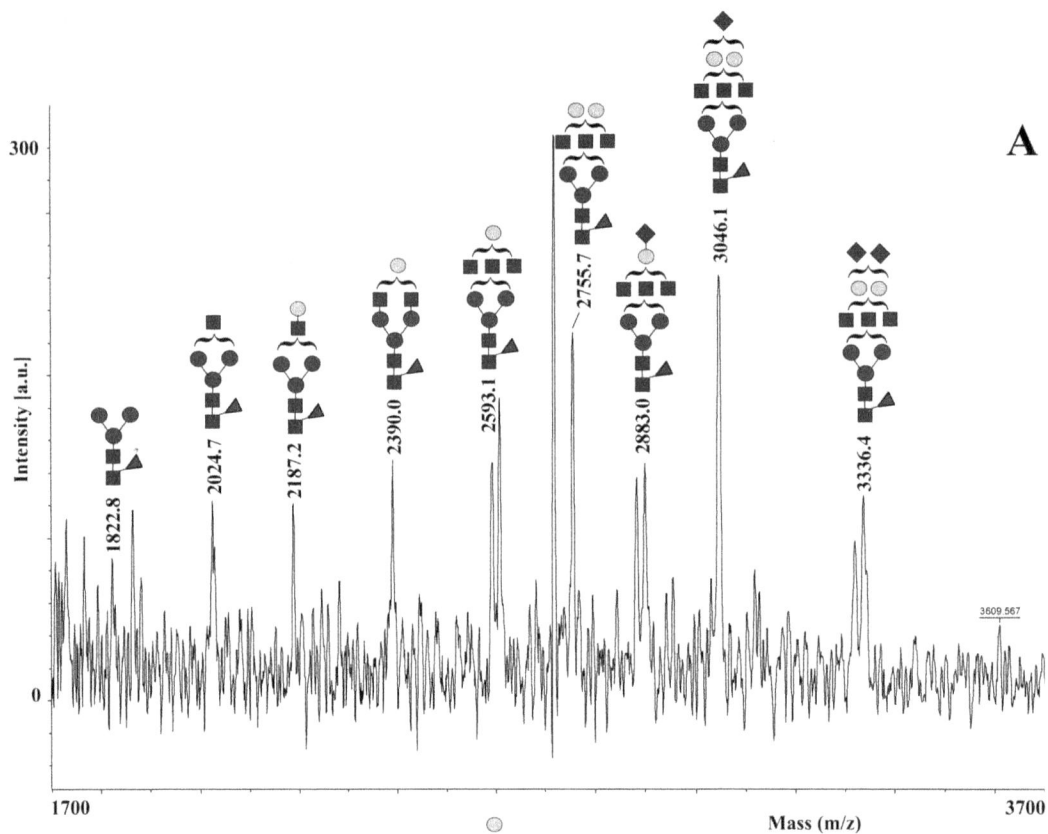

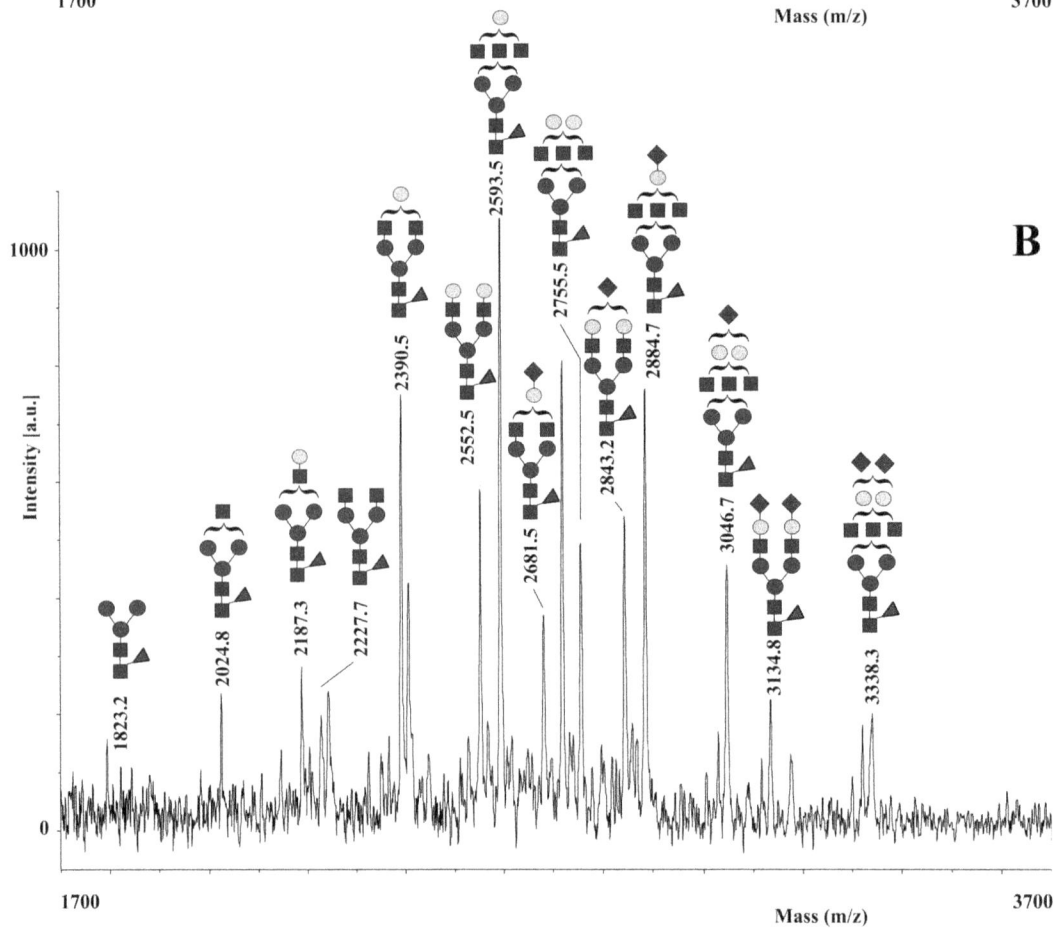

Figure 4. MALDI-TOF-MS analysis of the purified tryptic glycopeptides from rabOBP3 as obtained after HILIC enrichment and nanoLC separation. Spectra acquired in linear mode of the fractions eluting at 15 and 16 min are reported in panel A and B, respectively; shown are the mono-, bi- and tri-antennary complex-type glycan structures N-linked to Asn44 in peptide (44–50). ■, N-acetyl-glucosamine; ●, mannose; ○, galactose; ◀, fucose; ◆, N-acetyl-neuraminic acid.

previously isolated from the nasal epithelium [23]. Accordingly, we decided to rename UTE-7 as rabOBP3.

Post-translational modifications in rabOBP3

The blurred band and the discrepancy between the calculated (18 kDa) and apparent (23 kDa) molecular mass of the intact protein observed in SDS-PAGE, its broad MH^+ signal in MALDI-TOF-MS (data not shown) and the occurrence of two putative N-linked glycosylation sites (Asn29 and Asn44) in the corresponding amino acid sequence (as predicted by bioinformatic analysis) suggested that rabOBP3 could be a glycoprotein, similarly to what reported for pig SAL, horse EquC1 and some murine/rat MUPs [25,34,62]. To evaluate protein glycosylation and assign potential

modification site(s), a rabOBP3 sample resolved by SDS-PAGE was *in gel* reduced, alkylated with iodoacetamide and digested with trypsin. The corresponding peptide digest was then enriched for glycopeptides on a HILIC column and resolved by nanoLC into different fractions, which were then analyzed by MALDI-TOF-MS. Fractions eluting at 15 and 16 min showed a similar pattern of multiple signals in the mass spectrum (Figure 4A and B). On the basis of the measured mass values and known pathways of glycoprotein biosynthesis, all these peaks were assigned to peptide (44–50) having a pentasaccharide core N-linked to Asn44, and bearing mono-, bi- and tri-antennary complex glycan structures (theor. MH^+ values: m/z 1821.8, 2024.9, 2187.1, 2228.2, 2390.3, 2552.5, 2593.5, 2681.6, 2755.7, 2843.7, 2884.8, 3046.9, 3135.0

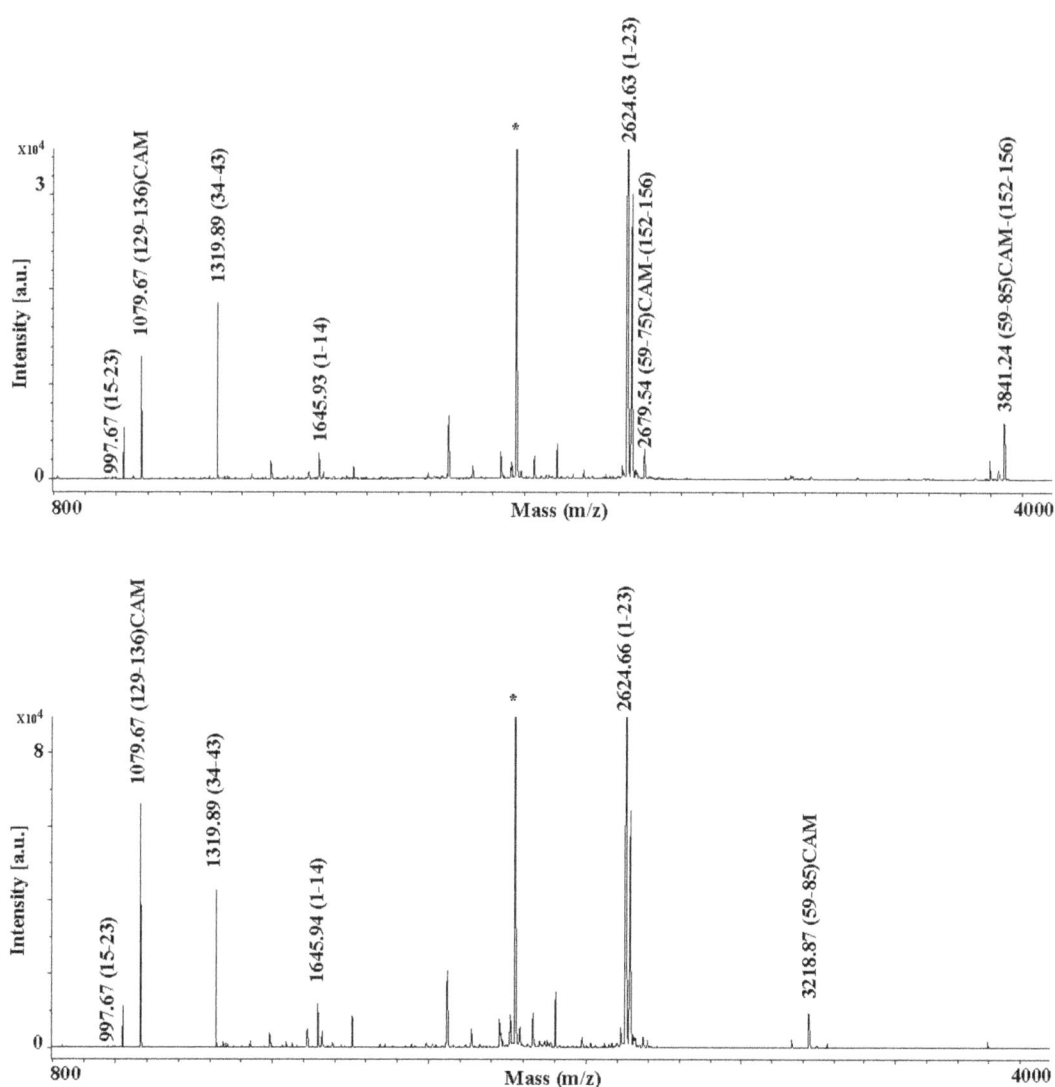

Figure 5. MALDI-TOF-MS analysis of the tryptic digest of rabOBP3 alkylated with iodoacetamide under denaturing, non-reducing conditions before (top) and following (bottom) treatment with dithiothreitol. Constant and variable signals are labelled in the spectra acquired in reflectron mode to highlight reduced and oxidized residues present under native conditions. Trypsin-derived peptides are indicated with an asterisk.

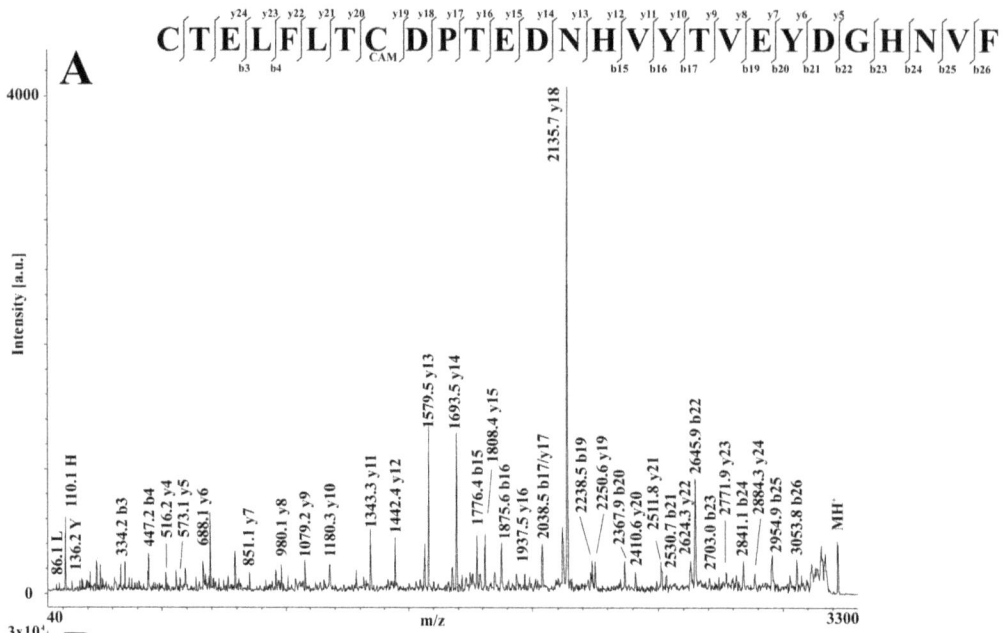

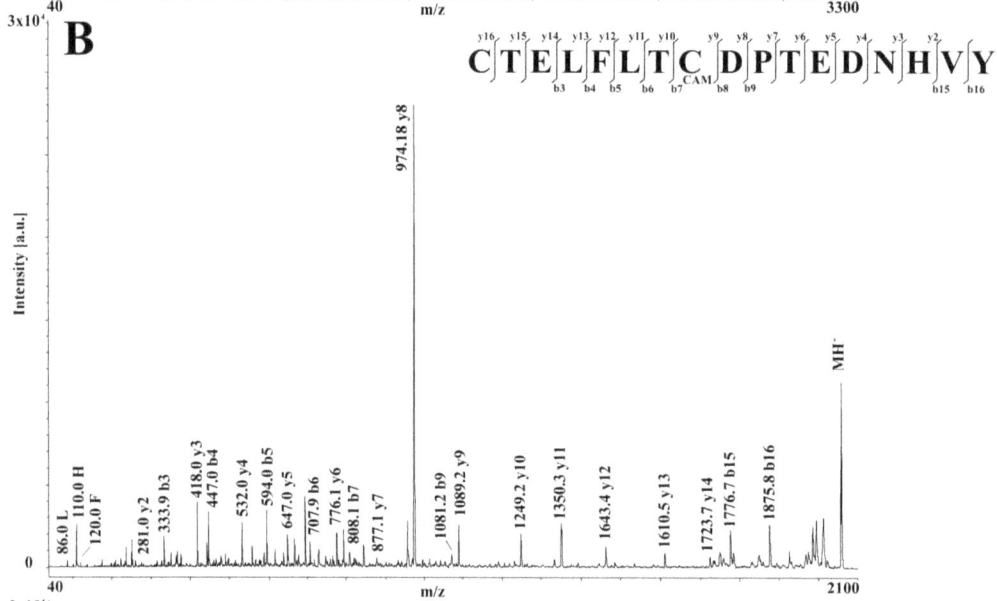

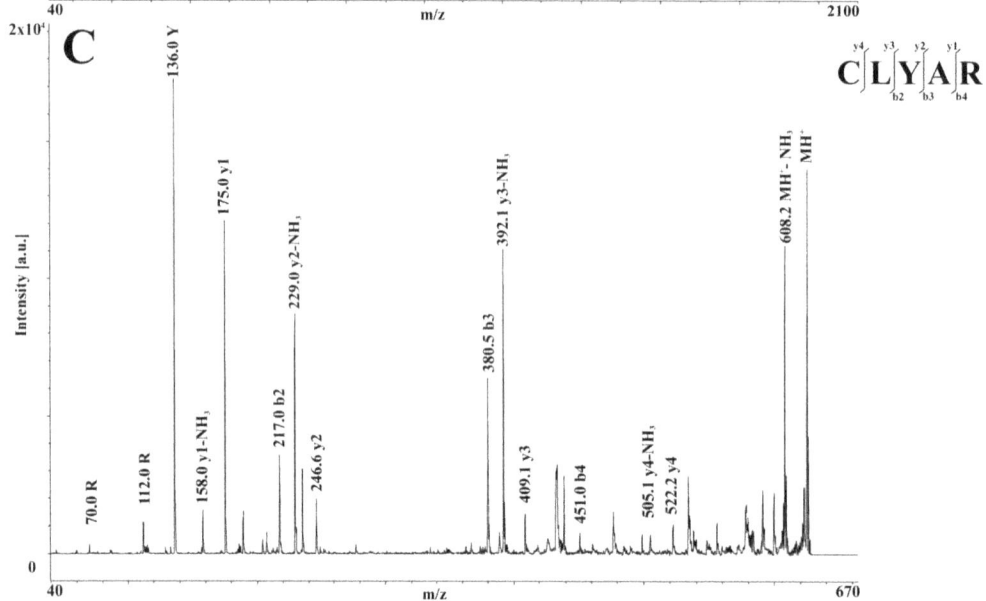

Figure 6. MALDI-TOF-TOF spectra of the disulfide-containing tryptic peptides from alkylated rabOBP3 following treatment with dithiothreitol. Fragmentation spectra of the peptides (59–85)CAM, (59–75)CAM and (152–156) are shown in panels A, B and C, respectively. In all cases, Cys residues originally involved in the S-S bond are present in a reduced status, the remaining ones occurring as carboxamidomethylated derivatives.

and 3338.2). After PNGase treatment, glycopeptides in both fractions collapsed to a unique component (peptide 44–50) having a MH$^+$ signal at m/z 784.08 (data not shown). MALDI-TOF-TOF-MS analysis of the deglycosylated peptide confirmed the expected Asn44>Asp conversion. Multiple signals associated with glycopeptides were also detected in the mass spectrum of the fractions eluting at 21 and 22 min. On the basis of measured mass values (exp. MH$^+$ values: m/z 3124.8, 3327.9, 3490.2, 3530.9, 3693.2, 3733.5, 3855.4, 3896.6, 3983.8, 4059.1, 4146.4, 4187.7, 4350.1, 4437.0, 4641.2 and 4932.9) and the relative intensities, these peaks were associated to peptide (34–50) having the same glycan structures reported in Figure 4 as N-linked to Asn44 (theor. MH$^+$ values: m/z 3123.3, 3326.5, 3488.7, 3529.7, 3691.9, 3732.9, 3854.0, 3895.0, 3983.1, 4057.2, 4145.3, 4186.3, 4348.4, 4436.5, 4639.7 and 4931.0). No signals related to the non-glycosylated peptide counterparts were detected in any LC fractions either from the entire protein digest or its glycopeptide-enriched portion, thus suggesting that rabOBP3 was completely modified at this site. On the other hand, no glycopeptides containing the other putative N-linked glycosylation site (Asn29) were observed in the tryptic digest or its HILIC eluate either before and after nanoLC separation; conversely, the corresponding non-glycosylated counterparts were always detected in both cases, thus demonstrating that no modification occurred at this site.

To evaluate protein thiol status and assign disulfide-bridged Cys residues, if present, rabOBP3 was treated with 1.1 M iodoacetamide under denaturing, non-reducing conditions and purified by size-exclusion chromatography. The alkylated protein was then digested with trypsin and split in two samples that were treated or not with DTT; Figure 5 shows the MALDI-TOF mass spectrum

of each sample. In addition to a number of common signals present in both spectra, the digest deriving from the protein not treated with DTT uniquely showed the presence of a clear MH$^+$ signal at m/z 3841.24, which was associated with the disulfide-containing peptides (59–85)CAM-(152–156) resulting from an aspecific cleavage at Phe85. A faint MH$^+$ peak at m/z 2679.54 was also observed; this signal was assigned to the smaller disulfide-containing peptide homologue (59–75)CAM-(152–156) derived from an aspecific hydrolytic event at Tyr85. Conversely, the digest treated with DTT showed the absence of the signals mentioned above and the exclusive occurrence of a MH$^+$ peak at m/z 3218.87, which was associated with the peptide (59–85)CAM. Due to its reduced mass value, no signal assigned to the peptide (152–156) was observed. These result confirmed the occurrence of one cysteine (Cys59 or Cys66) involved in a disulfide bond with Cys152 in the above-mentioned peptides, the remaining one being in a reduced status. On the other hand, both samples showed the presence of a MH$^+$ signal at m/z 1079.67, which derived from the peptide (129–136)CAM; the latter result demonstrate that rabOBP3 contains Cys133 as free thiol under native conditions.

To definitively assign the Cys residues involved in the protein S-S bond, disulfide-containing peptides (59–85)CAM-(152–156) and (59–75)CAM-(152–156) were then purified by nanoLC and reduced with DTT directly on the MALDI target. Resulting products showed MH$^+$ peaks at m/z 3220.2 and 2058.6, which were associated with the expected reduced peptides (59–85)CAM and (59–75)CAM, respectively, both having the Cys residue originally involved in the S-S bond in a reduced status and the remaining one as carboxamidomethylated species. In both cases, the occurrence of the reduced peptide (152–156) was also observed

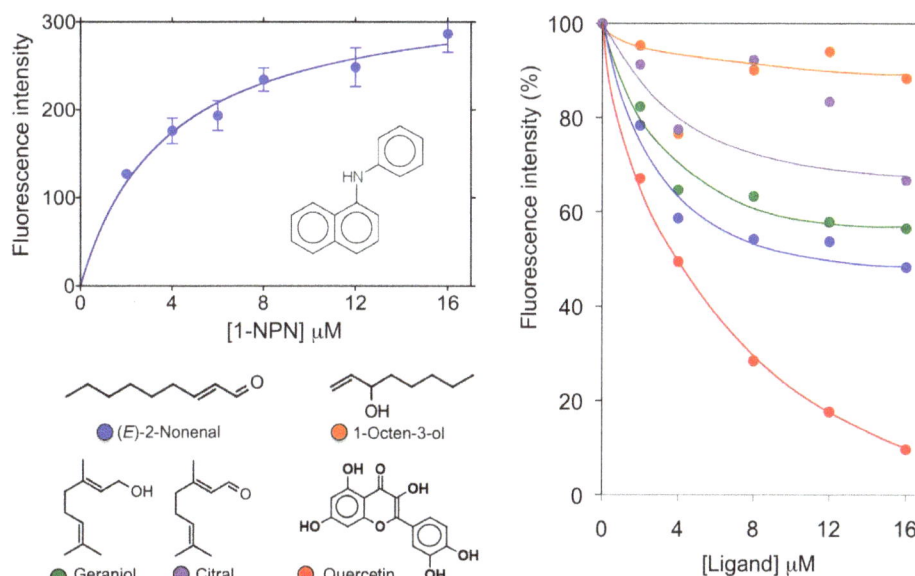

Figure 7. Binding of 1-NPN (left) and selected ligands (right) to rabOBP3 purified from seminal fluid and delipidated with dichloromethane. The protein binds the fluorescent probe 1-NPN with a dissociation constant of 3.8 μM (SD 0.9, n = 3). None of the ligands tested exhibited strong affinity to the protein, except quercetin, for which a physiological role does not seem plausible. Calculated dissociation constants are 2.2, 7.8 and 11.2 μM for quercetin, 2-nonenal and geraniol, respectively.

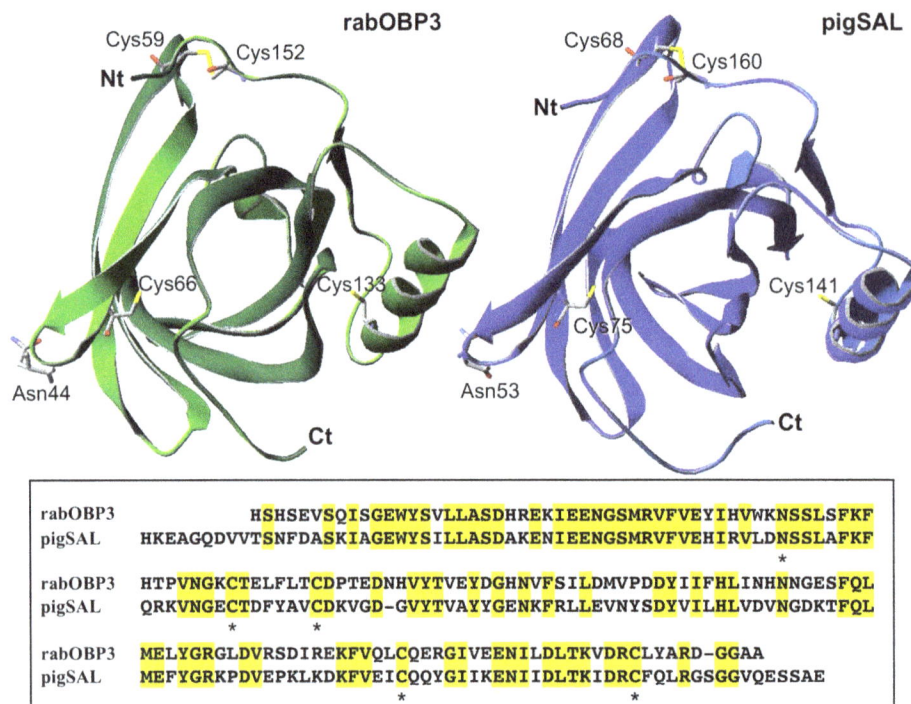

Figure 8. Three-dimensional model of rabOBP3 as built by using the crystallographic structure of pig SAL (Boar salivary lipocalin, PDB ID: 1 GM6) as a template [10]. Molecular model of rabOBP3 and pig SAL are shown in the left- and right-top panel, respectively. The corresponding sequence alignment is shown in the bottom panel, where conserved amino acids are highlighted in yellow. The conserved N-glycosylation site (Asn44), and oxidized (Cys59 and Cys152) and reduced (Cys66 and Cys133) residues are indicated by specific labelling (top) or asterisks (bottom).

in the corresponding MS spectra (exp. MH$^+$ signal at m/z 625.2). MALDI-TOF-TOF-MS analysis of the reduced peptides (59–85)CAM and (59–75)CAM finally assigned the thiol group to Cys59, definitively proving the existence of a disulfide bond in rabOBP3 linking together Cys59 and Cys152 (Figure 6).

Endogenous ligands of rabOBP3

Since pig SAL and murine/rat MUPs carry species-specific pheromones as endogenous ligands, we then searched for compounds that might be complexed with rabOBP3. Gas-chromatographic separation coupled with MS (GC-MS) analysis of a dichloromethane extract of the protein from rabbit seminal liquid showed the presence of several peaks, to none of which we could confidently assign a defined chemical structure.

Ligand-binding assays showed that rabOBP3 reversibly binds to the fluorescent probe N-phenyl-1-naphthylamine (1-NPN) with a dissociation constant of 3.8 μM (SD 0.9, n = 3). Competitive binding assays, performed with some common plant volatiles indicated significant, but modest affinity to 2-nonenal and geraniol. On the other hand, quercetin efficiently displaced 1-NPN from the complex, but is difficult to propose a role as a rabbit semiochemical for this compound (Figure 7).

Three-dimensional model of rabOBP3

Based on the significant (52%) sequence identity between rabOBP3 and pig SAL (Figure 8, bottom), a three-dimensional molecular model of the first protein was built up as deriving from the crystal structure of the latter (Boar salivary lipocalin, PDB ID: 1 GM6) (Figure 8, top). The good quality of this model was assessed by ANOLEA and GROMOS evaluations, which calculated small positive energy values for very few amino acids

scattered along the sequence. Although not fixed as initial structural constrains before the modelling procedure, a *post hoc* evaluation of the rabOBP3 model was in perfect agreement with the protein post-translational modifications determined in this study. In fact, Asn44 occurred at the most external position in a loop extending its side chain into the solvent, while Cys59 and Cys152 were present in the model with their S atoms at a distance compatible with the presence of a disulfide bridge (Figure 8). The latter result was not surprising, based on the high conservation of cystine moieties in rabOBP3, pig SAL, murine/rat MUPs, and other proteins [49–63]. As expected, the remaining cysteine residues (Cys66 and Cys133) occurred too far apart to be linked together, in a condition compatible with a reduced state.

Discussion

When the first OBP of vertebrates was discovered in the nasal tissue of the cow [11–12], its sequence similarity with urinary proteins of rodents immediately suggested a function in chemical communication for these polypeptides [64], which had been described several years earlier, but whose presence in the urine had represented an unsolved puzzle until then [65–66]. Since that time, the occurrence of proteins of the same class or even identical in olfactory organs and in secretions used in chemical communication has been well documented both in vertebrates and in insects. These polypeptides can be recognised among the family of OBPs on the basis of sequence similarity. Besides the urinary proteins of mouse and rat, OBPs of vertebrates include the boar salivary lipocalin SAL [30], the horse Equc1 (abundantly secreted in sweat) [25] and the hamster aphrodisin occurring in the vaginal discharge [49]. On the other hand, the human genome contains a pseudogene for a protein of this group, which presents a mutation at the donor site of the second intron, thus disrupting the corresponding ORF [67].

Insects OBPs have been reported in the sex organs. In particular, mosquito *Aedes aegypti* and lepidopteran *Helicoverpa armigera* OBPs, which also occur in the insect antennae, are produced in the male reproductive organ and are transferred to the female during mating. It has been shown that *H. armigera* OBP, when extracted from semen, is complexed with potential pheromones for the species and eventually is found on the surface of fertilised eggs [68]. In vertebrates, OBPs have been reported in reproductive organs: aphrodisin is secreted in the vaginal discharge of the hamster [47–48], while in humans the gene encoding an OBP is expressed in the prostate [69]. Data reported in this study suggest that also in the seminal liquid of the rabbit, OBPs might act as pheromone carriers. Unfortunately, information on rabbit pheromones is limited to the suckling pheromone, which directs pups towards the nipple [56–57]. Among the volatiles we have extracted from seminal rabOBP3, we were not able to identify any compound with confidence, thus suggesting that endogenous ligands of rabOBP3 might not be among common natural chemicals. In line with this consideration, preliminary competitive binding assays with common terpenoids and fatty acids excluded these compounds as protein endogenous ligands.

In conclusion, we propose that OBPs as pheromone carriers are likely present in the seminal fluid of other mammals. The isolation of OBPs in reproductive organs and the identification of their endogenous ligands could lead to the discovery of novel pheromones mediating behaviour between sexes, such as male competition, in mammals as it has been shown in some insect species. Besides the knowledge advancement in the biology of mammals, such information might suggest strategies to improve rearing conditions of economically important species, such as rabbit, cattle, pigs and horses.

Supporting Information

Figure S1 (**A**) PCR amplification of the gene encoding rabOBP3 in the prostate (P), as well as in male (mR) and female (fR) nasal respiratory tissue. All three samples gave amplification bands of around 500 bp, that were cloned and sequenced yielding the same sequence, reported in (**B**) with its translation. Similar experiments performed in the same conditions on uterus (Ut), uterine tubes (Tb) and ovaries (Ov) did not produce any amplification bands. (**C**) Alignment of the derived mature amino acid sequences of rabOBP3 cloned from nose and prostate, and compared with the sequence stored in the NCBI EST database as UTE-7 (entry EL341998). Mnose: male nasal tissue; Fnose: female nasal tissue; Prost: prostate.

Table S1   Results of a peptide mapping nanoLC-ESI-LI-MS/MS experiment on a tryptic digest of rabbit seminal OBP.

Acknowledgments

We thank Ms Olga Favilli of the Department of Veterinary Sciences, University of Pisa and Pampaloni Farm, Fauglia, Pisa, for help in the collection of rabbit semen.

Author Contributions

Conceived and designed the experiments: RM AN A. Scaloni PP. Performed the experiments: RM CD AN A. Serra A. Scaloni. Analyzed the data: RM CD A. Serra A. Scaloni PP. Contributed reagents/materials/analysis tools: AG A. Scaloni A. Serra PP. Wrote the paper: A. Scaloni PP.

References

1. Flower DR (1996) The lipocalin protein family: structure and function. Biochem J 318: 1–14
2. Flower DR (2000) Experimentally determined lipocalin structures. Biochim Biophys Acta 1482: 46–56
3. Monaco HL, Rizzi M, Coda A (1995) Structure of a complex of two plasma proteins: transthyretin and retinol-binding protein. Science 268: 1039–1041
4. Sawyer L, Kontopidis G (2000) The core lipocalin ß-lactoglobulin. Biochim Biophys Acta 1482: 136–148
5. Bianchet MA, Bains G, Pelosi P, Pevsner J, Snyder SH, et al. (1996) The three dimensional structure of bovine odorant-binding protein and its mechanism of odor recognition. Nat Struct Biol 3: 934–939
6. Tegoni M, Ramoni R, Bignetti E, Spinelli S, Cambillau C (1996) Domain swapping creates a third putative combining site in bovine odorant binding protein dimer. Nat Struct Biol 3: 863–867
7. Böcskei Z, Groom CR, Flower DR, Wright CE, Phillips EV, et al. (1992) Pheromone binding to two rodent urinary proteins revealed by X-ray crystallography. Nature 360: 186–188
8. Spinelli S, Ramoni R, Grolli S, Bonicel J, Cambillau C, et al. (1998) The structure of the monomeric porcine odorant binding protein sheds light on the domain swapping mechanism. Biochemistry 37: 7913–7918
9. Vincent F, Spinelli S, Ramoni R, Grolli S, Pelosi P, et al. (2000) Complexes of porcine odorant binding protein with odorant molecules belonging to different chemical classes. J Mol Biol 300: 127–139
10. Spinelli S, Vincent F, Pelosi P, Tegoni M, Cambillau C (2002) Boar Salivary Lipocalin: Three-dimensional X-Ray Structure and Androstenol/Androstenone Docking Simulations. Eur J Biochem 269: 2449–2456
11. Pelosi P, Pisanelli AM, Baldaccini NE, Gagliardo A (1981) Binding of 3H-2-isobutyl-3-methoxypyrazine to cow olfactory mucosa. Chem Senses 6: 77–85
12. Pelosi P, Baldaccini NE, Pisanelli AM (1982) Identification of a specific olfactory receptor for 2-isobutyl-3-methoxypyrazine. Biochem J 201: 245–248
13. Bignetti E, Cavaggioni A, Pelosi P, Persaud KC, Sorbi RT, et al. (1985) Purification and characterization of an odorant binding protein from cow nasal tissue. Eur J Biochem 149: 227–231
14. Pevsner J, Trifiletti RR, Strittmatter SM, Snyder SH (1985) Isolation and characterization of an olfactory receptor protein for odorant pyrazines. Proc Natl Acad Sci USA 82: 3050–3054
15. Pelosi P (1994) Odorant-binding proteins. Crit Rev Biochem Mol Biol 29: 199–228
16. Pelosi P (1996) Perireceptor events in olfaction. J Neurobiol 30, 3–19
17. Tegoni M, Pelosi P, Vincent F, Spinelli S, Campanacci V, et al. (2000) Mammalian odorant binding proteins. Biochim Biophys Acta 1482: 229–240
18. Dal Monte M, Andreini I, Revoltella R, Pelosi P (1991) Purification and characterization of two odorant binding proteins from nasal tissue of rabbit and pig. Comp Biochem Physiol 99B: 445–451
19. Marchese S, Pes D, Scaloni A, Carbone V, Pelosi P (1998) Lipocalins of boar salivary glands binding odours and pheromones. Eur J Biochem 252: 563–568
20. Pes D, Mameli M, Andreini I, Krieger J, Weber M, et al. (1998) Cloning and expression of odorant-binding proteins Ia and Ib from mouse nasal tissue. Gene 212: 49–55
21. Paolini S, Scaloni A, Amoresano A, Marchese S, Napolitano E, et al. (1998) Amino acid sequence post-translational modifications binding and labelling of porcine odorant-binding protein. Chem Senses 23: 689–698
22. Ganni M, Garibotti M, Scaloni A, Pucci P, Pelosi P (1997) Microheterogeneity of odorant-binding proteins in the porcupine revealed by N-terminal sequencing and mass spectrometry. Comp Biochem Physiol 117B: 287–291
23. Garibotti M, Navarrini A, Pisanelli AM, Pelosi P (1997) Three odorant-binding proteins from rabbit nasal mucosa. Chem Senses 22: 383–390
24. Pes D, Pelosi P (1995) Odorant-binding proteins of the mouse. Comp Biochem Physiol 112B: 471–479.
25. D'Innocenzo B, Salzano AM, D'Ambrosio C, Gazzano A, Niccolini A, et al. (2006) Secretory proteins as potential semiochemical carriers in the horse. Biochemistry 45: 13418–13428
26. Millery J, Briand L, Bezirard V, Blon F, Fenech C, et al. (2005) Specific expression of olfactory binding protein in the aerial olfactory cavity of adult and developing *Xenopus*. Eur J Neurosci 22: 1389–1399
27. Dal Monte M, Centini M, Anselmi C, Pelosi P (1993) Binding of selected odorants to bovine and porcine odorant binding proteins. Chem Senses 18: 713–721
28. Pevsner J, Hou V, Snowman AM, Snyder SH (1990) Odorant-binding protein characterization of ligand binding. J Biol Chem 265: 6118–6125
29. Hérent MF, Collin S, Pelosi P (1995) Affinities of nutty and green-smelling compounds to odorant-binding proteins. Chem Senses 20: 601–610

30. Loebel D, Marchese S, Krieger J, Pelosi P, Breer H (1998) Subtypes of odorant binding proteins: heterologous expression and assessment of ligand binding. Eur J Biochem 254: 318–324

31. Vincent F, Ramoni R, Spinelli S, Grolli S, Tegoni M, et al. (2004) Crystal structures of bovine odorant-binding protein in complex with odorant molecules. Eur J Biochem 271: 3832–3842

32. Pelosi P (1998) Odorant-binding proteins: structural aspects. Ann NY Acad Sci 855: 281–293

33. Pelosi P (2001) The role of perireceptor events in vertebrate olfaction. Cell Mol Life Sci 58: 503–509

34. Loebel D, Scaloni A, Paolini S, Fini C, Ferrara L, et al. (2000) Cloning, post-translational modifications, heterologous expression, ligand-binding and modelling of boar salivary lipocalin. Biochem J 350: 369–379

35. Pelosi P, Zhou J-J, Ban LP, Calvello M (2006) Soluble proteins in insect chemical communication. Cell Mol Life Sci 63: 1658–1676

36. Xu P, Atkinson R, Jones DN, Smith DP (2005) *Drosophila* OBP LUSH is required for activity of pheromone-sensitive neurons. Neuron 45: 193–200

37. Matsuo T, Sugaya S, Yasukawa J, Aigaki T, Fuyama Y (2007) Odorant-binding proteins OBP57d and OBP57e affect taste perception and host-plant preference in *Drosophila sechellia*. PLoS Biol 5: e118

38. Swarup S, Williams TI, Anholt RR (2011) Functional dissection of Odorant binding protein genes in *Drosophila melanogaster*. Genes Brain Behav 10: 648–657

39. Sun YF, De Biasio F, Qiao HL, Iovinella I, Yang SX, et al. (2012) Two Odorant-Binding Proteins Mediate the Behavioural Response of Aphids to the Alarm Pheromone (*E*)-ß-farnesene and Structural Analogues. PLoS One 7: e32759

40. Pevsner J, Hwang PM, Sklar PB, Venable JC, Snyder SH (1988) Odorant-binding protein and its mRNA are localized to lateral nasal gland implying a carrier function. Proc Natl Acad Sci USA 85: 2383–2387

41. Miyawaki A, Matsushita F, Ryo Y, Mikoshiba K (1994) Possible pheromone-carrier function of two lipocalin proteins in the vomeronasal organ. EMBO J 13: 5835–5842

42. Ohno K, Kawasaki Y, Kubo T, Tohyama M (1996) Differential expression of odorant-binding protein genes in rat nasal glands: implications for odorant-binding protein II as a possible pheromone transporter. Neuroscience 71: 355–366

43. Avanzini F, Bignetti E, Bordi C, Carfagna G, Cavaggioni A, et al. (1987) Immunocytochemical localization of pyrazine-binding protein in bovine nasal mucosa. Cell Tissue Res 247, 461–464.

44. Briand L, Eloit C, Nespoulous C, Bézirard V, Huet JC, et al. (2002) Evidence of an odorant-binding protein in the human olfactory mucus: location, structural characterization, and odorant-binding properties. Biochemistry 41, 7241–7252.

45. Cavaggioni A, Mucignat-Caretta C (2000) Major urinary proteins, alpha(2U)-globulins and aphrodisin. Biochim Biophys Acta 1482: 218–228

46. Cavaggioni A, Findlay JB, Tirindelli R (1990) Ligand binding characteristics of homologous rat and mouse urinary proteins and pyrazine binding protein of calf. Comp Biochem Physiol B 96: 513–520

47. Robertson DHL, Beynon RJ, Evershed RP (1993) Extraction characterisation and binding analysis of two pheromonally active ligands associated with major urinary protein of the house mouse (*Mus musculus*). J Chem Ecol 19: 1405–1416

48. Hurst JL, Payne CE, Nevison CM, Marie AD, Humphries RE, et al. (2001) Individual recognition in mice mediated by major urinary proteins. Nature 414: 631–634

49. Singer AG, Macrides F, Clancy AN, Agosta WC (1986) Purification and analysis of a proteinaceous aphrodisiac pheromone from hamster vaginal discharge. J Biol Chem 261: 13323–13326

50. Vincent F, Löbel D, Brown K, Spinelli S, Grote P, et al. (2001) Crystal structure of aphrodisin, a sex pheromone from female hamster. J Mol Biol 305: 459–469

51. Scaloni A, Paolini S, Brandazza A, Fantacci M, Marchese S, et al. (2001) Purification, cloning and characterisation of novel odorant-binding proteins in the pig. Cell Mol Life Sci 58: 823–834

52. Bacchini A, Gaetani E, Cavaggioni A (1992) Pheromone binding proteins in the mouse *Mus musculus*. Experientia 48: 419–421

53. Rouvinen J, Rautiainen J, Virtanen T, Zeiler T, Kauppinen J, et al. (1999) Probing the molecular basis of allergy Three-dimensional structure of the bovine lipocalin allergen Bos d2. J Biol Chem 274: 2337–2343

54. Hilger C, Kuehn A, Hentges F (2012) Animal lipocalin allergens. Curr Allergy Asthma Rep 12: 438–447

55. Mechref Y, Zidek L, Ma W-D, Novotny MV (2000) Glycosilated major urinary protein of the house mouse: characterization of its N-linked oligosaccharides. Glycobiology 10: 231–235

56. Virtanen T, Kinnunen T, Rytkönen-Nissinen M (2012) Mammalian lipocalin allergens—insights into their enigmatic allergenicity. Clin Exp Allergy 42: 494–504

57. Schaal B, Coureaud G, Langlois D, Giniès C, Sémon E, et al. (2003) Chemical and behavioural characterization of the rabbit mammary pheromone. Nature 424: 68–72

58. Charra R, Datiche F, Casthano A, Gigot V, Schaal B, et al. (2012) Brain processing of the mammary pheromone in newborn rabbits. Behav Brain Res 226: 179–188

59. Salzano AM, Novi G, Arioli S, Corona S, Mora D, et al. (2013) Mono-dimensional blue native-PAGE and bi-dimensional blue native/urea-PAGE or/SDS-PAGE combined with nLC-ESI-LIT-MS/MS unveil membrane protein heteromeric and homomeric complexes in *Streptococcus thermophilus*. J Proteomics 94: 240–261

60. Scaloni A, Monti M, Angeli S, Pelosi P (1999) Structural analysis and disulfide-bridge pairing of two odorant-binding proteins from *Bombyx mori*. Biochem Biophys Res Comm 266: 386–391

61. Picariello G, Ferranti P, Mamone G, Roepstorff P, Addeo F (2008) Identification of N-linked glycoproteins in human milk by hydrophilic interaction liquid chromatography and mass spectrometry. Proteomics 8: 3833–3847

62. Hilvo M, Baranauskiene L, Salzano AM, Scaloni A, Matulis D, et al. (2008) Biochemical characterization of CA IX, one of the most active carbonic anhydrase isozymes. J Biol Chem 283: 27799–27809

63. Perez-Miller S, Zou Q, Novotny MV, Hurley TD (2010) High resolution X-ray structures of mouse major urinary protein nasal isoform in complex with pheromones. Protein Sci 19: 1469–1479

64. Cavaggioni A, Sorbi RT, Keen JN, Pappin DJC, Findlay JBC (1987) Homology between the pyrazine-binding protein from nasal mucosa and major urinary proteins. FEBS Lett 212: 225–228

65. Finlayson JS, Asofsky R, Potter M, Runner CC (1965) Major urinary protein complex of normal mice: origin. Science 149: 981–982

66. Dinh BL, Tremblay A, Dufour D (1965) Immunochemical study of rat urinary proteins: their relation to serum and kidney proteins. J Immunol 95, 574–582

67. Zhang Z-D, Frankish A, Hunt T, Harrow J, Gerstein M (2010) Identification and analysis of unitary pseudogenes: historic and contemporary gene losses in humans and other primates. Genome Biology 11: R26

68. Sun YL, Huang LQ, Pelosi P, Wang CZ (2012) Expression in antennae and reproductive organs suggests a dual role of an odorant-binding protein in two sibling *Helicoverpa* species. PLoS One 7: e30040

69. Lacazette E, Gachon A-M, Pitiot G (2000) A novel human odorant-binding protein gene family resulting from genomic duplicons at 9q34: differential expression in the oral and genital spheres. Hum Mol Genetics 9: 289–301

10

Alpha-1-Antitrypsin: A Novel Human High Temperature Requirement Protease A1 (HTRA1) Substrate in Human Placental Tissue

Violette Frochaux[1], Diana Hildebrand[2], Anja Talke[3], Michael W. Linscheid[1], Hartmut Schlüter[2]*

1 Department of Chemistry, Humboldt-Universität zu Berlin, Berlin, Germany, **2** Department of Clinical Chemistry, University Medical Center Hamburg-Eppendorf, Hamburg, Germany, **3** Protealmmun GmbH, Berlin, Germany

Abstract

The human serine protease high temperature requirement A1 (HTRA1) is highly expressed in the placental tissue, especially in the last trimester of gestation. This suggests that HTRA1 is involved in placental formation and function. With the aim of a better understanding of the role of HTRA1 in the placenta, candidate substrates were screened in a placenta protein extract using a gel-based mass spectrometric approach. Protease inhibitor alpha-1-antitrypsin, actin cytoplasmic 1, tropomyosin beta chain and ten further proteins were identified as candidate substrates of HTRA1. Among the identified candidate substrates, alpha-1-antitrypsin (A1AT) was considered to be of particular interest because of its important role as protease inhibitor. For investigation of alpha-1-antitrypsin as substrate of HTRA1 synthetic peptides covering parts of the sequence of alpha-1-antitrypsin were incubated with HTRA1. By mass spectrometry a specific cleavage site was identified after met-382 (AIPM^{382} ↓ ^{383}SIPP) within the reactive centre loop of alpha-1-antitrypsin, resulting in a C-terminal peptide comprising 36 amino acids. Proteolytic removal of this peptide from alpha-1-antitrypsin results in a loss of its inhibitor function. Beside placental alpha-1-antitrypsin the circulating form in human plasma was also significantly degraded by HTRA1. Taken together, our data suggest a link between the candidate substrates alpha-1-antitrypsin and the function of HTRA1 in the placenta in the syncytiotrophoblast, the cell layer attending to maternal blood in the villous tree of the human placenta. Data deposition: Mass spectrometry (MS) data have been deposited to the ProteomeXchange with identifier PXD000473.

Editor: Mark Isalan, Imperial College London, United Kingdom

Funding: This work was supported by the BMBF (Bundesministerium für Forschung und Technologie. Grants: 031U216A, 0313694 A & 0313842A). The funders had no role in study design, data collection and analysis, decision to publish, or preparation of the manuscript.

Competing Interests: There were no competing interests (no employment, no consultancy, no patents, no products in development and no marketed products).

* Email: hschluet@uke.de

Introduction

Human HTRA1 belongs to the HtrA (high temperature requirement A) serine protease family. HtrA was first identified in Escherichia coli (E. coli) and is essential for the bacterial survival at high temperature [1]. The bacterial protease acts as a chaperone and degrades misfolded proteins at elevated temperatures [2]. Four human homologues of HtrA have been identified to date: HTRA1, HTRA2, HTRA3 and HTRA4. The human protease HTRA1 was first isolated from SV40-transformed fibroblast where it has been identified as a down-regulated gene [3]. HTRA1 contains a trypsin-like serine protease domain, a PDZ domain, an insulin-like growth factor binding protease (IGFBP) domain and a Kazal-type inhibitor domain [3]. Full-length HTRA1 has a molecular weight of ~50 kDa. Several ~30 kDa additional protein species of HTRA1 are known [4]. After binding a substrate, HTRA1 switches to an active conformation. In the active conformation, the catalytic triad is properly positioned for catalysis of the proteolytic reaction. Crystal structure of HTRA1 shows that HTRA1 crystallises as a trimer [5]. This oligomerization may play a role in proteolytic activity of the protease, as observed by E. coli HTRA [6].

Though its exact role is not well understood, HTRA1 seems to be involved in several pathologies, as rheumatoid arthritis [7], osteoarthritis [4], Alzheimer's disease [8], age-related macular degeneration [9] [10] [11] and some types of cancer [12] [13]. In ovarian cancer loss of HTRA1 expression may contribute to malignant phenotype whereas overexpression inhibits proliferation and cell growth.

Recently, several substrates of HTRA1 were identified and different hypotheses of its biological function were published. The tumour suppressor property of HTRA1 was correlated with its association to microtubule in cancer cells, which influences the cell motility [14,15]. Grau et al. demonstrated that HTRA1 degrades fibronectin which seems to lead to increasing mRNA and protein amounts of different matrix metallopeptidases (MMPs) in human synovial fibroblasts and finally to the degradation of the cartilage [7]. The same group suggests that HTRA1 is implicated in Alzheimers disease by degradation of amyloid β in the human brain, while the application of an HTRA1 inhibitor causes accumulation of amyloid β in astrocyte cells [8]. In the same way, Tennstaedt et al. suggest a protein quality control function for HTRA1 through degradation of aggregated tau [16]. HTRA1 was reported to have apoptotic properties. He et al. identified the substrate XIAP, a member of the "inhibitor of apoptosis family"

(IAP) and described the contribution of HTRA1 to chemoresistance [17]. The influence on apoptotic process may correlate with the tumour suppressor properties of the protease [15].

HTRA1 gene is expressed and translated in different tissues and organs. The highest levels of HTRA1 were found in the placenta [18]. Using in situ hybridization and immunohistochemistry methods the level of HTRA1 was found to be especially high in the third trimester of gestation [19]. Furthermore the level of HTRA1 in placenta with preeclampsia (PE) or trophoblastic diseases was reported to be deregulated compared with normal placenta [20–22].

To understand the functional roles of a protease, it is essential to elucidate its substrate repertoire [23]. As a proof that a defined protein is a substrate of a given protease, the purified candidate substrate should be proteolysed in the presence of the target protease. Using peptide libraries substrate specificities of proteases of interest can be determined. However, for the identification of a native substrate of a target protease the biological context is crucial. Many parameters contribute to the role of the protease in vivo, as tertiary and quaternary structure of the substrate or the cooperation of co-factors and ligands.

In this work, we have chosen a cell-free gel-based mass spectrometric approach to investigate the substrate repertoire of HTRA1 in the human placenta. The degradomics approach we used is fast and allows the identification of candidate substrates in the complex mixture of protein extracts from tissues. With this method several new candidate substrates of HTRA1 were identified, giving an insight into the biological function of the protease HTRA1 in the placenta.

Materials and Methods

Ethics Statement

In this work we have used the placentas of healthy women, who had given their written consent.

The use of this material has been approved by the ethics committee Ärztekammer Hamburg.

Protease and substrates

Recombinant human HTRA1 was expressed in insect cells and purified from insect cell culture supernatants (ProteaImmun GmbH, Berlin, Germany). Preparation was analysed by SDS-PAGE using coomassie blue staining ($\geq$70% of total protein). Protein bands were cut, digested in-gel with trypsin and analyzed via Mass spectrometry to confirm identity of the protease and to identify the impurities. All protein bands were found to be HTRA1 or fragments of HTRA1. A1AT was purified from pooled human plasma (ProteaImmun GmbH, Berlin, Germany), and purity was confirmed by SDS-PAGE using coomassie blue staining ($\geq$95% of total protein). β-Casein was purified from bovine Milk (Sigma), and purity was confirmed by SDS-PAGE using coomassie blue staining ($\geq$98% of total protein).

Preparation of a placenta protein extract from the placenta

Fresh placentas of healthy women, who had given their written consent, were cut into small pieces (at 4°C) and immediately frozen in liquid nitrogen. The frozen material was lyophilised, pulverised and homogenised by Ultra-Turrax at 4°C (IKA-Werke GmbH, Staufen) in phosphate buffered saline (PBS; 137 mM NaCl, 26 mM KCl, 14 mM KH_2PO_4, 80 mM Na_2HPO_4, pH 7.4; 4°C). After centrifugation (15,000 g, 4°C, 30 min) the supernatant was filtered by a 0.45 µm filter (Millipore Millex-HV Hydrophilic PVDF filter) followed by a second filtration using an Amicon Ultra-15 filter with a 15 kDa membrane. The total protein concentration of the resulting supernatant was determined using a Bradford assay [24].

Enrichment of HTRA1 using immunobeads, western blot of the placenta proteins

50 mg/ml placenta supernatant was diluted to a final concentration of 15 mg in 3 ml PBS. The placenta proteins were incubated with 10 µl anti-HTRA1-beads (ProteaDetect Extract HtrA1ProteaImmun GmbH, Berlin, Germany) for 3 hours at 37°C. After several washing steps the beads were incubated for 7 min at 95°C after addition of sample buffer without dithiothreitol (DTT) to elute the enriched HTRA1 protease. The eluent was loaded on a sodium dodecyl sulphate (SDS) gel. HtrA1 was detected by Western blot using the same monoclonal HTRA1 antibody as used in the anti-HTRA1-beads (ProteaImmun GmbH, Berlin, Germany).

Incubation of β-casein with HTRA1 in the Presence or absence of different protease-inhibitors followed by 1D-gel electrophoresis or HPLC analysis

A range of protease inhibitors (EDTA ethylenediaminetetraacetic acid, E64 [N-(trans-epoxysuccinyl)-L-leucine 4 guanidinobutylamide], pepstatinA, bestatin, leupeptin, AEBSF [4-(2-aminoethyl)benzenesulfonyl fluoride hydrochloride], and aprotinin, Sigma) were added separately to an incubation mixture containing 5 µg β-casein in 5 µl buffer (150 mM NaCl, 5 mM $CaCl_2$, 50 mM Tris-HCl, pH 7.4) and 0.5 µg recombinant human HTRA1 expressed in insect cells and purified from insect cell culture supernatants (0.2 µg/µl in protease buffer: 150 mM NaCl, 5 mM KCl, 50 mM imidazol, 50 mM Tris-HCl, 0.05% Brij-35, pH 7.5, ProteaImmun GmbH, Berlin, Germany) and incubated at 37°C. Aliquots from the incubation mixture were taken at different times and separated by SDS-Page as described by Laemmli [25] using 15% acrylamide final concentration. After electrophoresis the gels were stained with Coomassie blue. On the SDS-polyacrylamide gel electrophoresis (SDS-PAGE), proteolytic activity was registered as present, when no full-length β-casein was detected after 1 hour incubation. For validation, this experiment was repeated using high-performance liquid chromatography (HPLC) with UV detection for the analysis of the reaction products. Aliquots of the incubation mixture (150 pmol of the β-Casein) were injected on a reversed-phase (RP) column (Luna 00A-4041-C18; 5 µm, 150×0.5 mm; Phenomenex) driven by the HPLC system (Agilent 1200 system; Agilent Technologies, Waldbronn, Germany). Separation was performed using a binary mobile phase (eluent A: 94.9% deionized water, 5% acetonitrile, 0.1% formic acid (v/v/v); eluent B: 99.9% acetonitrile, 0.1% (v/v) formic acid) with a flow rate of 50 µl min^{-1}. Peptides were separated using a 60 min gradient as follow: 0–35 min 5% to 70% B; 35–37 min 70% B; 37–40 min 5% B; 40–60 min 70–5% B. MS detection was used to verify the identity of the β-casein peak of the UV-chromatogram. HPLC system was coupled with a Fourier transform ion cyclotron resonance mass spectrometer (FT-ICR-MS) (Finnigan LTQ FT ULTRA Thermo Fisher Scientific, Bremen, Germany). Electrospray ionization source (ESI) of the FT-ICR was working at 5.0 kV. Positive ionisation and a transfer capillary temperature of 275°C were applied.

Incubation of the placenta proteins with HTRA1

225 µg of the placenta proteins were incubated with 25 µg purified recombinant human HTRA1 expressed in insect cells in a final volume of 285 µl for 4 hours at 37°C (+HTRA1). As a

control 225 μg of the extracted placenta proteins were incubated with the same volume of protease buffer without HTRA1 using the same incubation conditions (−HTRA1). After incubation 2 ml cold acetone (−20°C) was added to each sample. After storing 1 hour at −20°C the precipitates were collected by centrifugation (20 min, 4°C, 15.000 g) and separated from the liquid phase. The pellets were washed 2 times with 90% acetone, 10% water at −20°C and centrifuged under same conditions. The pellets were finally dissolved in a sample buffer (9 M urea, 70 mM DTT, 2% ampholytes 2–4) for isoelectric focusing. For validation, incubation experiment analysed by 2-DE separation was performed in two independent replicates.

For further validation this experiment was repeated with addition of an inhibitor cocktail (P8340, sigma, concentrations of the inhibitors in the incubation solution are: AEBSF 1.04 mM, aprotinin 0.80 μM, bestatin 40 μM, E-64 14 μM, leupeptin 20 μM, pepstatin A 15 μM), or with addition of the protease inhibitor tissue inhibitor of metalloproteinase-1 (Timp-1, Sigma, ratio 10:1 total protein to inhibitor weight).

2-D-gel electrophoresis

The incubated samples (+HTRA1/−HTRA1) were subjected to 2D gel electrophoresis analysis (2-DE). Two dimensional gel electrophoresis was performed according to Proteome Factory's 2-D electrophoresis technique based on Klose and Kobalz [26]. 250 μg of total protein was applied to vertical rod gels (9 M urea, 4% acrylamide, 0.3% piperazine diacrylamide, 5% glycerol, 0.06% tetramethylethylenediamine (TEMED) and 2% carrier ampholytes (pH 2–11)) for isoelectric focusing (IEF) at 8820 Vh in the first dimension. After focusing, the IEF gels were incubated in equilibration buffer, containing 125 mM Trisphosphate (pH 6.8), 40% glycerol, 65 mM DTT, and 3% SDS for 10 minutes and subsequently frozen at −80°C. The second dimension SDS-PAGE gels (23×30×0.1 cm) were polymerized from 375 mM Tris-HCl buffer (pH 8.8), 15% acrylamide, 0.2% bisacrylamide, 0.1% SDS and 0.03% TEMED. After thawing, the equilibrated IEF gels were immediately applied to SDS-PAGE gels. Electrophoresis was performed using a two-step increase of current, starting with 15 min at 65 mA, followed by a run of 6 h at 140 mA, until the front reached the end of the gel. After electrophoresis, the analytical gels were stained with MS compatible silver nitrate (FireSilver staining kit PS-2001, Proteome Factory AG, Berlin, Germany). For image analysis the 2-DE gels used for comparison analysis were digitized at a resolution of 150 dpi using a PowerLook 2100XL with transparency adapter. Two-dimensional image analysis was performed using the Proteomweaver software (Definiens AG, Munich, Germany). For subsequent analysis selected protein spots were digested in-gel.

Trypsin in-gel digestion

For in-gel digestion selected protein spots were cut out, transferred to vials, and washed three times using 100–200 μl buffer (50% acetonitrile, and 50% 25 mM ammonium hydrogen carbonate (v/v)). Then the gel pieces were dehydrated using 50 μl acetonitrile. The supernatant was removed and the gel pieces were dried for 20 min at 37°C. 20 to 30 μl of modified trypsin (0.01 g/l, Promega) in 25 mM ammonium hydrogen carbonate was added subsequently and after 30 min 20 μl of 25 mM ammonium hydrogen carbonate buffer was added to each vial; trypsin digestion was carried out for 12 h at 37°C. After centrifugation, the supernatant was carefully removed, and the peptides were extracted twice from the gel by adding 50 μl of the following extraction solution: 50% acetonitrile, 50% formic acid (5%). Finally all pellets were extracted with 50 μl of acetonitrile. The supernatants were pooled, evaporated to dryness by a vacuum concentrator and redissolved in 20 μl 0.1% formic acid (v/v) for mass spectrometric analysis.

Mass spectrometry-based proteomic analysis

For nanoHPLC/ESI-MS/MS, a 1100 HPLC system (Agilent) was used. Separation was performed on a Zorbax 300 SB-C18 (150 mm×75 μm) column with a Zorbax 300 SB-C18 (0.3 mm×5 mm; Agilent Technologies; Waldbronn, Germany) enrichment column and a binary mobile phase water/acetonitrile gradient with a maximum flow rate of 0.30 μl min^{-1}. Peptides were separated using a 35 min gradient. The eluents used were as follows: A-94.9% deionized water, 5% acetonitrile, 0.1% formic acid (v/v/v); B-99.9% acetonitrile, 0.1% (v/v) formic acid. The separation column was coupled to a Finnigan LTQ FT ULTRA mass spectrometer (Thermo Fisher Scientific, Bremen, Germany) using a nanomate ESI interface (Advion) working at 1.7 kV. Positive ionisation and a transfer capillary temperature of 200°C were applied. Mass spectrometric detection was performed by a data-dependant method of acquisition controlled by Xcalibur 2.07 software version (Thermo Scientific, Bremen) where the five most intense precursor ions detected in the full MS scan (FT-ICR) were selected and fragmented in the Ion Trap by collision induced dissociation (CID) (35% energy, 4 amu mass isolation width). Proteome Discoverer 1.3 and the search engine SEQUEST (Thermo Scientific, Bremen) were used for data analysis. The MS/MS raw data were directly analysed with the Proteome Discoverer software using the following spectrum selector settings; minimum precursor mass 350 Da, maximum precursor mass 5000, total intensity threshold 100, minimum peak count 10, signal-to-noise threshold 5 (FT-only). The identification searching parameters were: tryptic digestion, 2 missed cleavages, deamidated (N), oxidation (M) and propionamide (C) dynamic modifications, precursor mass tolerance 3 ppm and fragment mass tolerance 0.5 Da. The database UniProtKB/Swiss-Prot homo sapiens (September 2013, 20'267 Proteins) was chosen for protein identification. A list of contaminants was removed from this database, namely keratin, type I and type II cytoskeletal and trypsin. A reverse database was used to prevent false positive identification. As discrimination criteria, protein identifications containing less than 2 peptides and SEQUEST scores lower than 40 were discarded. The mass spectrometry proteomics data included database and list of the contaminants have been deposited to the ProteomeXchange Consortium (http://proteomecentral.proteomexchange.org) via the PRIDE partner repository with the dataset identifier PXD000473 [27].

In vitro cleavage of pure plasma A1AT with HTRA1

10 μg A1AT purified from pooled human plasma (ProteaImmun GmbH, Berlin, Germany) solved in 10 μl PBS (137 mM NaCl, 26 mM KCl, 14 mM KH$_2$PO$_4$, 80 mM Na$_2$HPO$_4$, pH 7.4) was incubated with 0.5 μg HTRA1 (0.2 g/l) at 37°C over night (17 h). An aliquot was taken at different times. The incubation experiment was repeated three times and subjected to 2-D electrophoresis analysis as described above or to SDS-PAGE analysis (10% acrylamide final concentration). The gel was stained after electrophoresis with Coomassie blue. Selected protein spots were digested in-gel. The obtained peptides were analysed by LC-MS.

Incubation of A1AT with HTRA1 and β-casein

20 μg A1AT (1 g/l in PBS) was incubated with 0.4 μg HTRA1 (0.2 g/l) and 20 μg β-Casein (0.1 g/l in PBS) at 37°C. Aliquots

were taken at time 0 h, 30 min, 1 h and 3 h. The same experiment was repeated without A1AT, as positive control.

The aliquots were analysed by SDS Page using 15% acrylamide final concentration. The gel was stained after electrophoresis with Coomassie blue.

Incubation of A1AT peptides with HTRA1

5 μl of a 100 μM solution of the synthetic peptides A1AT(361–384), A1AT(379–400) and A1AT(396–418) (AG Kloetzel, Charité, Berlin) were incubated each separately with 12.5 μl HTRA1 (0.2 g/l) in a molar ratio 10:1 substrate to protease at 37°C. The same incubation experiment was performed without addition of HTRA1 as negative control. Aliquots from the incubation mixture were taken at different times and after dilution to 2 μM with 50% methanol, 50% formic acid (0.1%) analysed by matrix-assisted laser desorption/ionization (MALDI) Orbitrap MS (MALDI-Orbitrap-XL, Thermo Fisher Scientific, Bremen, Germany). A 3.5 mg/ml alpha-cyano-4-hydroxycinnamic acid solution in 50% methanol, 50% formic acid (0.1%) was used as matrix. The fragments were identified by molecular ions with an accuracy of 1–4 ppm and a laser energy of 8 μJ.

Results

Only full-length HTRA1-form in the placenta

To detect HTRA1 in the placenta protein extract, we enriched HTRA1 from the placenta proteins using immunobeads. The concentration of enriched HTRA1 was then estimated by Western blot using a monoclonal HTRA1 antibody and applying known amount of recombinant HTRA1 expressed in insect cells (Fig. 1). The monoclonal HTRA1 antibody recognizes the full-length HTRA1 at ~50 kDa, the ~37 kDa truncated monomer and the ~74 kDa dimer of recombinant truncated HTRA1. Without enrichment, only unspecific signals were found (Fig. 1 lane 5 and 6). After enrichment, one signal assigned to the full-length HTRA1 appeared on the Western blot (Fig. 1, lane 2 and 3). Approximately 150 ng full-HTRA1 was extracted from 15 mg placenta proteins. The same antibody against HTRA1 was tested for immunohistochemistry on placental tissue with the aim to show the co-localization of protease and candidate substrates, but this antibody was not compatible with formalin fixed tissue.

Screening for substrates of HTRA1 in placental tissue

For the identification of the candidate substrates of HTRA1 in the placenta we used a gel-based proteomics approach. The proteins for the screening experiments were extracted from human placental tissue as described in the experimental part. The placenta protein extract was incubated with or without addition of recombinant HTRA1 expressed in insect cells and potential substrates were identified by comparing the protein patterns of 2-DE separation followed by LC-MS analysis of the tryptic in-gel digest. The reduction in spot intensity on the 2-DE pattern on the gel representing the placenta protein extract incubated with HTRA1 compared with the 2-DE patterns of the control indicated intact substrates. The appearance of new spots indicated possible cleavage fragments (Fig. 2 and Fig. S1). Proteins were considered to be a potential substrate when the spot corresponding to the full-length product disappeared or was significantly reduced. The proteins underlying the spots A1 to A13 and B1 to B7 labeled in Figure 2 were chosen as potential substrates. These spots were cut, removed and the proteins were digested in-gel using trypsin. The obtained peptides were analysed by LC-MS and identified with protein data base with a search engine. With this approach we identified several potential substrates (Table 1, Fig. 2, for more

details see Table S1). Among the potential substrates, proteins of the cytoskeleton, proteins of the disulfide-isomerase family, several heat shock proteins, some proteins responsible for the protein biosynthesis and the protease inhibitor alpha-1-antitrypsin were identified (Table 1). The Spots B2–B6 corresponded to the added protease HTRA1 and to HTRA1 fragments. The degradation of HTRA1 after several hours of incubation at 37°C in the presence of other proteins (substrates or non-substrates) or in PBS buffer was regularly observed in this study and is probably an autocatalytic process.

Investigation of the proteolytic activity of HTRA1 in the presence of inhibitors

Many proteases are part of proteolytic cascades. If HTRA1 is integrated in such a cascade it may activate other proteases. These activated proteases will then digest further substrates which are not substrates of HTRA1 but could be misinterpreted as HTRA1 substrates. The identification of A1AT as a candidate substrate of HTRA1 raises the idea that the proteolytic activity of proteases will increase which are inhibited by A1AT, if the concentration of the latter decreases. To reduce the probability that other proteases than HTRA1 are responsible for degrading placenta tissue proteins, we looked for protease inhibitors which do not inhibit HTRA1. Therefore β-casein, a known substrate of HTRA1, was incubated with HTRA1 in the presence and in the absence of several inhibitors.

The protease activity of HTRA1 was not significantly inhibited by the serine protease inhibitors aprotinin, and AEBSF, the cysteine protease inhibitors leupeptin and E64, the metalloprotease inhibitor bestatin and the aspartic protease inhibitor pepstatin A (Fig. 3, Table 2). The chelator EDTA slowed slightly the proteolytic reaction, leading to the presumption, that metal ions like Ca^{2+} have a supporting effect on the catalysis of HTRA1 although HTRA1 is not a metalloprotease (Table 2).

Based on these results, placenta proteins were incubated with or without HTRA1 (control sample) for 4 hours at 37°C in the presence of a protease inhibitor cocktail (1% of the reaction volume) which contained AEBSF, E64, aprotinin, bestatin hydrochloride, leupeptin hemisulfate salt, and pepstatin A. Compared with the results of the first 2-DE analysis described above, the presence of the protease inhibitor cocktail did almost not change the 2-DE gel pattern of the protease treated sample and the same candidate substrates were identified (Fig. S2).

As the candidate substrate A1AT is known to be a substrate of several MMPs [28], [29], [30], we repeated the incubation experiment of placenta proteins in the presence or absence of HtrA1 and in presence of the MMP inhibitor Timp-1. Here again, there was no significant change in the 2-DE gel pattern compared to the first 2-DE data describe above (Fig. S3).

A1AT as target of HTRA1

Among the candidate substrates identified, A1AT was chosen for further validation experiments. A1AT was considered to be of particular interest because of its important role as protease inhibitor. A1AT is a high abundant plasma protein and is also expressed in the placenta in high concentrations [31]. Interestingly, A1AT is able to form a stable complex with HTRA1. This behaviour previously reported by Hu et al. [4], was verified in our study. On the 2-DE of the incubated sample of the placenta proteins and HTRA1, spots shifted from 55 kDa (A1AT) to approximately 120 kDa indicating the presence of the A1AT-HtrA1 complex (Fig. 4). This was confirmed by LC-MS analysis of the spot B7 after tryptic in-gel digestion.

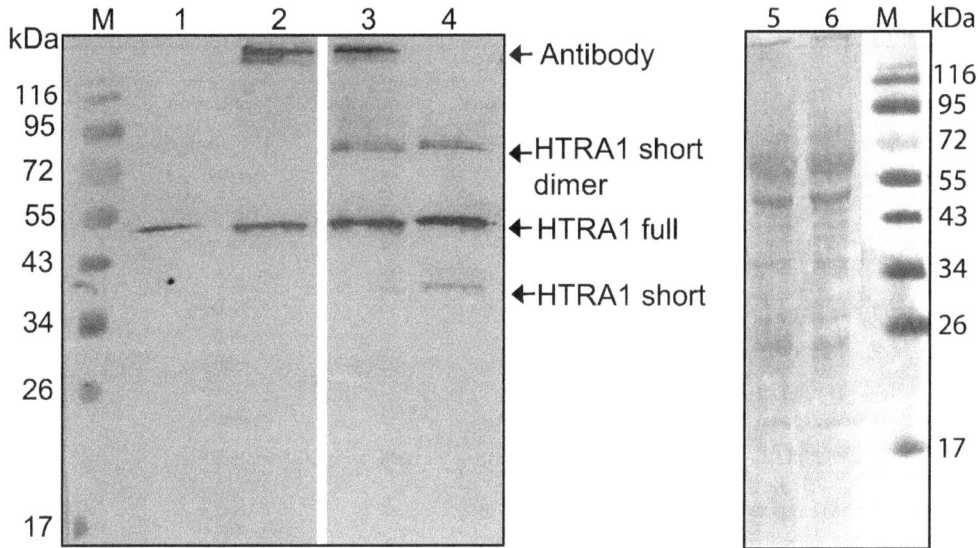

Figure 1. Western Blot analysis of placenta proteins after enrichment using immunobeads. The anti-HTRA1 antibody detected the 50 kDa (HTRA1 full) and the 37 kDa (HTRA1 short) recombinant HTRA1 forms expressed in insect cells (control: lane 1: 50 ng HTRA1 full, lane 4: 200 ng HTRA1 full). Lane 2 shows the placenta proteins after treatment with 10 μl immunobeads. Lane 3: Placenta proteins with 200 ng recombinant HTRA1 full after treatment with 10 μl immunobeads, to verify the binding capacity of the beads. Lane 5 and 6 show the placenta proteins before and after the extraction of HTRA1. Without enrichment, only unspecific signals were found. Based on the control, total amount of HTRA1 in lane 2 was estimated to 150 ng. The image is representative of two independent experiments.

A1AT is a highly glycosylated protein (asparagine 46, 83 and 247). Heterogeneity of the glycans (bi- and tri-antennary) gives rise to multiple protein species [32]. On the 2-DE several spots with pI shift of nearly 0.1 units and mass decrease of approximately 2 kDa are typical for A1AT [32] (Fig. 4, Fig. S6). The HTRA1 generated A1AT fragments show very similar species pattern with a decrease in the molecular weight of nearly 4 kDa compared to the full-length A1AT species. This similar pattern suggests that the glycosylation sites do not have an impact on the proteolytic action of HTRA1. Furthermore it can be assumed that HTRA1 cleaves the different A1AT species at one specific cleavage site.

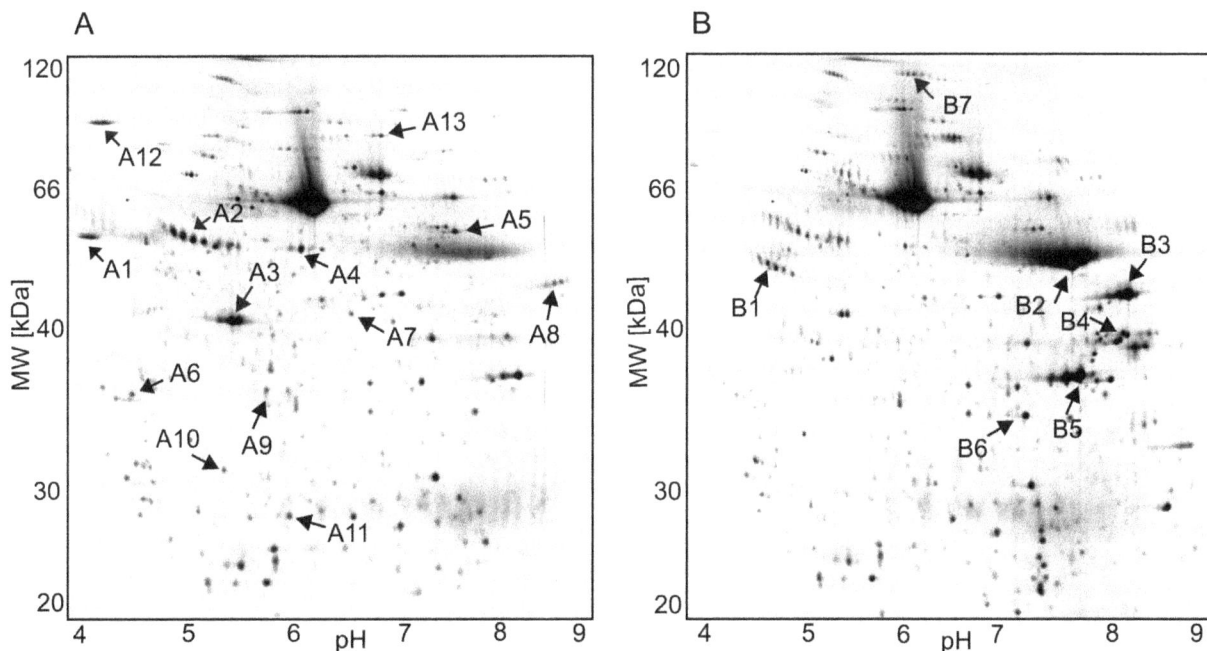

Figure 2. Details of the 2-D patterns of placental proteins incubated with HTRA1. The placenta proteins were incubated for 4 h at 37°C without (A) or with the protease HTRA1 (B). Numbers indicate protein spots that were subjected to tryptic in-gel digestion and LC-MS (A = disappeared spots, B = appeared spots). Details and accession no of the substrates can be found in Table 1. The shown images are representative of four independent experiments, one in the presence of the inhibitor Timp-1 and one in the presence of an inhibitor cocktail (Fig. S2 and S3).

Table 1. HTRA1 candidate substrates identified in the placenta protein.

Spot number	Protein name	Name	M [kDa]	localization	Accession number
A1	Protein disulfide-isomerase	PDIA1	57	Membrane, ER	P07237
A2, B1	Alpha-1-antitrypsin	A1AT	46	Secreted, ECM	P01009
A3	Actin cytoplasmic 1	ACTB	41	Cytoplasm,	P60709
	Actin, aortic smooth muscle	ACTA	42	cytoskeleton	P62736
A4	Protein disulfide-isomerase A3	PDIA3	56	ER	P30101
A5	Pyruvate kinase isozymes M1/M2	KPYM	58	Cytoplasm, nucleus	P14618
A6	Tropomyosin β-chain	TPM2	32	Cytoplasm, cytoskeleton	P07951
A7	Adenosyl homocysteinase	SAHH	47	Cytoplasm	P23526
A8	Elongation Factor –α 1	EF1A1	50	Cytoplasm, nucleus	P68104
A9	Estradiol 17-β-dehydrogenase 1	DHB1	35	Cytoplasm	P14061
A10	Chloride intracellular channel protein 1	CLIC1	26	Membrane, cytoplasm, nucleus	O00299
A11	Heat shock protein β-1	HSPB1	23	Cytoplasm, cytoskeleton	P04792
A12	Endoplasmin	ENPL	92	ER	P14625
A13	Elongation factor 2	EF2	95	Cytoplasm	P13639

ECM: extracellular matrix. 2D_A3: ACTB and ACTA are proteins of the actin family with very similar sequences.

To get more information about the cleavage site of HTRA1 in A1AT, the full A1AT and the fragment spots were analysed with MS, after trypsin digestion and LC separation. Compared with the full A1AT, peptides from the C-terminus were not detected in the fragment spots (Fig. 4). These results suggest a cleavage site within the C-terminal part of A1AT.

To validate the degradation of A1AT by HTRA1, purified A1AT from human plasma (ProteaImmun) was incubated with the protease in a ratio of protease to substrate of 1:10. The incubated sample and the control samples (time 0 h and overnight of incubation without HTRA1) were separated using 2-DE. Incubation generated cleavage products with a mass difference of nearly 4 kDa (Fig. 5). The incubation samples were also subjected to SDS-PAGE (Fig. S4). Selected protein spots of the SDS-PAGE were in-gel digested with trypsin and analysed by MS (spots 1D_A1, 1D_A2 and 1D_A3). Peptides of the C-terminal part of A1AT could be detected in the spot 1D_A1 but not in the spot 1D_A3 (cleavage product). A1AT and HTRA1 were both identified in spot 1D_A2. These results confirm that in the cleavage products of A1AT the C-terminal part is missing and that HTRA1 and A1AT are able to build a stable complex.

The protein A1AT is an inhibitor for a range of serine proteases. As HtrA1 is a serine protease, we tested if the presence of A1AT is affecting the activity of HTRA1. Therefore A1AT and HTRA1 were incubated in a ratio 50:1 followed by addition of the substrate β-Casein. The proteolytic activity of HTRA1 was monitored using SDS Page of aliquots of the incubation mixture taken at different times after the addition of β-Casein. Proteolysis of β-casein through the protease HtrA1 still takes place after 1 hour incubation. No significant difference was observed compared to the sample without A1AT (Fig. S5).

Mass spectrometric cleavage-site identification of HtrA1 in A1AT

To detect the exact cleavage site of HTRA1 in A1AT, three different peptides were synthetized. The peptides contained sequence part of A1AT where the cleavage site is supposed to be, according to the MS analysis of the A1AT-fragment spots of the 2-DE experiments. The synthetic peptides were: A1AT(361–384), A1AT(379–400) and A1AT(396–418). They were incubated each separately with HTRA1 in a molar ratio of protease to peptide of 1:10 at 37°C over night, and the proteolytic products were identified by MALDI-Orbitrap MS (Fig. 6). Incubation of the A1AT peptides without HTRA1 (negative control) generated no cleavage fragments (Fig. S7). Incubation of the peptide A1AT(361–384) with HTRA1 generated the fragment (361–382), thus cleavage occurred after met-382. The same specific cleavage site was found after incubation of the peptide A1AT(379–400) with the protease, as the fragment (383–400) was generated. Incubation of A1AT(396–418) produced as expected no cleavage fragments (Fig. S8).

Discussion

In this work the total amount of HTRA1 in human placenta tissue was estimated to be in a 10^{-5} mg range per 1 mg total protein using a western blot. This corresponds to nearly 0.001% of the protein in the placenta. Interestingly, only the full-length form of HTRA1 was detected on the western blot, although the monoclonal HTRA1 antibody is also able to recognise the truncated HTRA1 form. This result coincides to findings of Lorenzi et al. [21]. They describe a high amount of the full-length form in healthy human placenta and, in contrast to that, a total down-regulation of the full-length form and an up-regulation of a 30 kDa truncated form of HTRA1 in preeclampsia placenta. The authors suggested that the truncated HTRA1 form has a

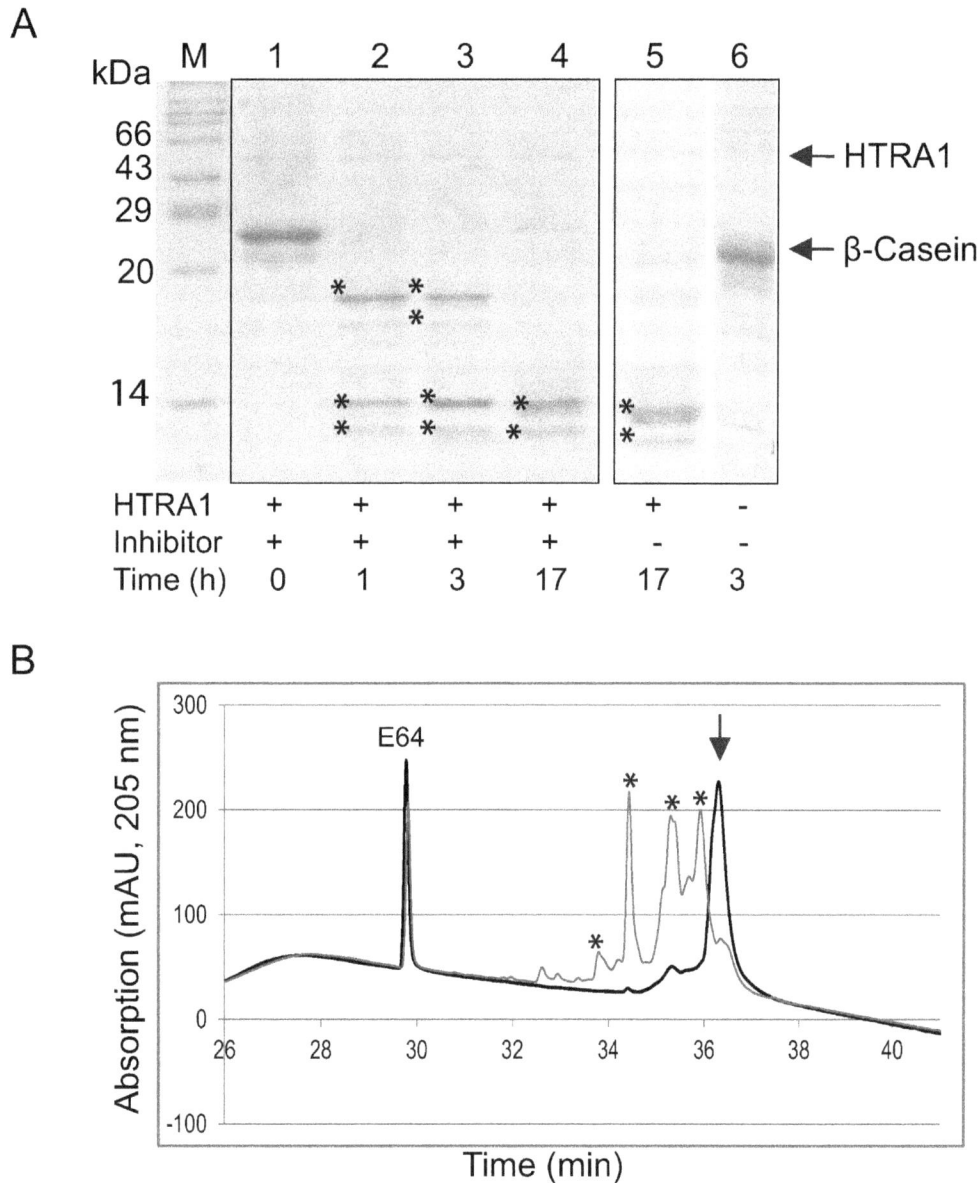

Figure 3. Influence of protease inhibitors on proteolytic activity of HTRA1, exemplified with E64. A: SDS-PAGE separation of the incubation products of HTRA1 and β-casein in the presence of E64. Lanes 1 to 4: 0 h, 1 h, 3 h and 17 h. Lane 5: incubation of HTRA1 and β-casein without E64, 17 h. Lane 6: incubation of β-casein without HTRA1 and without inhibitor for h. The proteolytic activity of HTRA1 in the presence of E64 is 100% compared to the control. The arrows indicate HTRA1 (above) and β-casein; the asterisks indicate the cleavage products of β-casein. M, marker. B: The proteolytic activity in the presence of E64 was monitored by HPLC analysis using UV detection. Incubation time 0 h (thick line) and 1 h (thin line). Height of the β-casein peak was used for relative quantification (mean of three independent replicates, Table 2). The asterisks indicate the cleavage products of β-casein. The arrow indicates β-casein. MS analysis was used to verify the identity of the β-casein peak (36.3 min).

physiological relevance and may be responsible for maldevelopment of the villi through degradation of fetal vessel fibronectin in placental villi.

High levels of HTRA1 were found in the third trimester of pregnancy compared to the first and second trimester [19]. HTRA1 is localized in the cytoplasm of the placenta cells and in the extracytoplasmic space of the stroma of placenta villi. In the first part of pregnancy HTRA1 is expressed in the syncytiotrophoblast and in the cytotrophoblast, the both layers surrounding placenta villi. In the third trimester, the expression is higher in the syncytiotrophoblast, the cell layer attending to maternal blood. The syncytiotrophoblast layer is continuously regenerated during

pregnancy by fusion of the underlying trophoblast cells, as its translation rate is low and the cells non-proliferative [33–35]. It is responsible for the transfer of oxygen and nutrients between the maternal blood and the fetus. Moreover, it is the place where many hormones required for fetal growth are synthetized. One of the identified potential substrates for HTRA1 in this study is the protease inhibitor alpha-1-antitrypsin (A1AT). The localization pattern of A1AT in the placenta shows similarity with the localization pattern of HTRA1. Both are located in the syncytiotrophoblast in mature placenta [19] [31]. Results of the Human Protein Atlas Project, publicly available online (www. proteinatlas.org) also show that HTRA1 and A1AT are both

Table 2. Influence of different protease inhibitors on the proteolytic activity of HTRA1.

Inhibitor	Concentration	β-casein after 1 h [%]
Control	-	21+/−4
EDTA	10 mM	33+/−5
E64	10 µM	25+/−2
Pepstatin A	700 mM	18+/−3
Aprotinin	800 nM	23+/−2
Bestatin	40 µM	25+/−3
Leupeptin	50 µM	23+/−2
AEBSF	0.2 mM	24+/−5

The protease inhibitors were added separately to an incubation mixture containing 5 µg β-casein and 0.5 µg HTRA1 at 37°C. Aliquots were taken at time 0 h and after 1 h incubation at 37°C and analysed by RP-HPLC. Percentage degradation of β-casein was calculated on the height basis from the UV chromatogram obtained at a wavelength of 205 nm, normalised to the height of the β-casein peak at time 0 h (mean of three replicates). As control, β-casein was incubated with HTRA1 without addition of an inhibitor. Compared with the control, protease activity of HTRA1 was not significantly inhibited by the inhibitors tested in this study. Only EDTA seems to slow slightly the proteolytic reaction.

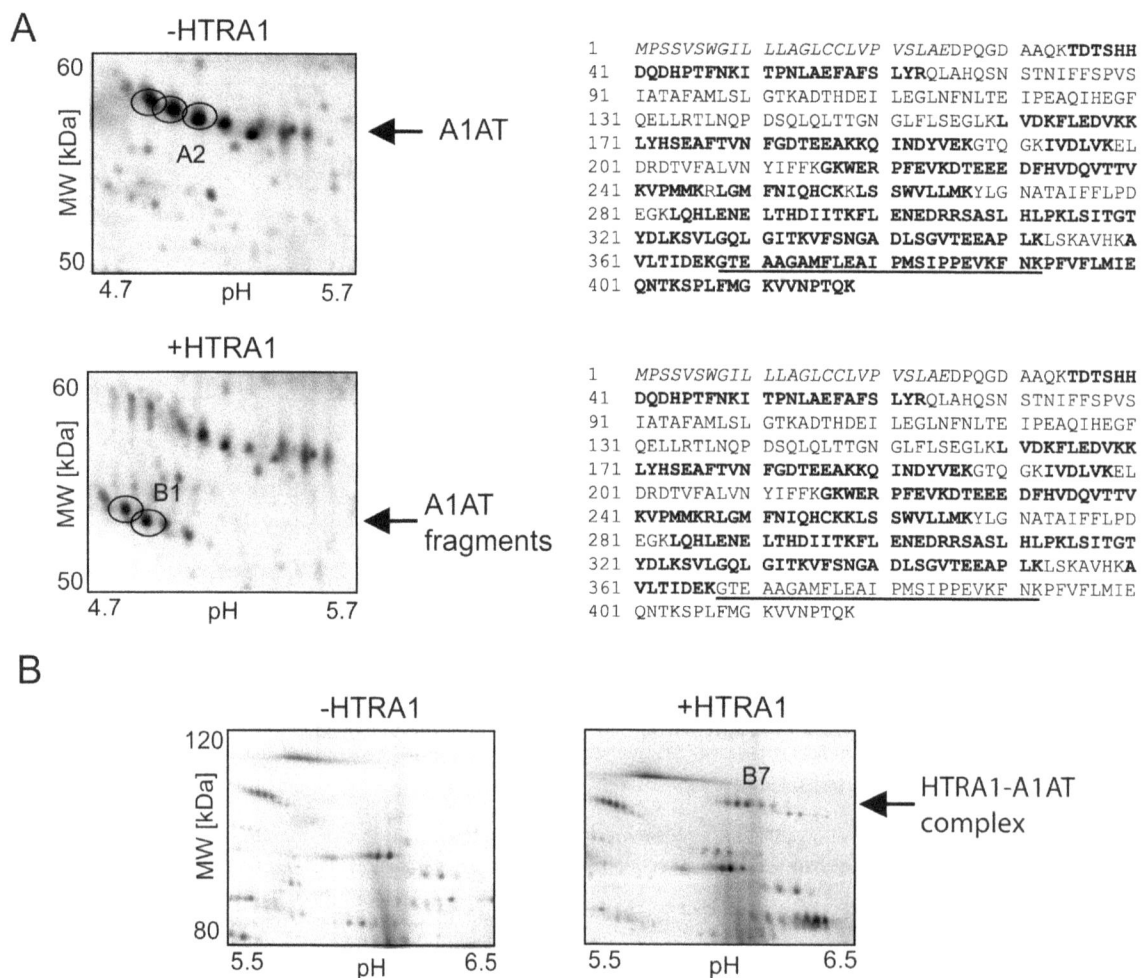

Figure 4. Details of the 2-D patterns of the candidate substrate A1AT incubated with HTRA1. Details of the 2-D pattern of the HTRA1 substrate A1AT in the placenta protein sample after incubation in the absence (−HTRA1) and in the presence (+HTRA1) of HTRA1 (A). The images are small sections of the Figure 2. The arrows indicate placenta A1AT (spots A2) and placenta A1AT fragments (Spots B1). On the right hand site: Sequences of A1AT identified by MS analysis (bold sequence) after tryptic in-gel digestion of the spots A2 and B1. The italic sequence indicates the signal peptide; the underlined sequence indicates the reaction center loop. (B) A1AT forms a stable complex with HTRA1 at high molecular weight range (spot B7); this was supported by MS measurements. Three replicates are shown in Figure S6.

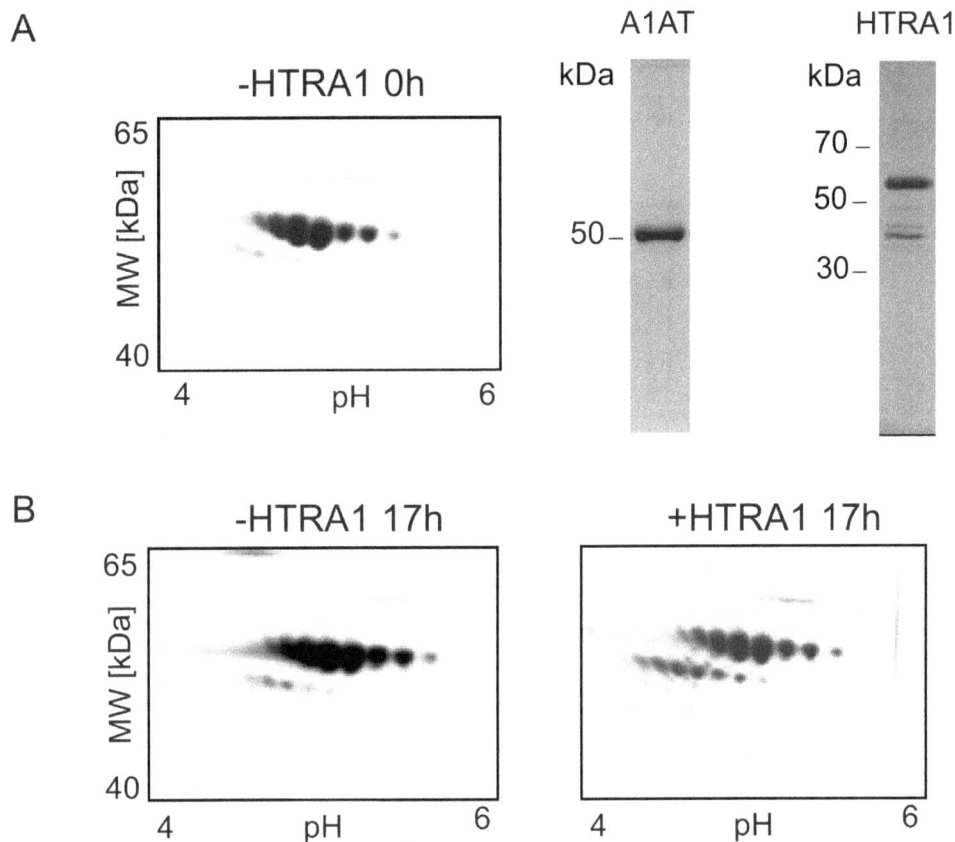

Figure 5. Validation of the candidate substrate A1AT of HTRA1. Purified A1AT from human plasma was incubated with HTRA1 at 37°C for 17 hours. (A) 2-D gel of A1AT in the absence of HTRA1 incubation time 0 h and SDS-PAGE of A1AT and HTRA1. The A1AT preparation contains only very small amount of cleaved A1AT. (B) Incubation over a period of 17 hours in the absence (left site) and presence (right site) of HTRA1. After incubation with HTRA1 a significant amount of A1AT was cleaved. The incubation of plasma A1AT with HTRA1 showed similar fragmentation patterns as the incubation of placental A1AT with HTRA1 (Fig. 4A).

located in the trophoblastic cells of the placenta [36,37]. This is important, as we used a homogenate of placental tissue containing proteins of different cell-types. The correlation of the localization pattern supports the assumption, that A1AT is a biological relevant substrate of HTRA1. As A1AT is a protease inhibitor, its inactivation through cleavage may have important consequences for the protease and protease inhibitor balance of the cell. Thus, A1AT was considered to be of particular interest and was chosen for further validation experiments.

Interestingly, incubation of placental A1AT with HTRA1 led to a limited proteolysis of A1AT. After cleavage, a decrease in molecular weight of nearly 4 kDa was observed. Similar results were obtained after incubation of purified A1AT from human plasma with HTRA1. A specific cleavage site was identified after met-382 (AIPM^{382} ↓ ^{383}SIPP) and the generated fragment is the residue C-36. The C-terminus of A1AT is crucial as it contains the reactive centre loop RCL (AA 368–392) which mediates the protease inhibitory property by binding to the target. The c-terminal part of A1AT seems to be particularly susceptible to proteolysis by a number of proteases ([38], MEROPS database, http://merops.sanger.ac.uk/index.shtml). For example matrix metallopeptidase-11 cleaves after ala-374 [28], thermolysin after met-375 [39] and periodontain after glu-378 [40]. The cleavage site after met-382 was reported for proteases such, as cathepsin L [41], elastase-2 [40] and chymotrypsin A [39]. Johnson et al. show that cleavage of A1AT after met-382 leads to the inactivation of

the protease inhibitor. Thus, the consequence of the cleavage of A1AT through HTRA1 will in a similar way leads to an inactivation of the protease inhibitor.

A1AT is a suicide protease-inhibitor with an extraordinary inhibitor mechanism. First, A1AT binds to the protease through residue of the RCL part. Then the protease cuts the protease inhibitor and a covalent acyl bond is built between A1AT and the protease. In response to the formation of this covalent bond A1AT changes its conformation causing a deformation of the protease, thereby inactivating it through formation of a very stable complex [42]. For instance the complex A1AT-chymotrypsin dissociates only after three to four days [43]. Finally, the covalent complex can be bound by receptors and degraded in lysosomes [44]. In contrast, the reaction of placental A1AT with HTRA1 leads not only to the formation of a protease-protease inhibitor complex, but also to the formation of cleavage fragments of A1AT within 4 hours. This leads to the assumption that either the A1AT-HTRA1 complex is not as stable as other A1AT-protease complexes, and dissociates after less than 4 hours into protease and the cleaved inactive protease inhibitor, or the reaction mechanism is a branched one, as already suggested by Patston et al. [45] for other protease inhibitors of the serpin family to which A1AT belongs [45–47]. It is challenging to predict how the reaction is regulated in the biological context and if A1AT is acting as a substrate or rather as an inhibitor for the protease HTRA1. Furthermore, the high concentration of HTRA1 used in our

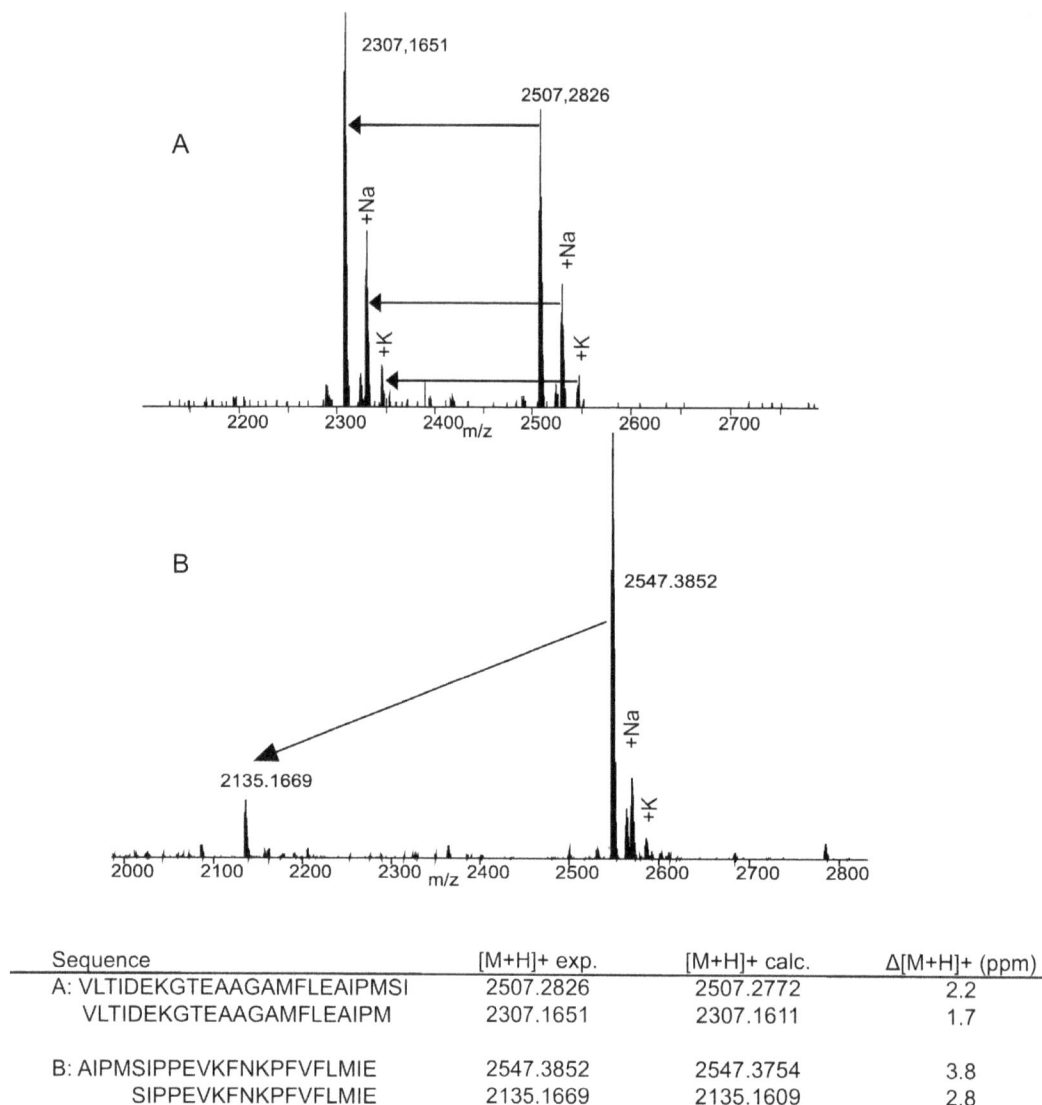

Sequence	[M+H]+ exp.	[M+H]+ calc.	Δ[M+H]+ (ppm)
A: VLTIDEKGTEAAGAMFLEAIPMSI	2507.2826	2507.2772	2.2
VLTIDEKGTEAAGAMFLEAIPM	2307.1651	2307.1611	1.7
B: AIPMSIPPEVKFNKPFVFLMIE	2547.3852	2547.3754	3.8
SIPPEVKFNKPFVFLMIE	2135.1669	2135.1609	2.8

Figure 6. MALDI-MS spectra after incubation of A1AT peptides with HTRA1. MALDI-Orbitrap-MS analysis as described in experimental procedures. A: Peptide A1AT(361–384) and generated fragment. Cleavage occurs after met381. B: Peptide A1AT(379–400) and generated fragment. Cleavage occurs again after met381. Further signals are sodium adduct (+22) and potassium adduct (+38). The image is representative of three independents experiments. The same incubation experiment was performed without addition of HTRA1 as negative control (Fig. S7).

experiments could influence the affinity of protease and protease-inhibitor in an artificial way. This remains to be further investigated in the future in a cellular context. Nevertheless, the consequences are the cleavage and inactivation of the protease inhibitor A1AT and this may have important consequences [48,49]. A disorder of the protease and protease inhibitor balance may be associated with pathophysiological events. Cleaved A1AT is also known to polymerize by β-strand linkage [50]. Accumulation of A1AT polymers can lead to severe diseases [51]. As the syncytiotrophoblast layer is directly attending to maternal blood, the inactivation of A1AT may have an impact on blood pressure regulation. The level of active proteases which are inhibited by A1AT could increase, as for instance neutrophil elastase, a protease participating in inflammation and responsible for reduced blood pressure through degradation of elastin in the arterial wall [49].

It was previously reported, that HTRA1 could be involved in apoptotic process, through degradation of proteins of the cytoskeleton as tubulin and microtubules, and of the apoptosis inhibitor XIAP [14,15] and to promote anoikis [52]. Since the protease inhibitor A1AT was reported to have an anti-apoptotic property possibly due to interaction with caspase-3 [53], the apoptosis supporting effect of HTRA1 may be also induced by diminishing the inhibitory property of A1AT.

The syncytiotrophoblast layer is formed from continuous fusion of the cytiotrophoblast. Most of the differentiated trophoblast cells are required to transport mRNA to the syncytium rather than for its growth [35]. Excessive syncytial formation is extruded in maternal blood as syncytial knots (accumulated apoptotic nuclei). This continuous turnover has to be tightly regulated. We suggest that HtrA1 may be involved in the regulation of apoptotic processes of the syncytiotrophoblast through inactivation of A1AT.

In this work, we have also identified two further members of the cytoskeleton, actin cytoplasmic 1 and tropomyosin beta as candidate substrates of HTRA1. Both are involved in cell motility and stability. Degradation of actin cytoplasmic 1 and tropomyosin beta by HTRA1 could play a role in cytoskeletal remodelling during apoptotic process, similar to the degradation of tubulin and microtubules.

Other candidate substrates identified here suggest that HTRA1 may be involved in protein biosynthesis process (EEF2, EEF1A1) and protein folding (HSPB1, PDIA1, PDIA3, HSP90B1), however, the significance of their degradation needs further investigation.

In conclusion, this study revealed several new candidate substrates of the protease HTRA1, which provides an insight into the function of HTRA1 in the placenta. Alpha-1-antitrypsin, actin cytoplasmic 1, tropomyosin beta chain and ten other proteins were identified as being degraded in the placenta protein sample after incubation with HtrA1. Furthermore we demonstrated that HTRA1 degrades purified plasma A1AT in vitro and, using synthetic peptides covering parts of the sequence of A1AT, we were able to identify a cleavage site within the reactive loop centre of the protease inhibitor. We postulate that HTRA1 may support trophoblast apoptosis, which is essential for the regenerating of the syncytiotrophoblast in the mature placenta and therefore for normal placentation through inactivation of the protease inhibitor A1AT. Main focus of our study was the investigation of potential substrates of HTRA1 in the protein repertoire of the placenta using a screening approach. However, in vivo evidence of the potential substrates identified in this study remains to be verified in cell experiments or animal models. Further studies are needed to confirm the relationship of HTRA1 and A1AT in vivo in the complex context of placenta development and growth.

Supporting Information

Figure S1   Colour coded picture of the 2-D gel electrophoresis separation of the placenta proteins. Colour coded picture of the two-dimensional protein separation of placenta proteins after 4 h incubation without (blue) and with the protease HTRA1 (orange). This picture was generated by overlaying the two 2-D gels presented in Figure 2 and by using artificial colours. Incubation time 4 hours.

Figure S2   Details of the 2-D patterns of placental proteins incubated with HTRA1 in the presence of the inhibitor cocktail P8340. The placenta proteins were incubated for 4 h at 37°C without (A) or with the protease HTRA1 (B) in the presence of the inhibitor cocktail P8340. The protein endoplasmin was not detected on both 2-DGE. Numbers (A = disappeared spots, B = appeared spots). Details and accession no of the substrates can be found in Table 1.

Figure S3   Details of the 2-D patterns of placental proteins incubated with HTRA1 in the presence of the inhibitor Timp-1. The placenta proteins were incubated 4 h at 37°C without (A) or with the protease HTRA1 (B) in the presence of the inhibitor Timp-1. Numbers (A = disappeared spots, B = appeared spots). Details and accession no of the substrates can be found in Table 1.

Figure S4   SDS-PAGE separation of the incubation products of A1AT with HTRA1. Incubation of A1AT in the absence (A) and in the presence of HTRA1 (B) in a ratio 1:10. Incubation time: 0 h and overnight. The arrows indicate A1AT, the A1AT-HTRA1 complex and the cleavage products of A1AT. This was confirmed

by MS-analysis (Table S1, spots 1D_A1, 1D_A2 and 1D_A3). M: marker, ON: overnight.

Figure S5   HTRA1 is not inhibited through A1AT. SDS-PAGE separation of the incubation products of HTRA1 and β-casein in the absence (A) and in the presence of A1AT (B) in a ration 1:50:50 after 0 h, 1 h and 2 h incubation time. The arrows indicate A1AT, HTRA1 and β-casein; the asterisks indicate the cleavage product of β-casein. M: marker. The image is representative of three independent replicates.

Figure S6   Details of the 2-D patterns of the candidate substrate A1AT incubated with HTRA1 three replicates shown. The placenta proteins were incubated 4 h at 37°C without (negative control Fig. A and C) or with the protease HTRA1 (Fig. B and D). A1AT forms a stable complex with HTRA1 at high molecular weight range (Fig. B, spot B7). No complex is observed when the placenta proteins are incubated without HTRA1 (A). A1AT species are significantly digested by HTRA1, new spot formations on the protease-treated sample are fragments of A1AT (Fig. D, spots B1). Control sample shows the A1AT species after incubation without HTRA1 (Fig. C). Details pictures lanes 1 and 2: two independent experiments. Details pictures lane 3: independent experiment in the presence of the inhibitor Timp-1.

Figure S7   MALDI-MS spectra after incubation of A1AT peptides without HTRA1 (negative control). MALDI Orbitrap-MS analysis as described in experimental procedures. Spectra of the peptides after incubation without HTRA1. A: Peptide A1AT(361–384) B: Peptide A1AT(379–400).

Figure S8   MALDI-MS spectra after incubation the peptide A1AT(396–418) with HTRA1. MALDI Orbitrap-MS analysis as described in experimental procedures. A: Peptide A1AT(396–418) after incubation with HtrA1. No cleavage fragment was generated. B: Peptide A1AT(396–418) after incubation without HTRA1 (negative control).

Table S1   List of candidate substrates of HTRA1 identified in the placenta. Spot 2D_A3: nearly same peptides were identified for the protein P60709 and P62736 (actin, aortic smooth muscle). 2D_A2 and 2D_A5: the protein P01019 (angiotensinogen) and P01857 (Ig gamma-1 chain C region) were also identified and stem probably from low abundant underlying spots. 2D_B7: the protein serum albumin stems from the huge protein spot at mass 66 kDa (Fig. 2). In spot 1D_A2 there is no serum albumin contamination. A1AT*: cleaved A1AT.

Acknowledgments

We want to acknowledge the Proteome Factory for the excellent 2-DE-Gel electrophoresis. We thank the PRIDE team for their support and for the possibility to deposit the MS data to the ProteomeXchange Consortium.

Author Contributions

Conceived and designed the experiments: HS MWL VF. Performed the experiments: VF AT DH. Analyzed the data: VF. Contributed reagents/materials/analysis tools: HS MWL AT. Wrote the paper: VF HS.

References

1. Lipinska B, Sharma S, Georgopoulos C (1988) Sequence analysis and regulation of the htrA gene of Escherichia coli: a sigma 32-independent mechanism of heat-inducible transcription. Nucleic Acids Res 16: 10053–10067.

2. Clausen T, Southan C, Ehrmann M (2002) The HtrA family of proteases: implications for protein composition and cell fate. Mol Cell 10: 443–455.

3. Zumbrunn J, Trueb B (1996) Primary structure of a putative serine protease specific for IGF-binding proteins. FEBS letters 398: 187–192.

4. Hu SI, Carozza M, Klein M, Nantermet P, Luk D, et al. (1998) Human HtrA, an evolutionarily conserved serine protease identified as a differentially expressed gene product in osteoarthritic cartilage. J Biol Chem 273: 34406–34412.

5. Truebestein L, Tennstaedt A, Monig T, Krojer T, Canellas F, et al. (2011) Substrate-induced remodeling of the active site regulates human HTRA1 activity. Nature structural & molecular biology 18: 386–388.

6. Krojer T, Sawa J, Schafer E, Saibil HR, Ehrmann M, et al. (2008) Structural basis for the regulated protease and chaperone function of DegP. Nature 453: 885–890.

7. Grau S, Richards PJ, Kerr B, Hughes C, Caterson B, et al. (2006) The role of human HtrA1 in arthritic disease. J Biol Chem 281: 6124–6129.

8. Grau S, Baldi A, Bussani R, Tian X, Stefanescu R, et al. (2005) Implications of the serine protease HtrA1 in amyloid precursor protein processing. Proc Natl Acad Sci U S A 102: 6021–6026.

9. An E, Sen S, Park SK, Gordish-Dressman H, Hathout Y (2010) Identification of novel substrates for the serine protease HTRA1 in the human RPE secretome. Invest Ophthalmol Vis Sci.

10. Zhang L, Lim SL, Du H, Zhang M, Kozak I, et al. (2012) High temperature requirement factor A1 (HTRA1) gene regulates angiogenesis through transforming growth factor-beta family member growth differentiation factor 6. J Biol Chem 287: 1520–1526.

11. Vierkotten S, Muether PS, Fauser S (2011) Overexpression of HTRA1 leads to ultrastructural changes in the elastic layer of Bruch's membrane via cleavage of extracellular matrix components. PLoS One 6: e22959.

12. Baldi A, De Luca A, Morini M, Battista T, Felsani A, et al. (2002) The HtrA1 serine protease is down-regulated during human melanoma progression and represses growth of metastatic melanoma cells. Oncogene 21: 6684–6688.

13. Chien J, Staub J, Hu SI, Erickson-Johnson MR, Couch FJ, et al. (2004) A candidate tumor suppressor HtrA1 is downregulated in ovarian cancer. Oncogene 23: 1636–1644.

14. Chien J, He X, Shridhar V (2009) Identification of tubulins as substrates of serine protease HtrA1 by mixture-based oriented peptide library screening. J Cell Biochem 107: 253–263.

15. Chien J, Ota T, Aletti G, Shridhar R, Boccellino M, et al. (2009) Serine protease HtrA1 associates with microtubules and inhibits cell migration. Mol Cell Biol 29: 4177–4187.

16. Tennstaedt A, Popsel S, Truebestein L, Hauske P, Brockmann A, et al. (2012) Human high temperature requirement serine protease A1 (HTRA1) degrades tau protein aggregates. J Biol Chem 287: 20931–20941.

17. He X, Khurana A, Maguire JL, Chien J, Shridhar V (2011) HtrA1 sensitizes ovarian cancer cells to cisplatin-induced cytotoxicity by targeting XIAP for degradation. Int J Cancer.

18. De Luca A, De Falco M, Severino A, Campioni M, Santini D, et al. (2003) Distribution of the serine protease HtrA1 in normal human tissues. J Histochem Cytochem 51: 1279–1284.

19. De Luca A, De Falco M, Fedele V, Cobellis L, Mastrogiacomo A, et al. (2004) The serine protease HtrA1 is upregulated in the human placenta during pregnancy. J Histochem Cytochem 52: 885–892.

20. Marzioni D, Quaranta A, Lorenzi T, Morroni M, Crescimanno C, et al. (2009) Expression pattern alterations of the serine protease HtrA1 in normal human placental tissues and in gestational trophoblastic diseases. Histol Histopathol 24: 1213–1222.

21. Lorenzi T, Marzioni D, Giannubilo S, Quaranta A, Crescimanno C, et al. (2009) Expression Patterns of Two Serine Protease HtrA1 Forms in Human Placentas Complicated by Preeclampsia with and without Intrauterine Growth Restriction. Placenta 30: 35–40.

22. Ajayi F, Kongoasa N, Gaffey T, Asmann YW, Watson WJ, et al. (2008) Elevated expression of serine protease HtrA1 in preeclampsia and its role in trophoblast cell migration and invasion. American journal of obstetrics and gynecology 199: 557 e551–510.

23. Overall CM, Blobel CP (2007) In search of partners: linking extracellular proteases to substrates. Nature reviews Molecular cell biology 8: 245–257.

24. Bradford MM (1976) A rapid and sensitive method for the quantitation of microgram quantities of protein utilizing the principle of protein-dye binding. Analytical biochemistry 72: 248–254.

25. Laemmli UK (1970) Cleavage of structural proteins during the assembly of the head of bacteriophage T4. Nature 227: 680–685.

26. Klose J, Kobalz U (1995) Two-dimensional electrophoresis of proteins: an updated protocol and implications for a functional analysis of the genome. Electrophoresis 16: 1034–1059.

27. Vizcaino JA, Cote RG, Csordas A, Dianes JA, Fabregat A, et al. (2013) The PRoteomics IDEntifications (PRIDE) database and associated tools: status in 2013. Nucleic acids research 41: D1063–1069.

28. Pei D, Majmudar G, Weiss SJ (1994) Hydrolytic inactivation of a breast carcinoma cell-derived serpin by human stromelysin-3. J Biol Chem 269: 25849–25855.

29. Nelson D, Potempa J, Travis J (1998) Inactivation of alpha1-proteinase inhibitor as a broad screen for detecting proteolytic activities in unknown samples. Analytical biochemistry 260: 230–236.

30. Knauper V, Reinke H, Tschesche H (1990) Inactivation of human plasma alpha 1-proteinase inhibitor by human PMN leucocyte collagenase. FEBS letters 263: 355–357.

31. Castellucci M, Theelen T, Pompili E, Fumagalli L, De Renzis G, et al. (1994) Immunohistochemical localization of serine-protease inhibitors in the human placenta. Cell Tissue Res 278: 283–289.

32. Mills K, Mills PB, Clayton PT, Mian N, Johnson AW, et al. (2003) The underglycosylation of plasma alpha 1-antitrypsin in congenital disorders of glycosylation type I is not random. Glycobiology 13: 73–85.

33. Potgens AJ, Drewlo S, Kokozidou M, Kaufmann P (2004) Syncytin: the major regulator of trophoblast fusion? Recent developments and hypotheses on its action. Human reproduction update 10: 487–496.

34. Huppertz B, Kingdom J, Caniggia I, Desoye G, Black S, et al. (2003) Hypoxia favours necrotic versus apoptotic shedding of placental syncytiotrophoblast into the maternal circulation. Placenta 24: 181–190.

35. Benirschke K, Kaufman P (2000) Pathology of the Human Placenta: Springer Verlag, New York.

36. Uhlen M, Oksvold P, Fagerberg L, Lundberg E, Jonasson K, et al. (2010) Towards a knowledge-based Human Protein Atlas. Nat Biotechnol 28: 1248–1250.

37. Ponten F, Jirstrom K, Uhlen M (2008) The Human Protein Atlas–a tool for pathology. J Pathol 216: 387–393.

38. Rawlings ND, Barrett AJ, Bateman A (2012) MEROPS: the database of proteolytic enzymes, their substrates and inhibitors. Nucleic acids research 40: D343–350.

39. Chang WS, Wardell MR, Lomas DA, Carrell RW (1996) Probing serpin reactive-loop conformations by proteolytic cleavage. Biochem J 314 (Pt 2): 647–653.

40. Nelson D, Potempa J, Kordula T, Travis J (1999) Purification and characterization of a novel cysteine proteinase (periodontain) from Porphyromonas gingivalis. Evidence for a role in the inactivation of human alpha1-proteinase inhibitor. J Biol Chem 274: 12245–12251.

41. Johnson DA, Barrett AJ, Mason RW (1986) Cathepsin L inactivates alpha 1-proteinase inhibitor by cleavage in the reactive site region. J Biol Chem 261: 14748–14751.

42. Huntington JA, Read RJ, Carrell RW (2000) Structure of a serpin-protease complex shows inhibition by deformation. Nature 407: 923–926.

43. Lobermann H, Lottspeich F, Bode W, Huber R (1982) Interaction of human alpha 1-proteinase inhibitor with chymotrypsinogen A and crystallization of a proteolytically modified alpha 1-proteinase inhibitor. Hoppe-Seyler's Zeitschrift fur physiologische Chemie 363: 1377–1388.

44. Perlmutter DH, Joslin G, Nelson P, Schasteen C, Adams SP, et al. (1990) Endocytosis and degradation of alpha 1-antitrypsin-protease complexes is mediated by the serpin-enzyme complex (SEC) receptor. J Biol Chem 265: 16713–16716.

45. Patston PA, Gettins P, Beechem J, Schapira M (1991) Mechanism of serpin action: evidence that C1 inhibitor functions as a suicide substrate. Biochemistry 30: 8876–8882.

46. Silverman GA, Bird PI, Carrell RW, Church FC, Coughlin PB, et al. (2001) The serpins are an expanding superfamily of structurally similar but functionally diverse proteins. Evolution, mechanism of inhibition, novel functions, and a revised nomenclature. J Biol Chem 276: 33293–33296.

47. Gettins PG (2000) Keeping the serpin machine running smoothly. Genome research 10: 1833–1835.

48. Mast AE, Enghild JJ, Pizzo SV, Salvesen G (1991) Analysis of the plasma elimination kinetics and conformational stabilities of native, proteinase-complexed, and reactive site cleaved serpins: comparison of alpha 1-proteinase inhibitor, alpha 1-antichymotrypsin, antithrombin III, alpha 2-antiplasmin, angiotensinogen, and ovalbumin. Biochemistry 30: 1723–1730.

49. Lisowska-Myjak B (2005) AAT as a diagnostic tool. Clinica chimica acta; international journal of clinical chemistry 352: 1–13.

50. Dunstone MA, Dai W, Whisstock JC, Rossjohn J, Pike RN, et al. (2000) Cleaved antitrypsin polymers at atomic resolution. Protein science: a publication of the Protein Society 9: 417–420.

51. Lomas DA, Evans DL, Finch JT, Carrell RW (1992) The mechanism of Z alpha 1-antitrypsin accumulation in the liver. Nature 357: 605–607.

52. He X, Ota T, Liu P, Su C, Chien J, et al. (2010) Downregulation of HtrA1 promotes resistance to anoikis and peritoneal dissemination of ovarian cancer cells. Cancer Res 70: 3109–3118.

53. Petrache I, Fijalkowska I, Zhen L, Medler TR, Brown E, et al. (2006) A novel antiapoptotic role for alpha1-antitrypsin in the prevention of pulmonary emphysema. American journal of respiratory and critical care medicine 173: 1222–1228.

11

Withanolide A Prevents Neurodegeneration by Modulating Hippocampal Glutathione Biosynthesis during Hypoxia

Iswar Baitharu[1,2], Vishal Jain[2], Satya Narayan Deep[2], Sabita Shroff[5], Jayanta Kumar Sahu[4], Pradeep Kumar Naik[1], Govindasamy Ilavazhagan[3]*

1 Department of Zoology, Guru Ghasidas Central University, Bilaspur, Chattishgarh, India, 2 Department of Neurobiology, Defence Institute of Physiology and Allied Sciences, Defense Research Development Organisation, Timarpur, Delhi, India, 3 Department of Research, Hindustan University, Chennai, Tamilnadu, India, 4 Department of Life Science, National Institute of Technology, Rourkela, India, 5 Department of Chemistry, Sambalpur University, Burla, India

Abstract

Withania somnifera root extract has been used traditionally in ayurvedic system of medicine as a memory enhancer. Present study explores the ameliorative effect of withanolide A, a major component of withania root extract and its molecular mechanism against hypoxia induced memory impairment. Withanolide A was administered to male Sprague Dawley rats before a period of 21 days pre-exposure and during 07 days of exposure to a simulated altitude of 25,000 ft. Glutathione level and glutathione dependent free radicals scavenging enzyme system, ATP, NADPH level, γ-glutamylcysteinyl ligase (GCLC) activity and oxidative stress markers were assessed in the hippocampus. Expression of apoptotic marker caspase 3 in hippocampus was investigated by immunohistochemistry. Transcriptional alteration and expression of GCLC and Nuclear factor (erythroid-derived 2)–related factor 2 (Nrf2) were investigated by real time PCR and immunoblotting respectively. Exposure to hypobaric hypoxia decreased reduced glutathione (GSH) level and impaired reduced gluatathione dependent free radical scavenging system in hippocampus resulting in elevated oxidative stress. Supplementation of withanolide A during hypoxic exposure increased GSH level, augmented GSH dependent free radicals scavenging system and decreased the number of caspase and hoescht positive cells in hippocampus. While withanolide A reversed hypoxia mediated neurodegeneration, administration of buthionine sulfoximine along with withanolide A blunted its neuroprotective effects. Exogenous administration of corticosterone suppressed Nrf2 and GCLC expression whereas inhibition of corticosterone synthesis upregulated Nrf2 as well as GCLC. Thus present study infers that withanolide A reduces neurodegeneration by restoring hypoxia induced glutathione depletion in hippocampus. Further, Withanolide A increases glutathione biosynthesis in neuronal cells by upregulating GCLC level through Nrf2 pathway in a corticosterone dependenet manner.

Editor: Antonio Paolo Beltrami, University of Udine, Italy

Funding: The study was funded by Defense Research Development Organisation. The funders had no role in study design, data collection and analysis, decision to publish, or preparation of the manuscript.

Competing Interests: The authors have declared that no competing interests exist.

* Email: govindasamyilavazhagan@gmail.com

Introduction

Prolonged exposure to hypobaric hypoxia at high altitude is known to cause hippocampal neurodegeneration leading to loss of memory and higher order brain dysfunctions [1–2]. Under hypoxic conditions, lower availability of oxygen at tissue level results in generation of superoxide radicals that subsequently generate hydroxyl and peroxynitrite radicals in a chain reaction [3]. The antioxidants and free radical scavenging enzyme system play a crucial role in quenching the free radicals generated as a byproduct of various biochemical reactions under normoxic condition. However, hypoxic exposure weakens the antioxidant defense mechanisms by causing alterations in activity of antioxidant enzymes like glutathione reductase and glutathione peroxidase [4–5]. The cumulative effect of impaired antioxidant system and increased free radical generation leads to lipid peroxidation, membrane damage, protein oxidation, DNA damage [6] and altered gene expression [7] that may finally culminate in cell death. The brain is vulnerable to oxidative stress because of its

high demand for oxygen, abundant fatty acids that are targets of lipid peroxidation, and lower antioxidant enzyme activities compared to other organs. Recent reports showed that hippocampal pyramidal neurons are more susceptible to oxidative stress induced damage compared to neurons at prefrontal cortex and cerebellum [8]. Administration of free radical quenchers like quercetin or antioxidant precursors such as N-acetyl cysteine has been reported to enhance cell viability in hypoxic stress [9–10].

Glutathione a tripeptide comprised of glutamate, cysteine and glycine, is a major antioxidant in the brain [11], with a concentration of approximately 2–3 mM. Glutathione is synthesized in cytosol by the consecutive action of the enzymes glutamate-cysteine ligase and glutathione synthetase which involve the utilization of ATP. Glutamate- cysteine ligase is the rate-limiting enzyme of GSH synthesis and is subjected to feedback inhibition by GSH [12]. Both enzymes are transcriptionally regulated by nuclear factor erythroid 2-related factor 2 (Nrf2), a redox-sensitive transcription factor member of the basic-leucine zipper family [13]. In response to oxidative stress, Nrf2 dissociates from its cytosolic inhibitor Keap1, translocates to the nucleus and binds to antioxidant-response elements (AREs) in the promoters of target genes. This leads to transcriptional induction of several cellular defense genes, including glutathione biosynthetic enzymes (glutathione cysteine ligase modifier subunit (GCLM) and catalytic subunit (GCLC) and GSH-dependent antioxidant enzymes (glutathione peroxidase 2, glutathione S-transferases and heme oxygenase-1) [14]. The Nrf2-mediated regulation of cellular antioxidant plays an important role in defense against oxidative stress [15]. Prophylactic Nrf2 activation by small molecules provide protection against a host of oxidative insults both *in vitro* as well as *in vivo*, including free radical donors and oxygen glucose deprivation (OGD), toxic levels of glutamate or N-methyl-D-aspartate (NMDA), neurotoxin or stroke-induced injury [16]. Exposure to hypobaric hypoxia depletes the neuronal glutathione in hippocampus [17]. Exogenous supplementation of GSH either through oral or intravenous route is hydrolyzed by γ-glutamyl-transpeptidase and is rapidly eliminated within seven minutes from general circulation. Comford et al. showed that only 0.5% of radiolabeled GSH administered by intra-carotid injection was detectable in brain extracts [18]. Although there are reports describing the existence of GSH transporters, glutathione generally doesn't cross the blood-brain-barrier [19]. Hence, compounds modulating the GSH biosynthesis play a much significant role in providing protection against oxidative insult compared to exogenous supplementation of GSH. Since a batteries of free radicals scavenging enzyme system depend directly on availability of GSH for detoxification of ROS, molecules capable of modulating glutathione biosynthesis could potentially protect free radicals mediated neurodegenration under hypoxic condition.

The root extract of *Withania somnifera* is used as a popular herbal drug in Ayurvedic medicine, and has been used traditionally as a tonic and nootropic agent. It facilitate cognitive function and augment mental retention capacity following diabetes, Aβ and scopolamine induced memory loss [20–21]. It is also known to augment cholinergic activity in hippocampus [22]. Recent reports from our laboratory showed that withanolide enriched extract of *Withania somnifera* root ameliorates hypoxia induced memory impairment by modulating corticosterone level in brain through Nitric oxide cyclooxygenase prostaglandin pathway [17]. Methanolic extract of Withania root demonstrate profound association with neurite extension and dendritic arborisation [23]. Treatment with withanolide A (WL-A), a major active constituents isolated from *Withania somnifera* root predominantly induces axonal outgrowth in normal cortical neurons [24]. Supplementation of withanolide enriched extract of *Withania somnifera* root restored hypoxia induced depleted antioxidant glutathione level and free radical scavenging enzyme system in brain [17]. Since glutathione is the major antioxidant in brain, modulation of its biosynthesis by withanolide A under hypoxic condition could ameliorate oxidative stress induced neurodegeneration and consequent memory dysfunction. In the present study, we investigated the effect of withanolide A on hippocampal glutathione biosynthesis during hypoxic exposure and its correlation with hypoxia induced neurodegeneration and memory dysfunction. The study further explores the possible mechanism underlying withanolide A mediated modulation of glutathione system during exposure to hypobaric hypoxia.

Materials and Methods

Ethics Statement

All the protocols followed in this experiment were approved by the Institutional Committee for Animal Care and Use (ICACU), Defense Institute of Physiology and Allied Sciences, New Delhi (Permit Number: DIP-12-250) following the guidelines of "Committee for the Purpose of Control and Supervision of Experiments on Animals" Govt. of India. Utmost care was taken to minimize suffering of animal during sampling. Sacrifice of the animals were done under sodium pentabarbitol anaesthesia.

Chemicals and reagents

Withanolide A (Cat # 74776; Purity ≥95%), corticosterone, buthionine sulfoximine (BSO) and metyrapone were procured from Sigma chemicals (Sigma-Aldrich, USA). Kits for estimation of glutathione reductase, glutathione peroxidase and superoxide dismutase activity were purchased from RANDOX (Randox laboratory, UK). Glutathione s transferase assay kit was procured from Caymen (Cayman, USA). All the primary and secondary antibodies used in the experiments were procured from abcam (abcam, USA). ATP chemiluminescence assay kit and EnzyChrom NADP+/NADPH Assay Kit was procured from Calbiochem (Calbiochem, San Diego, CA) and Bioassay System (Hayward, CA, USA) respectively. ABC staining kit for immunohistochemistry was purchased from the vectastain (Vector laboratory, USA). Superscript first strand cDNA synthesis kit and SYBR Green PCR Master Mix for real time analysis of GCLC and Nrf2 was purchased from Applied Biosystems (Applied Biosystems, Foster City, CA).

Animals

Adult male Sprague Dawley rats weighing 240–250 g were taken and maintained at 12 h light-dark cycle (lights on from 8:00 AM–8:00 PM) in the animal house of the institute. Food pellets (Lipton Pvt. Ltd., India) and water was given *ad libitum*. The temperature and humidity of the animal house was maintained at $25\pm2°C$ and $55\pm5\%$ respectively. All animal handling was performed between the time windows of 10.00 AM to 11.30 AM to avoid experimental deviations due to diurnal variations in corticosterone concentration.

Hypoxic exposure

Animals were exposed to a simulated altitude of 7600 m (25,000 ft, 282 mm Hg) in a specially designed animal decompression chamber where altitude could be maintained by reducing the ambient barometric pressure. Periodic evaluation of fluctuation in oxygen level arising from fresh air flush into the chamber was done using an oxygen sensor. The temperature and humidity in the chamber were maintained precisely at $25\pm2°C$ and $55\pm5\%$ respectively. The rate of ascent and descent to hypobaric conditions was maintained at 300 m/min as described previously [25–26]. The hypobaric hypoxic exposure was continuous for the stipulated period except for a 10–15 min interval each day for replenishment of food and water, drug administration and changing the cages housing the animals.

Experimental design

The study was performed in two phases. Phase I aimed at investigating the effect of withanolide A on glutathione level and GSH dependent free radical scavenging system in hippocampus following exposure to hypobaric hypoxia. Rats were screened using elevated plus maze and open field test to ensure that none of the animals selected for experimentation were having dysfunctions such as anxiety or locomotory problems. The selected rats were then divided into four groups randomly (n = 15/group) viz., normoxia, normoxia treated with withanolide A, hypoxia treated with vehicle (0.5% gum arabic solution) and hypoxic rats treated with withanolide A (Fig. 1). Alteration in level of reactive oxygen species, lipid peroxidation, GSH and activity of glutathione reductase, glutathione peroxidase, glutathione s transferase, superoxide dismutase and glutamyl cysteinyl ligase in hippocampal region were assessed. Changes in expression of glucocorticoid and mineralocorticoid receptor. corticosterone, ATP and NADPH level in hippocampal region was also estimated following hypoxic exposure.

Phase II study was conducted to explore the molecular mechanism underlying the modulatory effect of withanolide A on glutathione biosynthesis during exposure to hypobaric hypoxia and their effect on hypoxia induced neurodegeneration. Rats ($n = 80$) were divided into eight groups ($n = 10$/group) and drugs were administered as described in Table 1. The duration of hypoxic exposure was kept 7 days since neurodegeneration as well as the memory impairment was maximum on that day. Changes in apoptotic marker caspase 3 by immunohistochemistry and chromatin condensation by hoescht staining were evaluated in the CA3 region of hippocampus following hypoxic exposure and drugs administration.

Preparation of drug and pharmacological administration

Withanolide A was dissolved in 0.5% gum arabic solution and administered to rats orally by gavage using feeding cannula at a dose of 10 μmol/kg$^-$ BW (decided after dose optimization study). The Withanolide A feeding to rats was done for 21 days prior to and during exposure to hypobaric hypoxia for 7 days. Buthionine sulfoximine (4 mM/kg BW) was dissolved in phosphate buffer saline to a volume of 1 ml and administrered intraperitoneally (i.p) [27]. Metyrapone (50 mg/kg BW) was dissolved in the 40% polyethylene glycol (PEG) (w/v) in physiological saline. Metyrapone or an equivalent volume (1 ml) of the vehicle consisting of physiological saline and polyethylene glycol was injected intraperitoneally [28]. Corticosterone (40 mg/kg BW) was dissolved in peanut oil and injected subcutaneously [29] in a volume of 1 ml along with metyrapone during exposure to hypobaric hypoxia. Both the metyrapone as well as corticosterone administration was started from 3rd day and was continued till 7th day of hypoxic exposure. The drugs were administered once daily at 9:00 AM

when decompression chamber was opened to replace food and water.

Oxidative stress markers

On completion of the stipulated period of hypoxic exposure, rats were sacrificed and hippocampi were removed at 4–8°C in ice-cold 0.01 M phosphate buffer saline (PBS, pH 7.4). Tissue homogenates (10%) were prepared in 0.15 M KCl. The crude homogenates (250 μl) were taken for lipid peroxidation, GSH estimation and the remaining homogenates were centrifuged at 10,000 g for 30 min at 4–8°C. The supernatant was then collected and used for enzymatic estimations. The total protein content per 10 μl of each of the samples were estimated using bovine serum albumin as standard [30].

Estimation of reactive oxygen species

Reactive oxygen species mainly hydrogen peroxide (H_2O_2) and peroxinitrite ($ONOO^-$) in the hippocampal tissue were estimated spectrofluorimetrically using 2,7-dichlorofluorescein-diacetate (DCFHDA) as suggested by LeBel et al. [31] and modified by Myhre et al.. [32]. In brief, hippocampal homogenate (10%) was prepared in ice cold 0.15 M KCl and 1.494 ml of 0.1 M PBS (pH 7.4) was added to 25 μl of the crude homogenate followed by addition of 6 μl of DCFHDA (1.25 mM) [25]. The sample was then incubated for 15 min at 37°C in dark and readings were taken at 488 nm excitation and 525 nm emission. The readings were expressed as fluorescent units per mg of protein and converted to percentage by taking normoxic value as 100%.

Lipid peroxidation

Lipid peroxidation was measured by thiobarbituric acid test for malondialdehyde as per the method described by Das and Ratty [33] and modified by Colado et al.. [34]. Hippocampi were homogenized in 50 mM phosphate buffer, deproteinised with 40% trichloroacetic acid and 5 M hydrochloric acid. Thiobarbituric acid (2%) in 0.5 M sodium hydroxide was added to the deproteinised hippocampal sample. The reaction mixture was heated in a water bath at 90°C for 35 minutes and centrifuged at 12,000 g for 10 minutes. The pink chromogen formed was

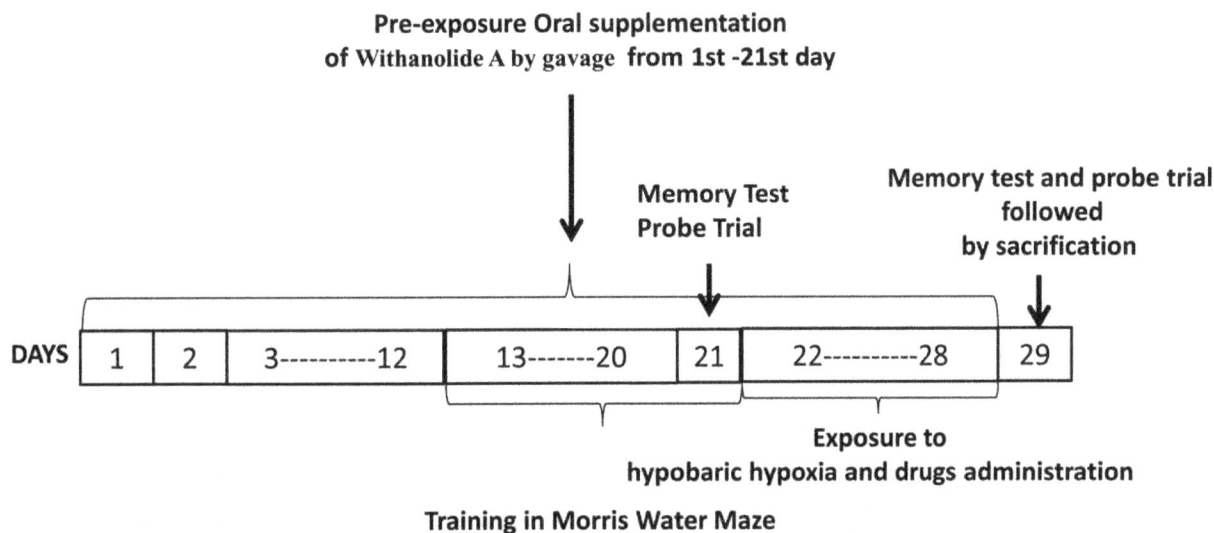

Figure 1. Showing the schedule of the training, probe trial and memory test in Morris Water Maze, supplementation of Withanolide A, administration of drugs and exposure to hypobaric hypoxia.

Table 1. Showing schedules and doses of drug administered during exposure to hypobaric hypoxia.

Groups	Description	Hypobaric Hypoxia	Intervention administered	Duration of treatment	Dose and mode of administration of drugs	Nature of durgs
Group I (n = 10)	Normoxia + Withanolide A	No	Withanolide A[#]	28 days	10 µmol kg^{-1} day^{-1} (Oral)	
Group II (n = 10)	Hypoxia + Vehicle	Yes(07days)	None	05 days	None	
Group III (n = 10)	Hypoxia + Withanolide A	Yes(07days)	Withanolide A[#]	28 days	10 µmol kg^{-1} day^{-1} (Oral)	
Group IV (n = 10)	Normoxia + BSO	No	BSO	05 days	4 mM/Kg BW (i.p) (Ando et al., 2009)	Glutathione Synthesis Inhibitor
Group V (n = 10)	Hypoxia + BSO	Yes(07days)	BSO[#]	05 days	4 mM/KgBW	Glutathione Synthesis Inhibitor
Group VI (n = 10)	Hypoxia + Withanolide A+ BSO	Yes(07days)	BSO[#] Withanolide A[#]	05 days 28 days	4 mM/KgBW 10 µmol kg^{-1} day^{-1} (Oral)	
Group VII (n = 10)	Hypoxia + Withanolide A+ Corticosterone	Yes(07days)	Withanolide A Corticosterone[#]	28 days 05 days	10 µmol kg^{-1} day^{-1} (Oral) 40 mg/Kg BW (i.p) (Smith Swintosky et al., 1996)	
Group VIII (n = 10)	Hypoxia + Metyrapone	Yes(07days)	Metyrapone*	05 days	50 mg/Kg (i.p) (Baitharu et al., 2011)	Corticosterone Synthesis Inhibitor
Group IX (n = 10	Hypoxia + Corticosterone + Metyrapone	Yes(07days)	Corticosterone*Metyrapone*	05 days 05 days	40 mg/Kg 50 mg/Kg	Corticosterone Synthesis Inhibitor

* Indicate the administration of drug started from 3rd day of hypoxic exposure.
# denotes the administration of Withanolide A started 21 days prior to hypoxic exposure and was continued during hypoxic exposure.

measured at 532 nm spectrophotometrically and expressed in mmol/mg protein. The results were then converted to percentage considering normoxic value as 100%.

Reduced glutathione

The reduced gluathone in 10% hippocampal tissue homogenate was measured as per the protocol followed by Hissin and Hilf [35]. In brief, 250 µl of the crude homogenates were taken, to which equal volume of 10% metaphosphoric acid was added. The mixture was then centrifuged at 10,000 g for 30 minutes at 4°C. The supernatants obtained were used for the estimation of GSH by incubation with o-pthaldehyde. Readings were taken spectro-fluorometrically at 350 nm excitation and 420 nm emission. The amount of GSH was calculated using a standard curve and expressed in mmol/mg of protein and converted to percentage taking normoxic value as 100%.

Glutathione Reductase and Glutathione Peroxidase activity

Glutathione reductase activity was measured as per the method described by Pinto and Bartley [36] and the values obtained were expressed in mmol of NADPH oxidized/min/g tissue. The glutathione peroxidase (GPx) (EC 1.11.1.9) activity was measured using glutathione peroxidase assay kit and the results obtained were expressed in U/mg protein and converted to percentages taking normoxic value as 100%.

Glutathione s transferase activity

The glutathione s transferase activity in the hippocampal tissue was estimated using glutathione s transferase activity assay kit as per the manufacturer protocol. Briefly, tissue was homogenized with ice cold 100 mM potassium phosphate buffer containing EDTA and centrifuged at 10,000 g for 15 minutes at 4°C. Supernatant was collected for assay. Optical density was measured spectrophotometrically (Molecular Devices, USA). The results obtained were expressed in U/mg protein and converted to percentages taking normoxic value as 100%.

Super oxide dismutase activity

The superoxide dismutase activity in the hippocampal tissue was estimated using RANDOX kit (RANDOX Laboratory Ltd.). The activity of the enzymes was expressed as U/mg protein and converted to percentages taking normoxic value as 100%.

ATP Level in hippocampus

The ATP content was determined using an ATP chemiluminescence assay kit (Calbiochem, San Diego, CA) as per the manufacturer's instructions. In brief, the tissue homogenate was treated with nuclear-releasing buffer for 5 min at room temperature with gentle shaking. To the tissue lysate, ATP monitoring enzyme was added and the luminescent reaction was immediately analyzed in a microplate reader (Spectra Max MII, Molecular Devices, Germany). The absolute ATP content was calculated by running an ATP standard curve with known ATP concentrations. Protein concentrations of samples were determined by Bradford assay (Bradford, 1976). The calculated total ATP concentration was expressed as nanomolar ATP/mg protein and converted to percentages taking normoxic value as 100%.

Estimation of NADP+/NADPH level in Hippocampus

The NADP+/NADPH ratio was determined by using the EnzyChrom NADP+/NADPH Assay Kit (ECNP-100) procured from BioAssay Systems, (BioAssay Systems, Hayward, CA, USA).

Briefly, Samples were homogenized with NADP+ extraction buffer for NADP+ determination and NADPH extraction buffer for NADPH determination separately. The tissue extracts was heated at 60°C for 5 min followed by addition of assay buffer and the opposite extraction buffer to neutralize the extracts. The mixture was spinned at 14,000 rpm for 5 min. Supernatant was used for NADP+/NADPH assays. Determination of both NADP+ and NADPH concentrations requires extractions from two separate samples. Calibration curve was prepared using NADP premix by mixing 1 mM standard and distilled water. Optical density (OD$_0$) was read for time "zero" at 565 nm and OD$_{30}$ after a 30 min incubation at room temperature. OD$_0$ was subtracted from OD$_{30}$ for the standard and sample wells and OD values were used to determine sample NADP+/NADPH concentration from the standard curve. The results thus obtained were converted to percentage considering normoxic values as 100%.

Estimation of corticosterone level in hippocampus by High Performance Liquid Chromatography

Levels of corticosterone was estimated in hippocampal tissue using high performance liquid chromatography (Waters, Milford, MS, USA). The extraction of corticosterone from hippocampal tissue was done with diethyl ether [37]. The ether evaporated tissue samples were reconstituted with 250 µl of methanol. 10 µl of the reconstituted sample was injected with the help of an auto sampler (Waters) to the HPLC system and resolved using C18 RP column with acetonitrile: Water: Glacial acetic acid (35: 65:05 v/ v) as solvent phase in isocratic condition. The flow rate of the mobile phase was maintained at 1 ml/min and detection of corticosterone fraction was done at 254 nm with a UV detector. The pressure in the column was maintained at 1800 psi and the samples were run for 30 minutes. A standard plot was prepared using corticosterone standard and methanol in the range of 10–1000 ng/ml by serial dilution. The standards were tested individually at different concentrations to record detection limit, retention time and peak area. Concentration of corticosterone was calculated from a standard plot of peak area of corticosterone versus concentration of corticosterone.

Determination of γ-GCL activity

γ-GCL activity was determined following the method described by Seelig et al.. [38]. Briefly, enzyme activity was determined at 37°C in reaction mixtures of 1.0 ml containing 100 mm Tris-HCl buffer (pH 8.2), 150 mm KCl, 5 mm ATP, 2 mm phosphoenol-pyruvate, 10 mm glutamate, 10 mm γ -aminobutyrate, 20 mm MgCl$_2$, 2 mm EDTA, 0.2 mm NADH, 17 µg pyruvate kinase, and 17 mg lactate dehydrogenase. The reaction was initiated by adding extract, and the rate of decrease in absorbance at 340 nm was monitored. Enzyme-specific activity was measured as micro-moles of NADH oxidized per minute per milligram protein. The results thus obtained were converted to percentage considering normoxic values as 100%.

Glucocorticoid receptor (GR), Mineralocorticoid receptor (MR), GCLC and Nrf2 in hippocampus by western blotting

The expression analysis of the proteins in the hippocampal region of the brain by western blotting was performed as described by Hota et al.. [25]. The hippocampi were dissected out at 4°C from the rat brain following decapitation and homogenized in ice-cold lysis buffer (0.01 M Tris–HCl, pH 7.6, 0.1 M NaCl, 0.1 M dithiothreitol, 1 mM EDTA, 0.1% NaN3, PMSF, Protease inhibitor cocktail). The homogenates were centrifuged at 10,000 g for 10 min at 4°C and the supernatants were used for protein expression analysis. SDS-PAGE (12%) was run in duplicates depending upon the molecular weight of the proteins of interest. Sample protein (50 mg) was resolved by SDS-PAGE and transferred to nitrocellulose membranes pre-soaked in transfer buffer (20% methanol, 0.3% Tris, 1.44% glycine in water) using a semidry transblot module (Bio-Rad). The transfer of the protein bands to the membrane was verified by Ponceau staining. The membranes were then blocked with 5% Blotto for 1 h, washed with PBST (0.01 M PBS, pH 7.4, 0.1% Tween 20) and probed overnight with polyclonal GR, MR, GCLC, GCLM and Nrf2 specific antibodies (abcam, USA). Subsequently, the membranes were washed with PBST thrice (10 min each) and were incubated with suitable secondary anti-IgG HRP conjugated antibody for 2–3 h. Chemiluminiscent peroxidase substrate kit was used to develop the membrane which were then stripped using stripping buffer (Bio-Rad) and probed for b-actin expression which was considered as loading control. The protein expression in each group was quantified by densitometric analysis.

Real-time polymerase chain reaction (PCR) of Nrf2 and γ-GCLC mRNA

Total RNA was extracted from the hippocampal tissues using TRIzol Reagent (Invitrogen) and reverse transcribed using the Superscript First-Strand Synthesis System (Invitrogen). To quantify the gene expression levels in the samples, real-time polymerase chain reaction was performed on an ABI Prism 7700 Sequence Detection System using SYBR Green PCR Master Mix (Applied Biosystems, Foster City, CA) and primers specific for the catalytic subunits of glutamate-cysteine ligase (GCL; EC 6.3.2.2) of rats; i.e., GCLC (forward primer, 5′-CTCTGCCTATGTGGTATTT-G-3′; reverse primer, 5′-TTGCTTGTAGTCAGGATGG-3′; amplicon size, 454 bp) and Nrf2 (forward primer, 5′ CGTGG-TGGACTTCTC TGCTACGTG GTG 3′; reverse primer, 5′ GGTCGGCATGCATTTGACTTCACAGTC 3′; amplicon size, 352 bp) and primers for β-actin to normalize the amount of mRNA in the samples (forward primer, 5′-TCTTCCAGCCTT-CCTTCC-3′; reverse primer, 5′-TAGAGCCACCAATCCA-CAC- 3′; amplicon size, 252 bp). The annealing temperature and the primer concentrations were optimized for amplification efficiency after validation of the dissociation curves and satisfactory separation of the PCR products on a 1.5% agarose gel. The optimal thermal cycle protocol for all the samples began with 10-min denaturation at 95°C, followed by 40 cycles of 95°C for 15 s, 62°C for 30 s, and 72°C for 45 s. The concentrations of the primers used for GCLC, Nrf2 and β-actin were 200, 260, and 200 μM, respectively. The relative amounts of mRNA for GCLC and Nrf2 in the drug treated groups versus the vehicle treated hypoxic and normoxic group were calculated as the relative expression ratios in comparison with β-actin and expressd in fold change.

Immunohistochemistry of caspase 3 in CA3 region of hippocampus

In brief, sections were washed in 0.1 M Phosphate Buffer Saline (pH 7.4) for 30 min, and treated with 0.3% hydrogen peroxide for 30 min to inhibit endogenous peroxidase activity. Sections were washed in 0.1 M PBS, permeabilized with 0.25% Triton X-100 and blocked with 1.5% goat serum for 3 h at room temperature. They were then incubated with rabbit anti-Caspase 3 (1:200) antibodies for 48 hours at 4°C. The sections were then washed and incubated with biotinylated goat anti-rabbit immunoglobulin (1:100) for 3 h at room temperature, washed with 0.1 M PBS (pH 7.4) and incubated with avidin-biotin complex tagged with

horse redox peroxidase enzyme prior to development with 3, 3′-diaminobenzidine (ABC kit, SantaCruz). Digital images were acquired using a light microscope (Olympus, Model BX 51, Japan) and the immunoreactive neurons for Caspase 3 were counted by Image Pro-Plus 5.1 software in six random fields of 0.1 mm^2.

Chromatin condensation by Hoechst staining

Chromatin condensation, which is an indicator of apoptosis, was studied by Hoechst 33342 staining. Hippocampal sections (15 μm thickness) were permeabilized in 0.1% triton and stained with Hoechst 33342 (10 mg/ml) for 30 min in the dark. The stained 6.04-mm sections of bregma were visualized using a blue filter in an Olympus BX-51 fluorescent microscope, and the brightly fluorescing cells were scored qualitatively.

Memory assessment by Morris Water Maze

Morris Water Maze was used to investigate the spatial reference memory of rodents [39]. An overhead camera and computer assisted tracking system with videomax software (Columbus Instruments, USA) was used to record the position of the rat in the maze. During reference memory task the rats were trained for a period of 8 days (sessions) followed by a probe trial on the ninth day. The platform position was kept fixed in one position throughout the training period. The rats were released randomly choosing any of the four quadrants as starting position and the starting position was changed in each release. The order of the starting position varied in every trial and no given sequence was repeated. The number of crossing over at the original platform position and the time (s) spent in the target quadrant were calculated. The rats were then exposed to hypobaric hypoxia for 7 days following which the reference memory was tested by a probe trial for 60 s and a single trial for memory test to locate the submerged platform. The amount of platform crossings and the time spent in the target quadrant during the probe trial and the latency as well as path length during the single trial was considered as measures for assessment of memory.

Statistical Analysis

Probe trial task in Morris Water Maze after exposure to hypoxia was analyzed using one-way Analysis of variance. Mean of latencies and pathlength for reference memory testing after exposure to hypobaric hypoxia was analyzed in similar manner since one trial was given for each task. The results of oxidative stress markers and other biochemical parameters are representations of six individual observations and presented as means ± SEM unless otherwise mentioned. Statistical analysis for multiple comparisons was done between normoxic group, hypoxic group and hypoxia with drug treated groups using one and two-way ANOVA wherever applicable. The post hoc analysis was done by Newman– Keul's test in all experimental groups wherever appropriate. Difference below or equal to the probability level (p≤0.05) was considered statistically significant.

Results

Withanolide A reduces oxidative stress in hippocampal region of the brain during exposure to hypobaric hypoxia

Exposure to hypobaric hypoxia significantly elevated the level of reactive oxygen species generation (F (3, 20) = 22.3, p≤0.05) and lipid peroxidation (F (3, 20) = 18.1, p≤0.05) along with significant reduction in GSH level (F (3, 20) = 21.8, p≤0.05) in hippocampus compared to normoxic group. Supplementation of withanolide A

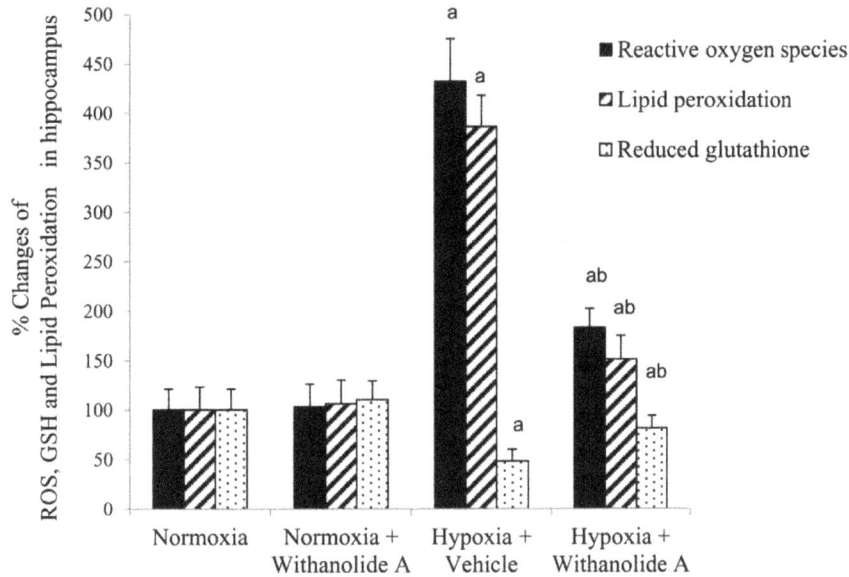

Figure 2. Effect of Withanolide A on oxidative stress markers. Administration of Withanolide A decreases the hypoxia induced elevated level of reactive oxygen species, lipid peroxidation and GSH in hippocampus. Data expressed as percentage change taking normoxic value as 100% and represents Mean ± SEM. 'a' denotes p≤0.05 vs. when compared to normoxic group and 'b' denotes p≤0.05 vs. when compared to 7 days hypoxic group treated with vehicle only.

21 days before and during exposure to hypobaric hypoxia for 7 days significantly decreased the free radicals level and lipid peroxidation and significantly increased GSH level in hippocampus compared to hypoxic vehicle treated group as shown in Fig. 2.

Withanolide A modulates GSH dependent free radicals scavenging system in hippocampus during hypobaric hypoxia

The activity of glutathione reductase (F $(3, 20) = 14.2$, p≤0.05), glutathione s transferase (F $(3, 20) = 11.3$, p≤0.05), and superoxide dismutase (F $(3, 20) = 09.3$, p≤0.05) was significantly decreased with concomittent increased glutathione peroxidase activity

following exposure to hypobaric hypoxia compared to normoxic group. Supplementation of withanolide A 21 days before and 7 days during exposure to hypobaric hypoxia significantly increased the activity of glutathione reductase, glutathione s transferase and superoxide dismutase with significant decrease in GPx activity compared to hypoxic group treated with vehicle only (Fig. 3).

Administration of Withanolide A during expoaure to hypobaric hypoxia alters ATP, NADPH level and GCLC activity in hippocampus

There was significant decrease of ATP level (F $(3, 20) = 16.1$, p≤0.05), NADPH level (F $(3, 20) = 12.3$, p≤0.05) and GCLC

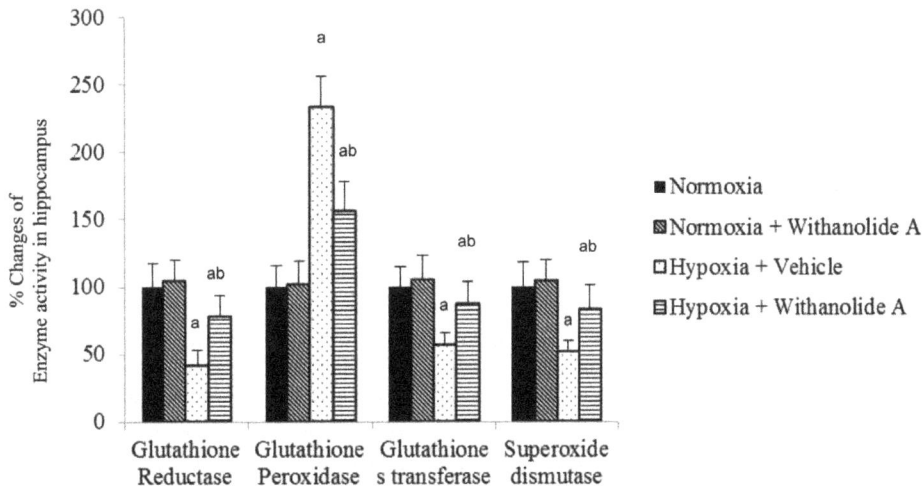

Figure 3. Effect of Withanolide A on free radical scavenging enzyme system. Withanolide A administration during hypoxic exposure increases hypoxia induced decreased activity of glutathione reductase, glutathione peroxidase, glutathione s transferase and superoxide dismutase in hippocampus. Data expressed as percentage change taking normoxic value as 100% and represents Mean ± SEM. 'a' denotes p≤0.05 vs. when compared to normoxic group and 'b' denotes p≤0.05 vs. when compared to 7 days hypoxic group treated with vehicle only.

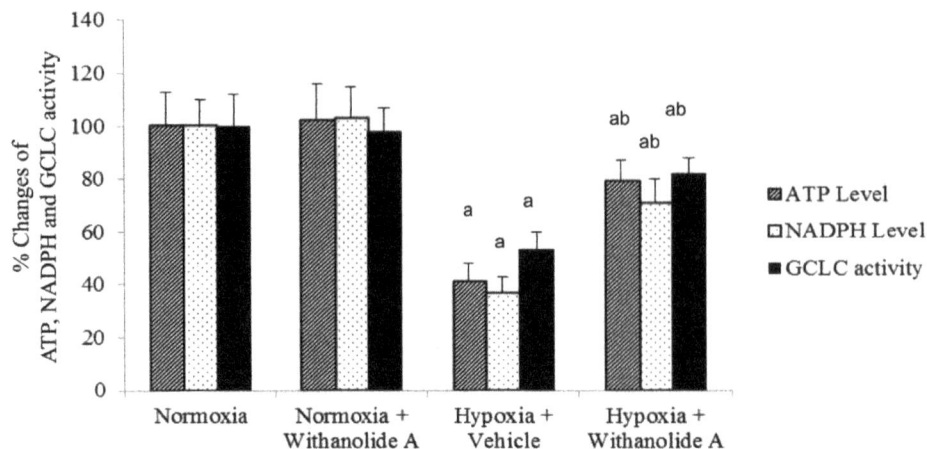

Figure 4. Restoration of ATP, NADPH and γ-glutamylcysteinyl ligase catalytic subunit activity in hippocampus following Withanolide A during hypoxia. Hypoxic exposure for 7 days decreases the hippocampal ATP, NADPH and γ-glutamylcysteinyl ligase catalytic subunit activity while administration of Withanolide A increases the ATP, NADPH and γ-glutamylcysteinyl ligase catalytic subunit activity in hippocampus. Data expressed as percentage change taking normoxic value as 100% and represents Mean ± SEM. 'a' denotes $p \leq 0.05$ vs. when compared to normoxic group and 'b' denotes $p \leq 0.05$ vs. when compared to 7 days hypoxic group treated with vehicle only.

activity $(F (3, 20) = 14.7, p \leq 0.05)$ in hippocampal region following exposure to hypobaric hypoxia compared to normoxic group. Admnistration of withanolide A 21 days prior to and during exposure to hypobaric hypoxia for 7 days significantly increased the ATP, NADPH and GCLC activity compared to hypoxic group treated with vehicle only as shown in Fig. 4.

Withanolide A modulates hippocampal corticosterone and its receptors during exposure to hypobaric hypoxia

Exposure to hypobaric hypoxia for 7 days significantly $(F (3, 20) = 26.5, p \leq 0.05)$ elevated the corticosterone level, glucocorticoid $F (3, 20) = 15.6, p \leq 0.05)$ and mineralocorticoid receptor $F (3, 20) = 21.4, p \leq 0.05)$ in hippocampus compared to normoxic group. Significant decrease in corticosterone level and glucocorticoid receptor in hippocampus was observed when administered with withanolide A 21 days prior to and during exposure to hypobaric hypoxia for 7 days compared to hypoxic group treated with vehicle only while no difference was noted in the mineralocorticoid receptor expression in hippocampus (Fig. S1 and S2).

Withanolide A provide neuroprotection during hypoxic exposure by modulating corticosterone level in hippocampus

Exposure to hypobaric hypoxia significantly $(F (4, 26) = 13.33, p \leq 0.05)$ increased the number of pycknotic cells in hippocampal region compared to normoxic group. Supplementation of withanolide A during exposure to hypobaric hypoxia significantly decreased the number of pycknotic cells in the CA3 region of hippocampus compared to hypoxic group treated with vehicle only. Administration of Withanolide A as well as metyrapone during exposure to hypobaric hypoxia significantly decreased the number of pycknotic neurons in the CA3 region of hippocampus compared to Hypoxic group treated with vehicle only as shown in Fig. S3.

Withanolide A restores hypobaric hypoxia induced memory impairment in rats

Exposure to hypobaric hypoxia significantly increased the latency as well as pathlength in Morris Water Maze during memory test compared to normoxic group. Supplementation of withanolide A during exposure to hypobaric hypoxia significantly decreased the

latency $(F (3, 56) = 9.13, p \leq 0.05)$ and pathlength $(F (3, 56) = 8.71, p \leq 0.05)$ during memory test compared to hypoxic group treated with vehicle only as shown in Fig. 5 i and ii. On the other hand, there was significant decrease in number of platform crossing $(F (3, 56) = 11.13, p \leq 0.05)$ as well as time spent in target quadrant during probe trial $(F (3, 56) = 8.13, p \leq 0.05)$ following exposure to hypobaric hypoxia compared to normoxic group. Administration of withanolide A 21 days prior to and during exposure to hypobaric hypoxia for 7 days significantly increased the number platform crossing and time spent in target quadrant during probe trial compared to hypoxic group trated with vehicle only (Fig. 5 iii and iv).

Withanolide A reverses neuronal apoptosis in CA3 region of hippocampus following hypoxic exposure and depletion of GSH

Exposure to hypobaric hypoxia significantly $(F (6, 36) = 11.33, p \leq 0.05)$ increased the number of apoptotic caspase 3 positive cells and hoescht positive cells $(F (6, 36) = 21.48, p \leq 0.05)$ in hippocampal region compared to normoxic group. Administration of buthionine sulfoximine during hypoxic exposure significantly increased the number of apoptotic cells and hoesct positive cells in hippocampus compared to hypoxic group treated with vehicle only. Though supplementation of withanolide A during exposure to hypobaric hypoxia significantly decreased the number of caspase positive cells as well as hoescht positive cells in the CA3 region of hippocampus compared to hypoxic group treated with vehicle only, combined administration of buthionine sulfoximine along with Withanolide A during hypoxic exposure significantly elevated the number of apoptotic cells in CA3 region of hippocampus compared to hypoxic group treated with Withanolide A only and vehicle treated group (Fig. 6 and 7 i and ii). Double labelling of apoptotic marker caspase 3 and neuronal marker NeuN further indicate the death of neurons in CA3 region hippocampus (Fig. S4).

Withanolide A maintains redox homeostasis by modulating glutathione biosynthesis in hippocampus through regulation of Nrf2 pathway and corticosterone signaling during hypoxic exposure

There was significant decrease in the expression of Nrf2 $(F (6, 22) = 17.36, p \leq 0.05)$ and γ-GCLC $(F (6, 22) = 21.13, p \leq 0.05)$ in

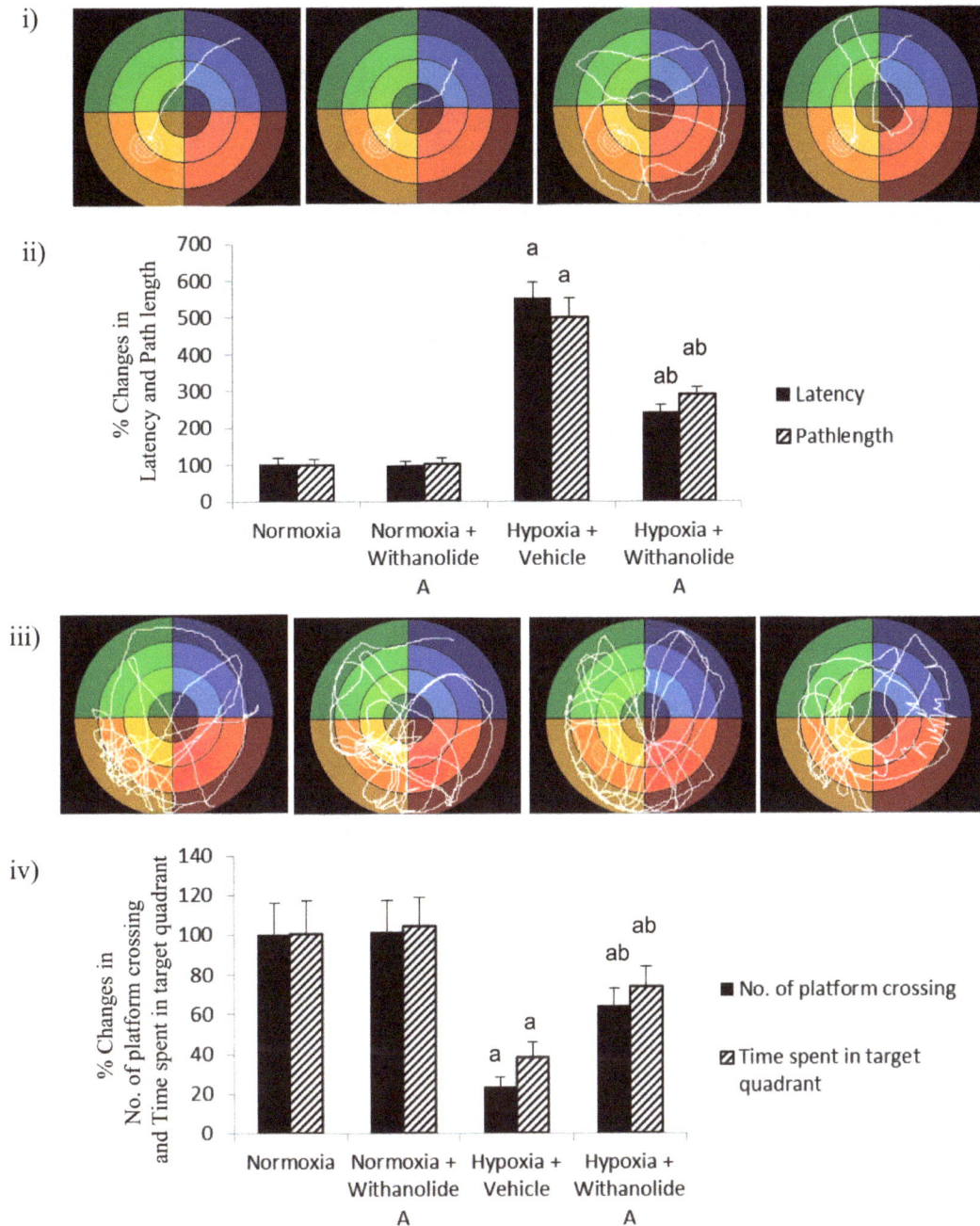

Figure 5. Amelioration of spatial memory function following Withanolide A administration during hypoxic exposure. Exposure to hypobaric hypoxia increased (i) path length and (ii) latency during spatial memory test but decreased (iii) number of platform crossing and (iv) time spent in target quadrant during probe trial when was reversed following withanolide A supplementation before and during exposure to hypobaric hypoxia. Data expressed as percentage change taking normoxic value as 100% and represents Mean ± SEM. 'a' denotes p≤0.05 vs. when compared to normoxic group and 'b' denotes p≤0.05 vs. when compared to 7 days hypoxic group treated with vehicle only.

hippocampus following exposure to hypobaric hypoxia for 7 days compared to normoxic group. Supplementation of withanolide A 21 days before and during exposure to hypoxic exposure significantly increased Nrf2 and γ-GCLC expression in hippocampus compared to hypoxic group treated with vehicle only. However, administration of corticosterone along with Withanolide Aduring exposure to hypobaric hypoxia significantly reduced the expression of Nrf2 as well as γ-GCLC in hippocampal region

compared to hypoxic group treated with withanolide A only. Further, administration of metyrapone during hypoxic exposure significantly increased the Nrf2 and γ-GCLC expression in hippocampus compared to withanolide A and corticosterone administered group. On the contrary, exogenous supplementation of corticosterone along with metyrapone significantly reduced the expression of Nrf2 and γ-GCLC in hippocampal region of brain

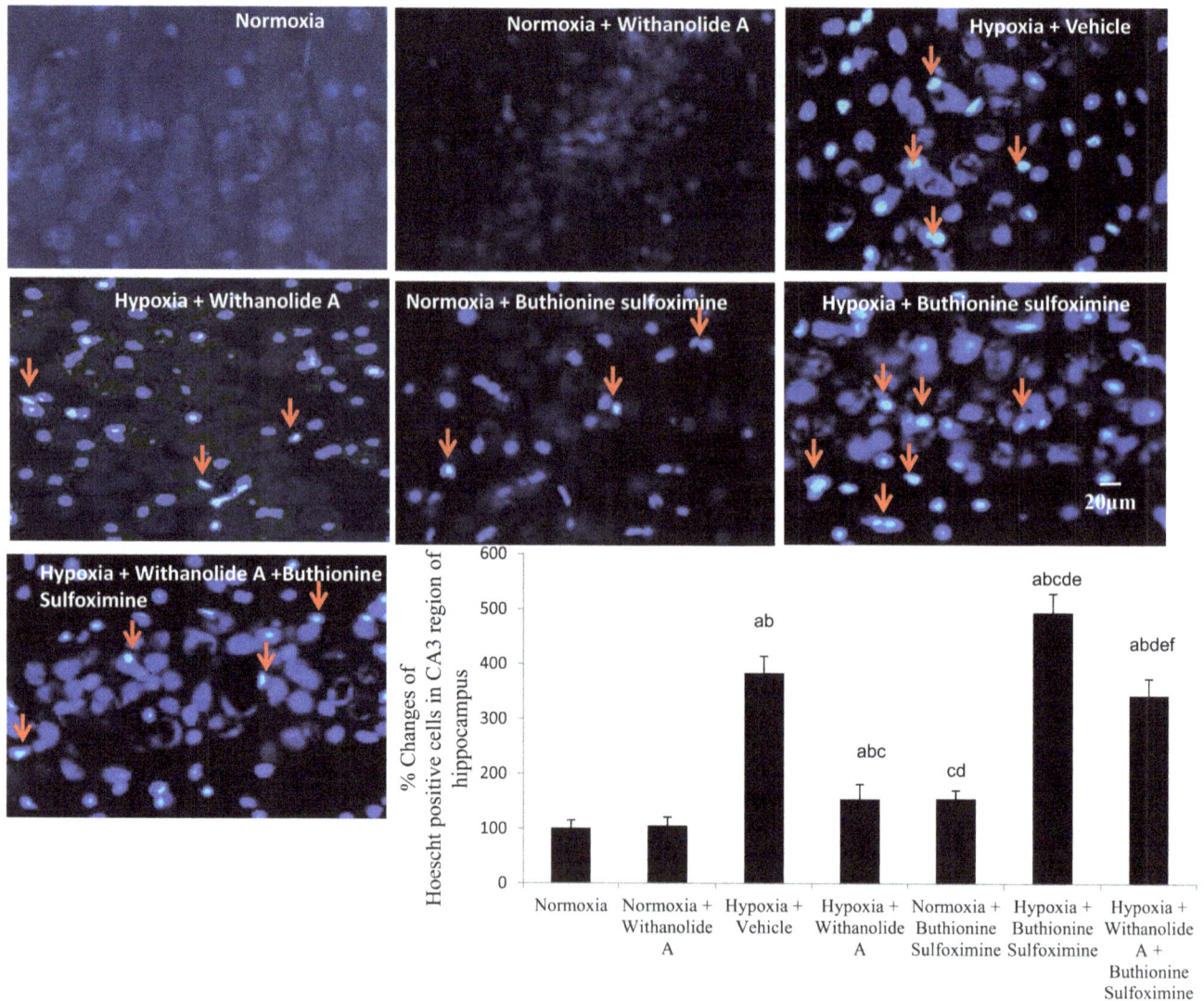

Figure 6. Modulation of hippocampal endogenous glutathione level by Withanolide A reduces hypoxia induced neurodegeneration. Withanolide A effectively decreases the number of degenerating neurons caused by hypoxic exposure. However, depletion of glutathione using buthionine sulfoximine during hypoxic exposure increases the number degenerating cells in hippocampus. Co-administration of withanolide A alongwith buthionine sulfoximine attenuates the neuroprotective effect of withanolide A during hypoxia. Data expressed as percentage change taking normoxic value as 100% and represents Mean ± SEM. 'a' denotes $p \leq 0.05$ vs. when compared to normoxic group. 'b' denotes $p \leq 0.05$ vs. when compared to normoxia + withanolide A group and 'c' denotes $p \leq 0.05$ vs. when compared to hypoxia + vehicle group, 'd' denotes $p \leq 0.05$ vs. when compared to hypobaric hypoxia + withanolide A, 'e' denotes $p \leq 0.05$ vs. when compared to normoxia + buthionine sulfoximine, and 'f' denotes $p \leq 0.05$ vs. when compared to hypobaric hypoxia + buthionine sulfoximine.

compared to only metyrapone treated group as shown in Fig. 8 i and ii.

Withanolide A mediated transcriptional regulation of Nrf2 and γ-GCLC during hypoxic exposure is corticosterone dependent

Exposure to hypobaric hypoxia for 7 days significantly decreased the Nrf2 (F $(6, 22) = 22.13$, $p \leq 0.05$) and γ-GCLC m-RNA level (F $(6, 22) = 17.43$, $p \leq 0.05$) in hippocampus compared to normoxic group. Supplementation of withanolide A 21 days prior to and during exposure to hypoxic exposure significantly upregulated Nrf2 and γ-GCLC expression in hippocampus compared to hypoxic group treated with vehicle only. However, administration of corticosterone along with withanolide A during exposure to hypobaric hypoxia significantly reduced the expres-

sion of Nrf2 as well as γ-GCLC m-RNA level in hippocampal region compared to only withanolide A treated hypoxic group. Further, administration of metyrapone during hypoxic exposure transcriptionally upregulted the Nrf2 and γ-GCLC expression in hippocampus significantly compared to withanolide A and corticosterone administered group. On the contrary, exogenous supplementation of corticosterone along with metyrapone significantly reduced the expression of Nrf2 and GCLC m-RNA level in hippocampal region of brain compared to only metyrapone treated group (Fig. 9).

Discussion

Hypobaric hypoxia induced prolonged elevation of corticosterone in hippocampus have been shown to cause enhanced oxidative stress, neurodegeneration and impairment of memory

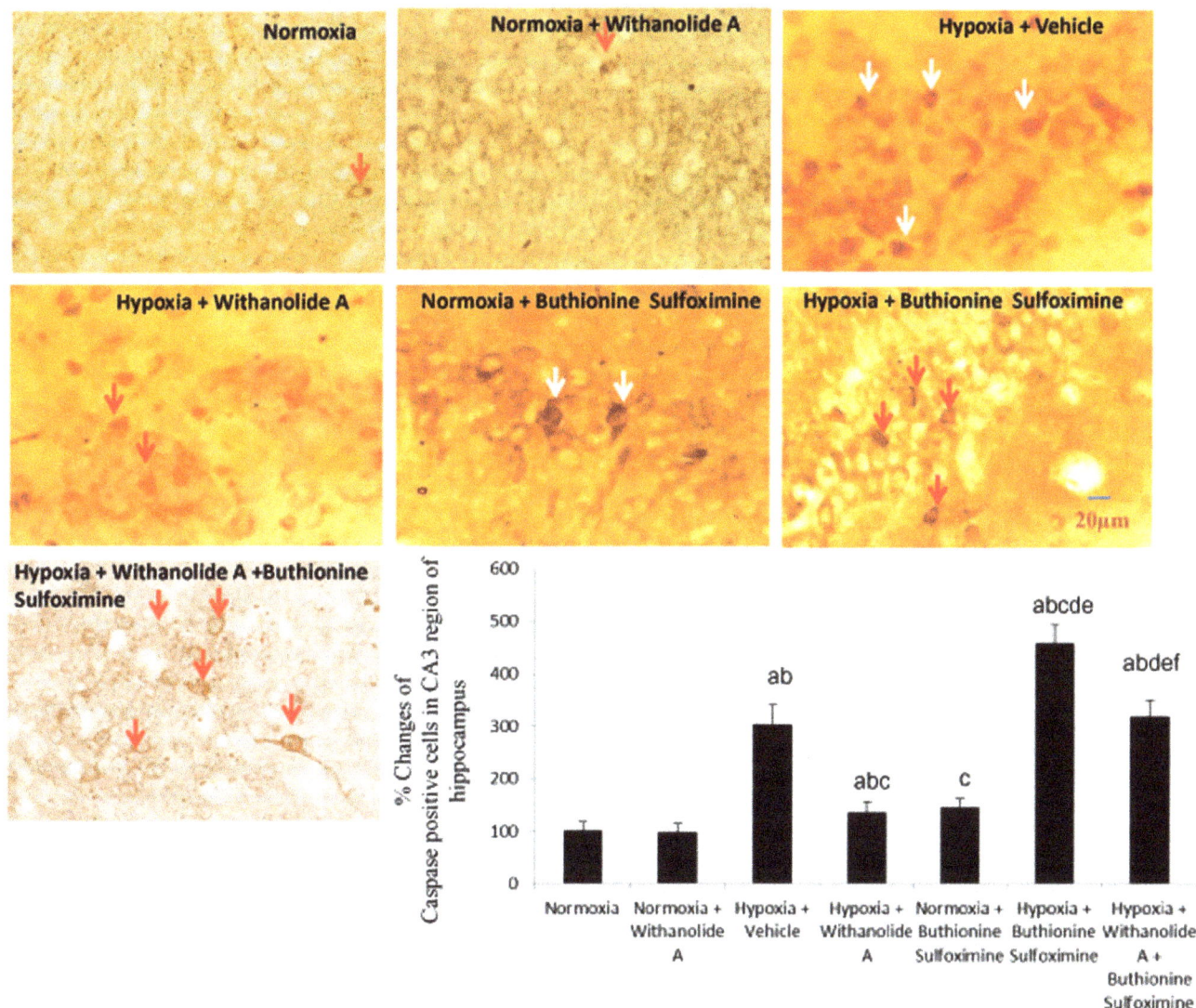

Figure 7. Withanolide A mediated restoration of endogenous glutathione level reverses hypoxia induced neuronal apoptosis in hippocampus. Administration of withanolide A prior to and during hypoxic exposure decreases number of apoptotic cells in CA3 region of hippocampus while glutathione depletion using buthionine sulfoximine elevates hypoxic neuronal apoptosis. Co-administration of withanolide A alongwith buthionine sulfoximine during hypoxic exposure enhances hypoxia induced apoptotic cells attenuating the nuroprotective effect of withanolide A. Data expressed as percentage change taking normoxic value as 100% and represents Mean ± SEM 'a' denotes p≤0.05 vs. when compared to normoxic group. 'b' denotes p≤0.05 vs. when compared to normoxia + withanolide A group and 'c' denotes p≤0.05 vs. when compared to hypoxia + vehicle group, 'd' denotes p≤0.05 vs. when compared to hypobaric hypoxia + withanolide A, 'e' denotes p≤0.05 vs. when compared to normoxia + buthionine sulfoximine, and 'f' denotes p≤0.05 vs. when compared to hypobaric hypoxia + buthionine sulfoximine.

consolidation and retrieval. Inhibition of corticosterone synthesis or blockade of glucocorticoid receptor during hypoxic exposure reduced neurodegeneration and ameliorated memory impairment [28,40]. However, use of synthetic inhibitors as prophylactic to prevent high altitude maladies may not be preferable because of their possible negative side effects. Recent studies from our laboratory demonstrate the prophylactic efficacy of withanolide enriched extract of *Withania somnifera* root in preventing hypoxia induced memory dysfunction by modulating corticosterone secretion and GSH level in hippocampal region of brain [17]. However, mechanism underlying such modulatory effect of withanolides on glutathione biosynthesis under hypoxic condition remains unexplored. Present investigation demonstrate that withanolide A causes augmented synthesis of glutathione in hippocampal neurons by upregulating key regulating enzyme for

glutathione biosynthesis γ-glutamyl cysteinyl ligase and Nrf2 and attenuate hypoxia induced hippocampal neurodegeneration. The study further indicate that glucocorticoid signaling play a pivotal role in modulating glutathione level in neuron by regulating expression of Nrf2 and GCLC during hypoxia.

Glutathione is the most abundant thiol-containing molecule and is crucial for neuroprotection in the brain which non-enzymatically reacts with superoxide [41], NO [42], ONOO− and hydroxyl radicals [43]. It is the major redox buffer that maintains intracellular redox homeostasis. Under conditions of oxidative stress, GSH can lead to reversible formation of mixed disulfides between protein thiol groups through S-glutathionylation, a process critical for preventing irreversible oxidation of proteins [44]. Thus, GSH modulates a variety of protein functions via S-glutathionylation. While cysteine itself has neurotoxic effects

i)

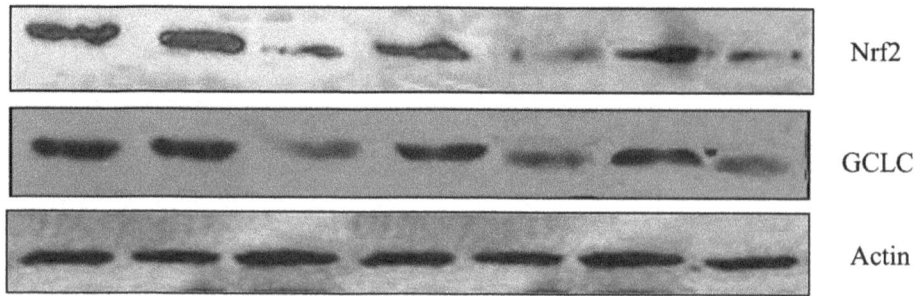

ii)

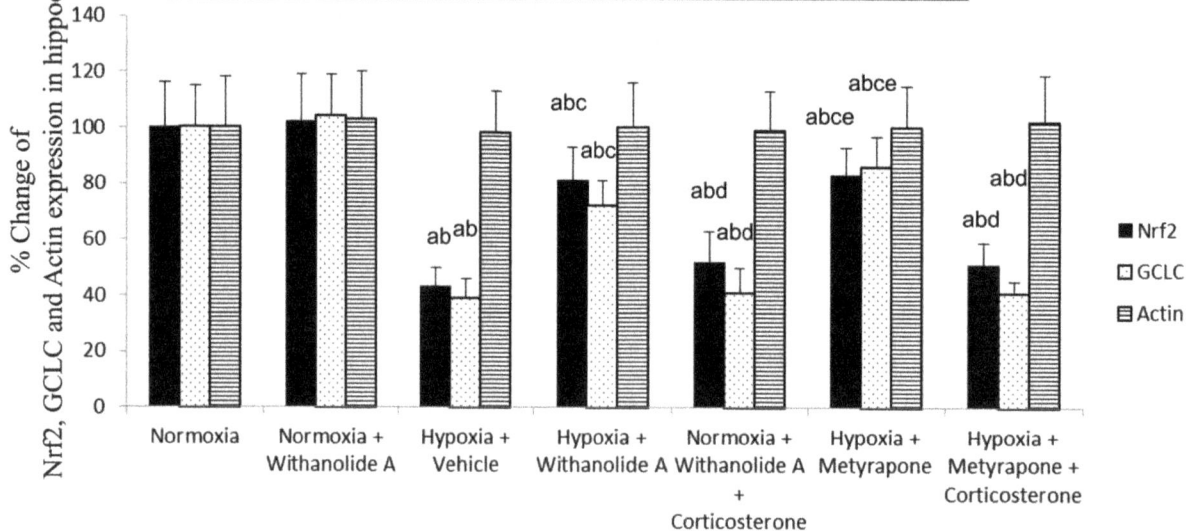

Figure 8. Withanolide A mediated elevation of hippocampal glutathione during hypoxia is corticosterone dependent. Withanolide A administration during hypoxic exposure upregulates Nrf2 and GCLC expression in hippocampus. Administration of Withanolide A alongwith exogenous corticosterone supplementation to the normoxic group decrease the Nrf2 as well as GCLC expression while inhibition of corticosterone synthesis using metyrapone reverses hypoxia induced downregulation of both Nrf2 and GCLC level. β-actin was used as a loading control. Data expressed as percentage change taking normoxic value as 100% and represents Mean ± SEM. 'a' denotes $p \leq 0.05$ vs. when compared to normoxic group. 'b' denotes $p \leq 0.05$ vs. when compared to normoxia + withanolide A group and 'c' denotes $p \leq 0.05$ vs. when compared to hypoxia + vehicle group, 'd' denotes $p \leq 0.05$ vs. when compared to hypobaric hypoxia + withanolide A, 'e' denotes $p \leq 0.05$ vs. when compared to hypobaric hypoxia + withanolide A + corticosterone and 'f' denotes $p \leq 0.05$ vs. when compared to hypobaric hypoxia + Metyrapone.

mediated by free radical generation, increasing extracellular glutamate and triggering overactivation of N-methyl-D-aspartate (NMDA) receptors [45], GSH is a non-toxic cysteine storage form with 10–100 times higher concentrations in mammalian tissues than cysteine [46]. Further, GSH can serve as a neuromodulator that bind to NMDA receptor via its γ-glutamyl moiety and is known to exert dual (agonistic/antagonistic) actions on neuronal responses [45]. Keeping in mind the neuroprotective effects exerted by GSH in neuronal system, it is expected that molecules modulating GSH synthesis could be of potent therapeutics importance to cure neurodegenerative disorders.

Oxygen scarcity causes impairment of electron transport chain in mitochondria owing to its pivotal role as electron sink. Incomplete reduction of oxygen in hypoxic condition results in elevated production of superoxide and hydroxyl radicals. Exposure to hypobaric hypoxia induces oxidative stress in brain [5,28]. Corroborating with previous findings, present study document an elevated level of free radicals and consequent lipid peroxidation following exposure to hypobaric hypoxia for 7 days along with a reduced level of endogenous antioxidant glutathione in hippocampus. The observed decrease in glutamylcysteinyl ligase activity, the key regulatory enzyme for gluatathione biosynthesis further

support the decreased level of glutathione in hippocampus under hypoxic condition. Administration of withanolide A before and during exposure to hypobaric hypoxia decreased the free radical level which further diminished the incidence of lipid peroxidation. Similar reports on several other stresses like diabetes and chronic food shock showed that *Withania somnifera* root extract administration during stress exposure increase GSH level, reduce reactive oxygen species generation and lipid peroxidation [47–48]. The anti-oxidative effect of *Withania somnifera* root extract could be attributed to the rich content of withanolides, flavonoids and other components with strong antioxidant potential [49]. Interestingly, withanolide A induced augmentation of endogenous antioxidant GSH level in hippocampus point towards the efficacy of plant components in modulating glutathione biosynthesis or its stabilization under hypoxic condition [50]. Increased GCLC activity following supplementation of withanolide A support the involvement of withanolide A in modulating glutathione biosynthesis under hypoxic condition. Similar modulation of glutathione synthesis and upregulation of GCLC activity by several molecules like adrenomedullin, flavanoids like butein and phloretin following exposure to stressors causing oxidative load have been shown to provide augmented neuroprotection [51–52].

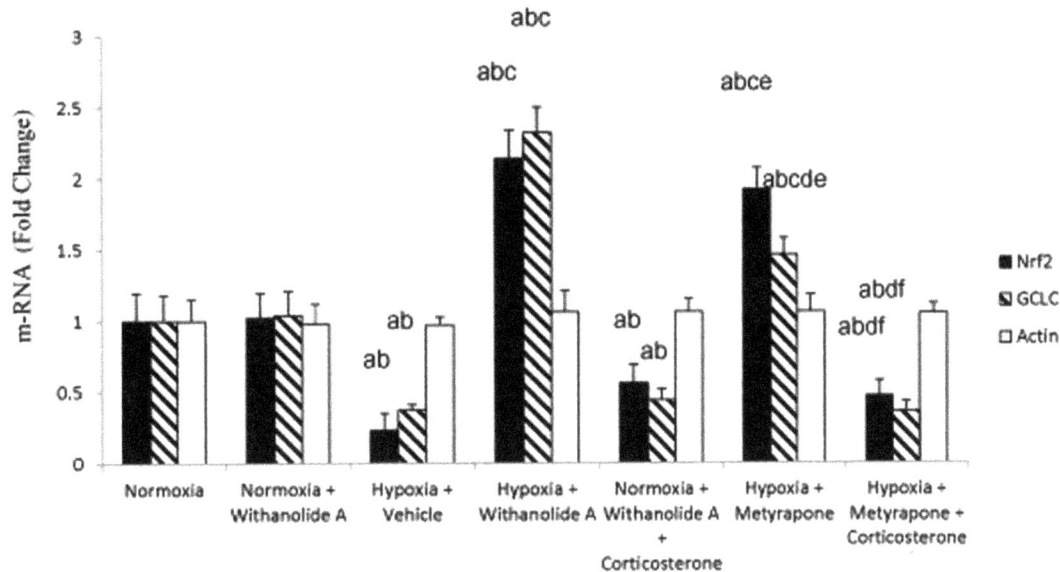

Figure 9. Withanolide A mediated transcriptional regulation of Nrf2 and GCLC expression depend on corticosterone signaling in hippocampus. Withanolide A administration during hypoxic exposure upregulates Nrf2 and GCLC expression in hippocampus. Administration of Withanolide A alongwith exogenous corticosterone supplementation to the normoxic group decrease the Nrf2 as well as GCLC expression while inhibition of corticosterone synthesis using metyrapone during hypoxia reverses hypoxia induced downregulation of both Nrf2 and GCLC level. β-actin was used as a loading control. Data expressed as percentage change taking normoxic value as 100% and represents Mean $\pm$ SEM. 'a' denotes $p \leq 0.05$ vs. when compared to normoxic group. 'b' denotes $p \leq 0.05$ vs. when compared to normoxia + withanolide A group and 'c' denotes $p \leq 0.05$ vs. when compared to hypoxia + vehicle group, 'd' denotes $p \leq 0.05$ vs. when compared to hypobaric hypoxia + withanolide A, 'e' denotes $p \leq 0.05$ vs. when compared to hypobaric hypoxia + withanolide A + corticosterone and 'f' denotes $p \leq 0.05$ vs. when compared to hypobaric hypoxia + metyrapone.

The brain contains high level of glutathione peroxidase which requires GSH for reduction of H_2O_2 and other peroxides [53,4]. In the present study, decreased glutathione reductase activity with concomittant increase in activity of glutathione peroxidase following exposure to hypobaric hypoxia causes elevated accumulation of oxidized glutathione in hippocampus. Regeneration of oxidized form of glutathione to its reduced form depends on availability of NADPH and ATP in neuronal system. On the other hand, exposure to hypobaric hypoxia decreased the ATP and NADPH level in hippocampus which could be a result of impaired mitochondrial functioning on account of reduced availability of oxygen and malfunctioning of the pentose phosphate pathway that generate NADPH and/or overutilization of NADPH under hypoxic condition [26]. Winterbourn et al.., (1994) reported similar decreased level of ATP and NADPH under ischemic and hypoxic condition in brain and other tissues [41,28]. Since synthesis of GSH is a ATP dependent process, decreased ATP level in hippocampus during hypoxic exposure could substantially influence the GSH level in hippocampus. Thus, it could be inferred that compromised glutathione biosynthesis and reduced availalbility of ATP and NADPH in hippocampal neuronal cells following exposure to hypobaric hypoxia impairs regeneration of glutathione from its oxidized state. Restoration of ATP and NADPH level, GCLC activity and augmented glutathione reductase activity following administration of withanolide A indicate the modulatory effect of withanolide A on glutathione biosynthesis and maintenance of neuronal redox potential in hippocampus.

Present study showed decreased activity of glutathione s transferase and superoxide dismutase activity in hippocampus following exposure to hypobaric hypoxia which was restored on supplementation of withanolide A. Elevation in glutathione s transferase and superoxide dismutase activity under hypoxic condition further supports the efficacy of withanolide A in modulating the free radicals scavenging enzyme system by maintaining neuronal GSH level. Similar elevated glutathione s transferase and superoxide dismutase activity following administration of *Withania somnifera* root extract and decrease oxidative stress induced neurodegeneration have been observed in restraint, ischemia and other moderate stress model [15,54].

Nitric oxide (NO) being a diffusible retrograde neurotransmitter and signaling molecule exerts multipronged effect on neuronal survivability through formation of peroxynitrite [55] and activation of cGMP cascade [56]. Previous report from our laboratory demonstrate that exposure to hypobaric hypoxia elevate nitric oxide level in hippocampus [5]. Nitric oxide can directly stimulate augmented synthesis and secretion of corticosterone by NO-COX-Prostaglandin pathway and enhance hypoxia induced neurodegeneration [17]. Since GSH serves as an endogenous NO reservoir by forming S-nitrosoglutathione (GSNO) [57], withanolide A mediated elevated synthesis of GSH can modulate the nitric oxide level in hippocampus during hypoxic exposure through formation of nitrosoglutathione, thereby attenuating the toxic effect nitric oxide. Studies showed similar protective effect of GSNO in the brain under oxidative stress conditions by regulating NO release and exerts different biological effects [58]. However, further studies are needed to unfold effect of GSNO in high altitude hypoxic condition.

Reduced glutathione regulates both apoptotic and necrotic cell death by modulating the expression/activity of caspases and other signaling molecules [59]. Depletion of GSH during stress enhance oxidative insult and causes neurodegeneration. Maiti et al.., reported that loss of memory following exposure to hypobaric hypoxia occurs due to neuronal apoptosis in hippocampal region

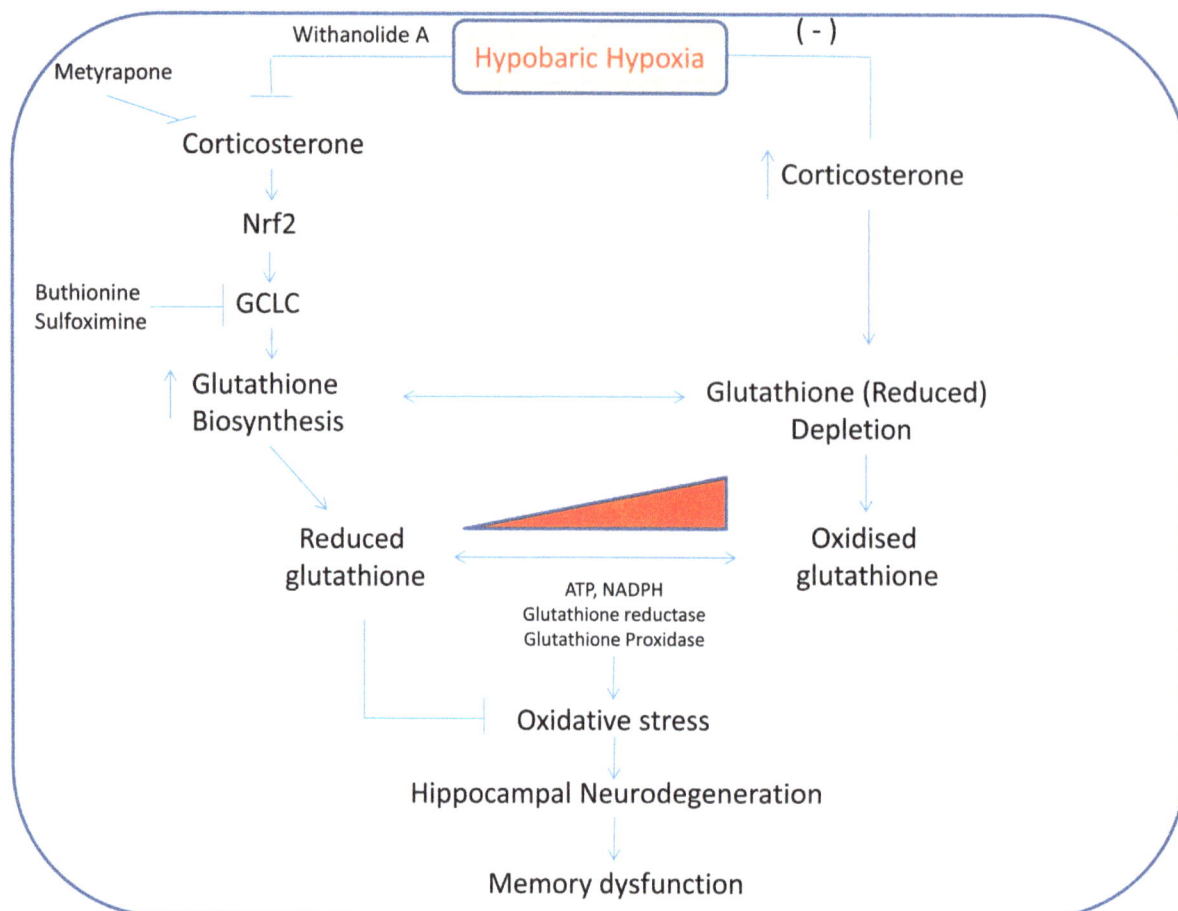

Figure 10. Schematic diagram showing Withanolide A mediated modulation of glutathione biosynthesis and neuroprotection during hypobaric hypoxia.

of the brain [60]. Present study showed that inhibition of key regulatory enzyme for glutathione biosynthesis GCLC using buthionine sulfoximine (BSO) during exposure to hypobaric hypoxia enhanced the hypoxia induced neurodegeneration as evident from elevated caspase positive and hoescht positive cells in the CA3 region of hippocampus. Similar studies showed that reduction of the brain glutathione content by buthionine sulfoximine (BSO), a specific inhibitor of GCLC enhances the toxic effects associated with elevated production of reactive oxygen species under ischemic condition [61–62] or treatment with 6-hydroxydopamine [63]. Decreased hoescht and caspase positive cells in CA3 region of hippocampus when administered with withanolide A indicate its neuroprotective effect under hypoxic condition. Interestingly, administration of buthionine sulfoximine along with withanolide A blunted its neuroprotective effect indicating the importance of glutathione biosynthesis in With-anolide A mediated neuroprotection. Similar studies also showed that GSH depletion in brain by treatment with buthionine sulfoximine leads to increased production of superoxide, hydroxyl radicals and H_2O_2 [64]. Decreased intracellular GSH on buthionine sulfoximine treatment worsen oxidative damage in hippocampus, while increased intracellular GSH by N-acetylcysteine (NAC) treatment ameliorated this damage [65,9]. Thus, the study supports the fact that intracellular GSH pool is important for limiting oxidative stress induced neuronal injury.

The key regulatory enzyme for glutathione biosynthesis γ-glutamylcysteinyl ligase is regulated by various transcription factors and environmental stimuli. Nuclear factor (erythroid-derived 2)-like 2 (Nrf2) is one of the major transcription factor that maintain the redox homeostasis in neuronal system. In the present study, prolonged exposure to hypobaric hypoxia down regulates the antioxidant regulatory transcription factor Nrf2 in the brain. Nrf2 in turn can down regulate the key regulatory enzyme GCLC for glutathione biosynthesis. In the present study, supplementation of withanolide A reversed hypoxia induced down regulation of Nrf2 in the hippocampal region of the brain with concomitant increased expression of GCLC. On the other hand, supplementation of corticosterone during exposure to hypobaric hypoxia down regulated the Nrf2 and GCLC in hippocampus. Interestingly, inhibition of corticosterone synthesis during hypoxic exposure using metyrapone upregulated Nrf2 as well as GCLC in hippocampus suggesting the regulatory role of corticosterone in Nrf2 mediated induction of GCLC expression. Further, administration of withanolide A along with exogenous supplementation of corticosterone blunted upregulation of Nrf2 and GCLC in hippocampal region indicating that Withanolide A upregulate Nrf2 and GCLC by decreasing the corticosterone level in hippocampus. Baitharu et al. showed that administration of withanolide A enriched extract of *withania somnifera* root extract during hypoxic exposure decreased the corticosterone level as well as the glucocorticoid receptor expression in hippocampus [66].

Supporting the present findings, similar studies by Kratschmar et al. (2012) demonstrate that glucocorticoids suppress cellular antioxidant defence capacity by impairing Nrf2-dependent antioxidant response [67]. Furthermore, combined administration metyrapone and corticosterone nullified the upregulation of Nrf2 as well as GCLC confirming the role of corticosterone or its receptor in regulating Nrf2 expression under hypoxic condition. While exposure to hypobaric hypoxia increased corticosterone level in hippocampus, activation of glucocorticoid receptors by corticosterone can supress Nrf2 expression resulting in decreased expression of GCLC. Supporting our findings, studies by Ki et al.. showed that activated glucocorticoid receptor modulates Nrf2 signaling and alters of Nrf2 target genes expression in brain through binding of glucocorticoid receptor to its glucocorticoid response element [68]. However, exact mechanisms involved in regulation of Nrf2 by glucocorticoid receptor in hypoxia need further investigation.

Conclusion

Present study demonstrate that in addition to its strong antioxidant property, withanolide A provide augmented neuroprotection by modulation of endogenous glutathione level in hippocampus during exposure to hypobaric hypoxia. Withanolide A increases glutathione biosynthesis in neuronal cells by upregulating GCLC level through Nrf2 pathway in a corticosterone dependenet manner (Figure 10). Since exogenous supplementation of GSH is not effective, modulation of glutathione biosynthesis by withanolide A could be of much therapeutic interest and can be used as a prophylactic to prevent/cure neurodegenerative disorders invoked by elevated oxidative insults in hypobaric hypoxia and other similar pathological condition.

Supporting Information

Exposure to hypobaric hypoxia elevates corticosterone level in hippocampal tissue while withanolide A administration attenuate cortcosterone elevation and maintain it in optimal level as shown Fig. S1. Withanolide A administration during hypoxic exposure decreases glucocorticoid receptor and increases mineralocorticoid receptor expression in hippocampus causing a receptor balance suitable for neuroprotection (Fig. S2). Persistent elevated corticosterone induces increased level of pycknosis in hippocampus while modulation the corticosterone level in hippocampus by corticosterone synthesis inhibitor metyrapone and withanoide A decreases the hypoxia induced elevated numbers of pycknotic cells (Fig. S3). Increased co-labelling of neuronal marker Neu N with apoptotic marker caspase 3 further indicate that the nature of cells undergoing apoptosis are neurons of hippocampal region (Fig. S4). Data expressed as percentage change taking normoxic value as 100% and represents Mean ± SEM. 'a' denotes p≤0.05 vs. when compared to normoxic group and 'b' denotes p≤0.05 vs. when compared to 7 days hypoxic group treated with vehicle only. This finding shows that elevated corticosterone or its downstream signaling play pivotal role in neuronal survivability and memory

functions in hypobaric hypoxic condition. Thus modulation of corticosterone could provide therapeutic strategy to reverse hypoxia induced physiological and pathological disorders.

Supporting Information

Figure S1 Withanolide A modulates corticosterone level in hippocampus during hypoxic exposure. Prolonged exposure to hypobaric hypoxia elevates hippocampal corticosterone level. Administration of withanolide A decreases the level of hippocampal corticosterone just above the normoxic level optimum for its protective effect. Data expressed as percentage change taking normoxic value as 100% and represents Mean ± SEM. 'a' denotes p≤0.05 vs. when compared to normoxic group and 'b' denotes p≤0.05 vs. when compared to 7 days hypoxic group treated with vehicle only.

Figure S2 Withanolide A modulates glucocorticoid and mineralocorticoid receptor expression in hippocampus during hypoxia. Withanolide A administration during hypoxic exposure decreases glucocorticoid receptor and increases mineralocorticoid receptor expression in hippocampus causing a receptor balance suitable for neuroprotection. Data expressed as percentage change taking normoxic value as 100% and represents Mean ± SEM. 'a' denotes p≤0.05 vs. when compared to normoxic group and 'b' denotes p≤0.05 vs. when compared to 7 days hypoxic group treated with vehicle only.

Figure S3 Optimal maintainance of corticosterone level using metyrapone and withanolide A during hypoxia provide neuroprotection in hippocampus. Withanolide A modulate the corticosterone level in hippocampus and decreases the hypoxia induced elevated pycknotic cells comparable to metyrapone. Data expressed as percentage change taking normoxic value as 100% and represents Mean ± SEM. 'a' denotes p≤0.05 vs. when compared to normoxic group and 'b' denotes p≤0.05 vs. when compared to 7 days hypoxic group treated with vehicle only.

Figure S4 Representative slides showing the double labelled neuronal cells with apoptotic marker caspase 3 and neuronal marker Neu N in the CA3 region of the hippocampus. Double labelled cells indicates the apoptotic neuronal cells.

Author Contributions

Conceived and designed the experiments: GI IB. Performed the experiments: IB SND VJ JKS SS. Analyzed the data: GI IB PKN. Contributed reagents/materials/analysis tools: SND IB GI VJ JKS SS. Wrote the paper: GI IB PKN.

References

1. Bahrke M, Hale BS (1993) Effect of altitude on mood, behavior and cognitive functioning. Sports Med 16: 97–125.
2. Baitharu I, Jain V, Deep SN, Sahu JK, Naik PK, et al. (2013) Exposure to hypobaric hypoxia and reoxygenation induces transient anxiety like behaviour in rat. J Behav Brain Sci 3: 591–602.
3. Won SJ, Kim DY, Gwag BJ (2002) Cellular and molecular pathways of ischemic neuronal death. J Biochem Mol Biol 35: 67–86.
4. Barker JE, Heales SJR, Cassidy A, Bolanos JP, Land JM, et al. (1996) Depletion of brain glutathione results in a decrease in glutathione reductase activity, an enzyme susceptible to oxidative damage. Brain Res 716: 118–122.

5. Maiti P, Singh SB, Sharma AK, Muthuraju S, Banerjee PK, et al. (2006) Hypobaric hypoxia induces oxidative stress in rat brain. Neurochem International 49: 709–716.
6. Moller P, Loft S, Lundby C, Olsen NV (2001) Acute hypoxia and hypoxic exercise induced DNA strand breaks and oxidative DNA damage in humans. FASEB J 15: 1181–1186.
7. Chandel NS, Maltepe E, Goldwasser E, Mathieu CE, Simon MC, et al. (1998) Mitochondrial reactive oxygen species trigger hypoxia-induced transcription. Proc Natl Acad Sci U S A. 95: 11715–11720.

8. Hota SK, Barhwal K, Singh SB, Ilavazhagan G (2007) Differential temporal response of hippocampus, cortex and cerebellum to hypobaric hypoxia: a biochemical approach. Neurochem International 51: 384–390.

9. Jayalakshmi K, Singh SB, Kalpana B, Sairam M, Muthuraju S, et al. (2007) N-acetyl cysteine supplementation prevents impairment of spatial working memory functions in rats following exposure to hypobaric hypoxia. Physiol Behav 92: 643–650.

10. Prasad J, Baitharu I, Sharma AK, Dutta R, Prasad D, et al. (2013) Quercetin reverses hypobaric hypoxia-induced hippocampal neurodegeneration and improves memory function in the rat. High Altitude Med Biol 14: 383–394.

11. Dringen R (2000) Glutathione metabolism and oxidative stress in neurodegeneration. European J Biochem 267: 4903.

12. Richman PG, Meister A (1975) Regulation of gamma-glutamylcysteine synthetase by nonallosteric feedback inhibition by glutathione. J Biol Chem 250: 1422–1426.

13. Nguyen T, Nioi P, Pickett CB (2009) The Nrf2-antioxidant response element signaling pathway and its activation by oxidative stress. J Biol Chem 284: 13291–13295.

14. Kensler TW, Wakabayashi N, Biswal S (2007) Cell survival responses to environmental stresses via the Keap1–Nrf2–ARE pathway. Ann Rev Pharmacol and Toxicol 47: 89–116.

15. Hussain S, Slikker W Jr, Ali SF (1996) Role of metallothionein and other antioxidants in scavenging superoxide radicals and their possible role in neuroprotection. Neurochem International 29: 145–152.

16. Shih AY, Imbeault S, Barakauskas V, Erb H, Jiang L, et al. (2005) Induction of the Nrf2-driven antioxidant response confers neuroprotection during mitochondrial stress in vivo. J Biol Chem 280: 22925–22936.

17. Baitharu I, Jain V, Deep SN, Hota KB, Hota SK, et al. (2013) *Withania somnifera* root extract ameliorates hypobaric hypoxia induced memory impairment in rats. J Ethnopharmacol 145: 431–441.

18. Cornford EM, Braun LD, Crane PD, Oldendorf WH (1978) Blood-brain barrier restriction of peptides and the low uptake of enkephalins. Endocrinol 103: 1297–1303.

19. Kannan R, Yi JR, Tang D, Li Y, Zlokovic BV, et al. (1996) Evidence for the existence of a sodium-dependent glutathione (GSH) transporter. Expression of bovine brain capillary mRNA and size fractions in Xenopus laevis oocytes and dissociation from gamma-glutamyltranspeptidase and facilitative GSH transporters. J Biol Chem 271: 9754–9758.

20. Dhuley JN (1998) Effect of ashwagandha on lipid peroxidation in stress-induced animals. J Ethnopharmacol 60: 173–178.

21. Naidu PS, Singh A, Kulkarni SK (2006) Effect of *Withania somnifera* root extract on reserpine-induced orofacial dyskinesia and cognitive dysfunction. Phytotherapy Res 220: 140–146.

22. Schliebs R, Liebmann A, Bhattacharya SK, Kumar A, Ghosal S, et al. (1997) Systemic administration of defined extracts from *Withania somnifera* (Indian Ginseng) and Shilajit differentially affects cholinergic but not glutamatergic and GABAergic markers in rat brain. Neurochem International 30: 181–190.

23. Tohda C, Kuboyama T, Komatsu K (2000) Dendrite extension by methanol extract of Ashwagandha (roots of *Withania somnifera*) in SK-N-SH cells. Neuroreport. 11: 1981–1985.

24. Kuboyama T, Tohda C, Zhao J, Nakamura N, Hattori M, et al. (2002) Axon- or dendrite-predominant outgrowth induced by constituents from Ashwagandha. Neuroreport 13: 1715–20.

25. Hota SK, Barhwal K, Baitharu I, Prasad D, Singh SB, et al. (2009) Bacopa monniera leaf extract ameliorates hypobaric hypoxia induced spatial memory impairment. Neurobiol Dis 34: 23–39.

26. Barhwal K, Hota SK, Baitharu I, Prasad D, Singh SB, et al. (2009) Isradipine antagonizes hypobaric hypoxia induced CA1 damage and memory impairment: complementary roles of L-type calcium channel and NMDA receptors. Neurobiol Dis 34: 230–244.

27. Ando D, Yamakita M, Kuriyama M, Yamagata Z, Koyama K (2009) Effects of glutathione depletion on hypoxia-induced erythropoietin production in rats. J Physiol Anthropol 28: 211–215.

28. Baitharu I, Deep SN, Jain V, Barhwal K, Malhotra AS, et al. (2011) Corticosterone synthesis inhibitor metyrapone ameliorates chronic hypobaric hypoxia induced memory impairment in rat. Behav Brain Res 228: 53–65.

29. Smith-Swintosky VL, Pettigrew LC, Sapolsky RM, Phares C, Craddock SD, et al. (1996) Metyrapone, an inhibitor of glucocorticoid production, reduces brain injury induced by focal and global ischemia and seizures. J Cerebral Blood Flow and Metabol 16: 585–598.

30. Bradford MM (1976) A rapid and sensitive method for the quantitation of microgram quantities of protein utilizing the principle of protein-dye binding. Anal Biochem 72: 248–254.

31. LeBel CP, Ali SF, McKee M, Bondy SC (1990) Organometal-induced increases in oxygen reactive species: the potential of 2,7-dichlorofluorescin diacetate as an index of neurotoxic damage. Toxicol Applied Pharmacol 104: 17–24.

32. Myhre O, Andersen JM, Aarnes H, Fonnum F (2003) Evaluation of the probes 2′, 7′-dichlorofluorescin diacetate, luminol, and lucigenin as indicators of reactive species formation. Biochem Pharmacol 65: 1575–1582.

33. Das NP, Ratty AK (1987) Studies on the effects of the narcotic alkaloids, cocaine, morphine, and codeine on nonenzymatic lipid peroxidation in rat brain mitochondria. Biochem Med Metab Biol 37: 258–264.

34. Colado MI, OShea E, Granaburn litvakdos R, Murray TK, Green AR (1997) In vivo evidence for free radical involvement in the degeneration of rat brain 5-HT following administration of MDMA and p-chloroamphetamine but not the degeneration following fenfluramine. Br J Pharmacol 121: 889–900.

35. Hissin PJ, Hilf R (1976) A fluorimetric method for determination of oxidized and GSH in tissue. Anal Biochem 74: 214–226.

36. Pinto RE, Bartley W (1969) The effect of age and sex on glutathione reductase and glutathione peroxidase activities and on aerobic glutathione oxidation in rat liver homogenates. Biochem J 112: 109–115.

37. Mishra A, Roy KP (2006) 2-OHE2-induced oocyte maturation involves steroidogenesis in cat fish. J Endocrinol 189: 341–353.

38. Seelig GF, Meister A (1984) γ-Glutamylcysteine synthetase. Interactions of an essential sulfhydryl group. J Biol Chem 259: 3534–3538.

39. Morris RGM (1984) Development of a water maze procedure for studying spatial learning the rat. J Neurosci Method 11: 47–60.

40. Baitharu I, Deep SN, Jain V, Prasad D, Ilavazhagan G (2013) Inhibition of glucocorticoid receptors ameliorates hypobaric hypoxia induced memory impairment in rat. Behav Brain Res 240: 76–86.

41. Winterbourn CC, Metodiewa D (1994) The reaction of superoxide with GSH. Arch Biochem Biophysics 314: 284–290.

42. Clancy RM, Levartovsky D, Leszczynska-Piziak J, Yegudin J, Abramson SB (1994) Nitric oxide reacts with intracellular glutathione and activates the hexose monophosphate shunt in human neutrophils: evidence for S-nitrosoglutathione as a bioactive intermediary. Proc Natl Acad Sci U S A 91: 3680–3684.

43. Bains JS, Shaw CA (1997) Neurodegenerative disorders in humans: the role of glutathione in oxidative stress-mediated neuronal death. Brain Res Rev 25: 335–358.

44. Giustarini D, Rossi R, Milzan A, Colombo R, Dalle-Donne I (2004) S-glutathionylation: from redox regulation of protein functions to human diseases. J Cell Mol Med 8: 201–212.

45. Janaky R, Ogita K, Pasqualotto BA, Bains JS, Oja SS, et al. (1999) Glutathione and signal transduction in the mammalian CNS. J Neurochem 73: 889–902.

46. Cooper AJ, Kristal BS (1997) Multiple roles of glutathione in the central nervous system. Biol Chem 378: 793–802.

47. Bhattacharya A, Ghosal S, Bhattacharya SK (2001) Anti-oxidant effect of Withania somnifera glycowithanolides in chronic footshock stress-induced perturbations of oxidative free radical scavenging enzymes and lipid peroxidation in rat frontal cortex and striatum. J Ethnopharmacol 74: 1–6.

48. Anwer T, Sharma M, Pillai KK, Khan G (2012) Protective effect of *Withania somnifera* against oxidative stress and pancreatic beta-cell damage in type 2 diabetic rats. Acta Polon Pharmacol 69: 1095–1101.

49. Parihar MS, Hemnani T (2003) Phenolic antioxidants attenuate hippocampal neuronal cell damage against kainic acid induced excitotoxicity. J Biosci 28: 121–128.

50. Gupta A, Gupta A, Datta M, Shukla GS (2000) Cerebral antioxidant status and free radical generation following glutathione depletion and subsequent recovery. Mol Cel Biochem 209: 55–61.

51. Kim JY, Yim JH, Cho JH, Kim JH, Ko JH, et al. (2006) Adrenomedullin regulates cellular glutathione content via modulation of gamma-glutamate-cysteine ligase catalytic subunit expression. Endocrinol 147: 1357–1364.

52. Yang YC, Lii CK, Lin AH, Yeh YW, Yao HT, et al. (2011) Induction of glutathione synthesis and heme oxygenase 1 by the flavonoids butein and phloretin is mediated through the ERK/Nrf2 pathway and protects against oxidative stress. Free Rad Biol Med 51: 2073–2081.

53. Blum J, Fridovich I (1985) Inactivation of glutathione peroxidase by superoxide radical. Arch Biochem Biophysics 240: 500–508.

54. Sharma R, Yang Y, Sharma A, Awasthi S, Awasthi YC (2004) Antioxidant role of glutathione S-transferases: protection against oxidant toxicity and regulation of stress-mediated apoptosis. Antioxidant Red Signal 6: 289–300.

55. Beckman JS, Beckman TW, Chen J, Marshall PA, Freeman BA (1990) Apparent hydroxyl radical production by peroxynitrite: implications for endothelial injury from nitric oxide and superoxide. Proc Natl Acad Sci USA 87: 1620–1624.

56. Miki N, Kawabe Y, Kuriyama K (1977) Activation of cerebral guanylate cyclase by nitric oxide. Biochem Biophysical Res Comm 75: 851–856.

57. Singh RJ, Hogg N, Joseph J, Kalyanaraman B (1996) Mechanism of nitric oxide release from S-nitrosothiols. J Biol Chem 271: 18596–18603.

58. Rauhala P, Lin AM, Chiueh CC (1998) Neuroprotection by S-nitrosoglutathione of brain dopamine neurons from oxidative stress. FASEB J 12: 165–173.

59. Garcia-Ruiz JC, Fernández-Checa C (2007) Redox regulation of hepatocyte apoptosis. J Gastroenterol Hepatol 22: S38–S42.

60. Maiti P, Singh SB, Mallick B, Muthuraju S, Ilavazhagan G (2008) High altitude memory impairment is due to neuronal apoptosis in hippocampus, cortex and striatum. J Chem Neuroanatom 36: 227–238.la

61. Mizui T, Kinouchi H, Chan PH (1992) Depletion of brain glutathione by buthionine sulfoximine enhances cerebral ischemic injury in rats. Am J Physiol 262: H313–317.

62. Wuellner U, Seyfried J, Groscurth P, Beinroth S, Winter S, et al. (1999) Glutathione depletion and neuronal cell death: the role of reactive oxygen intermediates and mitochondrial function. Brain Res 826: 53–62.

63. Pileblad E, Magnusson T, Fornstedt B (1989) Reduction of brain glutathione by L-buthionine sulfoximine potentiates the dopamine-depleting action of 6-hydroxydopamine in rat striatum. J Neurochem 52: 978–980.

64. Ferrari R, Ceconi C, Curello S, Cargnoni A, Alfieri O, et al. (1991) Oxygen free radicals and myocardial damage: protective role of thiol-containing agents. Am J Med 91: 95S–105S. Review.

65. Choy KH, Dean O, Berk M, Bush AI, van den Buuse M (2010) Effects of N-acetyl-cysteine treatment on glutathione depletion and a short-term spatial memory deficit in 2-cyclohexene-1-onetreated rats. European J Pharmacol 649: 224–228.

66. Bhatnagar M, Sharma D, Salvi M (2009) Neuroprotective effects of *Withania somnifera* dunal.: a possible mechanism. Neurochem Res 34: 1975–1983.

67. Kratschmar DV, Calabrese D, Walsh J, Lister A, Birk J, et al. (2012) Suppression of the Nrf2-Dependent Antioxidant Response by Glucocorticoids and 11b-HSD1-Mediated Glucocorticoid Activation in Hepatic Cells. PLoS ONE 7, e36774.

68. Ki SH, Cho IJ, Choi DW, Kim SG (2005) Glucocorticoid receptor (GR)-associated SMRT binding to C/EBPbeta TAD and Nrf2 Neh4/5: role of SMRT recruited to GR in GSTA2 gene repression. Mol Cel Biol 25: 4150–4165.

12

Differential Responses of the Antioxidant System of Ametryn and Clomazone Tolerant Bacteria

Leila Priscila Peters[1], Giselle Carvalho[1], Paula Fabiane Martins[1], Manuella Nóbrega Dourado[1], Milca Bartz Vilhena[1], Marcos Pileggi[2], Ricardo Antunes Azevedo[1]*

1 Departamento de Genética, Escola Superior de Agricultura Luiz de Queiroz, Universidade de São Paulo, Piracicaba, Brazil, 2 Departamento de Biologia Estrutural, Molecular e Genética, Universidade Estadual de Ponta Grossa, Ponta Grossa, Brazil

Abstract

The herbicides ametryn and clomazone are widely used in sugarcane cultivation, and following microbial degradation are considered as soil and water contaminants. The exposure of microorganisms to pesticides can result in oxidative damage due to an increase in the production of reactive oxygen species (ROS). This study investigated the response of the antioxidant systems of two bacterial strains tolerant to the herbicides ametryn and clomazone. Bacteria were isolated from soil with a long history of ametryn and clomazone application. Comparative analyses based on 16S rRNA gene sequences revealed that strain CC07 is phylogenetically related to *Pseudomonas aeruginosa* and strain 4C07 to *P. fulva*. The two bacterial strains were grown for 14 h in the presence of separate and combined herbicides. Lipid peroxidation, reduced glutathione content (GSH) and antioxidant enzymes activities were evaluated. The overall results indicated that strain 4C07 formed an efficient mechanism to maintain the cellular redox balance by producing reactive oxygen species (ROS) and subsequently scavenging ROS in the presence of the herbicides. The growth of bacterium strain 4C07 was inhibited in the presence of clomazone alone, or in combination with ametryn, but increased glutathione reductase (GR) and glutathione S-transferase (GST) activities, and a higher GSH concentration were detected. Meanwhile, reduced superoxide dismutase (SOD), catalase (CAT) and GST activities and a lower concentration of GSH were detected in the bacterium strain CC07, which was able to achieve better growth in the presence of the herbicides. The results suggest that the two bacterial strains tolerate the ametryn and clomazone herbicides with distinctly different responses of the antioxidant systems.

Editor: Guillermo López Lluch, Universidad Pablo de Olavide, Centro Andaluz de Biología del Desarrollo-CSIC, Spain

Funding: This work was funded by grant 09/54676-0 from the Fundação de Amparo à Pesquisa do Estado de São Paulo to RAA. The authors thank Conselho Nacional de Desenvolvimento Científico e Tecnológico (RAA and MBV) and the Fundação de Amparo à Pesquisa do Estado de São Paulo (LPP, PFM, GC and MND) for the fellowship and scholarships granted. The funders had no role in study design, data collection and analysis, decision to publish, or preparation of the manuscript.

Competing Interests: Dr. Paula F. Martins current address is: Souza Cruz S/A, Unidade Cachoeirinha, Avenida Frederico Augusto Ritter, 8000, Industrial, 94930-000 - Cachoeirinha, RS - Brazil. Dr. Paula F Martins was not an employee of the company when the study was being conducted. The Company is not involved in any way with this study.

* Email: raa@usp.br

Introduction

Pesticides are powerful tools in modern agriculture to minimize economic losses caused by weeds, insects and diseases, and to ensure adequate food production [1,2]. It is estimated that nearly 3 billion tons of pesticides are released into the environment each year [3], more than 35% of which are herbicides [4]. The intensive use of these xenobiotics in agroecosystems can result in the contamination of water and soils [5–7].

The herbicides ametryn and clomazone are widely employed in crops such as sugarcane, soybeans, corn, cotton, and are often detected in the environment [8–11]. In plants, ametryn toxicity is related to the blockage of the electron transport chain binding specifically to D1 proteins of photosystem II, thereby preventing photosynthesis [12]. In contrast, the mode of action of clomazone consists of inducing lipid peroxidation in cells by blocking carotenoid synthesis [13,14]. These two herbicides in the soil can affect microbial activity and induce a selection pressure, which

in turn allows the identification of tolerant microorganisms [15,16].

Previous studies have shown that many herbicides are redox-cycling agents able to alter the aerobic metabolism of microorganisms culminating in an oxidative stress condition [17–19]. This process induced by herbicides in bacteria results in the increased production and subsequent accumulation of reactive oxygen species (ROS), such as the superoxide radical ($O_2^{\bullet-}$), hydrogen peroxide (H_2O_2) and hydroxyl radical ($OH^{\bullet}$) [17]. These products of aerobic cell metabolism are toxic and may lead to enzyme inactivation, protein denaturation, lipid peroxidation and DNA mutation [20,21]. Therefore, any excess ROS that is produced has to be eliminated if a microbe is to survive [22].

Many bacteria can increase the rate of synthesis and accumulate non-enzymatic antioxidant compounds in response to excessive production of ROS (e.g. reduced glutathione (GSH) and ascorbic acid), as well as increase the activity of antioxidant enzymes [18,23]. The enzymes superoxide dismutase (SOD, EC 1.15.1.1) and catalase (CAT, 1.11.1.6) play crucial roles in the detoxification process of $O_2^{\bullet-}$ and H_2O_2, respectively [24–26]. Furthermore, the enzyme glutathione reductase (GR, EC 1.6.4.2) carries out the reduction of oxidized glutathione (GSSG), which is a fundamental reaction for maintaining the homeostasis between GSH/GSSG levels [27]. GSH is an antioxidant capable of directly neutralizing $OH^{\bullet}$ [28] and thus, is considered a key compound, in the stress tolerance process [29]. Glutathione S-transferase (GST, EC 2.5.1.18) is another important enzyme that is required for the degradation of pollutants, since it is primarily involved in cellular detoxification and redox biochemical mechanisms [30,31].

In this study we have examined the effects of the herbicides ametryn and clomazone on the antioxidant stress responses of two bacteria isolated from agricultural soils, previously treated with herbicides.

Materials and Methods

Ethics statement

The bacteria used in this work were isolated from soil samples collected in Fazenda Areão, Escola Superior de Agricultura Luiz de Queiroz (47°38′00″W; 22°42′30″S), Piracicaba, São Paulo State, Brazil. The location is an experimental area of the University and no specific permissions were required for sampling soils in this location. This field sampling did not involve or cause any harm to endangered or protected species.

Herbicides

Relevant characteristics of the two herbicides, ametryn (2-ethylamino-4-isopropylamino-6-methyl-thio-s-triazine) and clomazone (2-(2-chlorophenyl) methyl-4,4-dimethyl-3-isoxazolidinone), are listed in Table 1. Ametryn, which is a selective herbicide (Gesapax 500, Ciba Geigy) applied at 6 L ha^{-1} (3 kg ha^{-1}) to control narrow leaved weeds and broad leaved weeds, was used at 500 g L^{-1} (active ingredient-ai). Clomazone, which is a selective herbicide (Gamit 360 CS) that can be applied at a recommended dose of 1.8 L ha^{-1} (650 g ai ha^{-1}) in sugarcane and maize, was used at 360 g ai L^{-1}. However, ametryn and clomazone can be applied in combination in sugarcane at a recommended dose of 1 L ha^{-1} (300 g ai ha^{-1}), was used at 300 g ai L^{-1} and 1 L ha^{-1} (200 g ai ha^{-1}), was used at 200 g ai L^{-1}, respectively.

Bacterial strains isolation and growth conditions

The bacteria used in this work were isolated from soil samples collected in Fazenda Areão, Escola Superior de Agricultura Luiz de Queiroz (47°38′00″W; 22°42′30″S), Piracicaba, São Paulo State, Brazil. The soils were classified as Oxisol [32] of medium texture and had a history of ametryn and clomazone applications for five consecutive years.

The initial bacterial isolation was carried out using a plating technique with a serial dilution in 0.85% NaCl at concentrations of 10^{-3} and 10^{-5} inoculated in Minimal Salts medium containing 1.0 g (NH$_4$)$_2$SO$_4$, 1.0 g NaCl, 1.5 g K$_2$HPO$_4$, 0.5 g KH$_2$PO$_4$, and 0.2 g MgSO$_4$.7H$_2$O, per L of distilled water, at 30°C (pH 7.0) [33] in the absence and presence of the two herbicides. The concentrations of 25 mM ametryn, 9 mM clomazone and 20 mM of each herbicide were used based on the recommendations on the spray tank solution for each herbicide (5 g L^{-1} for ametryn, 1.8 g L^{-1} for clomazone and 5 g L^{-1} each, in combination).

The tolerant bacterial strains, CC07 and 4C07, were selected based on faster growth rates (compared to other bacteria isolates) and halo formation observed around the bacterial colony, indicating possible herbicide degradation, as observed by Nie et al. [33] and Martins et al. [34].

The bacterial strains were grown aerobically in nutrient Agar (Biobrás - Brazil) containing 5 g peptone, 3 g yeast extract and 15 g agar per L of distilled water, at 30°C (pH 7.0) both in the absence and presence of the herbicides. The herbicides concentrations were added as described above.

Bacterial identification

Bacterial DNA was extracted as previously described by Araújo et al. [35] and a partial sequence of the 16S rRNA gene was amplified with primers R1387 [36] and P027F [37]. PCR products were purified and sequenced with primers R1387, 519R and P027F for 16S rRNA (MegaBACE 1000). The sequences of the bacterial strains CC07 and 4C07 were retrieved from databases and used for alignment and phylogeny analyses [38,39] with MEGA 4.0 software package [40] based on the maximum parsimony (MP). The sequences obtained were deposited in GenBank^T under the accession numbers JX109938 and JX109935 for strains CC07 and 4C07, respectively.

Growth determination

Bacterial growth was monitored by measuring the number of colony-forming units (CFUs) mL^{-1} as described by Sangali and Brandelli [41]. Cultures inoculated with 0.1% of the original (Absorbance = 1.0 at 600 nm) were grown in 250 mL Erlenmeyer flasks containing 50 mL of nutrient medium and incubated in the dark on a rotary shaker (140 rpm) at 30°C for 14 h. The bacterial suspension was diluted to 10^{-6} in a saline solution containing 0.85% NaCl and then homogenized. At 2 h intervals the samples (20 µL) were loaded in triplicate for each treatment onto nutrient agar plates, which were further incubated at 30°C for 24 h. At the end of this period the CFUs were determined.

Table 1. Characteristics of the herbicides ametryn and clomazone.

	Ametryn	Clomazone
Manufacture	Syngenta, BASF, Bayer, Servatis, Sipcam Isagro Brasil	FMC Corporation
Agrochemical formulation	500 g ai L^{-1}, EC	360 g ai L^{-1}, EC
Molecular formula	C$_9$H$_{17}$N$_5$S	C$_{12}$H$_{14}$ClNO$_2$
Molecular weight	227.3	239.7
Vapor pressure	2.74×10^{-6} mm Hg at 25°C	1.4×10^{-4} mm Hg at 25°C
Water solubility	185 mg L^{-1} at 25°C	1,100 mg L^{-1} at 25°C

Physiological and biochemical measurements

Lipid peroxidation. Lipid peroxidation was determined by estimating the content of thiobarbituric acid reactive substance (TBARS) following the method of Heath and Packer [42]. Malondialdehyde (MDA) was monitored by measuring at 535 and 600 nm in a Perkin Elmer Lambda 40 spectrophotometer, and the concentration was calculated using an extinction coefficient of 155 mM^{-1} cm^{-1}.

Quantification of reduced glutathione (GSH). Bacterial cells (100 mg) were homogenized in 1.5 mL of 5% sulfosalicylic acid in a mortar and pestle at 4°C. The homogenate was centrifuged at 12,000×g for 20 min at 4°C. The content of GSH and GSSG was determined as described by Anderson [43] at 25°C in a mixture consisting of 1.75 mL 100 mM potassium phosphate buffer (pH 7.5) containing 0.5 mM ethylenediaminetetraacetic acid (EDTA) and 100 μL of 3 mM 5,5'-dithiobis(2-nitrobenzoic acid) (DTNB). The reaction was started by the addition of 250 μL of the bacterial cell homogenate. After 5 min, the absorbance for the determination of GSH was read at 412 nm using a spectrophotometer. A standard curve was prepared with known concentrations of GSH and the results were expressed in nmol g^{-1} FW.

Enzyme extraction and protein determination. Cultures were centrifuged at 12,000×g for 20 min at 4°C and the pellets macerated with liquid nitrogen in a mortar with a pestle. The extracts were homogenized (5:1, buffer volume: fresh weight) in 100 mM potassium phosphate buffer (pH 7.5) containing 1 mM EDTA, 3 mM DL-dithiothreitol (DTT) and 5% (w/w) polyvinyl-polypyrrolidone [44]. The homogenates were centrifuged at 12,000×g for 30 min at 4°C and the supernatants were stored in separate aliquots at −80°C prior to enzymatic analysis. The concentration of protein was determined by the method of Bradford [45] using bovine serum albumin as standard.

Polyacrylamide gel electrophoresis (PAGE). Electro-fphoretic analyses of antioxidant enzymes were carried out under non-denaturing condition in 12% polyacrylamide gels as described by Gratão et al. [44]. For denaturing SDS-PAGE, the gels were rinsed in distilled deionized water and incubated overnight in 0.05% Coomassie blue R-250 in a water/methanol/acetic acid 45:45:10 (v/v/v) solution and destained by successive washings in the same water/methanol/acetic acid 45:45:10 (v/v/v) solution.

SOD activity staining. SOD activity staining was carried out as described by Garcia et al. [46]. After non-denaturing PAGE separation, the gel was rinsed in distilled deionized water and incubated in the dark in 50 mM potassium phosphate buffer (pH 7.8) containing 1 mM EDTA, 0.05 mM riboflavin, 0.1 mM nitroblue tetrazolium, and 0.3% N,N,N',N'-tetramethylethylene-diamine. One unit of bovine liver SOD (Sigma, St. Louis, USA) was used as a positive control of activity. After 30 min, the gels were rinsed with distilled deionized water and then illuminated in water until the development of achromatic bands of SOD activity on a purple-stained gel. SOD isoenzyme characterization was performed as described by Azevedo et al. [47]. Briefly, SOD isoenzymes were distinguished by their sensitivity to inhibition by 2 mM potassium cyanide and 5 mM hydrogen peroxide. The relative intensities of the stained bands were determined by an ImageScanner III (GE Healthcare, Little Chalfont, UK) and the ImageQuant™ TL software (GE Healthcare, Uppsala, Sweden).

CAT activity staining. CAT activity following non-denaturing PAGE was determined as described by Boaretto et al. [48]. Gels were incubated in 0.003% hydrogen peroxide (H_2O_2) for 10 min and subsequently in a 1% (w/v) ferric chloride (FeCl$_3$) and 1% (w/v) potassium hexacyanoferrate III (K$_3$Fe(CN)$_6$) solution for

additional 10 min. One unit of bovine liver CAT (Sigma, St. Louis, USA) was used as a positive control.

GR activity staining. GR activity following non-denaturing PAGE was determined as described by Gomes-Junior et al. [49]. The gels were rinsed in distilled deionized water and incubated in the dark for 30 min at room temperature in the reaction solution contained 250 mM Tris (pH 7.5), 0.5 mM 3-(4,5-dimethyl-2-thiazolyl)-2,5-diphenyl-2H-tetrazolium bromide (MTT), 0.7 mM 2,6-dichloro-N-(4-hydroxyphenyl)-1,4-benzoquinoneimine sodium salt (DPIP), 3.4 mM GSSG (oxidized glutathione) and 0.5 mM NADPH. One unit of bovine liver GR (Sigma, USA) was used as a positive control of activity.

GR total activity determination. Total GR activity was assayed as described by Gratão et al. [44] at 30°C in a mixture consisting of 1.7 mL 100 mM potassium phosphate buffer (pH 7.5) containing 1 mM 5,5'-dithiobis(2-nitrobenzoic acid) (DTNB), 1 mM GSSG and 0.1 mM NADPH. The reaction was started by the addition of 50 μL of protein extract. The rate of reduction of oxidized glutathione was followed in a spectrophotometer by monitoring the change in absorbance at 412 nm for 1 min. GR activity was expressed as μmol min^{-1} mg^{-1} protein.

GST total activity determination. GST activity was assayed spectrophotometrically at 30°C in a mixture containing 900 μL 100 mM potassium phosphate buffer (pH 6.5), 25 μL 40 mM 1-chloro-2,4-dinitrobenzene (CDNB), 50 μL 1 mM GSH and 25 μL enzyme extract. The reaction mixture was followed by monitoring the increase absorbance at 340 nm over 5 min [50]. GST activity was expressed as μmol min^{-1} mg^{-1} protein.

Experimental design and statistical analysis. Total protein content and enzyme activity determinations were conducted on three replicates of each treatment, which were performed in a completely randomized design. The significance of the observed differences was verified by using a one-way analysis of variance (ANOVA) followed by the Tukey's test (p<0.05). All statistical analyses were carried out by using R software (URL http://www.r-project.org).

Results

Phylogenetic identity and bacterial growth

An almost-complete 16S rRNA gene sequence (1312 nts) was determined for both the CC07 and 4C07 strains and a phylogenetic tree was built up (Fig. 1). *Burkhloderia cariophilli* and *B. plantarii* were used as outgroups. Comparative analyses based on 16S rRNA gene sequences revealed that the CC07 strain is phylogenetically related to *Pseudomonas aeruginosa*, whereas the 4C07 strain exhibited homology with *P. fulva* (Fig. 1).

The growth of the strains in the presence of the herbicides is shown in Fig. 2. The two strains exhibited very distinct growth curves. Growth of both strains was not greatly affected in the presence of the herbicide ametryn, whilst clomazone strongly inhibited the growth of both strains. Strain CC07 exhibited a long (10 h) lag phase as an adaptation period before the exponential growth (12 h) (Fig. 2A), while strain 4C07 grew only for the first six hours. When the herbicides were used in combination, growth of the strain CC07 was only slightly inhibited, mainly during the early period, whereas for strain 4C07 there was considerable inhibition of growth, similar to that shown by clomazone alone.

Lipid peroxidation (MDA)

Lipid peroxidation was determined as the MDA content after 14 h of growth. Although similar trends in MDA content were detected in both strains (Fig. 3), statistically there was only a higher MDA content in strain CC07 in the presence of clomazone

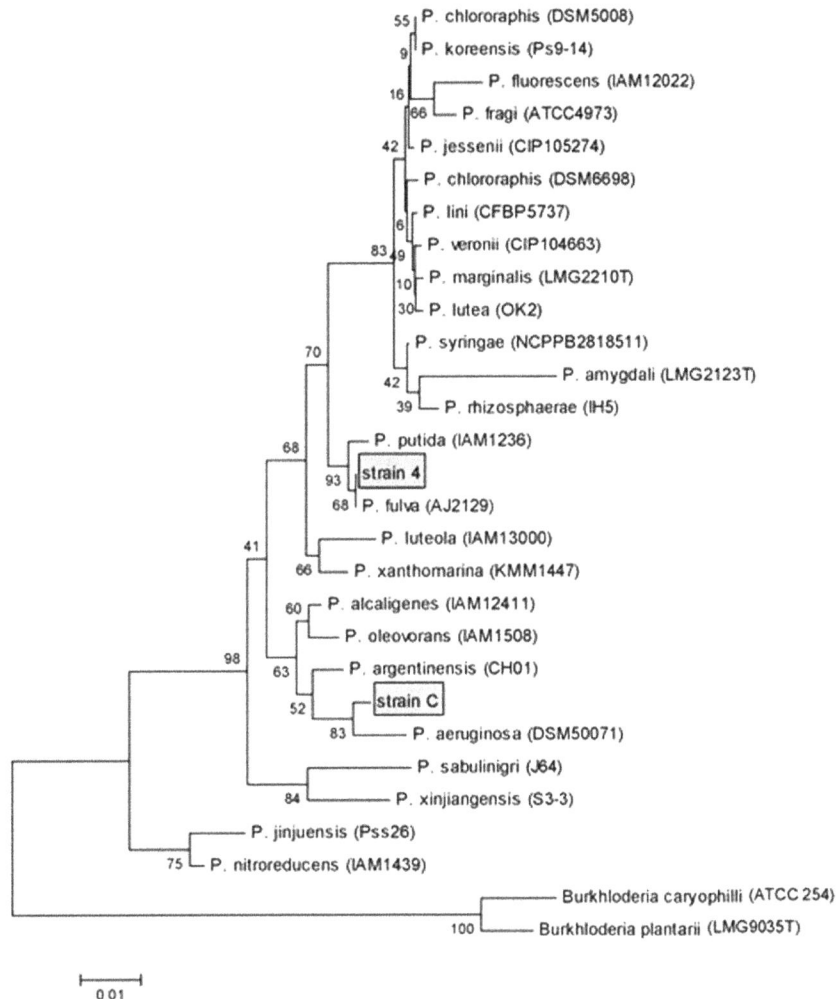

Figure 1. Maximum-parsimony phylogenetic tree constructed from the 16S rRNA gene. A total of 600 bp nucleotide of *Pseudomonas* spp. from RDP database were used. *Burkhloderia cariophilli* and *B. plantarii* served as outgroup. Bootstrap values were 1000 repetitions. Bars indicate the number of evolutionary steps with diverging sequences. Strains CC07 and 4C07 are shown inside the boxes.

(30.3%) and when exposed to the combination of the two herbicides (110.7%) (Fig. 3), when compared to growth in the control herbicide-free medium.

The content of reduced glutathione (GSH)

Different amounts of GSH were detected in the two bacterial strains, following exposure to the herbicides (Fig. 4). There was a significant increase in the GSH content of CC07 (16.5%) in the presence of ametryn, but GSH was reduced below control levels in the presence of clomazone or the herbicide mixture. Meanwhile in strain 4C07, the GSH content increased, following exposure to all the herbicide treatments (ametryn 39.4%, clomazone 60.7% and the mixture 50.5%), when compared to the untreated control (Fig. 4).

SDS-PAGE protein profile

Analysis following SDS-PAGE revealed clearly different protein profiles between the two bacterial strains, both with and without herbicide treatments (Fig. 5). There was a general reduction in intensity of the majority of the protein bands following electrophoresis of extracts of CC07 that had been subjected to herbicide treatment (Fig. 5, lanes 3, 4 and 5). In addition, a number of

protein bands varied in intensity or appearance/disappearance, depending on the treatment as indicated by the arrows in Fig. 5. For instance, a 225 kDa protein band (strain CC07) was reduced following treatment with the herbicide clomazone (Fig. 5, lane 4, band I), and a 58 kDa protein band was absent in strain 4C07 exposed to clomazone (Fig. 5, lane 8, band II), and greatly reduced following treatment by both herbicides (Fig.5, lane 9, band II), among other changes (arrows).

Effects of the herbicides on antioxidant enzymes

There were different responses in the antioxidant enzymes (SOD, CAT, GR and GST) isolated from the two bacterial strains, when treated with the herbicides alone or in combination. SOD activity was determined by activity staining following non-denaturing PAGE (Fig. 6). The analysis revealed that there were two distinct isoenzymes present in each bacterium, but that they had different electrophoretic mobilities (Fig. 6A and B). Both of the SOD isoenzymes isolated from strain CC07 were characterized as Mn/SODs (Fig. 6A, SOD I and II, lanes 1–4; and Fig. 6C 1, 2 and 3), whereas the two SOD isoenzymes, isolated from strain 4C07 comprised a Cu-Zn/SOD (I) and a Fe/SOD (II) (Fig. 6B, SOD I and II, lanes 5–8; and Fig. 6C 4, 5 and 6).

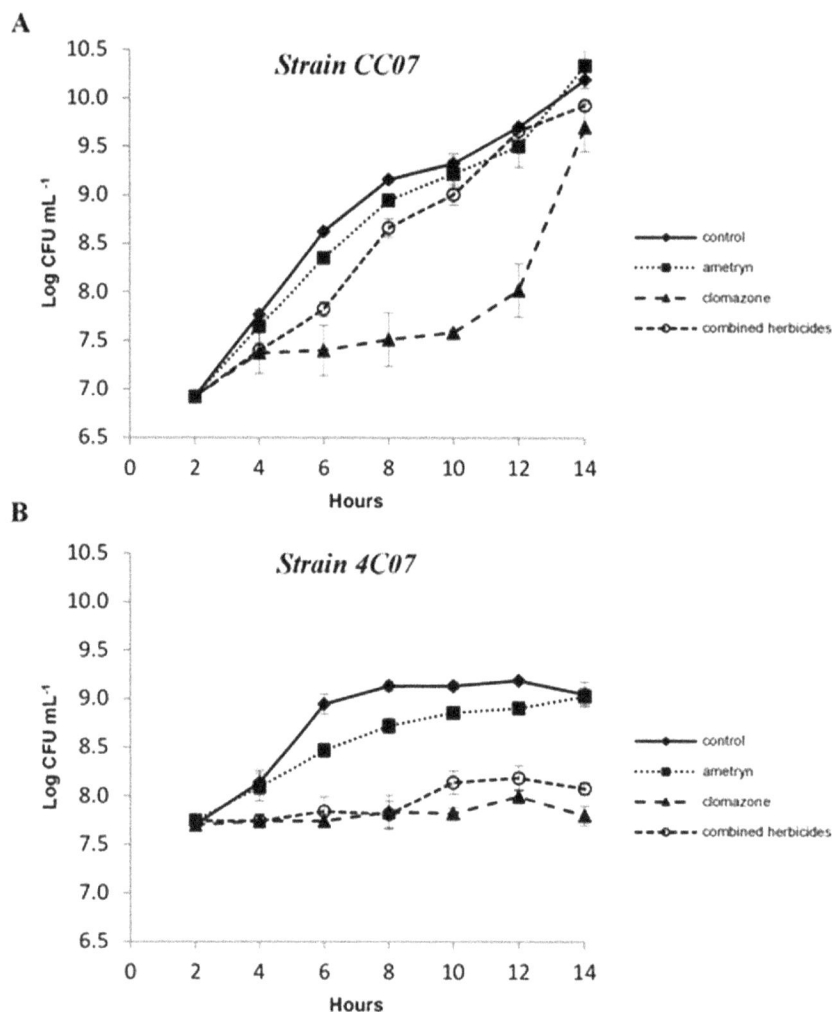

Figure 2. Growth curve of bacterial strains in the presence of 0 mM (control), 25 mM ametryn, 9 mM clomazone and 20 mM of each the herbicide. (A) Strain CC07 and (B) Strain 4C07. Values represent the means from three replicates ±SEM.

The activity of SOD I in strain CC07, was reduced following incubation with ametryn and the two combined herbicides, whilst SOD II was unaffected (Fig. 6A, lanes 2 and 4, respectively). A different SOD isoenzyme pattern was detected following native PAGE of extracts of 4C07, in which SOD II activity was increased following treatment with clomazone and the two combined herbicides (Fig. 6B, lanes 7 and 8, respectively), whereas the activity of SOD I was also stimulated in the presence of clomazone, but drastically inhibited following the combined herbicide treatment (Fig. 6B, lanes 7 and 8, respectively). Ametryn did not produce any major change in SOD I and II activity when compared to the control (Fig. 6B, lanes 5 and 6).

CAT activity staining following non-denaturing PAGE revealed the presence of three isoenzymes for strain CC07 (CAT I, II and III; Fig. 7, lanes 1–4) and four isoenzymes for strain 4C07 (CAT I, II, III and IV; Fig. 7, lanes 5–8), with CAT isoenzymes II (strain CC07), I (strain 4C07) and III (both bacteria) with the same relative mobility, suggesting they may be the same isoenzymes in both bacteria. The activity of CAT isoenzyme II was higher in 4C07 and was further increased when the strain was subjected to the combined herbicide treatment (Fig. 7, lane 8).

Furthermore, the CAT isoenzyme activity profiles were in a way similar to the enzyme pattern also observed for SOD (Fig. 6A and

B). For instance, in strain CC07, the activity of CAT I was reduced in the ametryn and combined herbicide treatments (Fig. 7, lanes 2 and 4) as observed for SOD I (Fig. 6, lanes 2 and 4). On the other hand, the activity of the CAT II isoenzyme was slightly reduced in the clomazone and combined herbicide treatments (Fig. 7, lanes 3 and 4), whereas the CAT III isoenzyme was unaltered in all treatments. For strain 4C07, apart from the unique CAT II isoenzyme (Fig. 7, lane 8), the activity of CAT isoenzyme I was clearly increased following treatment with clomazone and the combined herbicides (Fig. 7, lanes 7 and 8).

GR activity was determined as total specific activity (Fig. 8A and B) and by non-denaturing PAGE for isoenzyme identification (Fig. 8C). Total GR activity was not altered in strain CC07 regardless of the treatment (Fig. 8A), but was increased in strain 4C07 when exposed to clomazone (30.8% increase) and the combined herbicides (55.3% increase) (Fig. 8B). GR activity staining revealed the existence of 6 GR isoenzymes for strain CC07 (Fig. 8C, lanes 1–4), all present in all treatments tested, and up to 5 GR isoenzymes in the strain 4C07 depending on the treatment tested (Fig. 8C, lanes 5–8). In general, the GR isoenzyme activities were in accordance with the results obtained for total GR activity (Fig. 8A and B). Slight variations in band intensity were observed among the treatments, but all GR

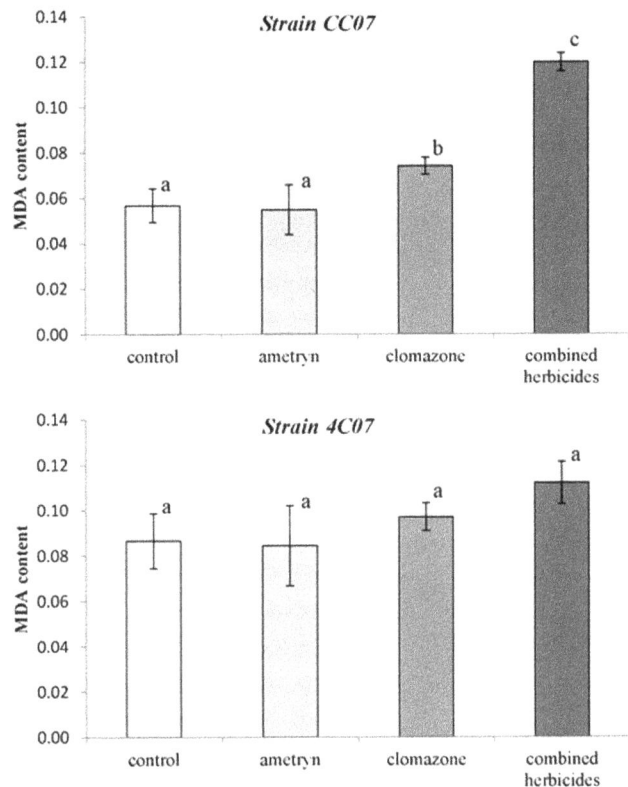

Figure 3. Lipid peroxidation (MDA content) of bacteria exposed to the herbicides. Values of MDA content (nmol g^{-1} fr. wt) represent the means from three replicates ±SEM. Means with different letters are significantly different ($P<0.05$) by one-way analysis of variance (ANOVA) and Tukey's test.

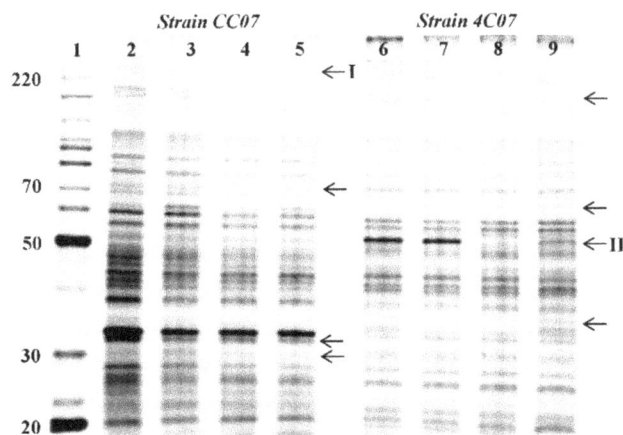

Figure 4. GSH content (nmol g^{-1} fr. wt) of bacteria exposed to the herbicides. Values represent the means from three replicates ±SEM. Means with different letters are significantly different ($P<0.05$) by one-way analysis of variance (ANOVA) and Tukey's test.

isoenzymes were present in the strain CC07 (Fig. 8C, lanes 1–4). On the other hand, there was increased total GR activity in strain 4C07, when subjected to clomazone and combined herbicide treatments, which is most likely due to the increased activity of GR isoenzyme II (Fig. 8C, lanes 7 and 8).

Total specific GST activity was also determined and the results revealed a distinct response for each bacterium when exposed to the herbicides (Fig. 9). Total GST activity increased by 51.5%, 41.5% and 105% when strain 4C07 was grown in the presence of ametryn, clomazone or the combination of herbicides, respectively (Fig. 9). On the other hand, there was a reduction in GST activity when strain CC07 was exposed to the herbicides ametryn or clomazone, but the activity was similar to the control when CC07 was grown in the presence of the combined herbicides (Fig. 9).

Discussion

Microorganisms present in soils must quickly adapt to environmental changes. This biochemical and physiological adaptation process play an important role in microbial survival, especially under stressful conditions [51]. Strains CC07 and 4C07 were selected for this study due to their ability to tolerate the high concentrations of herbicides tested and to the fact that bacteria of the *Pseudomonas* genus play a major role in degradation of xenobiotic compounds [52–55].

Generally, herbicides can alter the growth of degrading or tolerant bacteria [56]. Growth of the strains CC07 and 4C07 exhibited distinct responses when exposed to the herbicides

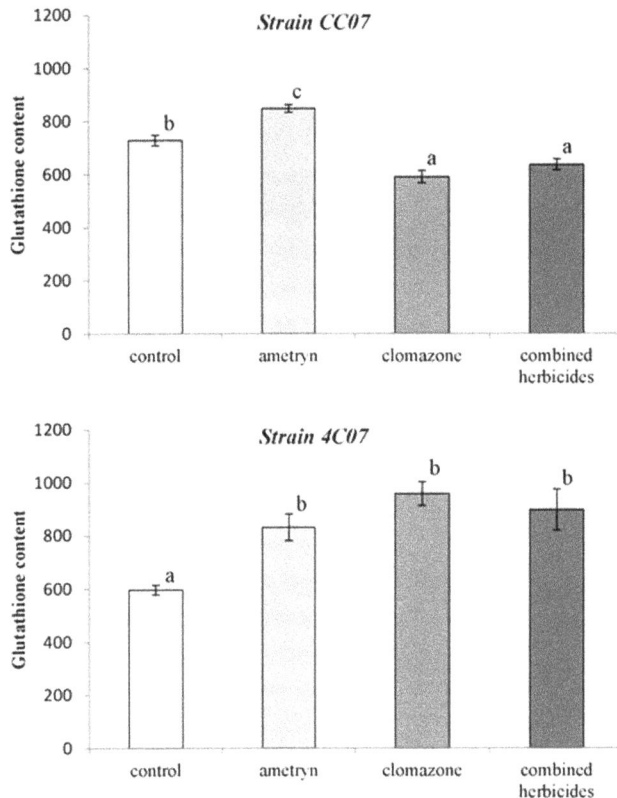

Figure 5. SDS-PAGE protein profiles of bacteria exposed to the herbicides. Lane 1, protein molecular mass markers (220 to 20 kDa). Lanes 2, 3, 4 and 5, strain CC07 grown in the presence of 0 mM (control), 25 mM ametryn, 9 mM clomazone and 20 mM of each the herbicide, respectively. Lanes 6, 7, 8 and 9, strain 4C07 grown in the presence of 0 mM (control), 25 mM ametryn, 9 mM clomazone and 20 mM of each the herbicide, respectively. Arrows indicate selected variations in intensity or appearance/disappearance depending on the treatment tested. I and II indicate protein bands of 225 kDa and 58 KDa that are depleted specifically in the presence of clomazone.

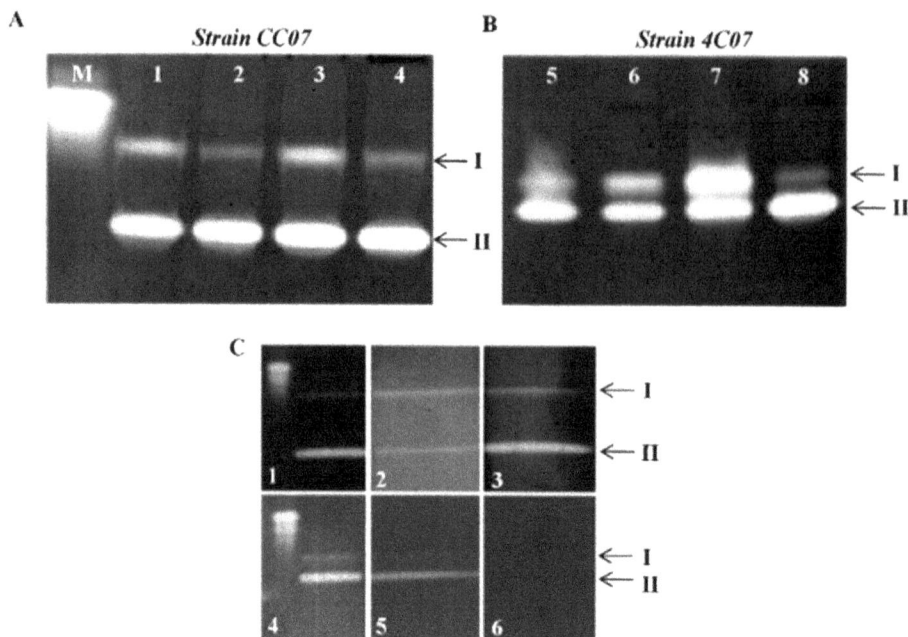

Figure 6. Activity staining for SOD following non-denaturing PAGE of extracts from cultured bacterial cells. (A) First lane is a bovine SOD standard, lanes 1, 2, 3 and 4 are strain CC07 grown in the presence of 0 mM (control), 25 mM ametryn, 9 mM clomazone and 20 mM of each herbicide, respectively. (B) Lanes 5, 6, 7 and 8 are strain 4C07 grown in the presence of 0 mM (control), 25 mM ametryn, 9 mM clomazone and 20 mM of each herbicide, respectively. Arrows indicate sequentially numbered SOD bands (I–II) that are independent of the bacterial strain. (C) Activity staining for SOD of strain CC07 (1, 2 and 3) and strain 4C07 (4, 5 and 6), used for classification of SOD isoenzymes. Lanes 1 and 4, control SOD activity. Lanes 2 and 5, SOD activity with 2 mM potassium cyanide treatment; lanes 3 and 6; 5 mM H_2O_2 treatment. Arrows indicate SOD bands that are sequentially numbered (I–II) according to Fig. 6A and B.

ametryn and clomazone, separately or in combination. The number of bacterial cells for all treatments was lower when compared to the control, indicating that the ROS generated by the herbicides and its metabolites can damage bacteria cells and decrease bacterial growth. However, clomazone was the treatment that markedly affected growth of both strains. These results may be associated with the higher toxicity of the clomazone molecule, due the presence of chloride (Table 1). García-Cruz et al. [57] demonstrated the toxicity of the chlorine atoms present in the herbicide 2,4-dichlorophenoxyacetic acid (2,4-D) and possible intermediates in bacterial biofilms. Another possible cause of the aggressiveness of clomazone towards bacteria cells may be due to

its mechanism of action, since it is an inhibitor of the deoxy-D-xylulose 5-phosphate (DXP) synthase enzyme [14,58], which plays a key role in isoprenoid biosynthesis and is also required for thiamine and pyridoxal phosphate production in prokaryotes [59].

Previous studies have shown that when bacteria are exposed to herbicides or their metabolites, a significant increase in ROS generation may occur and, consequently, induce an oxidative stress condition [22,23]. Lipid peroxidation is one of the best predictors of ROS level inducted under stress conditions [60], and can be used as a marker of oxidative stress [61]. An increased content of MDA was detected when strain CC07 was exposed to clomazone alone or in mixture with ametryn (Fig. 3), which

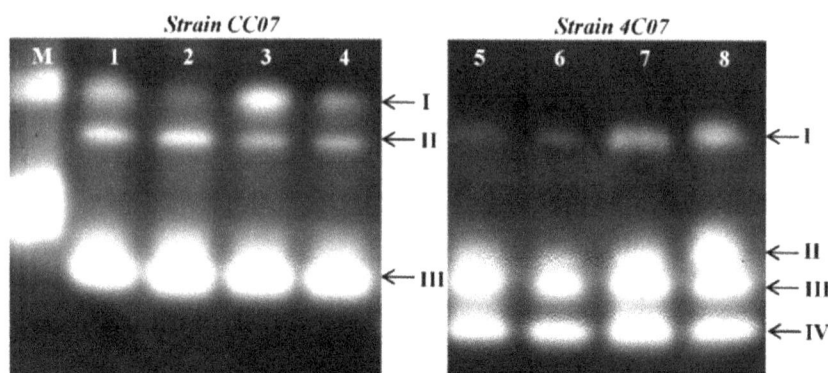

Figure 7. Activity staining for CAT following non-denaturing PAGE of extracts from cultured bacterial cells. Lane M is a bovine CAT standard, lanes 1, 2, 3 and 4, strain CC07 grown in the presence of 0 mM (control), 25 mM ametryn, 9 mM clomazone and 20 mM of each herbicide, respectively. Lanes 5, 6, 7 and 8, strain 4C07 grown in the presence of 0 mM (control), 25 mM ametryn, 9 mM clomazone and 20 mM of each herbicide, respectively. Arrows indicate sequentially numbered CAT bands for strains CC07 (I–III) and 4C07 (I–IV).

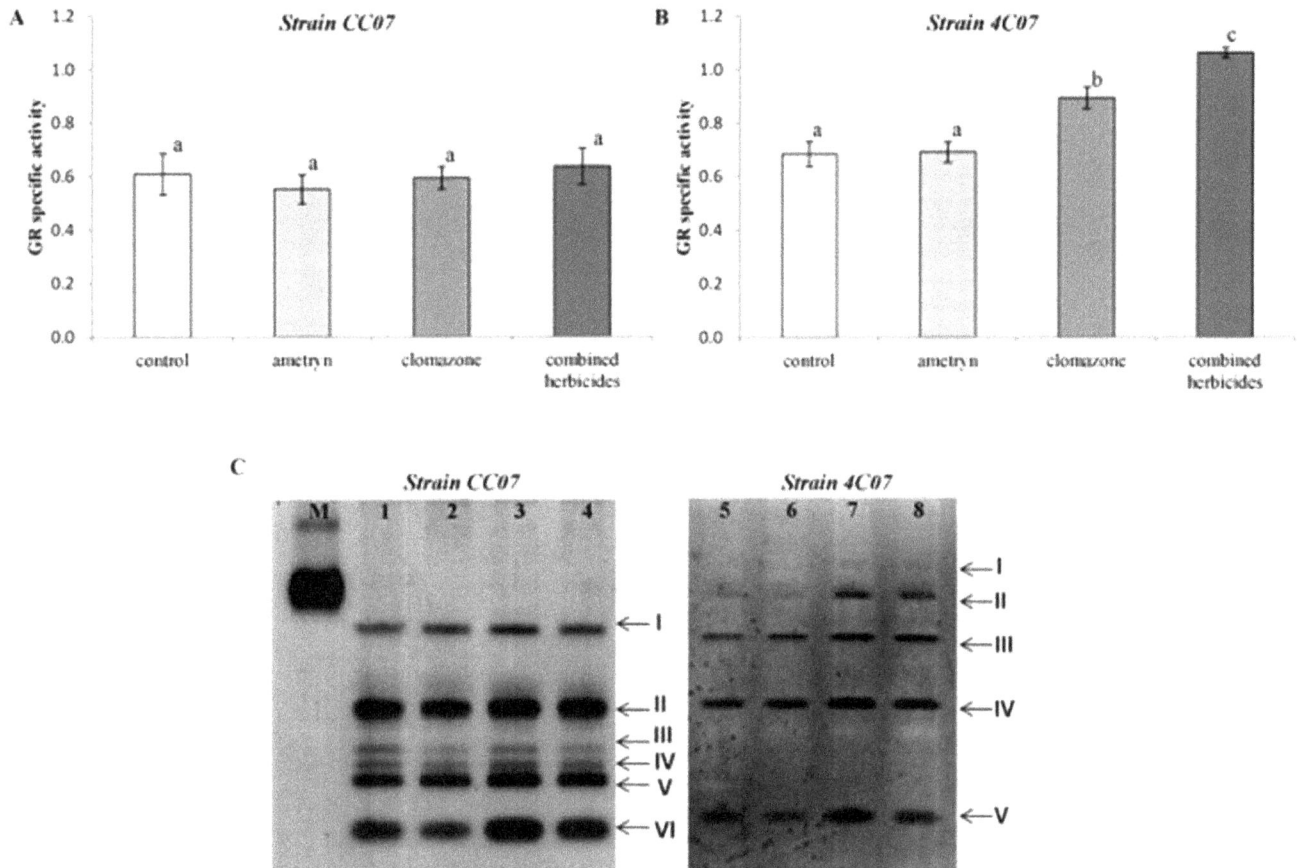

Figure 8. GR specific activity. (A) and (B) Specific activity of GR, expressed as μmol min^{-1} mg^{-1} protein. Values are the means of three replicates $\pm$SEM. Means with different letters are significantly different ($P<0.05$) by one-way analysis of variance (ANOVA) and Tukey's test. (C) Activity staining for GR following non-denaturing PAGE of extracts of cultured bacterial cells. Lane M is a bovine GR standard, lanes 1, 2, 3 and 4, strain CC07 grown in the presence of 0 mM (control), 25 mM ametryn, 9 mM clomazone and 20 mM of each herbicide, respectively. Lanes 5, 6, 7 and 8, strain 4C07 grown in the presence of 0 mM (control), 25 mM ametryn, 9 mM clomazone and 20 mM of each herbicide, respectively. Arrows indicate sequentially numbered GR bands (I–VI) that are independent of the bacterial strain.

suggests that the high concentration used in the combined treatment may have increased the amount of lipid peroxidation. Despite this, the CC07 strain was able to respond and tolerate the peroxidation stress, since following all the herbicide treatments, the growth observed after 14 h was very similar to that of the control (Fig. 2A). High concentrations of MDA have been found in strains of *Enterobacter asburiae* and *E. amnigenus* in the presence of the herbicides metolachlor and acetochlor, respectively [23], which correlated the oxidative stress with the mode of action of the herbicides tested, which involved the inhibition of the elongation of C18 and C16-fatty acids. On the other hand, changes in MDA content were not observed for strain 4C07 in the presence of the herbicides, suggesting that strain 4C07 could possess an effective antioxidant system to avoid the damage caused by ROS. Although no changes in MDA content were observed, the 4C07 strain exhibited limited growth, thus indicating sensitivity to the herbicides dose (s) in the medium.

To maintain ROS under the baseline levels, bacteria utilize a complex antioxidant defense system [20]. SOD is able to detoxify $O_2^{-\bullet}$, one of the two substrates of the Haber-Weiss reaction, which generates OH$^\bullet$ radicals, and is therefore plays a key role in the central defense mechanism of living organisms [62]. Following analysis of SOD activity by non-denaturing PAGE, two isoenzymes were identified in each strain, which were classified as

Mn-SOD (SODs I and II) for strain CC07, and as Cu/Zn-SOD (SOD I) and Fe-SOD (SOD II) for strain 4C07. Fe-SOD and Mn-SOD are present in the bacterial cytoplasm [63], whilst Cu/Zn-SOD isoenzymes are present in the periplasm and are more sensitive to endogenous oxidative stress. However, Cu/Zn-SODs are also responsible for protection against oxidative damage to DNA. Studies with mutants of *Escherichia coli* revealed that Cu/Zn-SOD activity increased bacterial resistance to the herbicide paraquat [64]. Thus, the presence of Cu/Zn-SOD in strain 4C07, in comparison to CC07, may have favored its antioxidant response to the $O_2^{-\bullet}$ increase in the presence of the herbicides. In contrast, following exposure to the herbicide quinclorac, there was an increase in both Fe-SOD and Mn-SOD activity in the bacterium *B. cepacia* WZ [17], which appeared to protect *B. cepacia* from the redox action of quinclorac [17].

Catalases are widely distributed and are considered important components of detoxification routes that prevent the formation of hydroxyl radicals [65]. Three active isoenzymes (CAT I, III and IV) were detected in extracts of the untreated strain 4C07, however in the presence of both herbicides a new isoenzyme (CAT II) was induced. This may have occurred due to bacterial sensitivity to both herbicides, thereby, the expression of a new isoform would contribute to maintain the CAT activity at a standard level.

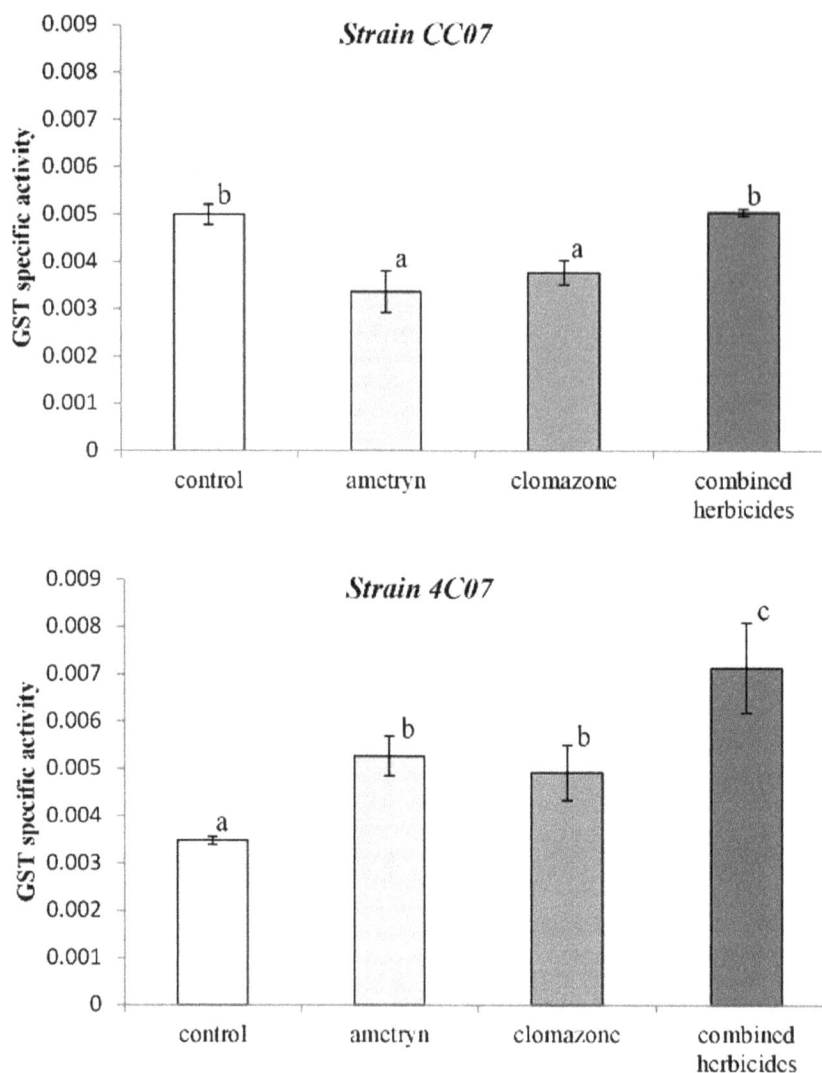

Figure 9. GST specific activity, expressed as units mg^{-1} protein. Values represent the means from three replicates ±SEM. Means with different letters are significantly different ($P<0.05$) by one-way analysis of variance (ANOVA) and Tukey's test.

By comparing the CAT and SOD results for strain CC07, it can be seen that the herbicide ametryn inhibited the activity of both enzymes, (isoenzymes SOD I and CAT I in particular), whilst the herbicide clomazone had a greater effect on the enzymes of strain 4C07. There is no consensus on the effect of herbicides on CAT and SOD; both enzymes are essential in *Bacillus subtilis* B19, *B. megaterium* and *E. coli* K12 for survival in the presence of the herbicide bensulfuron-methyl [19], and both enzymes also played an important role in copper resistance in *Amycolatopsis* spp species [66]. In contrast, Martins et al. [23] reported that SOD activity was not significantly changed when strains of *E. asburiae* and *E. amnigenus* were exposed to the herbicides metolachlor and acetochlor, respectively.

Reduced glutathione is the most abundant thiol in cells and its main function is to maintain the cytoplasm redox state in equilibrium [29]. There was a significant increase in the GSH content of strain CC07 in the presence of ametryn, possibly minimizing the occurrence of lipid peroxidation. In the presence of clomazone alone or in mixture with ametryn, however the GSH content was reduced. Hultberg [67] observed decreases in the concentration of GSH when the bacterium *P. fluorescens* was exposed to cadmium and suggested that GSH may be used as a

marker for the intensity of environmental stress. On the other hand, the GSH concentrations in strain 4C07 increased significantly in the presence of the herbicides. This result suggests a direct participation of GSH in ROS detoxification, particularly of the OH$^\bullet$ radical, consequently preventing lipid peroxidation and thus, contributing to the protection against oxidative stress. Generally in gram-negative bacteria, including *Pseudomonas*, the concentration of GSH is high [68] and it can react directly with a free radical resulting in a thiol radical, GS$^\bullet$, which reacts with another GS$^\bullet$ radical producing GSSG [69].

Under many intracellular stressful conditions the GSSG concentration increases due to GSH oxidation, however, a high GSH/GSSG rate is necessary to maintain the role of GSH as antioxidant and reducing agent [70], therefore, GR acts as a fundamental link between the two redox metabolites within a cell [29]. In strain CC07, total GR activity remained unchanged in the presence of the herbicides, which was reconfirmed with non-denaturing PAGE analysis. Veremeenko and Maksimova [68] showed that in *P. aurantiaca*, GSH and GR activity increased significantly in the presence of antibiotics. In this study, the increase in total GR activity in strain 4C07 in the presence of

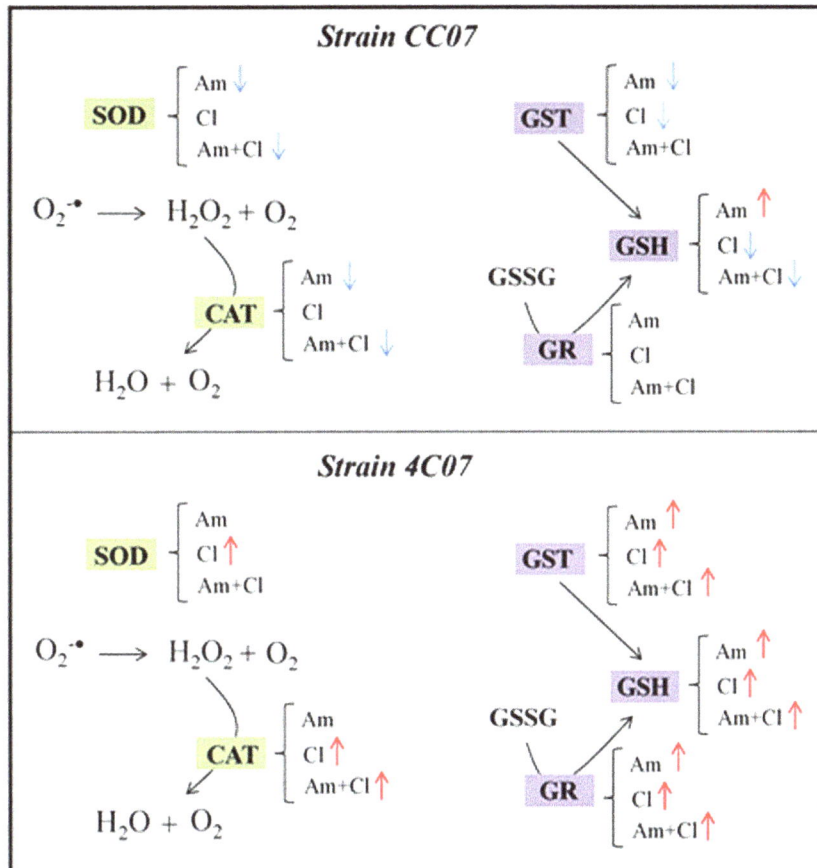

Figure 10. General view of the antioxidant system involving enzymatic and non-enzymatic components for the strains CC07 and 4C07 based on the data obtained in the present work. SOD (superoxide dismutase); CAT (catalase); GST (glutathione S-transferase); GR (glutathione reductase); GSSG (oxidized glutathione); GSH (reduced glutathione); Am (ametryn); Cl (clomazone); Am+Cl (combined herbicides). Blue arrows indicate decreases in enzymatic activity and glutathione content; red arrows indicate increases in enzymatic activity and glutathione content; the absence of symbols indicates no alterations. All changes are relative to the untreated control.

clomazone alone or in mixture with ametryn is associated with the increase in isoenzymes II and to a lesser extent I and V. These data suggest that the different isoenzymes may play a specific role in the antioxidant response of strain 4C07 to oxidative stress induced by the herbicides. Martins et al. [23] reported that when *E. asburiae* was exposed to the herbicide s-metolachlor (34 mM), two new GR isoforms were induced in response to the oxidative stress condition, suggesting a role for GR in herbicide tolerance. Similar responses were also found in plant species. For instance, in coffee cell suspension cultures subjected to the oxidative stress induced by heavy metals (cadmium, nickel and selenium) a new GR isoenzyme was induced during the first hours of stress. The authors also suggested that GR activity could be used as an early stress marker for this plant species [49,71,72].

GST is another enzyme with activity closely related to GSH. Its function is to conjugate GSH to xenobiotics, and therefore has a fundamental central role in detoxification [31]. GSTs are capable of detoxifying numerous classes of pesticides in bacterial cells including s-triazines and are involved in the first stage of the biodegradation of the herbicide atrazine, when the removal of the chlorine atom from the atrazine-GSH conjugate occurs [73]. The increase in GST activity in strain 4C07 may also be associated with the increase in GSH concentration, and could indicate the involvement of GST in the detoxification of the two herbicides.

Our results indicate that both strains are tolerant to the herbicides ametryn and clomazone or to their mixture. However, the herbicides caused an imbalance in the redox potential and metabolism of the bacterial cells. A summary of the key alterations observed are presented in Fig 10. The enzymes GR and GST, together with the antioxidant compound GSH, may play a major role in the tolerance of strain 4C07 to the herbicides ametryn and clomazone supplied individually or combined. In contrast, there was a decrease in the activity of enzymes SOD, CAT, GST and the GSH content in CC07, whilst GR activity remained unchanged. Compensatory mechanisms for the reduction of these enzymes activities and the depletion of GSH content may occur and account for the induction of another defense mechanism, since the strain CC07 managed to adjust its metabolism in response to the stress. These results indicate a different antioxidant response of the two bacterial strains to the herbicides; however, additional studies are required in order to understand the tolerance mechanisms. Nevertheless, strain CC07 grew at a higher rate, indicating that this bacterium was able to adapt better to the stressful environment, which could be useful for bioremediation strategies of environments contaminated with herbicides.

Author Contributions

Conceived and designed the experiments: LPP GC PFM RAA. Performed the experiments: LPP GC PFM MND. Analyzed the data: LPP GC PFM

MND MP RAA. Contributed reagents/materials/analysis tools: LPP GC
PFM MND RAA. Wrote the paper: LPP GC PFM MND MBV MP RAA.

References

1. Juraske R, Antón A, Castells F, Huijbregts MAJ (2007) PestScreen: A screening
approach for scoring and ranking pesticides by their environmental and
toxicological concern. Environ Int 33: 886–893.
2. Tejada M, Gomez I, Del Toro M (2011) Use of organic amendments as a
bioremediation strategy to reduce the bioavailability of chlorpyrifos insecticide in
soils. Effects on soil biology. Ecotox Environ Safe 74: 2075–2081.
3. Pimentel D, Peshin R, Dhawan AK (2009) Pesticides and pest control integrated
pest management: Innovation-development process. Netherlands: Springer. 83–
87 p.
4. Vercraene-Eairmal M, Lauga B, Saint Laurent S, Mazzella N, Boutry S, et al.
(2010) Diuron biotransformation and its effects on biofilm bacterial community
structure. Chemosphere 81: 837–843.
5. Mishra K, Sharma RC, Kumar S (2012) Contamination levels and spatial
distribution of organochlorine pesticides in soils from India. Ecotox Environ Safe
76: 215–225.
6. McKnight US, Rasmussen JJ, Kronvang B, Bjerg PL, Binning PJ (2012)
Integrated assessment of the impact of chemical stressors on surface water
ecosystems. Sci Total Environ 427–428: 319–331.
7. Pileggi M, Pileggi SAV, Olchanheski LR, da Silva PAG, Munoz Gonzalez AM,
et al. (2012) Isolation of mesotrione-degrading bacteria from aquatic
environments in Brazil. Chemosphere 86: 1127–1132.
8. Andrade SRB, Silva AA, Lima CF, D'Antonino L, Queiroz MELR, et al. (2010)
Lixiviação do ametryn em Argissolo Vermelho-Amarelo e Latossolo Vermelho-
Amarelo, com diferentes valores de pH. Planta Daninha 28: 655–663.
9. Cumming JP, Doyle RB, Brown PH (2002) Clomazone dissipation in four
tasmanian topsoils. Weed Sci 50: 405–409.
10. Monquero PA, Binha DP, Amaral LR, Silva PV, Silva AC, et al. (2008)
Lixiviação de clomazone + ametryn, diuron + hexazinone e isoxaflutole em dois
tipos de solo. Planta Daninha 26: 685–691.
11. Carlomagno M, Matho C, Cantou G, Sanborn JR, Last JA, et al. (2010) A
clomazone immunoassay to study the environmental fate of the herbicide in rice
(Oryza sativa). J Agr Food Chem 58: 4367–4371.
12. Vieira VC, Alves P, Picchi SC, Lemos MVF, Sena JAD (2010) Molecular
characterization of accessions of crabgrass (Digitaria nuda) and response to
ametryn. Acta Sci Agron 32: 255–261.
13. Yasuor H, TenBrook PL, Tjeerdema RS, Fischer AJ (2008) Responses to
clomazone and 5-ketoclomazone by Echinochloa phyllopogon resistant to
multiple herbicides in Californian rice fields. Pest Manag Sci 64: 1031–1039.
14. Ferhatoglu Y, Barrett M (2006) Studies of clomazone mode of action. Pestic
Biochem Physiol 85: 7–14.
15. Navaratna D, Elliman J, Cooper A, Shu L, Baskaran K, et al. (2012) Impact of
herbicide ametryn on microbial communities in mixed liquor of a membrane
bioreactor (MBR). Bioresource Technol 113: 181–190.
16. Mervosh TL, Sims GK, Stoller EW (1995) Clomazone fate in soil as affected by
microbial activity, temperature, and soil-moisture. J Agr Food Chem 43: 537–
543.
17. Lü ZM, Min H, Xia Y (2004) The response of Escherichia coli, Bacillus subtilis,
and Burkholderia cepacia WZ1 to oxidative stress of exposure to quinclorac.
J Environ Sci Health Part B Pestic Food Contam Agric Wastes 39: 431–441.
18. Lü ZM, Sang LY, Li ZM, Min H (2009) Catalase and superoxide dismutase
activities in a Stenotrophomonas maltophilia WZ2 resistant to herbicide pollution.
Ecotox Environ Safe 72: 136–143.
19. Lin X, Xu X, Yang C, Zhao Y, Feng Z, et al. (2009) Activities of antioxidant
enzymes in three bacteria exposed to bensulfuron-methyl. Ecotox Environ Safe
72: 1899–1904.
20. Gratão PL, Polle A, Lea PJ, Azevedo RA (2005) Making the life of heavy metal-
stressed plants a little easier. Funct Plant Biol 32: 481–494.
21. Monteiro CC, Carvalho RF, Gratão PL, Carvalho G, Tezotto T, et al. (2011)
Biochemical responses of the ethylene-insensitive Never ripe tomato mutant
subjected to cadmium and sodium stresses. Environ Exp Bot 71: 306–320.
22. Zhang Y, Meng DF, Wang ZG, Guo HS, Wang Y, et al. (2012) Oxidative stress
response in atrazine-degrading bacteria exposed to atrazine. J Hazard Mater
229: 434–438.
23. Martins PF, Carvalho G, Gratão PL, Dourado MN, Pileggi M, et al. (2011)
Effects of the herbicides acetochlor and metolachlor on antioxidant enzymes in
soil bacteria. Process Biochem 46: 1186–1195.
24. Gratão PL, Monteiro CC, Carvalho RF, Tezotto T, Piotto FA, et al. (2012)
Biochemical dissection of diageotropica and Never ripe tomato mutants to Cd-
stressful conditions. Plant Physiol Biochem 56: 79–96.
25. Monteiro CC, Rolão MB, Franco MR, Peters LP, Cia MC, et al. (2012)
Biochemical and histological characterization of tomato mutants. An Acad Bra
Cienc 84: 573–585.
26. Cia MC, Guimarães ACR, Medici LO, Chabregas SM, Azevedo RA (2012)
Antioxidant responses to water deficit by drought-tolerant and -sensitive
sugarcane varieties. Ann Appl Biol 161: 313–324.
27. Carvalho RF, Piotto FA, Schmidt D, Peters LP, Monteiro CC, et al. (2011) Seed
priming with hormones does not alleviate induced oxidative stress in maize
seedlings subjected to salt stress. Sci Agric 68: 598–602.
28. Lushchak VI (2011) Adaptive response to oxidative stress: Bacteria, fungi, plants
and animals. Comp Biochem Physiol Part C Toxicol Pharmacol 153: 175–190.
29. Masip L, Veeravalli K, Georgiou G (2006) The many faces of glutathione in
bacteria. Antioxid Redox Signaling 8: 753–762.
30. Ghelfi A, Gaziola SA, Cia MC, Chabregas SM, Falco MC, et al. (2011) Cloning,
expression, molecular modelling and docking analysis of glutathione transferase
from Saccharum officinarum. Ann Appl Biol 159: 267–280.
31. Allocati N, Federici L, Masulli M, Di Ilio C (2009) Glutathione transferases in
bacteria. Febs Journal 276: 58–75.
32. Camargo AO, Moniz AC, Jorge JA, Valadares JMA (2009) Métodos de análise
química, mineralógica e física de solos do instituto agronômico de Campinas.
Instituto Agronômico 106, 77.
33. Nie Z-J, Hang B-J, Cai S, Xie X-T, He J, et al. (2011) Degradation of
Cyhalofop-butyl (CyB) by Pseudomonas azotoformans strain QDZ-1 and cloning
of a novel gene encoding CyB-hydrolyzing esterase. J Agric Food Chem 59:
6040–6046.
34. Martins PF, Martinez CO, Carvalho G, Carneiro PIB, Azevedo RA, et al.
(2007) Selection of microorganisms degrading S-metolachlor herbicide. Braz
Arch Biol Technol 50: 153–159.
35. Araujo WL, Marcon J, Maccheroni W, Van Elsas JD, Van Vuurde JWL, et al.
(2002) Diversity of endophytic bacterial populations and their interaction with
Xylella fastidiosa in citrus plants. Appl Environ Microbiol 68: 4906–4914.
36. Heuer H, Krsek M, Baker P, Smalla K, Wellington EMH (1997) Analysis of
actinomycete communities by specific amplification of genes encoding 16S
rRNA and gel-electrophoretic separation in denaturing gradients. Appl Environ
Microbiol 63: 3233–3241.
37. Lane DJ, Pace B, Olsen GJ, Stahl DA, Sogin ML, et al. (1985) Rapid-
determination of 16S ribosomal RNA-sequences for phylogenetic analyses. Proc
Natl Acad Sci USA 82: 6955–6959.
38. Konstantinidis KT, Tiedje JM (2005) Towards a genome-based taxonomy for
prokaryotes. J Bacteriol 187: 6258–6264.
39. Saitou N, Nei M (1987) The neighbor-joining method - a new method for
reconstructing phylogenetic trees. Mol Biol Evol 4: 406–425.
40. Tamura K, Dudley J, Nei M, Kumar S (2007) MEGA4: Molecular evolutionary
genetics analysis (MEGA) software version 4.0. Molecular Mol Biol Evol 24:
1596–1599.
41. Sangali S, Brandelli A (2000) Feather keratin hydrolysis by a Vibrio sp strain kr2.
J Appl Microbiol 89: 735–743.
42. Heath RL, Packer L (1968) Photoperoxidation in isolated chloroplasts I. Kinetics
and stoichiometry of fatty acid peroxidation. Arch Biochem Biophys 125: 189–
198.
43. Anderson ME (1985) Dertermination of glutathione and glutathione disulfide in
biological samples. Methods Enzymol 113: 548–555.
44. Gratão PL, Monteiro CC, Antunes AM, Peres LEP, Azevedo RA (2008)
Acquired tolerance of tomato (Lycopersicon esculentum cv. Micro-Tom) plants to
cadmium-induced stress. Ann Appl Biol 153: 321–333.
45. Bradford MM (1976) A rapid and sensitive method for the quantitation of
microgram quantities of protein utilizing the principle of protein-dye binding.
Anal Biochem 72: 248–254.
46. Garcia JS, Gratão PL, Azevedo RA, Arruda MAS (2006) Metal contamination
effects on sunflower (Helianthus annuus L.) growth and protein expression in
leaves during development. J Agric Food Chem 54: 8623–8630.
47. Azevedo RA, Alas RM, Smith RJ, Lea PJ (1998) Response of antioxidant
enzymes to transfer from elevated carbon dioxide to air and ozone fumigation, in
the leaves and roots of wild-type and a catalase-deficient mutant of barley.
Physiol Plant 104: 280–292.
48. Boaretto LF, Carvalho G, Borgo L, Creste S, Landell MGA, et al. (2013) Water
stress reveals differential antioxidant responses of tolerant and non-tolerant
sugarcane genotypes. Plant Physiol Biochem 74: 165–175.
49. Gomes-Junior RA, Gratão PL, Gaziola SA, Mazzafera P, Lea PJ, et al. (2007)
Selenium-induced oxidative stress in coffee cell suspension cultures. Funct Plant
Biol 34: 449–456.
50. Zablotowicz RM, Hoagland RE, Locke MA, Hickey WJ (1995) Glutathione-S-
transferase activity and metabolism of glutathione conjugates by rhizosphere
bacteria. Appl Environ Microbiol 61: 1054–1060.
51. Mongkolsuk S, Dubbs J, Vattanaviboon P (2005) Chemical modulation of
physiological adaptation and cross-protective responses against oxidative stress in
soil bacterium and phytopathogen, Xanthomonas. J Ind Microbiol Biotechnol
32: 687–690.
52. Dwivedi S, Singh BR, Al-Khedhairy AA, Musarrat J (2011) Biodegradation of
isoproturon using a novel Pseudomonas aeruginosa strain JS-11 as a multifunctional bioinoculant of environmental significance. J Hazard Mater 185:
938–944.
53. Viegas CA, Costa C, André S, Viana P, Ribeiro R, et al. (2012) Does S-
metolachlor affect the performance of Pseudomonas sp. strain ADP as
bioaugmentation bacterium for atrazine-contaminated soils? PLoS ONE 7(5):
e37140.

54. Ramu S, Seetharaman B (2014) Biodegradation of acephate and methamidophos by a soil bacterium *Pseudomonas aeruginosa* strain Is-6. Environ Sci Heal B 49: 23–34.

55. aMattos MLT, Thomas R (1996) Degradation of the herbicide clomazone by *Pseudomonas fluorescens*. In: W Sand editor.Biodeterioration and Biodegradation.Weinheim: V C H Verlagsgesellschaft. pp.623–630.

56. Tironi SP, Belo AF, Fialho CMT, Galon L, Ferreira EA, et al. (2009) Efeito de herbicidas na atividade microbiana do solo. Planta Daninha 27: 995–1004.

57. García-Cruz U, Celis LB, Poggi P, Meraz M (2010) Inhibitory concentrations of 2,4D and its possible intermediates in sulfate reducing biofilms. J Hazard Mater 179: 591–595.

58. Ferhatoglu Y, Avdiushko S, Barrett M (2005) The basis for the safening of clomazone by phorate insecticide in cotton and inhibitors of cytochrome P450s. Pestic Biochem Physiol 81: 59–70.

59. Matsue Y, Mizuno H, Tomita T, Asami T, Nishiyama M, et al. (2010) The herbicide ketoclomazone inhibits 1-deoxy-D-xylulose 5-phosphate synthase in the 2-C-methyl-D-erythritol 4-phosphate pathway and shows antibacterial activity against *Haemophilus influenzae*. J Antibiot 63: 583–588.

60. Imlay JA (2008) Cellular defenses against superoxide and hydrogen peroxide. Annu Rev Biochem 77: 755–776.

61. Shao TJ, Yang GQ, Wang MZ, Lu ZM, Min H, et al. (2010) Reduction of oxidative stress by bioaugmented strain *Pseudomonas* sp HF-1 and selection of potential biomarkers in sequencing batch reactor treating tobacco wastewater. Ecotoxicology 19: 1117–1123.

62. Ryan KC, Johnson OE, Cabelli DE, Brunold TC, Maroney MJ (2010) Nickel superoxide dismutase: structural and functional roles of Cys2 and Cys6. J Biol Inorg Chem 15: 795–807.

63. Lushchak VI (2001) Oxidative stress and mechanisms of protection against it in bacteria. Biochemistry-Moscow 66: 476–489.

64. Goulielmos GN, Arhontaki K, Eliopoulos E, Tserpistali K, Tsakas S, et al. (2003) Drosophila Cu,Zn superoxide dismutase gene confers resistance to paraquat in *Escherichia coli*. Biochem Biophys Res Commun 308: 433–438.

65. Zeng HW, Cai YJ, Liao XR, Qian SL, Zhang F, et al. (2010) Optimization of catalase production and purification and characterization of a novel cold-adapted Cat-2 from mesophilic bacterium *Serratia marcescens* SYBC-01. Ann Microbiol 60: 701–708.

66. Dávila Costa JS, Albarracín VH, Abate CM (2011) Responses of environmental *Amycolatopsis* strains to copper stress. Ecotox Environ Safe 74: 2020–2028.

67. Hultberg M (1998) Rhizobacterial glutathione levels as affected by starvation and cadmium exposure. Curr Microbiol 37: 301–305.

68. Veremeenko EG, Maksimova NP (2010) Activation of the antioxidant complex in *Pseudomonas aurantiaca*-producer of phenazine antibiotics. Microbiology 79: 439–444.

69. Smirnova GV, Oktyabrsky ON (2005) Glutathione in bacteria. Biochemistry-Moscow+ 70: 1199–1211.

70. Foyer CH, López-Delgado H, Dat JF, Scott IM (1997) Hydrogen peroxide and glutathione associated mechanisms of acclamatory stress tolerance and signaling. Physiol Plant 100: 241–254.

71. Gomes-Junior RA, Moldes CA, Delite FS, Gratão PL, Mazzafera P, et al. (2006) Nickel elicits a fast antioxidant response in *Coffea arabica* cells. Plant Physiol Biochem 44: 420–429.

72. Gomes-Junior RA, Moldes CA, Delite FS, Pompeu GB, Gratão PL, et al. (2006) Antioxidant metabolism of coffee cell suspension cultures in response to cadmium. Chemosphere 65: 1330–1337.

73. Labrou NE, Karavangeli M, Tsaftaris A, Clonis YD (2005) Kinetic analysis of maize glutathione S-transferase I catalysing the detoxification from chloroacetanilide herbicides. Planta 222: 91–97.

13

Cadmium-Induced Hydrogen Sulfide Synthesis Is Involved in Cadmium Tolerance in *Medicago sativa* by Reestablishment of Reduced (Homo)glutathione and Reactive Oxygen Species Homeostases

Weiti Cui[1], Huiping Chen[2], Kaikai Zhu[1], Qijiang Jin[1], Yanjie Xie[1], Jin Cui[1], Yan Xia[1], Jing Zhang[1], Wenbiao Shen[1]*

1 College of Life Sciences, Laboratory Center of Life Sciences, Nanjing Agricultural University, Jiangsu Province, Nanjing, China, 2 Key Laboratory of Protection and Development Utilization of Tropical Crop Germplasm Resources, Hainan University, Haikou, China

Abstract

Until now, physiological mechanisms and downstream targets responsible for the cadmium (Cd) tolerance mediated by endogenous hydrogen sulfide (H_2S) have been elusive. To address this gap, a combination of pharmacological, histochemical, biochemical and molecular approaches was applied. The perturbation of reduced (homo)glutathione homeostasis and increased H_2S production as well as the activation of two H_2S-synthetic enzymes activities, including L-cysteine desulfhydrase (LCD) and D-cysteine desulfhydrase (DCD), in alfalfa seedling roots were early responses to the exposure of Cd. The application of H_2S donor sodium hydrosulfide (NaHS), not only mimicked intracellular H_2S production triggered by Cd, but also alleviated Cd toxicity in a H_2S-dependent fashion. By contrast, the inhibition of H_2S production caused by the application of its synthetic inhibitor blocked NaHS-induced Cd tolerance, and destroyed reduced (homo)glutathione and reactive oxygen species (ROS) homeostases. Above mentioned inhibitory responses were further rescued by exogenously applied glutathione (GSH). Meanwhile, NaHS responses were sensitive to a (homo)glutathione synthetic inhibitor, but reversed by the cotreatment with GSH. The possible involvement of cyclic AMP (cAMP) signaling in NaHS responses was also suggested. In summary, LCD/DCD-mediated H_2S might be an important signaling molecule in the enhancement of Cd toxicity in alfalfa seedlings mainly by governing reduced (homo)glutathione and ROS homeostases.

Editor: Ji-Hong Liu, Key Laboratory of Horticultural Plant Biology (MOE), China

Funding: This research was supported by the Fundamental Research Funds for the Central Universities (KYZ201316), the National Natural Science Foundation of China (grants no. 30971711, J1210056, J1310015), and the Priority Academic Program Development of Jiangsu Higher Education Institutions. The funders had no role in study design, data collection and analysis, decision to publish, or preparation of the manuscript.

Competing Interests: The authors have declared that no competing interests exist.

* Email: wbshenh@njau.edu.cn

Introduction

Cadmium (Cd) contamination is a non-reversible accumulation process, with the estimated half-life and high plant-soil mobility, thus resulting in a serious threat to human health through food chains. Normally, Cd exposure leads to the inhibition of plant growth, decrease of crop yield, and even plant cell death [1,2]. Indirectly stimulated generation of reactive oxygen species (ROS) that modify the antioxidant defence and bring out oxidative stress is ascribed to one of the Cd toxicities in plants, and therefore lipid peroxidation is considered as a hallmark of Cd exposure [3].

In plants, there are a lot of antioxidant defence mechanisms, which could keep the normally formed ROS at a low level and prevent them from exceeding toxic thresholds [3,4]. The glutathione (GSH) and ascorbate were subsequently recognized as the heart of the redox hub [5]. In plants, GSH is synthesized by two ATP-dependent steps: γ-glutamylcysteine (γ-EC) is synthesized from L-glutamate and L-cysteine by γ-glutamyl cysteine

synthetase (γ-ECS, also called as γ-GCS); and the second step, glycine is conjunct to γ-EC by glutathione synthetase (GS) [6,7]. In soybean and alfalfa plants, GSH homolog homoglutathione (hGSH) synthesized by homoglutathione synthetase (hGS) from β-alanine and γ-EC, is more abundant than GSH [8]. The rate of glutathione reductase (GR) reaction was the same with either oxidized glutathione (GSSG) or oxidized homoglutathione (hGSSGh) as the substrate [7]. Upon Cd exposure, it was confirmed that the rapid accumulation of peroxides and depletion of GSH and hGSH causes redox imbalance in *Medicago sativa* [9]. Subsequent experiments with comparing ten pea genotypes showing that, activities of ascorbate peroxidase (APX) decreased, but concentrations of GSH increased in the less Cd-sensitive genotypes [10].

Another sulphur-containing compound, hydrogen sulfide (H_2S), previously known as a toxic gas, has been progressively recognized as a gaseous signaling molecule with multiple functions in animals [11,12]. For example, H_2S has been revealed as a cytoprotectant and a regulator in various biological processes, such as oxidative stress suppression, smooth muscle relaxation, proliferation inhibition and apoptosis triggering [13–16]. Meanwhile, although previous reports observed that many plants can emit H_2S [17–19], there have been few studies on the physiological role of H_2S in *planta* during the last century.

In mammals, the majority of endogenous H_2S was produced by two enzymes, cystathionine β-synthase (CBS, EC 4.2.1.22) and cystathionine γ-lyase (CSE, EC 4.4.1.1), from L-cysteine [20]. Cysteine-degrading enzymes such as cysteine desulfhydrases are hypothesized to be involved in H_2S release in plants [21]. Previously, two specific desulfhydrases, L-cysteine desulfhydrase (LCD, EC 4.4.1.1; also called L-CDes or L-DES) and D-cysteine desulfhydrase (DCD, EC 4.4.1.15; also called D-CDes or D-DES), have been isolated and partially analyzed from *Arabidopsis thaliana* [22–24]. The LCD, which is considered as the most important enzyme with H_2S production in plants, shares a 100% sequence homolog with CSE in mammals [25]. By using sodium hydrosulfide (NaHS) as a H_2S donor, ample evidence further suggested that H_2S can protect plants against various stress-induced damage, such as salinity stress [26], drought [27–29], heavy metal exposure [30,31], and heat shock [32]. Additionally, H_2S can act as an inducer in several developmental processes, including adventitious root formation [33] and flower senescence [34]. However, exogenously applied H_2S donor without checking the kinetics of H_2S synthesis including corresponding metabolic enzyme activities or transcripts, may not fully replicate the function of endogenous H_2S in plants.

Cyclic AMP (adenosine 3′, 5′-cyclic monophosphate, cAMP) is a well-known second messenger playing important roles in many physiological processes. The cAMP is synthesized by adenylyl cyclase and broken down by cNMP phosphodiesterase. Dedioxyadenosine (DDA) and 1,3-diazinane-2,4,5,6-tetrone (alloxan) are well characterized as the inhibitors of adenylyl cyclase. Likewise, cNMP phosphodiesterase is sensitive to the inhibitor 1-methyl-3-(2-methylpropyl)-7*H*-purine-2,6-dione (IBMX) [35,36]. In animals, there is ample evidences to show H_2S-activited cAMP level or H_2S-regulated cAMP homeostasis [37,38]. It was found that H_2S acted via cAMP-mediated PI3K/Akt/p70S6K signal pathways to inhibit hippocampal neuronal apoptosis and protect neurons from OGD/R-induced injury [39]. However, the functions of cAMP signaling in H_2S-alleviated Cd stress in plants are still poorly understood.

Thus, the aim of this study was to investigate the signaling role of endogenous H_2S in the tolerance of *Medicago sativa* seedlings to Cd stress. For this purpose, we preliminarily investigated the synthesis of endogenous H_2S under Cd stress, which has not been fully performed. Furthermore, the effects of H_2S on GSH and hGSH metabolism, as well as ROS homeostasis were checked. Our results further indicated that Cd stress triggered endogenous H_2S production catalyzed by LCD/DCD pathways, and the elevated H_2S acts as a signal improving the homeostasis of GSH pool and keeping ROS under control, both of which finally contributed to Cd tolerance. Finally, the possible involvement of cAMP signaling in NaHS responses was also suggested.

Materials and Methods

Plant material, growth condition

Commercially available alfalfa (*Medicago sativa* L. Victoria) seeds were surface-sterilized with 5% NaClO for 10 min, and rinsed extensively in distilled water before being germinated for 1 d at 25°C in the darkness. Uniform seedlings were then selected and transferred to the plastic chambers and cultured with nutrient medium (quarter-strength Hoagland's solution) in the illuminating incubator (14 h light with a light intensity of 200 $\mu mol \cdot m^{-2} \cdot s^{-1}$, 25±1°C, and 10 h dark, 23±1°C). Five-day-old seedlings were then incubated in quarter-strength Hoagland's solution with or without varying concentrations of NaHS (Sigma-Aldrich; St Louis, MO, USA) or the other indicated chemicals (2 mM DL-propargylglycine (PAG), 1 mM GSH, 1 mM L-buthionine-sulfoximine (BSO), 50 μM 8-Br-cAMP (8Br), 200 μM alloxan (All), 1 mM DDA, and 500 μM IBMX) alone, or the combination of treatments for 6 h followed by the indicated time points of incubation in 200 μM CdCl$_2$. Seedlings without chemicals were used as the control (Con). The pH for both nutrient medium and treatment solutions was adjusted to 6.0.

After various treatments, above-ground parts and root tissues of seedlings were sampled immediately or flash-frozen in liquid nitrogen, and stored at −80°C for further analysis. Among these, above-ground parts and root tissues of 240 seedlings were respectively used for the determination of Cd contents. Seedling root tissues were also used for fresh weight determination (10 seedlings), thiobarbituric acid reactive substances (TBARS) content determination (120 seedlings), and other indicated tests (30 seedlings).

Determination of H_2S content, LCD and DCD activity

Hydrogen sulfide content was determined according to the method previously reported [19,34]. 100 mg of alfalfa seedling roots from 30 seedlings were ground under liquid nitrogen and extracted by 1 ml phosphate buffered saline (50 mM, pH 6.8) containing 0.1 M EDTA and 0.2 M ascorbic acid. After centrifugation at 13000 g for 15 min at 4°C, 400 μl of the supernatant was injected to 200 μl 1% zinc acetate and 200 μl 1 N HCl. After 30 min reaction, 100 μl 5 mM dimethyl-*p*-phenylene-diamine dissolved in 7 mM HCl was added to the trap followed by the injection of 100 μl 50 mM ferric ammonium sulfate in 200 mM HCl. After 15 min incubation at room temperature, the amount of H_2S was determined at 667 nm. Solutions with different concentrations of Na$_2$S were used in a calibration curve.

100 mg of alfalfa seedling roots from 30 seedlings were used for activity determination. The activities of LCD and DCD were determined as described by the methods previously reported [23,40]. L-cysteine desulfhydrase (LCD) activity was measured by the release of H_2S from L-cysteine in the presence of dithiothreitol (DTT). The formation of methylene blue was determined at 670 nm. To removal of the background, content of H_2S in the extracted protein solution was measured by same way with 50% trichloroacetic acid (TCA) instead of L-cysteine. The final LCD activity was calculated from the difference between the measured LCD activity and the background. D-cysteine desulfhydrase (DCD) activity was measured by the same method with following modifications: D-cysteine instead of L-cysteine, the pH of Tris-HCl was 8.0 rather than 9.0. Solutions with different concentrations of Na$_2$S were prepared, treated in the same way as the assay samples and were used for the quantification of enzymatically formed H_2S.

Determination of thiobarbituric acid reactive substances (TBARS), (h)GSH and (h)GSSG(h) contents

Lipid peroxidation was estimated by measuring the amount of TBARS as previously described [41]. About 400 mg of root tissues from 120 seedlings was ground in 0.25% 2-thiobarbituric acid (TBA) in 10% TCA using a mortar and pestle. After heating at 95°C for 30 min, the mixture was quickly cooled in an ice bath and centrifuged at 10,000×*g* for 10 min. The absorbance of the supernatant was read at 532 nm and corrected for unspecific turbidity by subtracting the absorbance at 600 nm. The concentration of lipid peroxides together with oxidatively modified proteins of plants were thus quantified in terms of TBARS amount using an extinction coefficient of 155 mM^{-1} cm^{-1} and expressed as nmol g^{-1} fresh weight (FW).

(h)GSH (GSH + hGSH) and (h)GSSG(h) (GSSG + hGSSGh) contents were measured by the 5,5′dithio-bis-(2-nitrobenzoic acid) (DTNB)-glutathione reductase (GR) recycling assay [41,42]. Frozen root tissues from 30 seedlings were homogenized in cold 5% 5-sulfosalicylic acid. The homogenate was centrifuged at 12,000×g for 20 min at 4°C and the supernatant was collected. Total glutathione ((h)GSH plus (h)GSSG(h)) was determined in the homogenates spectrophotometrically at 412 nm, using GR, DTNB, and NADPH. (h)GSSG(h) contents were determined by the same method in the presence of 2-vinylpyridine and (h)GSH contents were calculated from the difference between total glutathione and (h)GSSG(h).

Thiol analysis by reversed-phase HPLC

Low-molecular-weight thiols and their corresponding disulfides contents in root tissues from 30 seedlings were measured according to the methods previously reported [43–45], through derivatization with monobromobimane (mBBr) after reduction with DTT with or without previously blocked with N-ethylmaleimide (NEM), and separation by reversed-phase HPLC (Agilent Technologies, 1200 series Quaternary, Foster city, USA).

Histochemical analyses

Histochemical detection of lipid peroxidation and loss of plasma membrane integrity was performed with Schiff's reagent and with Evans blue described by previous reports [41,45].

Real-time quantitative RT-PCR analysis

Total RNA from root tissues of 30 seedlings was extracted using Trizol reagent (Invitrogen) according to the manufacturer's instructions. DNA-free total RNA (2 μg) from different treatments was used for first-strand cDNA synthesis in a 20-μL reaction volume containing 2.5 units of avian myeloblastosis virus reverse transcriptase XL (TaKaRa) and oligo dT primer.

Real-time quantitative RT-PCR reactions were performed with Mastercycler realplex^2 real-time PCR system (Eppendorf, Hamburg, Germany) using the SYBR *Premix Ex Taq* (TaKaRa) according to the user manual. The cDNA was amplified using primers (Table S1). The expression levels of the genes are presented as values relative to the corresponding control samples under the indicated conditions, with normalization of data to the geometric average of two internal control genes *MSC27* and *Actin2* [46].

Visualization of endogenous ROS by LSCM

Endogenous ROS was imaged using the fluorescent probe H$_2$DCFDA, and then scanned described by [45,47].

Statistical analysis

Values are means ± SD of three different experiments with three replicated measurements. Differences among treatments were analysed by one-way ANOVA, taking $P<0.05$ as significant according to Duncan's multiple range test.

Results

(h)GSH depletion and increased endogenous H$_2$S synthesis triggered by Cd stress

Considering alfalfa plants contain a thiol tripeptide homolog, hGSH, instead of or in addition to GSH [8,9], we detected the concentrations of GSH and hGSH. As shown in Table 1, the content of hGSH in alfalfa seedling roots under the control conditions, was about 8-fold higher than that of GSH. Similarly, hGSSGh is the main component of (h)GSSG(h) (total of hGSSGh and GSSG), because the GSSG content was almost negligible.

To further elucidate the correlation among GSH pool, H$_2$S and Cd tolerance, the time course of (homo)glutathione ((h)GSH; total of hGSH and GSH, and (h)GSSG(h)) contents, and H$_2$S synthesis were investigated in alfalfa seedling roots upon Cd stress. As expected, a decrease of (h)GSH content (especially hGSH) and an increase of (h)GSSH(h) (especially hGSSGh) level were progressively triggered by Cd stress within 12 h, thus leading to a decreased (h)GSH/(h)GSSH(h) ratio (12 h; Figure 1A-C), an important parameter for the intracellular redox status in *planta* upon Cd stress [3,45]. The ratio of hGSH/hGSSGh exhibited the similar tendency (Table 1). These results were consistent with the observed Cd toxicity, confirmed by the histochemical staining detecting the aggravated loss of plasma membrane integrity and lipid peroxidation with Evans blue and Schiff's reagent, increased TBARS content and growth stunt of seedling roots (Figure S1).

Because H$_2$S synthesis could be induced by oxidative stress and depletion of GSH both in animals and plants [48–50], we simultaneously investigated the production of H$_2$S in seedling roots after the exposure to Cd. Similar to the recent report [51], the production of H$_2$S was continuously increased after the exposure to Cd alone for 12 h (Figure 1D). The changes in activities of two H$_2$S synthetic enzymes LCD and DCD displayed similar tendencies (Figure 1E and F). Apparently, the reduced (homo)glutathione depletion and increased endogenous H$_2$S synthesis preceded Cd toxicity in alfalfa seedlings.

NaHS not only mimics intracellular H$_2$S content, but also alleviates Cd toxicity

Previous results revealed that the exogenously applied NaHS, a H$_2$S donor, alleviates Cd toxicity in bermudagrass seedlings [51]. Therefore, a preliminary work was carried out to compare the oxidative damage and growth performance of alfalfa seedlings upon Cd exposure with or without the indicated concentrations of NaHS pretreatment. Firstly, the results of histochemical staining and TBARS contents revealed that NaHS at 100 (in particular) and 500 μM was able to significantly decreased Cd-induced lipid peroxidation (Figure S1A and B). These beneficial roles were also supported by the changes of fresh weight of ten alfalfa seedling roots, showing that NaHS at 100 and 500 μM had the greatest effects on the alleviation of the inhibition of root growth caused by Cd stress (Figure S1C). The beneficial roles of 100 μM NaHS alone were also observed. Subsequent work confirmed that H$_2$S rather than other sulphur-containing derivatives and sodium exhibited the cytoprotective role in the improvement of Cd toxicity by using a series of sulphur- and sodium-containing chemicals including Na$_2$S, Na$_2$SO$_4$, Na$_2$SO$_3$, NaHSO$_4$, NaHSO$_3$, and NaAc, in comparison with the positive roles of NaHS (Figure S2).

Accordingly, we observed that the treatment with 100 μM NaHS for 3 h resulted in the enhancement of endogenous H$_2$S level in alfalfa seedling roots, which also mimicked a physiological response elicited by Cd alone for 12 h (Figure 2A). The addition of Cd to the NaHS-pretreated plants further strengthened the increased H$_2$S content. Therefore, 100 μM NaHS was used to mimic the physiological role of intracellular H$_2$S in the subsequent experiments.

Table 1. Concentrations of low molecular weight thiols and their disulfides, and hGSH/hGSSGh ratio in root tissues.

Treatment	cysteine (nmol g^{-1} FW)	cysteine disulfide (nmol g^{-1} FW)	γ-EC (nmol g^{-1} FW)	γ-EC disulfide (nmol g^{-1} FW)	GSH (nmol g^{-1} FW)	GSSG (nmol g^{-1} FW)	hGSH (nmol g^{-1} FW)	hGSSGh (nmol g^{-1} FW)	hGSH/hGSSGh
Con→Con	30±1 d	3.8±0.8 c	10±1 e	1.5±0.1	27±2 bc	0.2±1.9	252±16 b	28±2 c	8.86
Con→Cd	33±1 cd	5.7±0.8 b	14±2 d	1.7±0.1	21±2 c	0.2±1.4	112±13 f	33±1 bc	3.41
NaHS→Cd	40±2 c	4.0±0.6 c	18±2 bc	1.4±0.5	26±4 c	0.1±1.4	163±14 de	33±4 bc	4.89
NaHS→Con	34±2 cd	3.4±0.7 c	8±0 e	1.2±0.5	36±1 b	0.2±1.1	309±14 a	30±2 bc	10.23
NaHS + PAG→Cd	54±7 b	4.3±0.4 bc	21±3 b	1.2±0.5	29±5 bc	0.3±0.7	144±8 e	41±6 a	3.55
NaHS + PAG + GSH→Cd	65±8 a	4.7±1.6 bc	27±1 a	1.4±0.5	46±11 a	0.7±0.6	179±7 d	36±5 ab	4.91
PAG→Cd	52±6 b	7.4±0.7 a	21±1 b	1.6±0.2	29±1 bc	0.9±0.9	82±12 g	33±3 bc	2.41
PAG→Con	56±4 b	4.0±0.3 c	17±3 cd	1.2±0.5	29±2 bc	0.7±0.7	206±28 c	36±2 ab	5.67

Seedlings were pretreated with or without 100 μM NaHS, 2 mM PAG, 1 mM GSH, individual or combination for 6 h, and then exposed to 200 μM CdCl$_2$ for another 12 h. Values are means ± SD of three independent experiments with three replicates for each. Different letters within columns indicate significant differences ($P<0.05$) according to Duncan's multiple range test.

Changes of low molecular weight thiols and their disulfides as well as representative transcripts in response to NaHS

To determine the influence of H$_2$S at physiologically concentrations on (h)GSH depletion, GSH pool and corresponding metabolism associated genes were investigated. As shown in Figure 2B, the time-course analysis revealed that (h)GSH contents in seedling roots were significantly enhanced by the pretreatment with NaHS for 6 h, and remained high through 24 h of further incubation in the control solution. Meanwhile, NaHS pretreatment was able to slow down the decreased (h)GSH levels caused by Cd exposure. Changes of the (h)GSH/(h)GSSG(h) ratio also exhibited the similar tendencies (Figure 2C). Comparatively, Cd-induced cysteine and γ-EC (in particular), and cysteine disulfide contents were differentially strengthened or blocked by NaHS pretreatment, respectively (Table 1).

These results arises the question that, whether this increases in metabolites are, at least in part, duo to changes in the expression of genes involved in (h)GSH metabolism. Therefore, the expression of *ECS*, *GS*, and *GR1* genes, were analyzed by real-time RT-PCR. Results of Figure 3A and B revealed that the transcripts of *ECS*, *GS* and *GR1* (especially) in seedling roots approximately displayed a time-dependent increase during Cd stress for 24 h, while the transcriptional profiles of these genes in the control samples were relatively constant during the same period. The pretreatment with NaHS for 6 h in culture solution increased above transcripts, which were differentially strengthened by thereafter Cd stress.

NaHS-induced Cd tolerance, (h)GSH and ROS homeostases were sensitive to PAG, but rescued by GSH

To further verify the involvement of endogenous H$_2$S in Cd tolerance, DL-propargylglycine (PAG), an effective H$_2$S synthetic inhibitor [27], and GSH, applied individually and in combination, were used in the subsequent experiment. After 72 h exposure to Cd, the alfalfa seedlings displays severe growth inhibition both in roots and above ground parts, compared to control samples, both of which were improved by NaHS pretreatment (Figure 4A). By contrast, the improvement of seedling growth inhibition as well as the reestablishment of (h)GSH homeostasis triggered by NaHS were sensitive to PAG, but blocked by exogenously applied GSH (Figure 4, Figure S3A). An aggravated Cd toxicity in seedling growth inhibition was also observed when PAG was pretreated.

In an attempt to assess the potential role of endogenous H$_2$S in ROS homeostasis in Cd-stressed seedlings, ROS production was visualized by staining with H$_2$DCFDA. As expected, ROS in root tips with Cd alone were produced considerably, suggesting a perturbation in ROS homeostasis (Figure 5). However, the pretreatment with NaHS reduced the ROS abundance. Further results revealed that PAG pretreatment increased the H$_2$DCFDA fluorescence in Cd-stressed seedling roots, which was further blocked by the addition of GSH. The changes of TBARS content, an indictor of lipid peroxidation, exhibited the similar tendencies (Figure S3B).

Cd treatment caused the accumulation of Cd contents both in shoot and root (particularly) tissues (Figure S4). Similar to the previous reports [31], NaHS decreased Cd accumulation, which was significantly reversed by PAG, but was further blocked by the cotreatment with GSH.

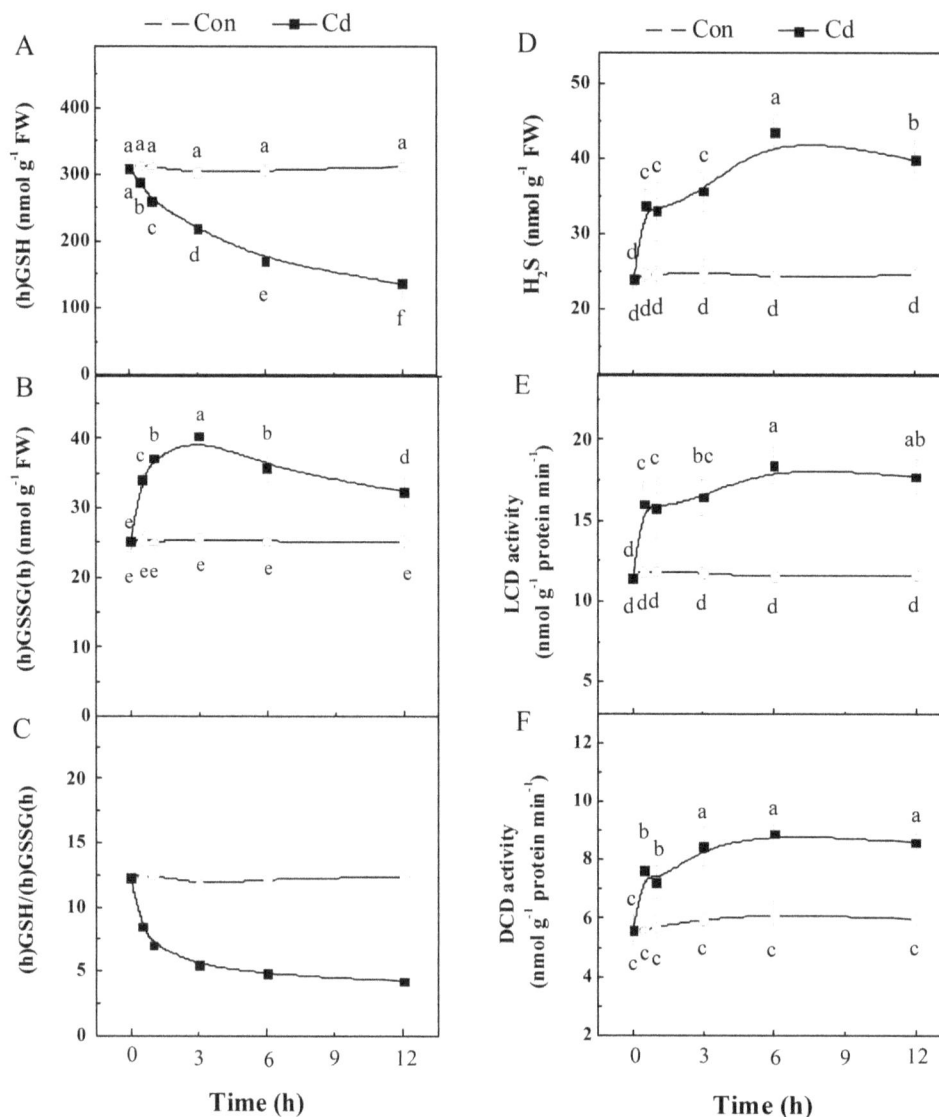

Figure 1. Time course changes of GSH pool and H₂S synthesis upon Cd stress. Upon 200 µM CdCl₂ treatment for 12 h, contents of (h)GSH (A), (h)GSSG(h) (B) and H₂S (D), the ratio of (h)GSH/(h)GSSG(h) (C), and the activities of LCD (E) and DCD (F) in root tissues were analyzed. Values are means ± SD of three independent experiments with three replicates for each. Bars denoted by the same letter did not differ significantly at $P<0.05$ according to Duncan's multiple range test.

Transcripts of representative antioxidant defense genes were sensitive to PAG, but rescued by GSH

Since ROS homeostasis was reestablished by NaHS in stressed conditions, the real-time RT-PCR test of corresponding genes involved in their metabolism, i.e. *Cu, Zn-SOD, APX1*, and *GPX* [3,5], were analysed. The results of Figure 6 revealed that in comparison with Cd alone samples, NaHS pretreatment followed by Cd exposure resulted in the enhancement in the transcript levels of *Cu, Zn-SOD, APX1*, and *GPX* in alfalfa seedling roots. The addition of PAG, however, significantly blocked the increases in the transcripts levels of these representative antioxidant enzymes induced by NaHS, all of which were reversed when GSH was added together with PAG.

NaHS responses were sensitive to a (h)GSH synthetic inhibitor, but reversed by the added GSH

The involvement of (h)GSH homeostasis in NaHS-induced cytoprotective against Cd stress were further investigated using a (h)GSH synthetic inhibitor and GSH applied exogenously. Pretreatment with NaHS, and ʟ-buthionine-sulfoximine (BSO) at 1 mM, a concentration expected to be effective [52], exhibited an aggravated Cd toxicity, which was confirmed by the severe growth stunt and TBARS overproduction, in comparison with Cd plus NaHS (Figure 7A and B). Similarly, NaHS-mediated reestablishment of (h)GSH homeostasis in Cd stressed alfalfa seedling roots was also perturbed by BSO (Figure 7C and D), which was confirmed by the significant decreased (h)GSH content and the ratio of (h)GSH/(h)GSSG(h), respect to Cd alone. By contrast, above BSO responses were sensitive to the addition of GSH when

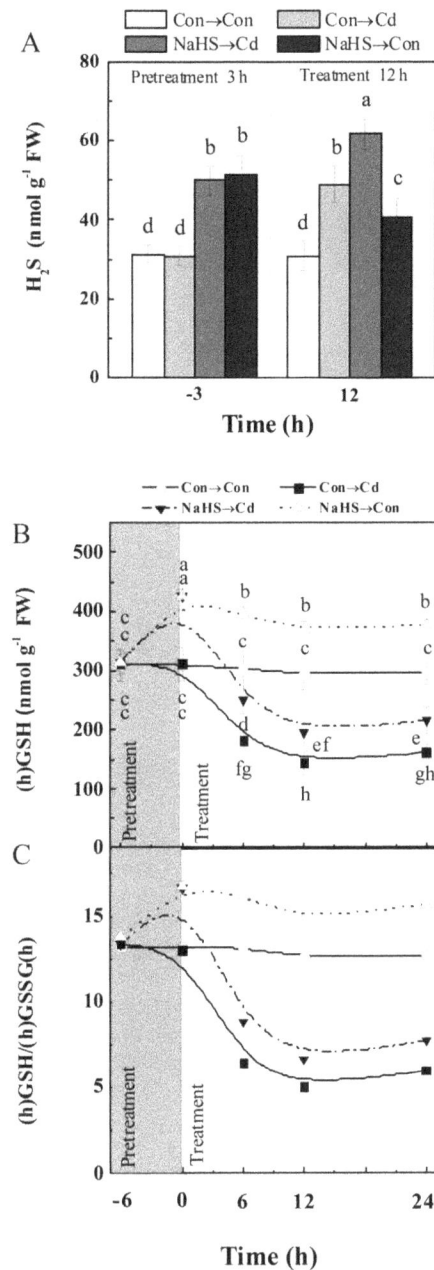

Figure 2. NaHS increased endogenous H₂S and (h)GSH contents, and the ratio of (h)GSH/(h)GSSG(h) upon Cd stress. Endogenous H₂S concentration in root tissues (A) was detected at 3 h after the beginning of 100 μM NaHS pretreatment (−3 h), and 200 μM CdCl₂ or chemical-free control treatments for 12 h (12 h). Meanwhile, contents of (h)GSH (B) and the ratio of (h)GSH/(h)GSSG(h) (C) in root tissues were detected at the indicated time points of treatments. Values are means ± SD of three independent experiments with three replicates for each. Bars denoted by the same letter did not differ significantly at P<0.05 according to Duncan's multiple range test.

applied together. Above results clearly indicated a requirement for (h)GSH homeostasis in NaHS-mediated alleviation of Cd toxicity.

cAMP signaling might be involved in NaHS responses

To testify the hypothesis that H₂S response is associated with cAMP signaling pathway, a pharmacological approach was used to manipulate endogenous cAMP. Results presented in Figure 8A

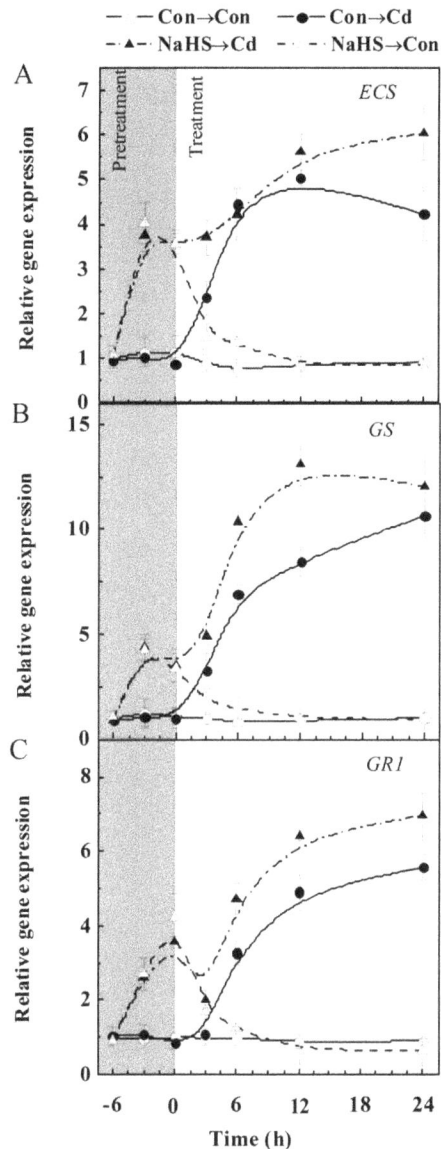

Figure 3. Time course of transcripts responsible for (h)GSH metabolism regulated by NaHS and Cd. Seedlings were pretreated with or without 100 μM NaHS for 6 h and then exposed to 200 μM CdCl₂ for another 24 h. The expression levels of ECS (A), GS (B) and GR1 (C) in root tissues analyzed by real-time RT-PCR are presented as values relative to the control at the beginning of pretreatment, normalized against expression of two internal reference genes in each sample. Values are means ± SD of three independent experiments with three replicates for each.

and B indicated that the pretreatment with 8-Br-cAMP, a membrane-permeable analogue of cAMP, alleviated Cd-induced decrease of fresh weight and increase of TBARS content in alfalfa seedling roots. Both of two adenylyl cyclase inhibitors, alloxan and DDA, blocked NaHS-alleviated Cd stress. Moreover, similar to the beneficial actions of 8-Br-cAMP (when was cotreated with PAG followed by Cd stress), a cNMP phosphodiesterase inhibitor IBMX also reversed the PAG responses in the aggravation of fresh weight loss and lipid peroxidation caused by Cd stress. Results from the real-time RT-PCR showed that 8-Br-cAMP and IBMX pretreatments followed by Cd stress, mimicked the effect of NaHS on GR1 up-regulation, regardless of whether PAG was added or

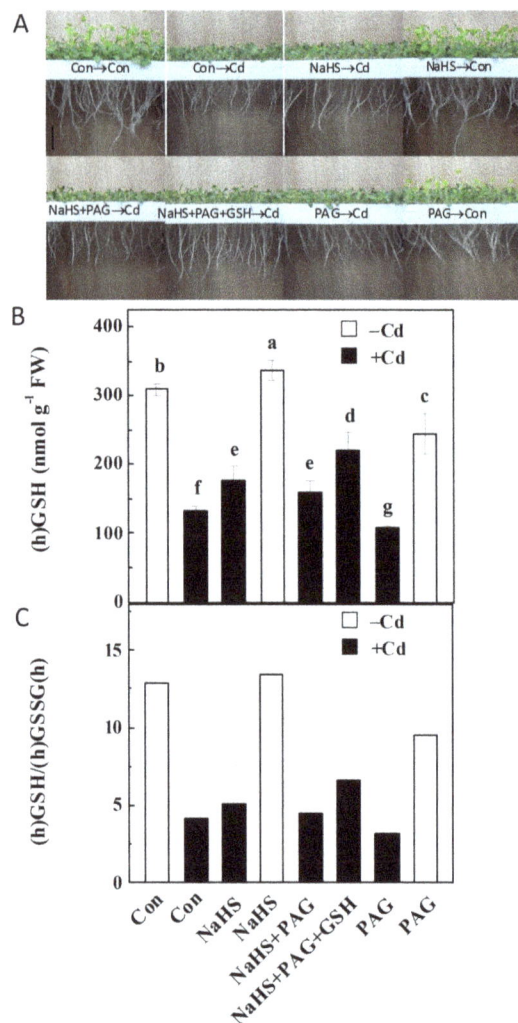

Figure 4. NaHS, PAG and GSH pretreatments differentially regulated seedling growth, (h)GSH content, and (h)GSH/ (h)GSSG(h) ratio. Corresponding phenotypes were photographed after 200 μM CdCl₂ treatment for 72 h, with or without 100 μM NaHS, 2 mM PAG, 1 mM GSH, individual or combination pretreatments for 6 h (A). Scale bar, 2 cm. Contents of (h)GSH (B), and the ratio of (h)GSH/ (h)GSSG(h) (C) in root tissues were also analyzed after 200 μM CdCl₂ treatment for 12 h, with or without 100 μM NaHS, 2 mM PAG, 1 mM GSH, individual or combination pretreatment for 6 h. Values are means ± SD of three independent experiments with three replicates for each. Bars denoted by the same letter did not differ significantly at $P<0.05$ according to Duncan's multiple range test.

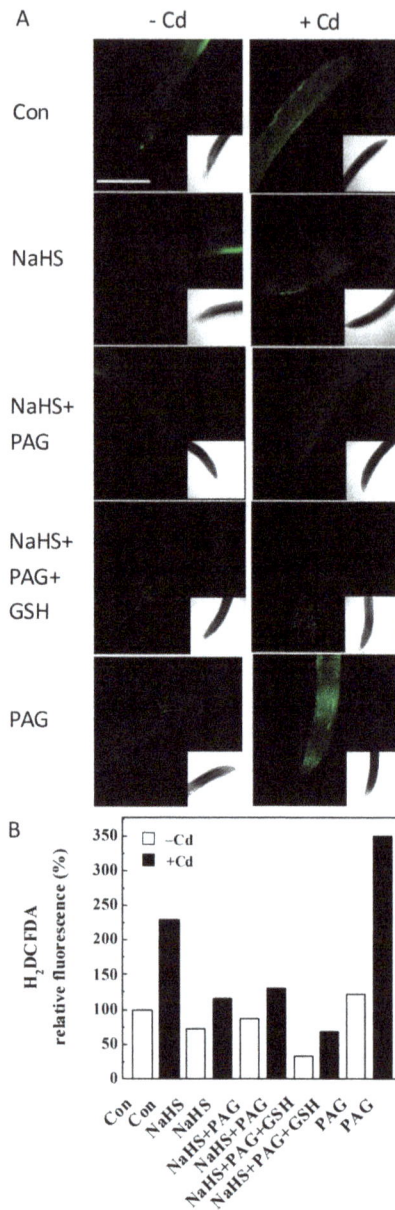

Figure 5. NaHS and GSH pretreatments alleviated Cd-induced ROS production, but blocked by PAG. LSCM results (A). Seedlings were pretreated with or without 100 μM NaHS, 2 mM PAG, 1 mM GSH, individual or combination for 6 h, and then exposed to 200 μM CdCl₂ for another 6 h. After various treatments, the roots were respectively stained with H₂DCFDA, then washed thoroughly to removal extra dye and immediately photographed by LSCM. Scale bar, 0.5 mm. The relative DCF fluorescence intensity in the corresponding roots (B).

not (Figure 8C). Two inhibitors alloxan and DDA partially blocked NaHS plus Cd-induced *GR1* transcripts. A similar tendency was found in the changes in *GPX* transcripts (Figure 8F). Results presented in Figure 8D and E further revealed the negative effects of adenylyl cyclase inhibitors on the transcripts of *Cu, Zn-SOD* and *APX1* in NaHS-pretreated seedling roots upon Cd, in comparison with the positive responses of 8-Br-cAMP and IBMX in the presence or absence of PAG.

Discussion

Although H₂S is a hazardous gaseous molecule with a strong odor of rotten eggs, it has been described as an important regulator with a variety of biological roles in animals and recently

in plants [11–16,25–34,53–56]. Moreover, recent works on *Populus euphratica* cells [57] and bermudagrass seedlings [51], demonstrated that exogenously applied NaHS, a H₂S donor, resulted in an enhanced Cd tolerance in these species. However, possible physiological mechanisms and downstream targets responsible for the observed Cd tolerance triggered by intracellular H₂S remain elusive. In this report, we discovered endogenous H₂S production in response to Cd stress, and further provided evidence demonstrating a requirement of (h)GSH and ROS homeostases, at least partially, in the intracellular H₂S-meaited plant adaptation

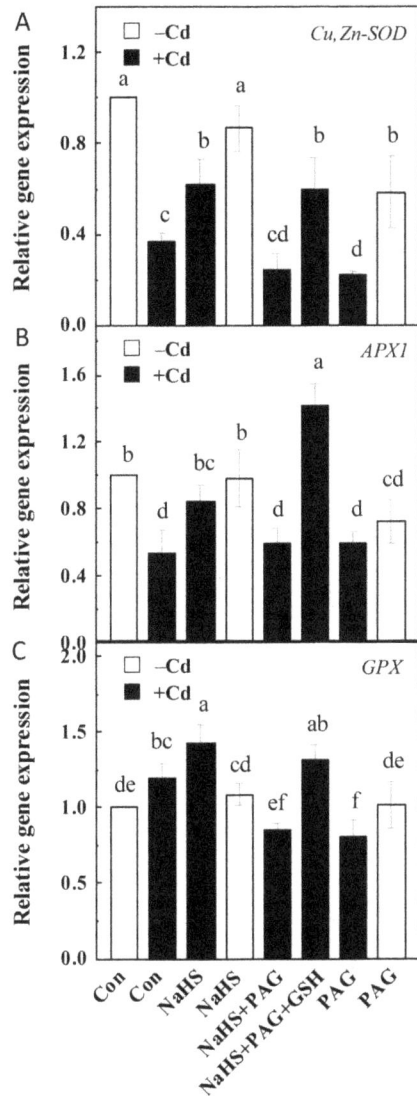

Figure 6. Transcripts of *Cu, Zn-SOD*, *APX1*, and *GPX* regulated by NaHS, PAG, GSH and Cd. Seedlings were pretreated with or without 100 μM NaHS, 2 mM PAG, 1 mM GSH, individual or combination for 6 h, and then exposed to 200 μM CdCl$_2$ for another 12 h. The expression levels of *Cu,Zn-SOD* (A), *APX1* (B), and *GPX* (C) transcripts in root tissues analyzed by real-time RT-PCR are presented as values relative to the control, normalized against expression of two internal reference genes in each sample. Values are means ± SD of three independent experiments with three replicates for each. Bars denoted by the same letter did not differ significantly at *P*<0.05 according to Duncan's multiple range test.

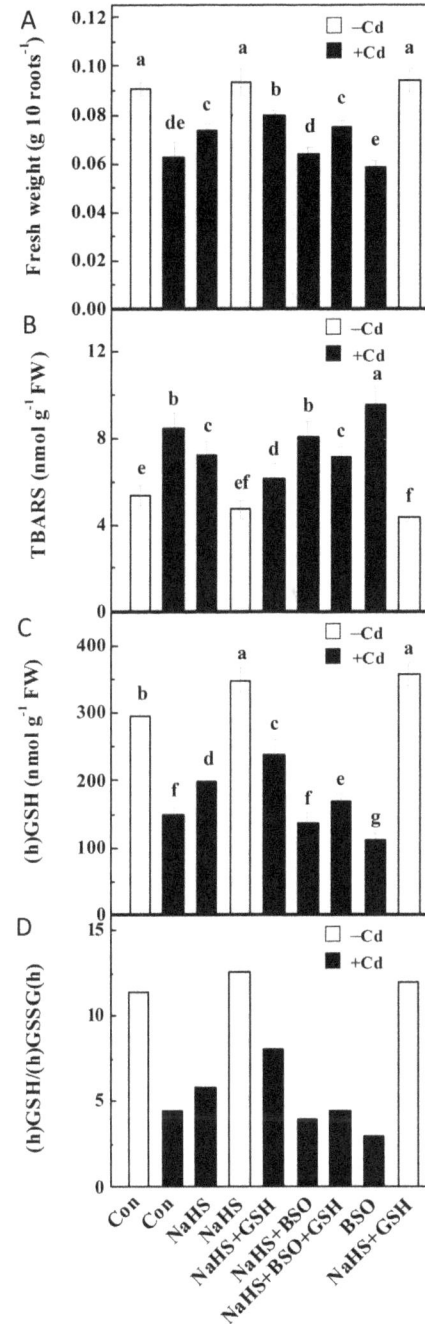

Figure 7. NaHS, GSH and BSO pretreatments differentially regulated seedling growth, TBARS accumulation, (h)GSH content, and (h)GSH/(h)GSSG(h). Fresh weight of 10 roots (A), TBARS accumulation (B), (h)GSH contents (C), and (h)GSH/(h)GSSG(h) ratio (D) in root tissues were determined after seedlings were pretreated with or without 100 μM NaHS, 1 mM GSH, 1 mM BSO, individual or combination for 6 h, and exposed to 200 μM CdCl$_2$ for 72 h (A), 24 h (B) and 12 h (C and D). Values are means ± SD of three independent experiments with three replicates for each. Bars denoted by the same letter did not differ significantly at *P*<0.05 according to Duncan's multiple range test.

against Cd toxicity. Therefore, our results presented in this work are vital for both fundamental and applied plant biology.

Endogenous H$_2$S production in response to Cd stress: the possible involvement of LCD/DCD

In animals, it was previously reported that diverse stress-inducing stimuli could result in the production of H$_2$S, including oxidative stress [49], depletion of cysteine (or its derivatives) [58] and glutathione [50]. Recent work in Arabidopsis [25] and bermudagrass seedlings [51] reported drought- and Cd-induced H$_2$S production. Because the signal compound H$_2$S is very reactive [53], the rapid regulation of the activity of H$_2$S biosynthetic

enzymes seems essential to fulfill H$_2$S-depenent functions. In this work, we further showed that Cd-triggered endogenous H$_2$S production might be related to LCD/DCD pathways (Figure 1D-F), since the similar increasing changes in the levels of intracellular

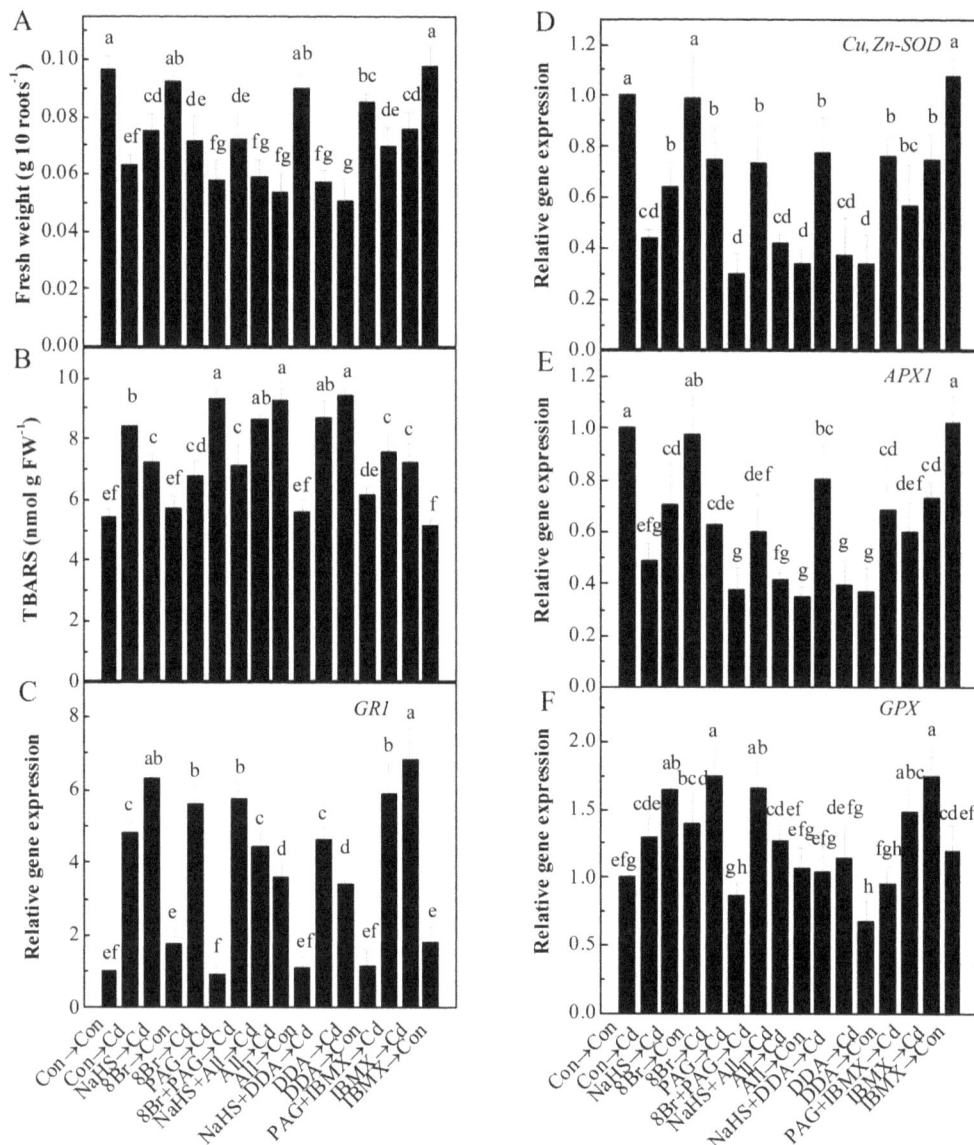

Figure 8. cAMP pathway might be involved in H₂S-alleviated Cd toxicity. Fresh weight of 10 roots (A), TBARS accumulation (B), *GR1* (C), *Cu,Zn-SOD* (D), *APX1* (E), and *GPX* (F) gene expression in alfalfa seedling roots upon Cd stress. Seedlings were pretreated with or without 100 μM NaHS, 50 μM 8-Br-cAMP (8Br), 2 mM PAG, 200 μM alloxan (All), 500 μM IBMX, 1 mM DDA alone, or the combination of treatments for 6 h, and then exposed to 200 μM CdCl₂ for 72 h (A), 24 h (B) and 12 h (C–F). The expression levels of corresponding genes analyzed by real-time RT-PCR are presented as values relative to the control, normalized against expression of two internal reference genes in each sample. Values are means ± SD of three independent experiments with three replicates for each. Bars denoted by the same letter did not differ significantly at *P*<0.05 according to Duncan's multiple range test.

H₂S as well as LCD/DCD activities were observed in the seedling roots of alfalfa challenged with Cd for 12 h. Meanwhile, similar to previous reports in wheat [30], bermudagrass [51], *Spinacia oleracea* seedlings [59], and strawberry plants [60], NaHS-induced H₂S production in alfalfa plants was also observed (Figure 2A).

In plants, both LCD and DCD are hypothesized to be involved in intracellular H₂S synthesis [21,27]. Several LCD/DCD candidates have been cloned and partially analyzed from the model plant *Arabidopsis* to *Brassica napus* [24,61]. Our above findings are consistent with those reported by Bloem et al. [40], in which they found that *Brassica napus* was able to react to *Pyrenopeziza brassicae* infection with a greater potential to release H₂S, which was reflected by an increasing LCD activity with fungal infection. More recently, auxin-induced DES-mediated

H₂S generation was also found to be involved in lateral root formation in tomato seedlings [62]. In view of the fact that all H₂S synthetic enzymes are not fully elucidated, our results suggested that LCD/DCD pathways might be, at least partially, related to Cd-induced H₂S production in alfalfa seedlings. In a future study, the role of other enzymatic and non-enzymatic pathways-mediated induction of H₂S synthesis in alfalfa seedlings upon Cd stress need be further elucidated.

The mechanism underlying the role of intracellular H₂S in the alleviation of Cd toxicity: reestablishment of reduced (homo)glutathione and ROS homeostases

Ample evidence revealed a clear relationship between metal stress and redox homeostasis and antioxidant capacity

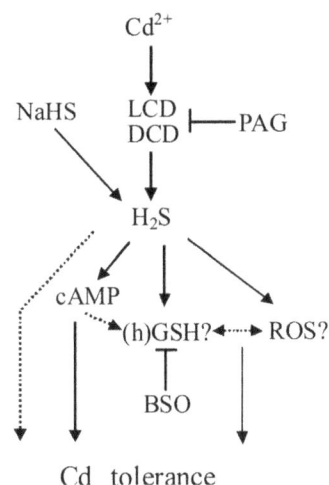

Figure 9. Simplified scheme of mechanisms involved in Cd tolerance by LCD/DCD-produced H_2S-modulated (h)GSH and ROS homeostases. Abbreviations: NaHS, sodium hydrosulfide; PAG, DL-propargylglycine; LCD, L-cysteine desulfhydrase; DCD, D-cysteine desulfhydrase; H_2S, hydrogen sulfide; ROS, reactive oxygen species; (h)GSH, reduced (homo)glutathione; BSO, L-buthionine-sulfoximine; cAMP, cyclic AMP. The dashed line denotes possible signaling cascade. T bars, inhibition.

[3,9,63–66]. Also, GSH could function as a heavy metal-ligand and an antioxidant [5,67]. In plants, H_2S serves as a signal as well as a novel antioxidant in hormonal and defense responses against abiotic stress [53,60]. Genetic evidence further revealed that the GSH deficiency mutant *pad2-1* shows the more oxidized redox state in contrast to wild type [68]. Arabidopsis mutants deficient in phytochelatins (PCs) and GSH biosynthesis respectively, *cad1* and *cad2*, are consequently more sensitive to Cd [6,69,70], that showed the crucial role of PCs, especially their precursor GSH in responding to Cd challenge. In the assays described here, as expected, when alfalfa seedling plants were upon Cd exposure, (h)GSH homeostasis is altered, which is reflected by the fact that the concentrations of reduced GSH and hGSH dropped (Table 1, Figure 1A–C), possible as a consequence of initiated PCs biosynthesis [71,72]. Similarly, a low ratio of (h)GSH/(h)GSSH(h), an important redox index related to Cd tolerance in alfalfa plants [45], was also observed. These changes thereafter cause redox imbalance and in turn Cd toxicity (Figures S1A, S3 and S4; Figures 4A and 5).

Our further experiments provide strong evidence to support the existence of a causal relationship between the endogenous H_2S signal and the alleviation of Cd toxicity in alfalfa seedlings partly by reestablishment of (h)GSH and ROS homeostases, which might be associated with the cAMP pathway. This conclusion is based on several pieces of evidence: (i) increased H_2S metabolism as well as the perturbation of (h)GSH homeostasis in alfalfa seedling roots are two early responses to the exposure of Cd (Figure 1, Table 1). These changes were consistent with the phenotypes of Cd toxicity (Figure 4A, Figure S1C). (ii) Application of a H_2S-releasing compound NaHS (also called as H_2S donor), not only mimics intracellular H_2S content triggered by Cd, but also alleviates Cd toxicity (Figures 2 and 4). Consistently, we also detected reestablishment of (h)GSH homeostasis, which was reflected by a higher (h)GSH content and ratio of (h)GSH/(h)GSSG(h) upon Cd stress. The observed Cd tolerance might be due to the available (h)GSH by the up-regulation of (h)GSH synthesis related genes, *ECS* and *GS* (Figure 3A and B), as well as *GR1* (Figure 3C), because besides

the synthesis of PCs, availability of GSH and concerted activity of GR seem to play a important role for plants to combat oxidative stress and Cd toxicity [7,72,73]. While, the inhibition of H_2S production caused by its synthetic inhibitor PAG blocked NaHS-induced Cd tolerance and reestablishment of (h)GSH and ROS homeostases, the latter of which was confirmed by the histochemical staining detecting the alleviation of plasma membrane integrity and lipid peroxidation, decreased ROS content and up-regulation of *Cu,Zn-SOD*, *APX1* and *GPX* transcripts, as well as declined TBARS level (Table 1, Figures 2–6, and Figures S1, S3 and S4). (iii) Above mentioned PAG responses were further rescued by exogenously applied GSH (Table 1, Figures 4–6). (iv) NaHS responses were sensitive to a (h)GSH synthetic inhibitor, but reversed by the added GSH (Figure 7), both of which suggesting a requirement of (h)GSH homeostasis for NaHS cytoprotective roles; and (v) Previous reports in animals showed H_2S-activited cAMP level or H_2S-regulated cAMP homeostasis [37,38]. Here, we found that two adenylyl cyclase inhibitors, alloxan and DDA, blocked the beneficial responses conferred by NaHS in alfalfa seedlings subjected to Cd stress (Figure 8). On the contrary, an analogue of cAMP 8-Br-cAMP and a cNMP phosphodiesterase inhibitor IBMX mimicked the effects of NaHS on the alleviation of Cd toxicity as well as the regulation of (h)GSH homeostasis and ROS metabolism (*GR1*, *Cu,Zn-SOD*, *APX1*, and *GPX*, etc). Above pharmacological evidence indicated the involvement of cAMP signaling in NaHS responses. Additionally, NaHS-triggered cytoprotective roles were confirmed to act as a H_2S-dependent fashion (Figure S2). Above results clearly established a casual link between intracellular H_2S in the alleviation of Cd toxicity and reestablishment of (h)GSH and ROS homeostases.

Conclusions

In summary, our pharmacological, histochemical, biochemical and molecular evidence suggested that the intracellular H_2S was able to ameliorate Cd toxicity in alfalfa seedlings at least partly by reestablishment of (h)GSH and ROS homeostases. Figure 9 illustrates a simplified scheme of mechanisms involved in Cd tolerance by LCD/DCD-produced H_2S-modulated (h)GSH and ROS homeostases, since 1) LCD/DCD-produced H_2S acts as a signal triggered by Cd to regulated (h)GSH metabolisms; 2) both (h)GSH and ROS homeostases could be reestablished by H_2S and further linked to Cd tolerance; 3) cAMP signaling pathway might be related to NaHS-triggered Cd tolerance, partially through the regulation of GSH homeostasis and ROS metabolism. Taking into account that H_2S participates in stressful responses and developmental process, our study therefore may extend our understanding of the complex system integrating environmental and developmental signals.

Supporting Information

Figure S1 NaHS pretreatment alleviates Cd toxicity.

Figure S2 H_2S or HS^-, but not other compounds derived from NaHS contribute to NaHS responses.

Figure S3 Effects of NaHS, PAG and GSH pretreatments on the fresh weight (A) and TBARS concentrations (B) in alfalfa seedling roots upon Cd stress.

Figure S4 Effects of NaHS, PAG and GSH pretreatments on Cd concentrations in alfalfa seedlings upon Cd stress.

Table S1   The sequences of primers for real-time RT-PCR.

Author Contributions

Conceived and designed the experiments: WS. Performed the experiments: WC KZ QJ. Analyzed the data: WC Y. Xie WS. Contributed reagents/materials/analysis tools: HC JC Y. Xia JZ. Wrote the paper: WC HC WS.

References

1. Gao J, Sun L, Yang X, Liu JX (2013) Transcriptomic analysis of cadmium stress response in the heavy metal hyperaccumulator *Sedum alfredii* hance. PLoS ONE 8(6): e64643.
2. Ye Y, Li Z, Xing D (2013) Nitric oxide promotes MPK6-mediated caspase-3-like activation in cadmium-induced *Arabidopsis thaliana* programmed cell death. Plant Cell Environ 36: 1–15.
3. Sharma SS, Dietz KJ (2009) The relationship between metal toxicity and cellular redox imbalance. Trends Plant Sci 14: 43–50.
4. Tkalec M, Štefanić PP, Cvjetko P, Šikić S, Pavlica M, et al. (2012) The effects of cadmium-zinc interactions on biochemical responses in tobacco seedlings and adult plants. PLoS ONE 9(1): e87582.
5. Foyer CH, Noctor G (2011) Ascorbate and glutathione: the heart of the redox hub. Plant Physiol 155: 2–18.
6. Cobbett CS, May MJ, Howden R, Rolls B (1998) The glutathione-deficient, cadmium-sensitive mutant, *cad2-1*, of *Arabidopsis thaliana* is deficient in γ-glutamylcysteine synthetase. Plant J 16: 73–78.
7. Cruz de Carvalho MH, Brunet J, Bazin J, Kranner I, d' Arcy-Lameta A, et al. (2010) Homoglutathione synthetase and glutathione synthetase in drought-stressed cowpea leaves: expression patterns and accumulation of low-molecular-weight thiols. J Plant Physiol 167: 480–487.
8. Matamoros MA, Moran JF, Iturbe-Ormaetxe I, Rubio MC, Becana M (1999) Glutathione and homoglutathione synthesis in legume root nodules. Plant Physiol 121: 879–888.
9. Ortega-Villasante C, Rellán-Álvarez R, Del Campo FF, Carpena-Ruiz RO, Hernández LE (2005) Cellular damage induced by cadmium and mercury in *Medicago sativa*. J Exp Bot 56: 2239–2251.
10. Metwally A, Safronova VI, Belimov AA, Dietz KJ (2005) Genotypic variation of the response to cadmium toxicity in *Pisum sativum* L. J Exp Bot 56: 167–178.
11. Abe K, Kimura H (1996) The possible role of hydrogen sulfide as an endogenous neuromodulator. J Neurosci 16: 1066–1071.
12. Li L, Rose P, Moore PK (2011) Hydrogen sulfide and cell signaling. Annu Rev Pharmacol Toxicol 51: 169–187.
13. Hosoki R, Matsuki N, Kimura H (1997) The possible role of hydrogen sulfide as an endogenous smooth muscle relaxant in synergy with nitric oxide. Biochem Biophys Res Commun 237: 527–531.
14. Du J, Hui Y, Cheung Y, Bin G, Jiang H, et al. (2004) The possible role of hydrogen sulfide as a smooth muscle cell proliferation inhibitor in rat cultured cells. Heart Vessels 19: 75–80.
15. Yang G, Sun X, Wang R (2004) Hydrogen sulfide-induced apoptosis of human aorta smooth muscle cells via the activation of mitogen-activated protein kinases and caspase-3. FASEB J 18: 1782–1784.
16. Kimura Y, Goto Y, Kimura H (2010) Hydrogen sulfide increases glutathione production and suppresses oxidative stress in mitochondria. Antioxid Redox Signal 12: 1–13.
17. Wilson LG, Bressan RA, Filner P (1978) Light-dependent emission of hydrogen sulfide from plants. Plant Physiol 61: 184–189.
18. Hällgren JE, Fredriksson SÅ (1982) Emission of hydrogen sulfide from sulfur dioxide-fumigated pine trees. Plant Physiol 70: 456–459.
19. Sekiya J, Schmidt A, Wilson LG, Filner P (1982) Emission of hydrogen sulfide by leaf tissue in response to L-cysteine. Plant Physiol 70: 430–436.
20. Wang R (2002) Two's company, three's a crowd: can H2S be the third endogenous gaseous transmitter? FASEB J 16: 1792–1798.
21. Papenbrock J, Riemenschneider A, Kamp A, Schulz-Vogt HN, Schmidt A (2007) Characterization of cysteine-degrading and H2S-releasing enzymes of higher plants – from the field to the test tube and back. Plant Biol 9: 582–588.
22. Léon S, Touraine B, Briat JF, Lobréaux S (2002) The *AtNFS2* gene from *Arabidopsis thaliana* encodes a NifS-like plastidial cysteine desulfhurase. Biochem J 366: 557–564.
23. Riemenschneider A, Wegele R, Schmidt A, Papenbrock J (2005) Isolation and characterization of a D-cysteine desulfhydrase protein from *Arabidopsis thaliana*. FEBS J 272: 1291–1304.
24. Álvarez C, Calo L, Romero LC, García I, Gotor C (2010) An *O*-acetylserine(thiol)lyase homolog with L-cysteine desulfhydrase activity regulates cysteine homeostasis in Arabidopsis. Plant Physiol 152: 656–669.
25. Jin Z, Shen J, Qiao Z, Yang G, Wang R, et al. (2011) Hydrogen sulfide improves drought resistance in *Arabidopsis thaliana*. Biochem Biophys Res Commun 414: 481–486.
26. Wang Y, Li L, Cui W, Xu S, Shen W, et al. (2012) Hydrogen sulfide enhances alfalfa (*Medicago sativa*) tolerance against salinity during seed germination by nitric oxide pathway. Plant Soil 351: 107–119.
27. García-Mata C, Lamattina L (2010) Hydrogen sulphide, a novel gasotransmitter involved in guard cell signalling. New Phytol 188: 977–984.
28. Lisjak M, Srivastava N, Teklic T, Civale L, Lewandowski K, et al. (2010) A novel hydrogen sulfide donor causes stomatal opening and reduces nitric oxide accumulation. Plant Physiol Biochem 48: 931–935.

29. Zhang H, Jiao H, Jiang CX, Wang SH, Wei ZJ, et al. (2010) Hydrogen sulfide protects soybean seedlings against drought-induced oxidative stress. Acta Physiol Plant 32: 849–857.
30. Zhang H, Hu LY, Hu KD, He YD, Wang SH, et al. (2008) Hydrogen sulfide promotes wheat seed germination and alleviates oxidative damage against copper stress. J Integr Plant Biol 50: 1518–1529.
31. Li L, Wang Y, Shen W (2012) Roles of hydrogen sulfide and nitric oxide in the alleviation of cadmium-induced oxidative damage in alfalfa seedling roots. Biometals 25: 617–631.
32. Li ZG, Gong M, Xie H, Yang L, Li J (2012) Hydrogen sulfide donor sodium hydrosulfide-induced heat tolerance in tobacco (*Nicotiana tabacum* L) suspension cultured cells and involvement of Ca2+ and calmodulin. Plant Sci 185–186: 185–189.
33. Lin YT, Li MY, Cui WT, Lu W, Shen WB (2012) Haem oxygenase-1 is involved in hydrogen sulfide-induced cucumber adventitious root formation. J Plant Growth Regul 31: 519–528.
34. Zhang H, Hu SL, Zhang ZJ, Hu LY, Jiang CX, et al. (2011) Hydrogen sulfide acts as a regulator of flower senescence in plants. Postharvest Biol Tec 60: 251–257.
35. Ma W, Qi Z, Smigel A, Walker RK, Verma R, et al. (2009) Ca2+, cAMP, and transduction of non-self perception during plant immune responses. Proc Natl Acad Sci U S A 106: 20995–21000.
36. Jin XC, Wu WH (1998) Involvement of cyclic AMP in ABA- and Ca2+-mediated signal transduction of stomatal regulation in *Vicia faba*. Plant Cell Physiol 40: 1127–1133.
37. Kimura H (2000) Hydrogen sulfide induces cyclic AMP and modulates the NMDA receptor. Biochem Biophys Res Commun 267: 129–133.
38. Lu M, Liu YH, Ho CY, Tiong CX, Bian JS (2012) Hydrogen sulfide regulates cAMP homeostasis and renin degranulation in As4.1 and rat renin-rich kidney cells. Am J Physiol Cell Physiol 302: C59–C66.
39. Shao JL, Wan XH, Chen Y, Bi C, Chen HM, et al. (2011) H2S protects hippocampal neurons from anoxia-reoxygenation through cAMP-mediated PI3K/Akt/p70S6K cell-survival signaling pathways. J Mol Neurosci 43: 453–460.
40. Bloem E, Riemenschneider A, Volker J, Papenbrock J, Schmidt A, et al. (2004) Sulphur supply and infection with *Pyrenopeziza brassicae* influence L-cysteine desulphydrase activity in *Brassica napus* L. J Exp Bot 55: 2305–2312.
41. Cui W, Gao C, Fang P, Lin G, Shen W (2008) Alleviation of cadmium toxicity in *Medicago sativa* by hydrogen-rich water. J Hazard Mater 260: 715–724.
42. Smith IK (1985) Stimulation of glutathione synthesis in photorespiring plants by catalase inhibitors. Plant Physiol 79: 1044–1047.
43. Herschbach C, Pilch B, Tausz M, Rennenberg H, Grill D (2002) Metabolism of reduced and inorganic sulphur in pea cotyledons and distribution into developing seedlings. New Phytol 153: 73–80.
44. Meyer AJ, Brach T, Marty L, Kreye S, Rouhier N, et al. (2007) Redox-sensitive GFP in *Arabidopsis thaliana* is a quantitative biosensor for the redox potential of the cellular glutathione redox buffer. Plant J 52: 973–986.
45. Cui W, Li L, Gao Z, Wu H, Xie Y, et al. (2012) Haem oxygenase-1 is involved in salicylic acid-induced alleviation of oxidative stress duo to cadmium stress in *Medicago sativa*. J Exp Bot 63: 5521–5534.
46. Vandesompele J, De Preter K, Pattyn F, Poppe B, Van Roy N, et al. (2002) Accurate normalization of real-time quantitative RT-PCR data by geometric averaging of multiple internal control genes. Genome Biol 3: research0034.
47. Kováčik J, Babula P, Klejdus B, Hedbavny J, Jarošová M (2014) Unexpected behavior of some nitric oxide modulators under cadmium excess in plant tissue. PLos ONE 9(3): e91685.
48. Rennenberg H, Filner P (1982) Stimulation of H2S emission from pumpkin leaves by inhibition of glutathione synthesis. Plant Physiol 69: 766–770.
49. Kwak WJ, Kwon GS, Jin I, Kuriyama H, Sohn HY (2003) Involvement of oxidative stress in the regulation of H2S production during ultradian metabolic oscillation of *Saccharomyces cerevisiae*. FEMS Microbiol Lett 219: 99–104.
50. Sohn HY, Kum EJ, Kwon GS, Jin I, Adams CA, et al. (2005) *GLR1* plays an essential role in the homeodynamics of glutathione and the regulation of H2S production during respiratory oscillation of *Saccharomyces cerevisiae*. Biosci Biotechnol Biochem 69: 2450–2454.
51. Shi H, Ye T, Chan Z (2014) Nitric oxide-activated hydrogen sulfide is essential for cadmium stress response in bermudagrass (*Cynodon dactylon* (L). Pers.). Plant Physiol Biochem 74: 99–107.
52. Rüegsegger A, Schmutz D, Brunold C (1990) Regulation of glutathione synthesis by cadmium in *Pisum sativum* L. Plant Physiol 93: 1579–1584.
53. Lisjak M, Teklic T, Wilson ID, Whiteman M, Hancock JT (2013) Hydrogen sulfide: environmental factor or signalling molecule? Plant Cell Environ 36: 1607–1616.

54. García-Mata C, Lamattina L (2013) Gasotransmitters are emerging as new guard cell signaling molecules and regulators of leaf gas exchange. Plant Sci 201–202: 66–73.

55. Kimura Y, Kimura H (2004) Hydrogen sulfide protects neurons from oxidative stress. FASEB J 18: 1165–1167.

56. Li ZG, Gong M, Liu P (2012) Hydrogen sulfide is a mediator in H_2O_2-induced seed germination in *Jatropha Curcas*. Acta Physiol Plant 34: 2207–2213.

57. Sun J, Wang R, Zhang X, Yu Y, Zhao R, et al. (2013) Hydrogen sulfide alleviates cadmium toxicity through regulations of cadmium transport across the plasma and vacuolar membranes in *Populus euphratica* cells. Plant Physiol Biochem 65: 67–74.

58. Sohn HY, Kuriyama H (2001) The role of amino acids in the regulation of hydrogen sulfide production during ultradian respiratory oscillation of *Saccharomyces cerevisiae*. Arch Microbiol 176: 69–78.

59. Chen J, Wu FH, Wang WH, Zheng CJ, Lin GH, et al. (2011) Hydrogen sulphide enhances photosynthesis through promoting chloroplast biogenesis, photosynthetic enzyme expression, and thiol redox modification in *Spinacia oleracea* seedlings. J Exp Bot 62: 4481–4493.

60. Christou A, Manganaris GA, Papadopoulos I, Fotopoulos V (2013) Hydrogen sulfide induces systemic tolerance to salinity and non-ionic osmotic stress in strawberry plants through modification of reactive species biosynthesis and transcriptional regulation of multiple defence pathways. J Exp Bot 64: 1953–1966.

61. Xie Y, Lai D, Mao Y, Zhang W, Shen W, et al. (2013) Molecular cloning, characterization, and expression analysis of a novel gene encoding L-cysteine desulfhydrase from *Brassica napus*. Mol Biotechnol 54: 737–746.

62. Fang T, Cao Z, Li J, Shen W, Huang L (2014) Auxin-induced hydrogen sulfide generation is involved in lateral root formation in tomato. Plant Physiol Biochem 76: 44–51.

63. Dawood M, Cao F, Jahangir MM, Zhang G, Wu F (2012) Alleviation of aluminum toxicity by hydrogen sulfide is related to elevated ATPase, and suppressed aluminum uptake and oxidative stress in barley. J Hazard Mater 209–210: 121–128.

64. Jin CW, Mao QQ, Luo BF, Lin XY, Du ST (2013) Mutation of *mpk6* enhances cadmium tolerance in *Arabidopsis* plants by alleviating oxidative stress. Plant Soil 371: 387–396.

65. Lagorce A, Fourçans A, Dutertre M, Bouyssiere B, Zivanovic Y, et al. (2012) Genome-wide transcriptional response of the archaeon *Thermococcus gammatolerans* to cadmium. PLoS ONE 7(7): e41935.

66. Thapa G, Sadhukhan A, Panda SK, Sahoo L (2012) Molecular mechanistic model of plant heavy metal tolerance. Biometals 25: 489–505.

67. Dixit P, Mukherjee PK, Ramachandran V, Eapen S (2013) Glutathione transferase from *Trichoderma virens* enhances cadmium tolerance without enhancing its accumulation in transgenic *Nicotiana tabacum*. PLoS ONE 6(1): e16360.

68. Dubreuil-Maurizi C, Vitecek J, Marty L, Branciard L, Frettinger P, et al. (2011) Glutathione deficiency of the Arabidopsis mutant *pad2-1* affects oxidative stress-related events, defense gene expression, and the hypersensitive response. Plant Physiol 157: 2000–2012.

69. Howden R, Andersen CR, Goldsbrough PB, Cobbett CS (1995) A cadmium-sensitive, glutathione-deficient mutant of *Arabidopsis thaliana*. Plant Physiol 107: 1067–1073.

70. Howden R, Goldsbrough PB, Andersen CR, Cobbett CS (1995) Cadmium-sensitive, *cad1* mutants of *Arabidopsis thaliana* are phytochelatin deficient. Plant Physiol 107: 1059–1066.

71. Grill E, Löffler S, Winnacker EL, Zenk MH (1989) Phytochelatins, the heavy-metal-binding peptides of plants, are synthesized from glutathione by a specific γ-glutamylcysteine dipeptidyl transpeptidase (phytochelatin synthase). Proc Natl Acad Sci USA 86: 6838–6842.

72. Mishra S, Srivastava S, Tripathi RD, Govindarajan R, Kuriakose SV, et al. (2006) Phytochelatin synthesis and response of antioxidants during cadmium stress in *Bacopa monnieri* L. Plant Physiol Biochem 44: 25–37.

73. Verbruggen N, Juraniec M, Baliardini C, Meyer CL (2013) Tolerance to cadmium in plants: the special case of hyperaccumulators. Biometals 26: 633–638.

14

Immunogenicity Evaluation of a Rationally Designed Polytope Construct Encoding HLA-A*0201 Restricted Epitopes Derived from *Leishmania major* Related Proteins in HLA-A2/DR1 Transgenic Mice: Steps toward Polytope Vaccine

Negar Seyed[1], Tahereh Taheri[1], Charline Vauchy[2,3,4], Magalie Dosset[2,3,4], Yann Godet[2,3,4], Ali Eslamifar[5], Iraj Sharifi[6], Olivier Adotevi[2,3,4,7], Christophe Borg[2,3,4,7], Pierre Simon Rohrlich[2,3,4,8], Sima Rafati[1]*

1 Molecular Immunology and Vaccine Research Lab, Pasteur Institute of Iran, Tehran, Iran, **2** INSERM U1098, Unité Mixte de Recherche, Besançon, France, **3** Etablissement Français du Sang de Bourgogne Franche-Comté, Besançon, France, **4** Université de Franche-Comté, Besançon, France, **5** Department of Electron Microscopy and Clinical Research, Pasteur Institute of Iran, Tehran, Iran, **6** School of Medicine, Leishmaniasis Research Center, Kerman University of Medical Sciences, Kerman, Iran, **7** CHRU de Besançon, Service d'Oncologie, Besançon, France, **8** CHRU de Besançon, Service de pédiatrie, Besançon, France

Abstract

Background: There are several reports demonstrating the role of CD8 T cells against *Leishmania* species. Therefore peptide vaccine might represent an effective approach to control the infection. We developed a rational polytope-DNA construct encoding immunogenic HLA-A2 restricted peptides and validated the processing and presentation of encoded epitopes in a preclinical mouse model humanized for the MHC-class-I and II.

Methods and Findings: HLA-A*0201 restricted epitopes from LPG-3, *Lm*STI-1, CPB and CPC along with H-2Kd restricted peptides, were lined-up together as a polytope string in a DNA construct. Polytope string was rationally designed by harnessing advantages of ubiquitin, spacers and HLA-DR restricted Th1 epitope. Endotoxin free pcDNA plasmid expressing the polytope was inoculated into humanized HLA-DRB1*0101/HLA-A*0201 transgenic mice intramuscularly 4 days after Cardiotoxin priming followed by 2 boosters at one week interval. Mice were sacrificed 10 days after the last booster, and splenocytes were subjected to *ex-vivo* and *in-vitro* evaluation of specific IFN-γ production and *in-vitro* cytotoxicity against individual peptides by ELISpot and standard chromium-51(^{51}Cr) release assay respectively. 4 H-2Kd and 5 HLA-A*0201 restricted peptides were able to induce specific CD8 T cell responses in BALB/C and HLA-A2/DR1 mice respectively. IFN-γ and cytolytic activity together discriminated LPG-3-P1 as dominant, *Lm*STI-1-P3 and *Lm*STI-1-P6 as subdominant with both cytolytic activity and IFN-γ production, *Lm*STI-1-P4 and LPG-3-P5 as subdominant with only IFN-γ production potential.

Conclusions: Here we described a new DNA-polytope construct for *Leishmania* vaccination encompassing immunogenic HLA-A2 restricted peptides. Immunogenicity evaluation in HLA-transgenic model confirmed CD8 T cell induction with expected affinities and avidities showing almost efficient processing and presentation of the peptides in relevant preclinical model. Further evaluation will determine the efficacy of this polytope construct protecting against infectious challenge of *Leishmania*. Fortunately HLA transgenic mice are promising preclinical models helping to speed up immunogenicity analysis in a human related mouse model.

Editor: Clive M. Gray, University of Cape Town, South Africa

Funding: This work was financially supported by a grant from Pasteur Institute of Iran (PhD scholarship), National Science Foundation of Iran (grant no. 87020176) and INSERM UMR 1098, Besancon, France. The funders had no role in study design, data collection and analysis, decision to publish, or preparation of the manuscript.

Competing Interests: The authors have declared that no competing interests exist.

* Email: sima-rafatisy@pasteur.ac.ir

Introduction

Cutaneous, Visceral and Mucocutaneous leishmaniasis are three main features of a vector-born parasitic disease caused by *Leishmania* genus and transmitted by sandfly bite [1]. Leishmaniasis can be transmitted in many tropical and subtropical countries, and is found in parts of about 98 countries on 5 continents. Different forms of the disease predominate in different regions of the world. Countries like Morocco, Nepal, India, China, Iraq and Bangeladesh are mostly involved with visceral leishmaniasis while others like Algeria, Syria, Iran, Tunisia, Afghanistan,

Pakistan and Saudi Arabia are involved with cutaneous form. Brazil, is almost exclusively involved with all three forms of the disease at a very high incidence rate [2].

Current control relies on chemotherapy to alleviate the disease and on vector control to reduce transmission. A few drugs are available for chemotherapy but facing problems such as high toxicity, variable efficacy, inconvenient treatment schedules, costs and drug resistance [3]. Vector control has also appeared extremely difficult due to sand fly generalization and adaption to many different micro-landscapes [4]. Thus an effective vaccination would be of great interest to control this expanding disease. Unfortunately despite all efforts made using different vaccination strategies [5,6,7], no protective vaccine for human is available to control the disease except for a multi-protein vaccine namely LEISH-F(F1, F2, F3) which is still in clinical trial and has not entered the market yet [8,9,10].

Leishmania is an obligatory intracellular parasite residing and proliferating inside macrophages as ultimate host cells. Therefore with no doubt IFN-γ plays a vital role in controlling the infection since it induces the signal for nitric oxide production by macrophages. Nitric oxide is a nitrogen metabolite that inhibits parasite survival [11,12]. Consensually CD4+ Th1 cells have been considered the main IFN-γ providers in *Leishmania* specific response, but today's knowledge also remarks the CD8+ cytotoxic T cells (Tc1) role in this scenario [13,14], especially in controlling secondary *Leishmania (L.) major* infection. [15,16,17]. There was an unresolved paradigm around the role of these cells controlling primary infection [18,19,20] but Belkaid's elegant experiment with low rather than high dose inoculation finally shed light on this enigma. Intradermal low-dose (100–1000) metacyclic challenge with *L. major* (resembling the natural infection transmitted by sandfly bite) in C57BL/6 mice depleted of CD8+ T cells successfully established a progressive infection defeating the immune system [21]. Later on, Uzonna *et al.* delineated a transient Th2 response at early stages of low dose challenge that was modified and diverted to Th1 only in the presence of IFN-γ producing CD8+ T cells and not in CD8+ T cell depleted mice [22].

Besides their IFN-γ production [23], cytolytic activity of CD8+ T cells has also been under question [24,25,26,27,28]. On one hand the massive proliferation of the parasite in non-ulcerative nodules from patients suffering from diffuse cutaneous leishmaniasis and post Kala-Azar dermal leishmaniasis has been ascribed to CD8+ T cell exhaustion due to long lasting infection [29,30]. On the other hand, the parasite-free pathologic lesions of patients suffering from mucosal leishmaniasis have been ascribed to hyperactivity of CD8+ T cells at involved tissue [31,32]. Whether the cytolytic activity is responsible for parasite eradication directly by apoptosis or indirectly by disrupting parasite infected macrophages is unclear.

Besides all other vaccination strategies, today protective and therapeutic peptide-based vaccine concept has drawn attraction in the field of intracellular infections [33,34,35] and cancer [36,37] where multi-CD8 cytotoxic T cell responses are crucial mediators of immunity. Since the evidence continues to pile up about CD8+ T cells role [38,39,40,41,42], peptide vaccine might open a new way in the battle over leishmaniasis.

In our previous study six known proteins from *L. major* were screened for best HLA-A2 binding 9 mer peptides by immunoinformatics tools. A few peptides from *L. major* Stress Inducible Protein-1 (*Lm*STI-1) and Lipophosphoglycan Protein-3 (LPG-3) were then shown to be immunogenic stimulating PBMC from cutaneous leishmaniasis recovered individuals [43]. Since DNA constructs are well known for CD8+ T cell stimulation, we

rationally designed a DNA construct encoding previously selected peptides to evaluate the processing, presentation and immunogenicity of various encoded epitopes in a preclinical mouse model humanized for HLA class I and II molecules (HLA-DRB1*0101/HLA-A*0201). These preclinical models are precious tools in hand to study the immunogenicity of *in-silico* selected peptides for vaccination purposes in humans.

Material and Methods

Ethics statement

Transgenic animals, homozygous for all modified genetic characters, were bred at the IBCT animal facility of INSERM UMR1098, Besançon, France (authorization number D-25-056-7). The protocol including maintenance, anesthesia (under standard isoflurane inhalation) and euthanasia (via cervical dislocation) was reviewed and approved by the competent authority at INSERM UMR1098 for compliance with the French and European Regulations on Animal Welfare (protocol number 10005).

Design of the polytope construct

HLA-A2 restricted epitopes from 4 different *Leishmania* proteins (CPB and CPC, 5 peptides, *Lm*STI-1, 4 peptides and LPG-3, 4 peptides) along with 4 epitopes presented in H-2Kd allele context from BALB/c mice were included. Peptides were lined up together to satisfy two basic criteria: least junctional peptides (neoepitope, generated by the juxtaposition of two authentic epitopes) and most cleaved peptides of interest out of the polytope sequence. To this end all possible arrangements (24 different arrangements) without spacer and with 4 different spacers: AAA (3 alanine residues) [44,45], AAY (2 alanine and one tyrosine residues) [46,47], K (lysine) [48,49] and AD (one alanine and one aspartic acid residues) [50] (a total of 120 different arrangements for each protein) were analysed with two common on-line algorithms for proteasome cleavage: NetCTL 1.2 [51] and nHLApred [52]. This approach was followed for peptide arrangements of each individual protein and then the most preferable arrangements satisfying both basic criteria were selected to be combined again in a longer polytope. Using NetCTL 1.2 a few combinations were selected. To finalise the selection process we checked for the hydrophobicity pattern of the different combinations. Final arrangement flanking with additional sequences as, *Hin*dIII restriction site, kozak sequence and mouse ubiquitin sequence at amino-terminal (N-terminal) and Tetanus Toxoid universal Th1 epitope (TT$_{830}$) with *Bam*HI restriction site at carboxy-terminal (C-terminal) was synthesized by BIOMATIK company (Canada). 921 base pair (bp) long sequence (PT) was codon optimized for optimal expression in mouse cells and was received as pUC57-PT (Codon Optimization was done using BIOMATIK proprietary software. 15% cut off was used for codon efficiency and any codon below 15% was removed except for positions with strong secondary structures. Secondary structure was checked using a build in M-fold module (Internal ribosomal binding sites were removed). Table 1 summarizes the HLA restriction and immunoinformatics characteristics of individual peptides in the construct (6 HLA-A2 restricted and 4 H-2Kd restricted peptides marked in bold were analysed in this study).

Cloning pathway

To meet the objectives of this study, pUC57-PT (BIOMATIK, Canada) was subject to a few cloning steps. 918 bp long fragment digested by *Hin*dIII – *Bam*HI (Roche, Germany) enzymatic reaction was directly cloned into pEGFP-N3 plasmid (Clontech,

Table 1. Characteristics of *in silico* predicted *L. major* specific CD8[+] T cell 9-mer peptides included in the polytope construct.

Protein (GeneDB accession number)	Peptide sequence	HLA Restriction	Scores predicted by online immunoinformatics software						
			SYFPEITHI (1)	BIMAS (2)	EpiJen (3)#	RANKpep/Proteasome cleavage (4)	nHLAPred (5)	NetCTL (6)	Multipred (7)
CPB (LmjF08.1080)	LMLQAFEVV	HLA-A*0201	22	1617	+	72/−	1	1.255	MB[a]
	QLNHGVLLV		28	159	+	73/+	1	1.055	MB
	LLTGYPVSV		28	118	+	91/−	1	1.284	MB
CPC (LmjF29.0820)	FLGGHAVKL		27	98	+	73/+	0.97	1.097	MB
	LLATTVSGL		29	83	+	90/+	1	1.137	MB
LmsTI-1 (LmjF08.1110)	**LLMLQPDYV* (P4)**		23	1179	+	68/+	1	1.027	MB
	ALQAYDEGL (P6)		24	10	+	63/+	0.93	1.218	MB
	QLDEQNSVL (P3)		22	14	+	64/+	1	0.791	MB
	YMEDQRFAL (P2)		21	108	+	72/+	0.99	1.102	MB
LPG-3 (LmjF29.0760)	**LLLLGSVTV (P1)**		30	437	+	86/+	1	1.023	MB
	FLVGDRVRV		25	319	+	80/+	1	1.191	MB
	MLDILVNSL (P5)		28	33	+	76/+	1	1.142	MB
	MTAERVLEV		25	15	+	74/+	1	1.181	HB[b]
LmjF25.0150	**AYSVSASSL (Kd1)**	**H-2Kd**	28	2880	-	-	-	-	-
LmjF14.0650	**SYETGSSTL (Kd2)**		25	2400	-	-	-	-	-
LmjF29.2650	**FYQEAAELL (Kd3)**		27	2400	-	-	-	-	-
LmjF29.0867	**SYSSLVSAL (Kd4)**		28	2880	-	-	-	-	-

* Peptides analyzed in HLA transgenic mice (P1–P6) and BALB/c mice (Kd1–Kd4) are in bold.
[a] Moderate binder.
[b] High binder.
Peptide falling over pre-set threshold.
(1) http://www.syfpeithi.de/bin/MHCServer.dll/EpitopePrediction.htm.
(2) http://www-bimas.cit.nih.gov/molbio/hla_bind/.
(3) http://www.jenner.ac.uk/EpiJen/.
(4) http://immunax.dfci.harvard.edu/Tools/rankpep.html.
(5) http://www.imtech.res.in/raghava/nhlapred/comp.html.
(6) http://www.cbs.dtu.dk/services/NetCTL/.
(7) http://antigen.i2r.a-star.edu.sg/multipred.
CPB (Cathepsin L-like Protease or Type I Cysteine Proteinase), CPC (Cathepsin L-like Protease or Type I Cysteine Proteinase), LmsTI-1 (*L.major* Stress Indussible Protein), LPG-3 (Lipophosphoglycan Biosynthetic Protein).

USA) digested with the same enzymes. Polytope sequence was inserted in-frame upstream to the EGFP sequence generating pEGFP-PT. The polytope sequence in tandem with EGFP was digested out of the pEGFP-PT by *Bgl* II – *Not* I restriction enzymes (Roche, Germany) and sub-cloned into pLEXSY-neo-2 plasmid (Jena Biosciences, Germany) making pLEXSY-PT-EGFP. Eventually polytope sequence was amplified with a set of primers (forward: CGCAAGCTTACCATGCAGATTTTCG and Reverse: AATGGATCCCTACACCAGCAGCACGCC, MWG, Germany) with *Hin*dIII and *Bam*HI restriction sites (underlined) and stop codon on reverse primer (bold). The PCR product was directly cloned into pcDNA3.1(+) vector (Invitrogen, USA) which was digested with *Hin*dIII and *Bam*HI. Recombinant pcDNA-PT was next subject to sequencing.

Cell lines

African green monkey kidney fibroblast-like cell line, COS-7 (ATCC CRL-1651), was cultured in RPMI-1640 medium (Sigma, Germany) supplemented with 10% inactivated fetal calf serum (Gibco, USA), 2 mM L-Glutamine, 10 mM HEPES and 50 µg/ml Gentamicin (all from Sigma, Germany). CT26, an undifferentiated colon carcinoma cell line from BALB/c origin (ATCC CRL-2638) was cultured in DMEM – Glutamax (+) medium supplemented with 1% penicillin/streptomycin antibiotic mixture and 10% inactivated fetal bovine serum (All from Gibco, USA). RMA/s cells, Transporter associated with antigen processing (TAP) deficient cells from C57BL/6 origin transfected with HLA-A*0201 were cultured in RPMI-1640 Glutamax (+) medium supplemented with 1% penicillin/streptomycin antibiotic mixture and 10% inactivated fetal bovine serum (FBS).

HLA-Transgenic mice

The humanized HLA-DRB1*0101/HLA-A*0201 were kindly provided by F.A. Lemonier, Pasteur Institute of Paris and were bred in animal facility of INSERM UMR1098. These mice are double knockouts of H-2 class I and class II genes and thus express only human MHC molecules (H-2 class I ($\beta 2m^O$)/class II (IA β^O)-KO). These C57BL/6 background mice express human alpha-1/alpha-2 domains of HLA-A*0201 heavy chain and murine alpha-3 domain covalently linked to human beta-2 microglobulin domain.

In-vitro evaluation of polytope expression using COS-7 cells

Recombinant pEFGP-PT was purified (Qiagen midi-plasmid purification kit, Germany) and transiently transfected into COS-7 cells by means of linear Polyethylenimine 25 KDa (LINPEI.25-Polysciences, USA), as previously described [53]. Briefly, freshly prepared DNA/PEI complex in HBS buffer with 18 µl of PEI at NrE (Number of Equivalents) = 7 and 5 µg of plasmid DNA was mixed with COS-7 cells (5×10^4 cell/well in opti-MEM serum free medium (Invitrogen, USA)) at 70% confluency. In this experiment pEGFP-N3 plasmid without insert was used as control. Six hours later, cells were washed and incubated for further 18 hours. Green fluorescence of GFP expressing cells was detected both by fluorescent microscope (Nikon-Japan) and by flow cytometry (Becton Dickinson FACScalibur, USA) after 24 hours.

Stable transfection of *Leishmania tarentolae* with pLEXSY-PT-EGFP

Stable transfection of *L. tarentolae* was performed as previously described [54]. Briefly, *L. tarentolae* parasites at logarithmic growth phase were suspended in electroporation buffer at a total number of 4×10^7 cells/400 µl. Pre-cooled parasite suspension

mixed with 15 µg of purified linear pLEXSY-PT-GFP (Promega gel extraction kit, USA) in electroporation cuvette, was electroporated (Gene Pulser Xcell, Bio Rad, USA), was recovered in M199 (Sigma, Germany) -10% FCS selection antibiotic free medium and was subsequently plated on Nobel agar plates supplemented with or without 30 µg/ml of G-418 (Gibco, USA). Resistant clones were sub-cultured with 200 µg/ml of G-418 for three consecutive weeks in logarithmic growth phase. Resistant parasites were subject to complementary assays at DNA and RNA levels. Integration of the linear plasmid encoding polytope-EGFP fragment into the chromosome was subsequently confirmed with a set of forward (F3001: TATTCGTTGTCAGATGGCGCAC) and reverse (A1715: GATCTGGTTGATTCTGCCAGTAG) primers (MWG, Germany) each specific for chromosomal and plasmid DNA respectively. Complementary PCRs (polymerase chain reactions) with sets of primers specific for different integrated genes confirmed polytope integration. At the RNA level, after RNA extraction (QIAGEN RNeasy extraction kit-Germany) and following oligo-dT reverse transcriptional cDNA amplification, expression of the polytope was confirmed by EGFP sequence specific primers (EGFP-F: ATGATATCAAGATCTATGGTGAGCAAGGGC, EGFP-R: GCTCTAGATTAGGTACCCTTGTACAGCTCGTC).

Polytope degradation assay using fluorescent *L. tarentolae*

Resistant parasites under high drug concentration (200 µg/ml) were washed and treated with Proteasome Inhibitor MG132 (Biotrend, Germany) at two different concentrations: 5 µM and 10 µM. Un-treated cells were used as control. After three hours of incubation at 26°C, cells were washed and re-suspended in PBS buffer for microscopic and flow cytometric analysis.

Mouse Immunization and splenocyte isolation protocol

Endotoxine free recombinant pcDNA-PT was prepared by QIAGEN EndoFree Plasmid Mega Kit, (QIAGEN, Germany) for immunization. Eight weeks-old male HLA-DRB1*0101/HLA-A*0201 transgenic mice were humanly anesthetized and injected with 100 µl diluted Cardiotoxin (Latoxan, France) 4 days before DNA immunization (6.8 µg per mouse). 100 µg Purified plasmid DNA was inoculated bilaterally in hamstring muscles of each anesthetized mice followed by 2 boosters with one week interval each. Ten days after the last booster, mice were humanly euthanized. Dissected spleen from each individual mouse was split into single cell suspension of splenocytes (BD Bioscienses 70 µm cell strainer) and depleted of red blood cells by ACK lyses buffer (Sigma, France). RBC free suspension was washed and re-suspended in x-vivo 15 medium (Lonza, France) for further incubation (3 hours) prior to immunoassays. Both experiments in BALB/c and transgenic mice were repeated twice.

IFN-γ ELISpot assay

Ex-vivo ELISpot was conducted as previously described [55,56]. Briefly, splenocytes from immunized mice were incubated at 2×10^5 cells per well in duplicates in ELISpot Anti-IFN-γ coated plates in presence of the relevant or control peptides (Proimmune, Ltd-UK). Plates were incubated for 16 to 18 hours at 37°C, and spots were revealed following the manufacturer's instructions (GenProbe-France). Spot-forming cells (SFC) were counted using C.T.L. Immunospot system (Cellular Technology Ltd. Germany). Stimulations resulting in spots 2 times the negative control (un-stimulated cells) and more than 10 were considered positive. Anti-HLA DR antibody (clone L243, provided by Bernard Maillère laboratory, France) was used to confirm HLA class I restricted response.

Short term CTL line generation

12×10^6 splenocytes per well from each individual mouse were plated in 6 well plates (Greiner Bio-one, France) in RPMI-1640 supplemented with 10% fetal bovine serum and 50 μM β-mercaptoethanol. Individual peptides used for *in-vitro* stimulation were supplied at 5 μg/ml final concentration. To amplify responding clones' frequency, IL-2 was added at 100 U/ml final concentration and cells were incubated for three more days. Peptide specific cytotoxic activity of $CD8^+$ T cell lines generated during *in-vitro* culture was further assessed in CTL assay.

Chromium 51(^{51}Cr) release assay (CTL assay)

Cytolytic activity was tested in a standard 4-hour ^{51}Cr release assay [57]. RMA/s target cells loaded with individual peptides at a 10 μg/ml final concentration and labeled with ^{51}Cr radioactive isotope were co-cultured with short term CTL lines making three different effector-to-target ratios. Supernatant fluids from all stimulation conditions were harvested in 96 well Lumaplates (PerkinElmer, USA). Plates were air dried (overnight) and radioactivity was measured in 1450 MicroBeta (Wallac, Finland) the following day. Results are expressed as the mean of triplicates in % of specific lyses: [(experimental − spontaneous release)/(total − spontaneous release)] ×100. An irrelevant HLA-A*0201 restricted 9 mer peptide derived from human telomerase was used as negative control.

Statistical analysis

Data were analyzed based on variance difference significance. In the case of P-value<0.05, un-paired t-test was used and in the case of P-value>0.05, Mann-Whitney U non-parametric test was used. P-value less than 0.05 was considered significant in each test.

Results

Rational design of a polytope construct

To design a construct able to stimulate *Leishmania* specific CD8 T cell responses, 13 selected peptides from our previous study plus 4 H-2Kd control restricted peptides were arranged in tandem. Figure 1 depicts the final arrangement of peptides. Besides spacers we had inserted additional extensions as Alanine/Arginine/Tyrosine (ARY) into flanking area of each H-2Kd determinant to increase the peptide affinity for TAP molecule [47,50,58]. As illustrated in Figure 2-A (the results of NetCTL analysis) and Figure 2-B (the results of nHLApred analysis), the final arrangement was chopped into desired peptides (top of the list) resulting in least junctional peptides. Since a protein is ubiquitinated only if it carries degradation signals [59,60] and a polytope is devoid of these signals, the final arrangement was additionally supported with an ubiquitin sequence upstream of the polytope with a G76A substitution to protect deubiquitination by hydrolases. Ubiquitination efficiency was further evaluated by a GFP expressing *Leishmania* species.

Furthermore, TT_{830} sequence, a universal Th1 epitope from tetanus toxoid, was integrated downstream of the polytope sequence to meet the prerequisite of naïve CD8 T cell response activation [61,62]. Th1 peptide was separated by ARY extension since cytoplasmic Th1 epitope must gain access to secretory cavity to appear in the context of HLA class II molecules [63,64,65]. TT_{830} was chosen as helper epitope due to lack of sufficient knowledge on human HLA class II restricted *Leishmania* peptides.

Further testing of the polytope construct devoid of TT_{830} will show whether inclusion of such an epitope is necessary.

Hydrophobic/hydrophilic nature of the protein is an additional point which is necessary to be taken into account for efficient expression. Figure S1 depicts the hydrophobicity pattern of the final arrangement analysed by Expasy-Protscale (http://web.expasy.org/protscale/). N-terminal region of the polytope was quite hydrophilic due to ubiquitin sequence. 921 bp long sequence, cloned in pUC57, was subject to a few cloning steps summarized in Figure S2.

Polytope expression was confirmed in transiently transfected mammalian COS-7 cells

COS-7 cells were transiently transfected with pEGFP-PT to verify the expression of the polytope in eukaryotic cells recognizing CMV promoter. As shown in Figure 3, 24% of pEGFP-PT transfected COS-7 cells were GFP positive. Since the polytope sequence was inserted upstream of the EGFP sequence, fluorescense detection directly correlated with polytope expression. pEGFP-N3 transfected cells with 32% positivity were used as control.

Proteasome degradation was obviously ubiquitin dependant

In this study we established a GFP-based instead of radioactive degradation detection method [66] to evaluate the efficacy of ubiquitination. In this system, homologous recombination directly inserted the polytope-EGFP (PT-EGFP) sequence with N-terminal ubiquitin into the ribosomal DNA region (r-DNA) of *L. tarentolae* where the high expression levels meet the demands for ribosomal assembly. The plasmid integration and expression was confirmed at both DNA (Figure S3) and RNA level (Figure S4). Stably transfected *L. tarentolae* parasites with pLEXSY-PT-EGFP were cultured three consecutive weeks under G-418 pressure (200 μg/ml) for logarithmic expansion. Recombinant parasites shining green were barely detectable before treatment with a small three-peptide inhibitor which easily disseminates into the cell and transiently disturbs the proteasome function in a competitive manner in contrast to control pLEXSY-EGFP transfected *L. tarentolae* (data not shown). One prominent character of this system was easy handling of parasite in a rather simple medium for weeks in a logarithmic phase to assure a low level of GFP expression. Recombinant clones were then treated with MG132 to transiently stop degradation process. Parasites shining green were easily detectable by flow cytometry and fluorescent microscope monitoring within three hours of treatment (Figure 4). The effect of MG132 treatment on pLEXSY-PT-EGFP transfected clones was compared to another clone transfected with an ubiquitin free construct. The level of expression before and after treatment with proteasome inhibitor roughly differed in latter case compared to ubiquitinated construct (Figure S5). So it was firmly established that ubiquitination successfully directed the polytope to cytoplasmic degradation right after synthesis.

Polytope expression was confirmed by DNA immunization in BALB/c mice

Four H-2Kd restricted epitopes (previously introduced by Domunteil *et al.* [67]) were used in a control experiment in BALB/c mice. Figure 5A shows individual IFN-γ responses of 4 immunized mice against individual peptides detected by *ex-vivo* ELISpot assay following DNA-DNA prime-boost immunization. The response against all 4 peptides appeared positive and statistically significant ($p < 0.05$) compared to un-stimulated control

MQIFVKTLTGKTITLEVEPSDTIENVKAKIQDKEGIPPDQQRL
IFAGKQLEDGRTLSDYNIQKESTLHLVLRLRGA AD YMEDQRF
AL AD ALQAYDEGL AD LLMLQPDYV AD QLDEQNSVL AD FLVGD
RVRV AD LLLLGSVTV AD MLDILVNSL K MTAERVLEV AD ARY LLT
GYPVSV AD LLATTVSGL AAY FLGGHAVKL AAA LMLQAFEWV AD
ARY AYSVSASSL AD ARY SYETGSSTL AD ARY FYQEAAELL AD ARY
SYSSLVSAL AD ARY QYIKANSKFIGITEL AD ARY QLNHGVLLV

Figure 1. Final arrangement of the polytope construct used for immunization. 13 HLA-A*0201 plus 4 H-2Kd control restricted peptides were arranged in tandem with spacers for accurate proteosomal cleavage (in red). Additional N-terminal ubiquitin sequence (bigger box) for proteosomal degradation, and C-terminal TT_{830} epitope (smaller box) for CD8 T cell response enhancement were also included. Peptides depicted in green are from *Lm*STI-1, peptides in blue from LPG-3, peptides in gray from CPB/CPC and peptides in yellow are H-2Kd restricted (AYS = Kd1, SYE = Kd2, FYQ = Kd3, SYS = Kd4).

cells. As shown in figure 5B, Kd4 stimulated a significant but weaker IFN-γ production compared to the other three examined peptides. These results confirmed that the long polytope was properly expressed, translated and chopped into peptides by proteasome cleavage leading to peptide presentation and priming of naïve $CD8^+$ T cells.

Leishmania specific CD8 T cells were induced against HLA-A*0201 restricted peptides in HLA transgenic mice

To evaluate the *in-vivo* immunogenicity of selected peptides, HLA-A2 Transgenic mice were immunized with polytope DNA construct. Immune reactivity was assessed by IFN-γ secretion from peptide stimulated splenocytes. Figure 6 illustrates the results for *ex-vivo* stimulation (16 hrs of culture) of splenocytes from each individual mice against individual HLA-A*0201 restricted peptides at 5 µg/ml/peptide final concentration. IFN-γ producing cells were enumerated by ELISpot assay. 10 mice out of 11 responded to 1-3 peptide out of 6 (Figure S6). LPG-3-P1 (P1: LLLLGSVTV) elicited a dominant response compared to *Lm*STI-1-P3 (P3: QLDEQNSVL), *Lm*STI-1-P4 (P4: LLMLQPDYV), LPG-3-P5 (P5: MLDILVNSL) and *Lm*STI-1-P6 (P6: ALQAYDEGL). *Lm*STI-1-P2 (P2: YMEDQRFAL) at this stage provoked no response. The response against P1, P4 and P5 was statistically significant compared to negative control peptide ($p<0.05$).

To further elucidate the immune response under different stimulation conditions, splenocytes were stimulated *in-vitro* with individual peptides (5 µg/ml/peptide) along with IL-2 for one week. Splenocytes from 4 mice out of 5 responded to 2–4 peptides out of 6 (Figure S7). As shown in Figure 7A, an elevated response was detected against peptides P3 and P6. The response against P1 was detected as strong as before, but the response against peptide P2, P4 and P5 was barely affected.

Next the splenocytes were stimulated with a higher concentration of each peptide (10 µg/ml/peptide) during *in-vitro* stimulation. Splenocytes from 5 mice out of 6 responded to 3–6 peptides out of 6 (Figure S8). As shown in Figure 7B an elevated response was detected against all six peptides including P2. The response against P1, P4 and P6 was statistically significant ($p<0.05$) (with 6 mice out of 6 responding to the relevant peptides). The response was $CD8^+$ T cell restricted since CD4-T cell blockade by anti-HLA-DR, did not influence the outcome (Figure 7C).

Therefore, we could consider P1 as a dominant high affinity peptide since it elicited potent IFN-γ response both *ex-vivo* and *in-vitro* even at low peptide concentration. The remaining tested peptides could be considered as subdominant regarding IFN-γ response since the response could be raised by *in-vitro* IL-2 stimulation (P3 and P6) and/or by increasing the level of peptide concentration (P2, P3, P4, P5, P6).

Specific CD8-T cell lymphocytes displayed cytotoxic activity against peptide loaded target cells

The cytolytic activity of the T cell clones against all six peptides was investigated after one week stimulation with 5 µg/ml/peptide and 100 U/ml IL-2. RMA/s cells loaded with relevant peptides served as targets. All the results were compared to a negative control. P1, P3 and P6 induced specific lysis by CTL lines regarding P1 as the most potent inducer. P2, P4 and P5 provoked no cytolysis at all. Figure 8 illustrates the percent of specific lysis of targets loaded by P1, P3 and P6 by T cell clones from individual mice at 3 different effector-to-target (E/T) ratios and Table 2 compares the results at 30:1 E/T ratio. Thereby 3 out of 6 peptides were considered as Tc1 type cells with both IFN-γ production potential and cytolytic activity. This makes the construct quite attractive for vaccination against *Leishmania* infectious challenge.

Discussion

Today it is believed that $CD8^+$ T cells take a significant part in immunity against leishmaniasis. So polytope vaccines might turn a hope in Leishmaniasis control. Therefore in this study we decided to evaluate the immunogenicity of a DNA polytope construct encompassing previously determined immunogenic peptides [43] in a relevant preclinical mouse model. As indicated in our previously published paper, we examined the immune response of CL recovered HLA-A2+ individuals against different peptide pools and included in our DNA construct all peptides which had detectable stimulatory effect on PBMCs (CPB and CPC -5 peptides, *Lm*STI-1 -4 peptides and LPG-3 - 4 peptides). A weaker response to CPB/CPC peptide pool was detected which was implied to be more potentially restricted to other closely related super-types besides A2. In this study we only focused on 6 out of 8 remaining A2 restricted peptides (based on their scores and importance) to check the immunogenicity in HLA-A2 transgenic animals.

A polytope DNA construct was provided sticking to basics of rational design such as inserting spacers between adjacent peptides [68] and targeting the polytope to proteasome enzymatic degradation [69]. We selected DNA immunization instead of peptide immunization (which is deeply dependent on modulators of immunogenicity) or polytope string (which is more immunogenic but less cost effective) because DNA constructs are well appreciated as efficient for stimulating both T-helper-1(Th-1) and T-cytotoxic-1(Tc-1) responses. We first assessed the

A

```
NetCTL-1.2 predictions using MHC supertype A2. Threshold 0.750000

169 ID Sequence pep LLTGYPVSV aff  0.7526 aff_rescale  1.1219 cle 0.9674 tap  0.2500 COMB  1.2795 <-E
204 ID Sequence pep LMLQAFEWV aff  0.7746 aff_rescale  1.1547 cle 0.2471 tap  0.6120 COMB  1.2224 <-E
 90 ID Sequence pep ALQAYDEGL aff  0.6911 aff_rescale  1.0302 cle 0.6932 tap  0.9470 COMB  1.1815 <-E
123 ID Sequence pep FLVGDRVRV aff  0.6892 aff_rescale  1.0273 cle 0.8841 tap  0.1780 COMB  1.1689 <-E
155 ID Sequence pep MTAERVLEV aff  0.6879 aff_rescale  1.0254 cle 0.7231 tap  0.4580 COMB  1.1568 <-E
145 ID Sequence pep MLDILVNSL aff  0.6417 aff_rescale  0.9566 cle 0.9676 tap  0.7060 COMB  1.1371 <-E
180 ID Sequence pep LLATTVSGL aff  0.6305 aff_rescale  0.9398 cle 0.9396 tap  0.8560 COMB  1.1236 <-E
101 ID Sequence pep LLMLQPDYV aff  0.6580 aff_rescale  0.9808 cle 0.7525 tap  0.2640 COMB  1.1069 <-E
192 ID Sequence pep FLGGHAVKL aff  0.6214 aff_rescale  0.9263 cle 0.9634 tap  0.6640 COMB  1.1040 <-E
 79 ID Sequence pep YMEDQRFAL aff  0.6186 aff_rescale  0.9222 cle 0.9364 tap  0.7730 COMB  1.1013 <-E
294 ID Sequence pep QLNHGVLLV aff  0.6177 aff_rescale  0.9209 cle 0.9453 tap  0.2010 COMB  1.0727 <-E
134 ID Sequence pep LLLLGSVTV aff  0.5707 aff_rescale  0.8508 cle 0.9616 tap  0.2170 COMB  1.0059 <-E
293 ID Sequence pep YQLNHGVLL aff  0.5457 aff_rescale  0.8135 cle 0.8510 tap  1.1600 COMB  0.9992 <-E
  7 ID Sequence pep TLTGKTITL aff  0.5203 aff_rescale  0.7756 cle 0.9788 tap  1.0180 COMB  0.9734 <-E
205 ID Sequence pep MLQAFEWVA aff  0.6431 aff_rescale  0.9587 cle 0.2428 tap -0.4530 COMB  0.9725 <-E
135 ID Sequence pep LLLGSVTVA aff  0.5834 aff_rescale  0.8696 cle 0.7182 tap -0.4460 COMB  0.9550 <-E
263 ID Sequence pep SLVSALADA aff  0.5306 aff_rescale  0.7909 cle 0.2780 tap -0.3310 COMB  0.8161 <-E
259 ID Sequence pep YSYSSLVSA aff  0.4894 aff_rescale  0.7295 cle 0.5005 tap -0.1960 COMB  0.7948 <-E
 65 ID Sequence pep STLHLVLRL aff  0.4024 aff_rescale  0.5999 cle 0.9486 tap  0.8370 COMB  0.7841 <-E
112 ID Sequence pep QLDEQNSVL aff  0.4064 aff_rescale  0.6058 cle 0.9535 tap  0.6720 COMB  0.7824 <-E
199 ID Sequence pep KLAAALMLQ aff  0.4686 aff_rescale  0.6985 cle 0.0675 tap  0.0850 COMB  0.7129
102 ID Sequence pep LMLQPDYVA aff  0.3925 aff_rescale  0.5852 cle 0.8066 tap -0.3440 COMB  0.6889
 97 ID Sequence pep GLADLLMLQ aff  0.4718 aff_rescale  0.7033 cle 0.0300 tap -0.4150 COMB  0.6871
```

B

ALLELE: HLA-A*0201

Threshold for .5 % with score: .5

Prediction method	Rank	Sequence	Residue No.	Peptide Score
ANNs+QM	1	STLHLVLRL	65	1.000
ANNs+QM	2	QLDEQNSVL	112	1.000
ANNs+QM	3	FLVGDRVRV	123	1.000
ANNs+QM	4	LLLLGSVTV	134	1.000
ANNs+QM	5	MLDILVNSL	145	1.000
ANNs+QM	6	LLTGYPVSV	169	1.000
ANNs+QM	7	LLATTVSGL	180	1.000
ANNs+QM	8	AYSVSASSL	218	1.000
ANNs+QM	9	YMEDQRFAL	79	0.990
ANNs+QM	10	FYQEAAELL	246	0.980
ANNs+QM	11	FLGGHAVKL	192	0.970
ANNs+QM	12	SYSSLVSAL	260	0.960
ANNs+QM	13	ALQAYDEGL	90	0.930
ANNs+QM	14	SYETGSSTL	232	0.470
ANNs+QM	15	RLIFAGKQL	42	0.290
ANNs+QM	16	ARYAYSVSA	215	0.180
ANNs+QM	17	HAVKLAAAL	196	0.130
ANNs+QM	18	KESTLHLVL	63	0.100

Figure 2. Immunoinformatic prediction of proteosomal cleavage pattern. A. NetCTL cleavage pattern of the final selected polytope sequence. This diagram shows the output frm NetCTL software after proteosomal cleavage with a threshold over 0.75 to discriminate between binders and non-binders. Junctional peptides have negative scores for TAP binding. Peptides scored over threshold are enclosed in a box. B. Immunoproteasome cleavage pattern of the final selected polytope sequence predicted by nHLApred. This diagram shows the output frm nHLApred software after immunoproteosomal cleavage with a threshold over 0.5 to discriminate between binders an non-binders. Peptides scored over threshold are enclosed i a box. First Peptide in the box was an ubiquitin derived peptide.

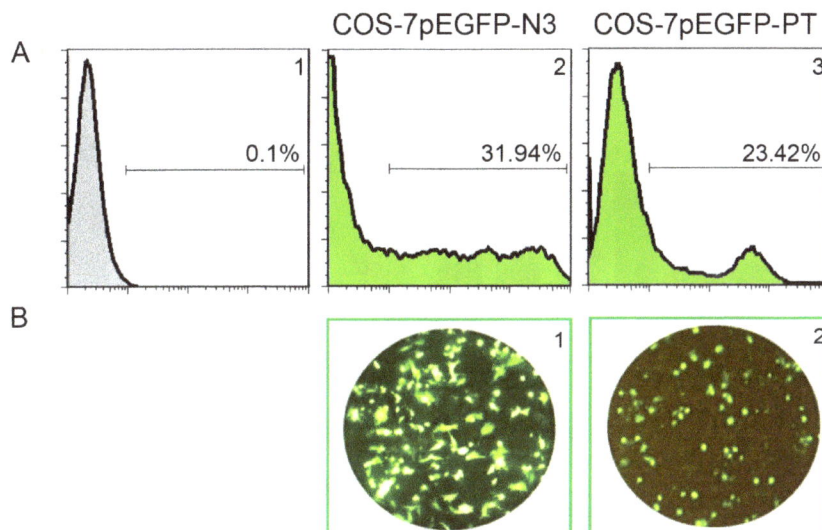

Figure 3. *In-vitro* evaluation of polytope expression using COS-7 cells. Recombinant pEFGP-PT was transiently transfected into COS-7 cells by means of linear Polyethylenimine 25 KDa. A1 represents COS-7 cells without any transfection, A2 represents COS-7 cells transfected with pEGFP-N3 as positive control and A3 represents COS-7 cells transfected with pEGFP-PT. Panel B represents the corresponding microscopic feature of each condition, B1 represents positive control and B2 represents COS-7 cells transfected with pEGFP-PT after 24 hours.

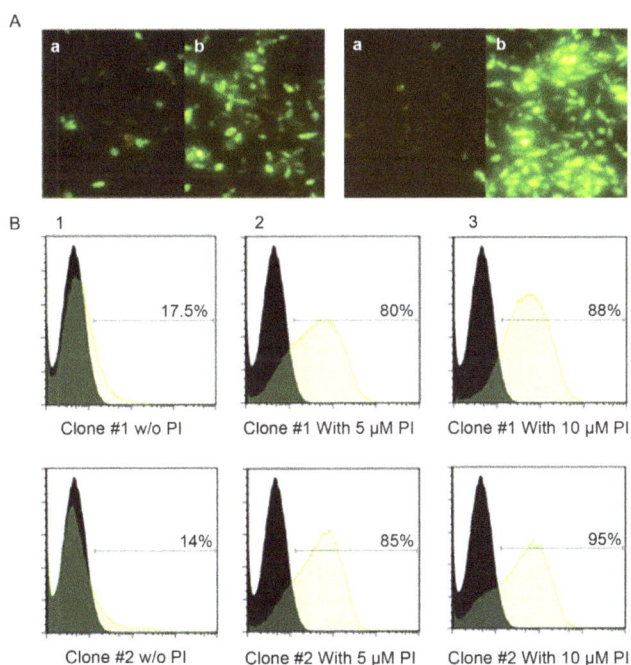

Figure 4. Transfected *Leishmania* parasites before and after treatment with proteasome inhibitor. Stable transfected parasites harboring polytope sequence were generated and used for evaluation of ubiquitinated polytope expression and degradation. A. Shows the fluorescent microscope patterns in 2 microscopic fields. Pale colored parasites before MG132 treatment turned sharp green in the presence of MG132 inhibitor. Part "a" in each field reflects before and part "b" after treatment with 10 μM MG132 for 3 hours. B. Illustrates fluorescent intensity of 2 different transfected clones (Clone#1 and clone #2) before and after treatment with MG132. Column 1: before treatment, column 2: treatment with 5 μM and column 3: treatment with 10 μM of MG132. Numbers on each plot represent the GFP positive population which drastically increases after MG132 treatment. PI: proteasome inhibitor.

immunogenicity of the polytope construct in BALB/c mice where 4 out of 4 H-2Kd restricted peptides previously characterized by Dumonteil et al. were shown to be immunogenic [67]. In their experiment, the peptides were directly inoculated into BALB/c mice with adjuvant and ranked in Kd2>Kd3>Kd1>Kd4 order based on ELISA detected IFN-γ production. Our results were totally concordant confirming that all 4 peptides were quite accessible for the immune system after proteasome degradation. In this DNA construct the H-2Kd restricted peptides were inserted downstream to the polytope. Therefore it was inferred that the polytope was fully expressed till the end point and was chopped into desired peptides. This result further encouraged us to proceed with HLA-humanized mice which harbor less CD4$^+$ and CD8$^+$ T cell amounts than conventional mice strains [70].

The moderate efficacy of many vaccine trials primarily reported as protective in wild-type animal models is partly explained by the different influence that human and animal MHC have on the outcome of the immune response [71]. Humanised transgenic mice models expressing human HLA instead of mouse MHC, fill this gap between human and mice. These preclinical models have shown promising results despite subtle differences in antigen processing machinery including proteasome cleavage and TAP molecules affinity for peptides [72], since the immunological hierarchy is approximately the same in both models and about 80% of peptides immunogenic in one are also immunogenic in the other [73].

In our previous study we had focused on peptides in a pool to screen the immunogenicity by human PBMCs. As reported, the response against *Lm*STI-1 peptide pool (4 peptides) appeared in higher frequency and lower potency quite contrary to LPG-3 peptide pool (4 peptides) with higher potency and lower frequency. Here we studied the individual response of 4 peptides in *Lm*STI-1 pool and 2 out of 4 in LPG-3 stimulating pool in a DNA construct. Almost all peptides were able to induce specific CD8 T cell responses *in-vivo*, indicating the adequacy of HLA-A2 transgenic mice T cell repertoire. P1, derived from LPG-3, dominantly induced Tc1 type responses. P3, P4, P6 (derived from *Lm*STI-1)

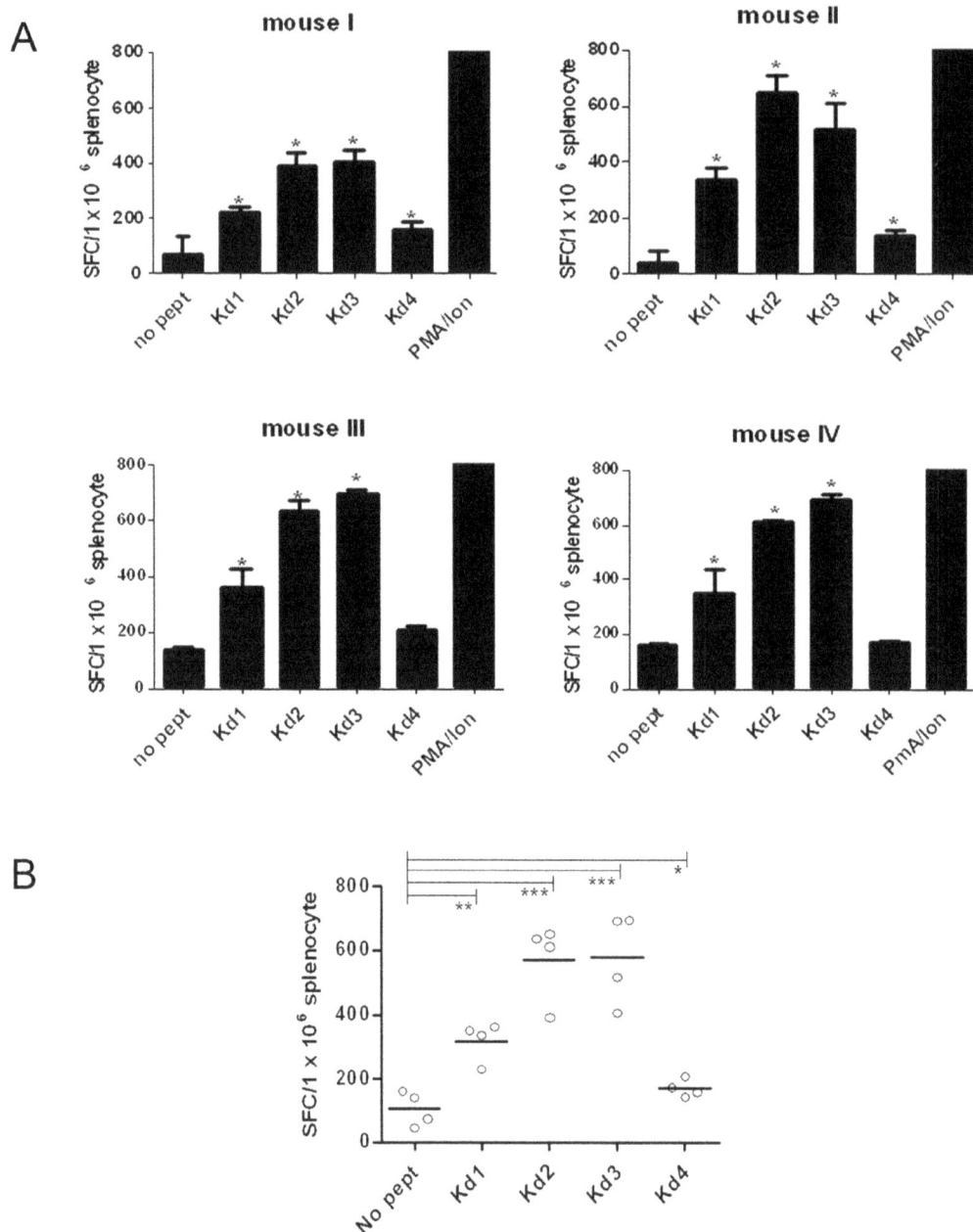

Figure 5. Balb/c response against 4 H-2Kd restricted peptides. 4 mice were immunized with polytope construct three times with one week interval and sacrificed 10 days after the last booster. Splenocytes from individual mice were *in-vitro* re-stimulated by representative peptides (Kd1-4) of Balb/c and specific IFN-γ production was evaluated by *ex-vivo* ELISPOT assay. A. Representative of two experiments. Columns, mean of spots from duplicate wells for each mice from one representative experiment; bars, SD. Stimulations resulting in spots two times the negative control (unstimulated cells) and more than 10 were considered positive (stars). B. Statistical analysis of consolidated data from 4 mice against each individual peptide. The response against all 4 peptides appeared statistically significant compared to un-stimulated control cells ($p<0.05$) with an exception for Kd4 which was subdominant in comparison to the rest. Horizontal lines represent the mean value. SFC: Spot Forming Cells.

and P5 (the other peptide form LPG-3 group) appeared subdominant. P2 was rather a weak immunogen. In this study P1 from LPG-3 appeared quite dominant regarding IFN-γ production and cytolytic activity compared to peptides from *Lm*STI-1. The level of expression of LPG-3 declines in the amastigote stage of the parasite but *Lm*STI-I is constitutively expressed both in promastigote and amastigote stage of parasite with even higher expression in amastigote stage. So the level of

response in human experiment shows a good correlation between affinity, final avidity and immunogenicity of the peptides [55,74].

We concluded that the hierarchy observed between different peptides is in fact a function of peptide affinity but not the levels of expression or proteasome cleavage. Fig-1 clarifies our declaration since the P1 peptide appears quite at the middle of the sequence so there is no superiority for level of expression. Also as shown in Fig-2A and 2B, the P1 peptide has no superiority to other peptides regarding its position in the list. More importantly all of the 6

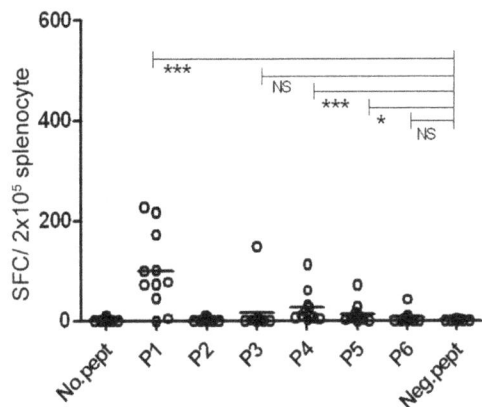

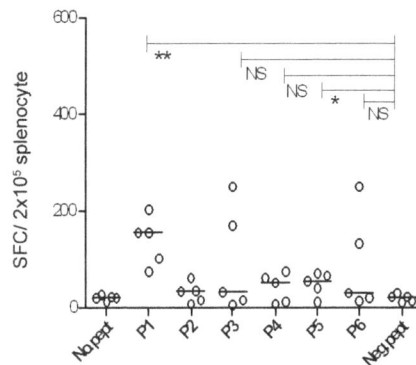

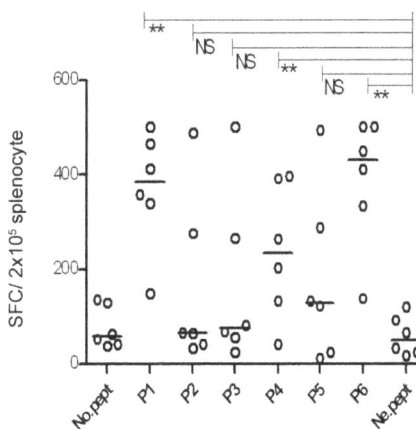

Figure 6. *Ex-vivo* **evaluation of the specific response against six peptides (5 μg/ml/peptide) in HLA A2/DR1 mice.** A total of 11 mice in two rounds of experiments were immunized with polytope construct three times with one week interval and sacrificed 10 days after the last booster. Splenocytes from individual mice were *in-vitro* re-stimulated by representative peptides (P1–P6) of HLA-A2 and specific IFN-γ production was evaluated by *ex-vivo* ELISPOT assay. Each dot represents mean of duplicate wells for each individual mice response against each peptide. Neg.pept (negative control peptide) represents a 9 mer HLA-A*0201 restricted peptide from human telomerase. Horizontal lines represent the mean value. No.pept = no peptide stimulation. NS: not-significant.

peptides examined are separated by "AD" as spacer (Fig-1). Even subtle differences between human and HLA transgenic mice in TAP-peptide affinity (as described by Pascolo *et al.* [72]) is of less importance because HLA-A2 peptide restriction is less TAP dependant than other HLAs. The main positive point for this peptide is that the score predicted by SYFPEITHI immunoinformatics software is higher than the rest. Some studies confine the selection on SYFPEITHI scores over 24 [75,76] to stringently select dominant peptides restricted to HLA-A*0201. But here we showed that scores over 20 appear quite satisfactory because helped us predict valuable low affinity peptides. Whether a construct composed of both dominant and subdominant epitopes effectively protects against *Leishmania* challenge is open to further investigations [77,78].

To end with a proper arrangement of 17 peptides (Thirteen HLA-A*0201 restricted peptides and four H-2Kd restricted ones), all possible combinations with or without spacers were examined by immunoinformatic methods. In our experience, arrangements with AAA, AAY and AD spacers had a better performance regarding peptide processing compared to spacer free arrangements and arrangements with lysine (K) as spacer. This is in agreement with the "P1" premise. First experiments with polytope constructs appeared as string of beads without any flanking sequences between each two determinants with detectable immunogenicity for comprising peptides. However in some cases, the immunogenicity was apparently a function of peptide position regarding flanking sequences [79]. The influence of the P1 amino acid (the first residue next to the C-terminus of the peptide) on processing efficiency was elucidated by further investigation. Where some results indicated a preference for natural flanking sequences for proteasome processing [80] others showed that C-terminal flanking of epitopes with alanine increased the epitope processing and recognition by T cells [81,82].

In our study "AD" was more satisfying as spacer regarding the least junctional peptide criteria in comparison to other two spacers. This might be rather hard to discuss since it is fully

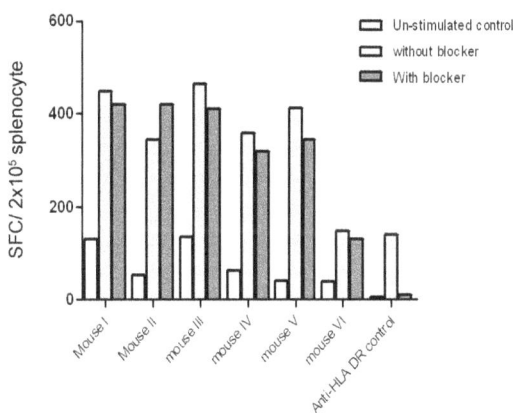

Figure 7. *In vitro* **evaluation of the specific response against six peptides in HLA A2/DR1 mice after one week stimulation (with 100 u/ml IL-2).** A. Splenocytes from a total of 5 mice immunized with polytope construct three times with one week interval and sacrificed 10 days after the last booster were re-stimulated by representative peptides (5 μg/ml/peptide) of HLA-A2. Specific IFN-γ production was evaluated by *ex-vivo* ELISPOT assay. Each dot represents mean of duplicate wells for each individual mice response against each peptide. B. Splenocytes were stimulated with 10 μg/ml/peptide instead of 5 μg/ml/peptide. Neg.pept (negative control peptide) represents a 9 mer HLA-A*0201 restricted peptide from human telomerase. Horizontal lines represent the mean value. No.pept = no peptide stimulation. NS: not-significant. C. P1 stimulation of splenocytes from B. along with Anti-HLA-DR (L243) blocker antibody. The response was CD8+ T cell restricted since CD4-T cell blockade by anti-HLA-DR, did not influence the outcome. L243 definitely lowers a potent CD4 T cell response in human PBMC against a TERT derived universal MHC class II restricted cancer peptide (UCP).

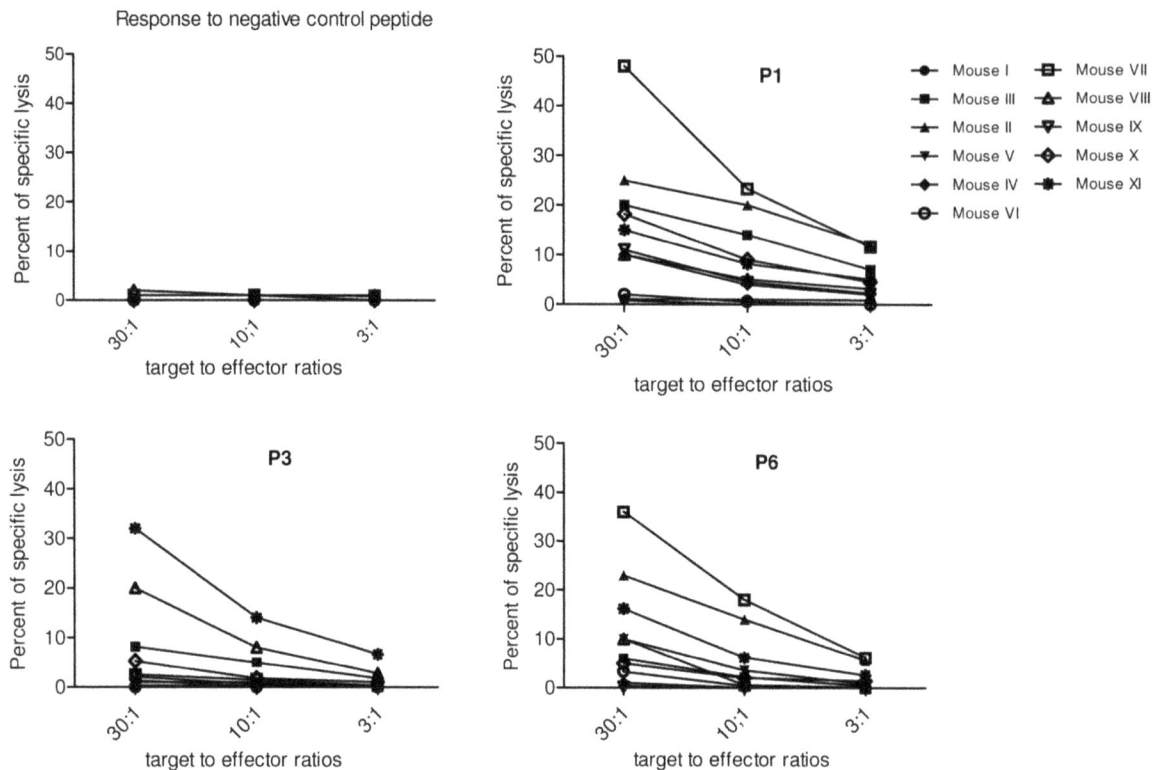

Figure 8. Cytotocic T cell response against RMA/s target cells loaded with individual peptides. Cytolytic activity was tested in a standard 4-hour ^{51}Cr release assay. RMA/s target cells loaded with individual peptides at a 10 µg/ml final concentration and labeled with ^{51}Cr radioactive isotope were co-cultured with short term CTL lines making three different effector-to-target ratios. Results are expressed as the mean of triplicates in % of specific lyses: [(experimental − spontaneous release)/(total − spontaneous release)] ×100. A 9-mer HLA-A0201 restricted peptide from human telomerase was used as negative control.

dependant on the peptide composition at one side and prediction methods and the algorithms they rely on, on the other side. Since flanking spacers and their nature is rather a controversial subject, immunoinformatic helps compare different options simultaneously saving time and energy but it should be kept in mind that *in-silico* prediction is still at its infancy and needs to grow up with more and more accurate data feeding.

Another important point was C-terminal glycine substitution of ubiquitin molecule (G76) with an alanine moiety to keep off hydrolytic enzymes. Protein-ubiquitin complex is rather unstable and readily disassembles by hydrolytic activity [66]. It has been shown that G76 substitutions simply guarantee ubiquitination [83,84]. This way the polytope is efficiently targeted to further ubiquitination and final degradation right after synthesis. In our experience G76A substitution worked efficiently to stabilize the

complex as shown by the difference before and after MG132 treatment. There are some reports for C-terminal insertion and efficient processing [50], but N-terminal conjugates attract more [85,86]. This could be simply explained by the fact that ubiquitin molecules bind other proteins and also other ubiquitin residues by their C-terminal glycine.

Here we investigated the efficiency of a polytope DNA construct sticking to the optimal designation criteria in the literature including: N-terminal G76A ubiquitin sequence, proteosomal cleavage considerations and C-terminal universal Tetanus Toxoid T-helper epitope to minimize the cost of study with transgenic animals. But it will be invaluable to further assess the different construct designations with many different options for processing and presentation like N-terminal signal sequence instead of Ubiqitine or *Leishmania* derived HLA-class II restricted peptides

Table 2. Percent of specific lysis of targets loaded by P1, P3 and P6 by T cell clones from individual mice at 30:1 E/T effector to target ratio.

Protein	Peptide sequence	R/T[a]	Specific lysis (%)[b] of responders
LPG-3	LLLLGSVTV (P1)	8/11	20, 25, 10, 48, 10, 11, 18, 15
*Lm*STI-1	QLDEQNSVL (P3)	4/11	8.2, 20, 5.3, 32
	ALQAYDEGL (P6)	7/11	6, 23, 10, 36, 10, 5, 16.2

[a]Responder to tested mice.
[b]Specific lysis at 30:1 effector to target ratio.

instead of TT_{830} and evaluate the level of protection conferred by *Leishmania* challenge in transgenic mice.

In our recent studies we focused on *in-silico* prediction of CD8 stimulating peptides from well known proteins of *Leishmania* and extended the study to experimental *in-vitro* and *in-vivo* evaluations. It was eventually proved that this procedure could result in potential CD8 stimulating peptides identification. To our knowledge, these studies for the first time report HLA class I restricted peptides from *Lm*STI-1 and LPG-3 proteins of *Leishmania*, immunogenic both in human and relevant animal model. Previous works focused on membrane associated proteins such as gp63 [87] and KmP11 [88] and reverse predicted peptides from *Leishmania* genome regarding membrane associated ORFs [89,90]. These kinds of antigens are postulated to get access to HLA class I system more efficiently. However Dumonteil *et al.* screened out immunogenic H-2Kd peptide epitopes from *L. major* proteome including 8272 annotated proteins without localization considerations [67]. *Lm*STI-1, recently detected in *L. donovani* secretome [91], is an intracellular protein [92] and one of the immunogenic components of Leish-F vaccine. LPG-3 is a highly immunogenic protein [93,94] which resembles mammalian endoplasmic reticulum chaperone *GRP94* required for phosphoglycan synthesis and localizes to endoplasmic reticulum of *Leishmania* [95]. Reiner *et al.* have recently characterized members of *L. donovani* secretome with a variety of subcellular localizations. In their study nearly one-third of *Leishmania* secreted proteins in exosomes were predicted to be cytoplasmic including ribosomal, nuclear, mitochondrial and glycosomal proteins but not endoplasmic reticulum proteins. Many proteins appear in the secretome without harboring a conventional N-terminal signal sequence as *Lm*STI-1 [96]. Therefore it could be a valuable approach to characterize novel protein antigens with CD8 T cell activating properties, through wide screening of *Leishmania* genome re-focusing on all possible ORFs with or without annotated function and without considering sub-cellular localization.

Our results were promising since at the end we were able to identify at least one dominant high affinity peptide out of 13 previously *in-silico* predicted ones. The rest of the peptides were subdominant with lower affinities but from a very effective vaccine candidate as *Lm*STI-1 and could be manipulated in further studies to altered peptides with even higher affinity. We cannot ignore other potential peptides that could have been missed by *in-silico* predictions or overlooked, but deeply believe that these kinds of studies in high-through-put scales could in fact speed up peptide screening and identification, both CD4 and CD8 stimulating ones, to build up peptide libraries for different *Leishmania* species. This is a prerequisite for further studies in both ways as polytope vaccination and new antigen hunting for *Leishmania* and is hardly achievable by classical peptide selection methods. Obviously we need more accurate prediction tools and more sensitive *in-vitro* detection tests to reach this end. Fortunately HLA transgenic mice are applicable preclinical models helping to speed up immunogenicity analysis in a human related mouse model.

Supporting Information

Figure S1 Kate and Dolite analysis (Protscale) of hydrophobic profile of the final polytope arrangement. This aanlysis was used to finally select between different combinations and shows the pattern of the final selected polytope sequence. No hydrophobic patch was detected specially at the N-terminal region to hinder translation.

Figure S2 Cloning pathway. 921 bp long polytop (PT) sequence, was codon optimized for optimal expression in mice and received in pUC57 (pUC57-PT). pEGFP-PT was used to confirm the expression of the sequence in mammalian cells by CMV promoter, pLEXSY-PT-EGFP was used to confirm the stability of the expressed polytope and pcDNA-PT was used for inoculations.

Figure S3 Representative PCR reactions used to confirm the plasmid integration into the genome from one transfected Leishmania tarentolae clone. A. Schematic representation of genome sequence after plasmid integration into rDNA *ssu* region. **B.** Full set of PCR reactions followed to confirm the integration at DNA level. Lane 1 and 6: 1 kb DNA ladder marker, lane 2: EGFP fragment (727 bp), lane 3: Polytope fragment (929 bp), lane 4: Polytope-EGFP fragment (1665 bp), lane 7: *SSU* fragment (1070 bp) and lane 8: EGFP-SSU (3000 bp). Lane 7 points to the most important reaction with 2 primers specific for a chromosomal sequence and plasmid sequence with *ssu* fragment in between. Lane 5 refers to un-transfected cells confirmed with F3001/A1715 PCR reaction.

Figure S4 RNA expression evaluation with a set of primers specific for EGFP. Lane 1: Fermentas 1 Kb ladder marker, lane 2 and 3: RT-PCR reaction from 2 transfected clones, lane 3: un-transfected cells.

Figure S5 Effect of MG132 treatment on ubiquitinated and non-ubiquitinated constructs. A. un-transfected parasite, **B.** EGFP transfected parasite, **C** and **D**, *L. tarentolae* transfected with ubiqitinated construct (pLEXSY-PT-EGFP). **E** and **F**, *L. tarentolae* transfected with un-ubiquitinated construct (pLEXSY-A2-EGFP). Expression level of EGFP before and after treatment with proteasome inhibitor roughly differs for un-ubiqitinated protein quite contrary to ubiquitinated protein. Numbers on each plot represent GFP positive population PI: proteasome inhibitor.

Figure S6 Ex-vivo response of individual mice against six peptides (5 µg/ml/peptide) in HLA A2/DR1 mice. A total of 11 mice in two rounds of experiments were immunized with polytope construct three times with one week interval and sacrificed 10 days after the last booster. Splenocytes from individual mice were *in-vitro* re-stimulated by representative peptides (P1-P6) of HLA-A2 and specific IFN-γ production was evaluated by *ex-vivo* ELISPOT assay. Each column represents the mean of duplicate wells stimulated with each peptide. Numbers on each plot show the number of peptides with positive response for each mouse. Peptide stimulations resulting in spots two times the negative control (Neg.pept) and more than 10 were considered positive (stars). Neg.pept (negative control peptide) represents a 9mer HLA-A*0201 restricted peptide from human telomerase.

Figure S7 In vitro evaluation of the specific response against six peptides in HLA A2/DR1 mice after one week stimulation (with 100 u/ml IL-2). Splenocytes from a total of 5 mice immunized with polytope construct three times with one week interval and sacrificed 10 days after the last booster were re-stimulated by representative peptides (5 µg/ml/peptide) of HLA-A2. Specific IFN-γ production was evaluated by *ex-vivo* ELISPOT assay. Each column represents mean of duplicate wells for each individual mice response against each peptide. Numbers on each plot show the number of peptides with positive response for each mouse. Peptide stimulations resulting in spots two times the negative control (Neg.pept) and more than 10 were considered positive (stars).

Figure S8 *In vitro* **evaluation of the specific response against six peptides in HLA A2/DR1 mice after one week stimulation with 100 u/ml IL-2 and higher concentration of peptides.** Splenocytes were stimulated with 10 µg/ml/ peptide instead of 5 µg/ml/peptide. Numbers on each plot show the number of peptides with positive response for each mouse. Peptide stimulations resulting in spots two times the negative control (Neg.pept) and more than 10 were considered positive (stars).

Acknowledgments

Authors from Pasteur Institute of Iran wish to sincerely acknowledge INSERM UMR1098, Besançon, France for their collaboration in *in-vivo* assays of this project. Sincere thanks goes to Prof. Philippe Saas, director at INSERM UMR1098, whose concern and comprehension provided the opportunity for this project to be accomplished. Special thank goes also to Patricia Letondal for her crucial suggestions and expertise, also Myriam David, Sarah Odrion and Jean-René Pallandre for their helps and concerns. We thank Dominique Paris for helping us in animal unit and also deeply thank Prof. François Lemonnier for HLA transgenic mice.

Author Contributions

Conceived and designed the experiments: NS PSR SR. Performed the experiments: NS TT CV MD YG. Analyzed the data: NS YG OA CB PSR SR. Contributed reagents/materials/analysis tools: AE IS OA CB PSR. Wrote the paper: NS PSR SR. Critical review of the manuscript: CV YG CB.

References

1. Organization WH (2010) Control of the leishmaniases. World Health Organization technical report series: xii.
2. Alvar J, Velez ID, Bern C, Herrero M, Desjeux P, et al. (2012) Leishmaniasis worldwide and global estimates of its incidence. PLoS One 7: e35671.
3. Croft SL, Olliaro P (2011) Leishmaniasis chemotherapy–challenges and opportunities. Clin Microbiol Infect 17: 1478–1483.
4. Kishore K, Kumar V, Kesari S, Dinesh DS, Kumar AJ, et al. (2006) Vector control in leishmaniasis. Indian J Med Res 123: 467–472.
5. Kedzierski L (2010) Leishmaniasis Vaccine: Where are We Today? J Glob Infect Dis 2: 177–185.
6. Modabber F (2010) Leishmaniasis vaccines: past, present and future. Int J Antimicrob Agents 36 Suppl 1: S58–61.
7. Okwor I, Mou Z, Liu D, Uzonna J (2012) Protective immunity and vaccination against cutaneous leishmaniasis. Front Immunol 3: 128.
8. Beaumier CM, Gillespie PM, Hotez PJ, Bottazzi ME (2013) New vaccines for neglected parasitic diseases and dengue. Translational Research 162: 144–155.
9. Velez ID, Gilchrist K, Martinez S, Ramirez-Pineda JR, Ashman JA, et al. (2009) Safety and immunogenicity of a defined vaccine for the prevention of cutaneous leishmaniasis. Vaccine 28: 329–337.
10. Nascimento E, Fernandes DF, Vieira EP, Campos-Neto A, Ashman JA, et al. (2010) A clinical trial to evaluate the safety and immunogenicity of the LEISH-F1+ MPL-SE vaccine when used in combination with meglumine antimoniate for the treatment of cutaneous leishmaniasis. Vaccine 28: 6581–6587.
11. Mougneau E, Bihl F, Glaichenhaus N (2011) Cell biology and immunology of Leishmania. Immunol Rev 240: 286–296.
12. Nylen S, Gautam S (2010) Immunological perspectives of leishmaniasis. J Glob Infect Dis 2: 135–146.
13. Herath S, Kropf P, Muller I (2003) Cross-talk between CD8(+) and CD4(+) T cells in experimental cutaneous leishmaniasis: CD8(+) T cells are required for optimal IFN-gamma production by CD4(+) T cells. Parasite Immunol 25: 559–567.
14. Ruiz JH, Becker I (2007) CD8 cytotoxic T cells in cutaneous leishmaniasis. Parasite Immunol 29: 671–678.
15. Muller I (1992) Role of T cell subsets during the recall of immunologic memory to Leishmania major. Eur J Immunol 22: 3063–3069.
16. Muller I, Kropf P, Etges RJ, Louis JA (1993) Gamma interferon response in secondary Leishmania major infection: role of CD8+ T cells. Infect Immun 61: 3730–3738.
17. Muller I, Kropf P, Louis JA, Milon G (1994) Expansion of gamma interferon-producing CD8+ T cells following secondary infection of mice immune to Leishmania major. Infect Immun 62: 2575–2581.
18. Huber M, Timms E, Mak TW, Rollinghoff M, Lohoff M (1998) Effective and long-lasting immunity against the parasite Leishmania major in CD8-deficient mice. Infect Immun 66: 3968–3970.
19. Overath P, Harbecke D (1993) Course of Leishmania infection in beta 2-microglobulin-deficient mice. Immunol Lett 37: 13–17.
20. Wang ZE, Reiner SL, Hatam F, Heinzel FP, Bouvier J, et al. (1993) Targeted activation of CD8 cells and infection of beta 2-microglobulin-deficient mice fail to confirm a primary protective role for CD8 cells in experimental leishmaniasis. J Immunol 151: 2077–2086.
21. Belkaid Y, Von Stebut E, Mendez S, Lira R, Caler E, et al. (2002) CD8+ T cells are required for primary immunity in C57BL/6 mice following low-dose, intradermal challenge with Leishmania major. J Immunol 168: 3992–4000.
22. Uzonna JE, Joyce KL, Scott P (2004) Low dose Leishmania major promotes a transient T helper cell type 2 response that is down-regulated by interferon gamma-producing CD8+ T cells. J Exp Med 199: 1559–1566.
23. Nateghi Rostami M, Keshavarz H, Edalat R, Sarrafnejad A, Shahrestani T, et al. (2010) CD8+ T cells as a source of IFN-gamma production in human cutaneous leishmaniasis. PLoS Negl Trop Dis 4: e845.
24. Barral-Netto M, Barral A, Brodskyn C, Carvalho EM, Reed SG (1995) Cytotoxicity in human mucosal and cutaneous leishmaniasis. Parasite Immunol 17: 21–28.
25. Faria DR, Souza PE, Duraes FV, Carvalho EM, Gollob KJ, et al. (2009) Recruitment of CD8(+) T cells expressing granzyme A is associated with lesion progression in human cutaneous leishmaniasis. Parasite Immunol 31: 432–439.
26. Machado P, Kanitakis J, Almeida R, Chalon A, Araujo C, et al. (2002) Evidence of in situ cytotoxicity in American cutaneous leishmaniasis. Eur J Dermatol 12: 449–451.
27. Rogers KA, Titus RG (2004) Characterization of the early cellular immune response to Leishmania major using peripheral blood mononuclear cells from Leishmania-naive humans. Am J Trop Med Hyg 71: 568–576.
28. Russo DM, Chakrabarti P, Higgins AY (1999) Leishmania: naive human T cells sensitized with promastigote antigen and IL-12 develop into potent Th1 and CD8(+) cytotoxic effectors. Exp Parasitol 93: 161–170.
29. Hernandez-Ruiz J, Salaiza-Suazo N, Carrada G, Escoto S, Ruiz-Remigio A, et al. (2010) CD8 cells of patients with diffuse cutaneous leishmaniasis display functional exhaustion: the latter is reversed, in vitro, by TLR2 agonists. PLoS Negl Trop Dis 4: e871.
30. Joshi T, Rodriguez S, Perovic V, Cockburn IA, Stager S (2009) B7-H1 blockade increases survival of dysfunctional CD8(+) T cells and confers protection against Leishmania donovani infections. PLoS Pathog 5: e1000431.
31. Faria DR, Gollob KJ, Barbosa J Jr, Schriefer A, Machado PR, et al. (2005) Decreased in situ expression of interleukin-10 receptor is correlated with the exacerbated inflammatory and cytotoxic responses observed in mucosal leishmaniasis. Infect Immun 73: 7853–7859.
32. Gaze ST, Dutra WO, Lessa M, Lessa H, Guimaraes LH, et al. (2006) Mucosal leishmaniasis patients display an activated inflammatory T-cell phenotype associated with a nonbalanced monocyte population. Scand J Immunol 63: 70–78.
33. Cong H, Mui EJ, Witola WH, Sidney J, Alexander J, et al. (2010) Human immunome, bioinformatic analyses using HLA supermotifs and the parasite genome, binding assays, studies of human T cell responses, and immunization of HLA-A*1101 transgenic mice including novel adjuvants provide a foundation for HLA-A03 restricted CD8+T cell epitope based, adjuvanted vaccine protective against Toxoplasma gondii. Immunome Res 6: 12.
34. Geluk A, van den Eeden SJ, Dijkman K, Wilson L, Kim HJ, et al. (2011) ML1419c peptide immunization induces Mycobacterium leprae-specific HLA-A*0201-restricted CTL in vivo with potential to kill live mycobacteria. J Immunol 187: 1393–1402.
35. Mudd PA, Martins MA, Ericsen AJ, Tully DC, Power KA, et al. (2012) Vaccine-induced CD8+ T cells control AIDS virus replication. Nature 491: 129–133.
36. Brinkman JA, Fausch SC, Weber JS, Kast WM (2004) Peptide-based vaccines for cancer immunotherapy. Expert Opin Biol Ther 4: 181–198.
37. Perez SA, von Hofe E, Kallinteris NL, Gritzapis AD, Peoples GE, et al. (2010) A new era in anticancer peptide vaccines. Cancer 116: 2071–2080.
38. Colmenares M, Kima PE, Samoff E, Soong L, McMahon-Pratt D (2003) Perforin and gamma interferon are critical CD8+ T-cell-mediated responses in vaccine-induced immunity against Leishmania amazonensis infection. Infect Immun 71: 3172–3182.
39. Gurunathan S, Stobie L, Prussin C, Sacks DL, Glaichenhaus N, et al. (2000) Requirements for the maintenance of Th1 immunity in vivo following DNA vaccination: a potential immunoregulatory role for CD8+ T cells. J Immunol 165: 915–924.
40. Mendez S, Belkaid Y, Seder RA, Sacks D (2002) Optimization of DNA vaccination against cutaneous leishmaniasis. Vaccine 20: 3702–3708.
41. Mendez S, Gurunathan S, Kamhawi S, Belkaid Y, Moga MA, et al. (2001) The potency and durability of DNA- and protein-based vaccines against Leishmania

major evaluated using low-dose, intradermal challenge. J Immunol 166: 5122–5128.

42. Stager S, Rafati S (2012) CD8(+) T cells in leishmania infections: friends or foes? Front Immunol 3: 5.

43. Seyed N, Zahedifard F, Safaiyan S, Gholami E, Doustdari F, et al. (2011) In silico analysis of six known Leishmania major antigens and in vitro evaluation of specific epitopes eliciting HLA-A2 restricted CD8 T cell response. PLoS Negl Trop Dis 5: e1295.

44. Oseroff C, Sette A, Wentworth P, Celis E, Maewal A, et al. (1998) Pools of lipidated HTL-CTL constructs prime for multiple HBV and HCV CTL epitope responses. Vaccine 16: 823–833.

45. Toes RE, Hoeben RC, van der Voort EI, Ressing ME, van der Eb AJ, et al. (1997) Protective anti-tumor immunity induced by vaccination with recombinant adenoviruses encoding multiple tumor-associated cytotoxic T lymphocyte epitopes in a string-of-beads fashion. Proc Natl Acad Sci U S A 94: 14660–14665.

46. Velders MP, Weijzen S, Eiben GL, Elmishad AG, Kloetzel PM, et al. (2001) Defined flanking spacers and enhanced proteolysis is essential for eradication of established tumors by an epitope string DNA vaccine. J Immunol 166: 5366–5373.

47. Huebener N, Fest S, Strandsby A, Michalsky E, Preissner R, et al. (2008) A rationally designed tyrosine hydroxylase DNA vaccine induces specific antineuroblastoma immunity. Mol Cancer Ther 7: 2241–2251.

48. Li X, Yang X, Jiang Y, Liu J (2005) A novel HBV DNA vaccine based on T cell epitopes and its potential therapeutic effect in HBV transgenic mice. Int Immunol 17: 1293–1302.

49. Pinchuk I, Starcher BC, Livingston B, Tvinnereim A, Wu S, et al. (2005) A CD8+ T cell heptaepitope minigene vaccine induces protective immunity against Chlamydia pneumoniae. J Immunol 174: 5729–5739.

50. Bazhan SI, Karpenko LI, Ilyicheva TN, Belavin PA, Seregin SV, et al. (2010) Rational design based synthetic polyepitope DNA vaccine for eliciting HIV-specific CD8+ T cell responses. Mol Immunol 47: 1507–1515.

51. Larsen MV, Lundegaard C, Lamberth K, Buus S, Brunak S, et al. (2005) An integrative approach to CTL epitope prediction: a combined algorithm integrating MHC class I binding, TAP transport efficiency, and proteasomal cleavage predictions. Eur J Immunol 35: 2295–2303.

52. Bhasin M, Raghava GP (2007) A hybrid approach for predicting promiscuous MHC class I restricted T cell epitopes. J Biosci 32: 31–42.

53. Doroud D, Zahedifard F, Vatanara A, Najafabadi AR, Taslimi Y, et al. (2011) Delivery of a cocktail DNA vaccine encoding cysteine proteinases type I, II and III with solid lipid nanoparticles potentiate protective immunity against Leishmania major infection. J Control Release 153: 154–162.

54. Bolhassani A, Taheri T, Taslimi Y, Zamanilui S, Zahedifard F, et al. (2011) Fluorescent Leishmania species: development of stable GFP expression and its application for in vitro and in vivo studies. Exp Parasitol 127: 637–645.

55. Adotevi O, Mollier K, Neuveut C, Cardinaud S, Boulanger E, et al. (2006) Immunogenic HLA-B*0702-restricted epitopes derived from human telomerase reverse transcriptase that elicit antitumor cytotoxic T-cell responses. Clin Cancer Res 12: 3158–3167.

56. Adotevi O, Mollier K, Neuveut C, Dosset M, Ravel P, et al. (2010) Targeting human telomerase reverse transcriptase with recombinant lentivector is highly effective to stimulate antitumor CD8 T-cell immunity in vivo. Blood 115: 3025–3032.

57. Rohrlich PS, Cardinaud S, Firat H, Lamari M, Briand P, et al. (2003) HLA-B*0702 transgenic, H-2KbDb double-knockout mice: phenotypical and functional characterization in response to influenza virus. Int Immunol 15: 765–772.

58. Burgevin A, Saveanu L, Kim Y, Barilleau E, Kotturi M, et al. (2008) A detailed analysis of the murine TAP transporter substrate specificity. PLoS One 3: e2402.

59. Mogk A, Schmidt R, Bukau B (2007) The N-end rule pathway for regulated proteolysis: prokaryotic and eukaryotic strategies. Trends Cell Biol 17: 165–172.

60. Sewell DA, Shahabi V, Gunn GR 3rd, Pan ZK, Dominiecki ME, et al. (2004) Recombinant Listeria vaccines containing PEST sequences are potent immune adjuvants for the tumor-associated antigen human papillomavirus-16 E7. Cancer Res 64: 8821–8825.

61. Cho HI, Celis E (2012) Design of immunogenic and effective multi-epitope DNA vaccines for melanoma. Cancer Immunol Immunother 61: 343–351.

62. Iurescia S, Fioretti D, Fazio VM, Rinaldi M (2012) Epitope-driven DNA vaccine design employing immunoinformatics against B-cell lymphoma: a biotech's challenge. Biotechnol Adv 30: 372–383.

63. Dani A, Chaudhry A, Mukherjee P, Rajagopal D, Bhatia S, et al. (2004) The pathway for MHCII-mediated presentation of endogenous proteins involves peptide transport to the endo-lysosomal compartment. J Cell Sci 117: 4219–4230.

64. Lich JD, Elliott JF, Blum JS (2000) Cytoplasmic processing is a prerequisite for presentation of an endogenous antigen by major histocompatibility complex class II proteins. J Exp Med 191: 1513–1524.

65. Tewari MK, Sinnathamby G, Rajagopal D, Eisenlohr LC (2005) A cytosolic pathway for MHC class II-restricted antigen processing that is proteasome and TAP dependent. Nat Immunol 6: 287–294.

66. Rodriguez F, Zhang J, Whitton JL (1997) DNA immunization: ubiquitination of a viral protein enhances cytotoxic T-lymphocyte induction and antiviral protection but abrogates antibody induction. J Virol 71: 8497–8503.

67. Herrera-Najera C, Pina-Aguilar R, Xacur-Garcia F, Ramirez-Sierra MJ, Dumonteil E (2009) Mining the Leishmania genome for novel antigens and vaccine candidates. Proteomics 9: 1293–1301.

68. Babe LM, Chen Y, Chesnut R, DeYoung LM, Huang MT, et al. (2009) Optimized multi-epitope constructs and uses thereof. Google Patents.

69. Fu F, Li X, Lang Y, Yang Y, Tong G, et al. (2011) Co-expression of ubiquitin gene and capsid protein gene enhances the potency of DNA immunization of PCV2 in mice. Virol J 8: 264.

70. Pajot A, Schnuriger A, Moris A, Rodallec A, Ojcius DM, et al. (2007) The Th1 immune response against HIV-1 Gag p24-derived peptides in mice expressing HLA-A02.01 and HLA-DR1. Eur J Immunol 37: 2635–2644.

71. Pajot A, Michel ML, Fazilleau N, Pancre V, Auriault C, et al. (2004) A mouse model of human adaptive immune functions: HLA-A2.1-/HLA-DR1-transgenic H-2 class I-/class II-knockout mice. Eur J Immunol 34: 3060–3069.

72. Pascolo S (2005) HLA class I transgenic mice: development, utilisation and improvement. Expert Opin Biol Ther 5: 919–938.

73. Firat H, Garcia-Pons F, Tourdot S, Pascolo S, Scardino A, et al. (1999) H-2 class I knockout, HLA-A2.1-transgenic mice: a versatile animal model for preclinical evaluation of antitumor immunotherapeutic strategies. Eur J Immunol 29: 3112–3121.

74. Sette A, Vitiello A, Reherman B, Fowler P, Nayersina R, et al. (1994) The relationship between class I binding affinity and immunogenicity of potential cytotoxic T cell epitopes. J Immunol 153: 5586–5592.

75. Gomez-Nunez M, Pinilla-Ibarz J, Dao T, May RJ, Pao M, et al. (2006) Peptide binding motif predictive algorithms correspond with experimental binding of leukemia vaccine candidate peptides to HLA-A*0201 molecules. Leuk Res 30: 1293–1298.

76. Elkington R, Walker S, Crough T, Menzies M, Tellam J, et al. (2003) Ex vivo profiling of CD8+-T-cell responses to human cytomegalovirus reveals broad and multispecific reactivities in healthy virus carriers. J Virol 77: 5226–5240.

77. Dominguez MR, Silveira EL, de Vasconcelos JR, de Alencar BC, Machado AV, et al. (2011) Subdominant/cryptic CD8 T cell epitopes contribute to resistance against experimental infection with a human protozoan parasite. PLoS One 6: e22011.

78. Im EJ, Hong JP, Roshorm Y, Bridgeman A, Letourneau S, et al. (2011) Protective efficacy of serially up-ranked subdominant CD8+ T cell epitopes against virus challenges. PLoS Pathog 7: e1002041.

79. Livingston BD, Newman M, Crimi C, McKinney D, Chesnut R, et al. (2001) Optimization of epitope processing enhances immunogenicity of multiepitope DNA vaccines. Vaccine 19: 4652–4660.

80. Neisig A, Roelse J, Sijts AJ, Ossendorp F, Feltkamp MC, et al. (1995) Major differences in transporter associated with antigen presentation (TAP)-dependent translocation of MHC class I-presentable peptides and the effect of flanking sequences. J Immunol 154: 1273–1279.

81. Eggers M, Boes-Fabian B, Ruppert T, Kloetzel PM, Koszinowski UH (1995) The cleavage preference of the proteasome governs the yield of antigenic peptides. J Exp Med 182: 1865–1870.

82. Del Val M, Schlicht HJ, Ruppert T, Reddehase MJ, Koszinowski UH (1991) Efficient processing of an antigenic sequence for presentation by MHC class I molecules depends on its neighboring residues in the protein. Cell 66: 1145–1153.

83. Imai T, Duan X, Hisaeda H, Himeno K (2008) Antigen-specific CD8+ T cells induced by the ubiquitin fusion degradation pathway. Biochem Biophys Res Commun 365: 758–763.

84. Rodriguez F, Whitton JL (2000) Enhancing DNA immunization. Virology 268: 233–238.

85. Chen JH, Yu YS, Chen XH, Liu HH, Zang GQ, et al. (2012) Enhancement of CTLs induced by DCs loaded with ubiquitinated hepatitis B virus core antigen. World J Gastroenterol 18: 1319–1327.

86. Chou B, Hiromatsu K, Okano S, Ishii K, Duan X, et al. (2012) Antiangiogenic tumor therapy by DNA vaccine inducing aquaporin-1-specific CTL based on ubiquitin-proteasome system in mice. J Immunol 189: 1618–1626.

87. Rezvan H, Rees R, Ali S (2012) Immunogenicity of MHC Class I Peptides Derived from Leishmania mexicana Gp63 in HLA-A2.1 Transgenic (HHDII) and BALB/C Mouse Models. Iran J Parasitol 7: 27–40.

88. Basu R, Roy S, Walden P (2007) HLA class I-restricted T cell epitopes of the kinetoplastid membrane protein-11 presented by Leishmania donovani-infected human macrophages. J Infect Dis 195: 1373–1380.

89. John L, John GJ, Kholia T (2012) A reverse vaccinology approach for the identification of potential vaccine candidates from Leishmania spp. Appl Biochem Biotechnol 167: 1340–1350.

90. Schroeder J, Aebischer T (2011) Vaccines for leishmaniasis: from proteome to vaccine candidates. Hum Vaccin 7 Suppl: 10–15.

91. Silverman JM, Reiner NE (2011) Leishmania exosomes deliver preemptive strikes to create an environment permissive for early infection. Frontiers in cellular and infection microbiology 1.

92. Santarém N, Silvestre R, Tavares J, Silva M, Cabral S, et al. (2007) Immune response regulation by leishmania secreted and nonsecreted antigens. BioMed Research International 2007.

93. Abdian N, Gholami E, Zahedifard F, Safaee N, Rafati S (2011) Evaluation of DNA/DNA and prime-boost vaccination using LPG3 against Leishmania major infection in susceptible BALB/c mice and its antigenic properties in human leishmaniasis. Exp Parasitol 127: 627–636.

94. Pirdel L, Hosseini AZ, Kazemi B, Rasouli M, Bandehpour M, et al. (2012) Cloning and Expression of Leishmania infantum LPG3 Gene by the Lizard Leishmania Expression System. Avicenna J Med Biotechnol 4: 186–192.

95. Descoteaux A, Avila HA, Zhang K, Turco SJ, Beverley SM (2002) Leishmania LPG3 encodes a GRP94 homolog required for phosphoglycan synthesis implicated in parasite virulence but not viability. EMBO J 21: 4458–4469.

96. Silverman JM, Chan SK, Robinson DP, Dwyer DM, Nandan D, et al. (2008) Proteomic analysis of the secretome of Leishmania donovani. Genome Biol 9: R35.

15

Transcriptomics of Desiccation Tolerance in the Streptophyte Green Alga *Klebsormidium* Reveal a Land Plant-Like Defense Reaction

Andreas Holzinger[1]*, Franziska Kaplan[1], Kathrin Blaas[1], Bernd Zechmann[2], Karin Komsic-Buchmann[3], Burkhard Becker[3]

1 University of Innsbruck, Functional Plant Biology, Innsbruck, Austria, 2 Baylor University, Center for Microscopy and Imaging, Waco, Texas, United States of America, 3 University of Cologne, Botanical Institute, Biocenter, Cologne, Germany

Abstract

Background: Water loss has significant effects on physiological performance and survival rates of algae. However, despite the prominent presence of aeroterrestrial algae in terrestrial habitats, hardly anything is known about the molecular events that allow aeroterrestrial algae to survive harsh environmental conditions. We analyzed the transcriptome and physiology of a strain of the alpine aeroterrestrial alga *Klebsormidium crenulatum* under control and strong desiccation-stress conditions.

Principal Findings: For comparison we first established a reference transcriptome. The high-coverage reference transcriptome includes about 24,183 sequences (1.5 million reads, 636 million bases). The reference transcriptome encodes for all major pathways (energy, carbohydrates, lipids, amino acids, sugars), nearly all deduced pathways are complete or missing only a few transcripts. Upon strong desiccation, more than 7000 transcripts showed changes in their expression levels. Most of the highest up-regulated transcripts do not show similarity to known viridiplant proteins, suggesting the existence of some genus- or species-specific responses to desiccation. In addition, we observed the up-regulation of many transcripts involved in desiccation tolerance in plants (e.g. proteins similar to those that are abundant in late embryogenesis (LEA), or proteins involved in early response to desiccation ERD), and enzymes involved in the biosynthesis of the raffinose family of oligosaccharides (RFO) known to act as osmolytes. Major physiological shifts are the up-regulation of transcripts for photosynthesis, energy production, and reactive oxygen species (ROS) metabolism, which is supported by elevated cellular glutathione content as revealed by immunoelectron microscopy as well as an increase in total antiradical power. However, the effective quantum yield of Photosystem II and CO_2 fixation decreased sharply under the applied desiccation stress. In contrast, transcripts for cell integrative functions such as cell division, DNA replication, cofactor biosynthesis, and amino acid biosynthesis were down-regulated.

Significance: This is the first study investigating the desiccation transcriptome of a streptophyte green alga. Our results indicate that the cellular response is similar to embryophytes, suggesting that embryophytes inherited a basic cellular desiccation tolerance from their streptophyte predecessors.

Editor: Wagner L. Araujo, Universidade Federal de Vicosa, Brazil

Funding: This work was supported by FWF grant P24242-B16 (http://www.fwf.ac.at/). The funders had no role in study design, data collection and analysis, decision to publish, or preparation of the manuscript.

Competing Interests: The authors have declared that no competing interests exist.

* Email: Andreas.Holzinger@uibk.ac.at

Introduction

Poikilohydric plants such as algae do not have protective structures (such as a waxy cuticula) and cannot actively regulate the transpiration rate (e.g. by stomata), which can easily lead to desiccation under water-limited conditions. Desiccation tolerance of these organisms can be defined as the ability to survive drying to ~10% remaining water content, which is equivalent to ~ 50% relative air humidity (RH) at 20°C [1,2]. Aeroterrestrial green algae are naturally exposed to this stress, which they must tolerate in order to survive and propagate under these conditions (for a recent summary see [3]). The mechanisms to tolerate dehydration might involve the formation of permanent stages (e.g. akinetes, zygospores; e.g. [4]), but dehydration might also be tolerated in the vegetative state. Desiccation-tolerant species of green algae have been found in phylogenetically distinct lineages, e.g. in the so-called lichen algae, Trebouxiophyceae (e.g. [5,6]), but also in streptophyte green algae, which represents the sister lineage to land plants (e.g. [7–11]).

A primary target of dehydration in green algae is photosynthesis (for a summary see [3]). Early studies reported an effect on CO_2 exchange in the Trebouxiophyte *Apatococcus lobatus* [12]. While the most favourable carbon assimilation was measured in *A. lobatus* at ~98% RH, even at RH 90%, 50% of the maximum CO_2 uptake still occurred. The lower limit of carbon assimilation was observed at 68% RH [12]. More recently, Gray et al. [13] used ambient air at 25% RH to desiccate various strains of algae isolated in deserts and their aquatic relatives from the Chlorophyceae (e.g. *Bracteacoccus*, *Scenedesmus*, *Chlorogonium*) and Trebouxiophyceae (e.g. *Chlorella*, *Myrmecia*). They found that the desert algae could survive 4 weeks of desiccation when dried in darkness, and recovered photosynthetic quantum yield within 1 h of rehydration. Resurrection kinetics were also studied in other desiccation-tolerant Chlorophyta (e.g. *Desmococcus*, *Apatococcus*, *Trebouxia*), where green biofilm samples underwent prolonged desiccation (up to 80 days) and upon rehydration recovered about half of the initial value prior to desiccation [6]. In searching for the mechanisms behind this tolerance, Wieners et al. [14] observed desiccation-induced non-radiative dissipation in isolated green lichen algae (*Trebouxia*). Dynamic photoinhibition has been confirmed for several species of desert and aquatic algae (e.g. *Klebsormidium*, *Cylindrocystis*, *Stichococcus*; [15]). Photoprotective mechanisms were also found in the investigated algal strains; however, lineage-specific modifications occurred [15].

In *Klebsormidium crenulatum* [16–18] and a related *Klebsormidium* species [19–20], the optimum quantum yield decreased to zero after the cells were desiccated at ambient room temperature and RH; the exact percentage decrease was not determined, but recovered to different extents quickly after rehydration [16,19]. Differences in desiccation tolerance might be also important for intra-generic ecological differentiation. For example, Škaloud and Rindi [21] suggested the existence of an ecological differentiation in *Klebsormidium,* according to different water-holding capacities of the substrates. However, on the global scale populations of *Klebsormidium* were described as genetically homogenous, and the local genotypes may be caused by ecological factors such as habitat differentiation [22]. Phylogenetic relationships within the genus *Klebsormidium* were recently elucidated (e.g. [20–21,23,24]), and the strain of *K. crenulatum* investigated here has been characterized by its *rbc*L gene [18] and is belonging to clade F according to [24].

Although some information is available on physiological adaptations to desiccation in green algae, no molecular approach has been used so far to understand desiccation tolerance in these organisms. Proteome analyses of the Trebouxiophycean green alga *Asterochloris erici*, a lichen-forming organism, found that only a few proteins increase upon desiccation [25]. It was not possible to identify changes in protein content involved in the light reactions of photosynthesis; however, dehydration led to a decrease in proteins associated with the Calvin cycle, indicating a reduction in carbon fixation. Interestingly, *A. erici* is still able to maintain initial values of maximum and actual quantum yield until 20% relative water content is reached [5]. Presumably, the loss of certain proteins does not halt carbon fixation during drying [25].

Deep sequencing (Illumina RNA-seq) provides a fast and reliable approach to generate large expression databases for functional genomic analysis, and is particularly suitable for non-model species with unsequenced genomes (e.g. [26–28]). While the analysis of the plant transcriptomes has produced an unexpected abundance of data on transcript identity in vascular plants, including resurrection plants (e.g. [29–31]), little information is yet available on transcriptomes in algae. In the unsequenced microalga *Chlorella vulgaris*, triacylglycerol biosynthetic pathways were analyzed by a transcriptome approach [32]. In *C. vulgaris* Illumina sequencing produced 27 million reads, Velvet and Oasis program suites were utilized for *de novo* transcriptome assembly,

yielding 29,237 transcripts with an average length of 970 b. The results suggested an up-regulation of fatty-acid and triacylglycerol-biosynthetic mechanisms under nitrogen limitation [32]. A study using brown algal kelp *Saccarina latissima* identified transcriptional changes upon temperature and light stress; 32% of the genes showed an altered expression under the experimental conditions used [33]. Heinrich et al. [33] also found that high temperatures had a stronger effect on gene expression than low temperatures. The molecular processes of acclimation included the adjustment of the primary metabolism, induction of several ROS scavengers and regulation of heat-shock proteins (HSPs) [33]. A combination of high temperatures and high light intensities proved to be most harmful to this kelp species. Another study on the kelp *Saccharina japonica* investigated the effect of blue light on the transcriptome [34]. Their RNA-seq analysis yielded more than 70,000 non-redundant unigenes with an average length of 538 bp, and 40.2% of these could be mapped to transcript-encoding regions. Upon blue light exposure, ~7,800 unigenes were up-regulated and ~3,850 down-regulated. The feasibility of next-generation sequencing in the green algae *Coleochaete* and *Spirogyra* has been shown by Timme and Delwiche [35], and more than 30 transcriptomes were sequenced in the 1KP Project (http://onekp.com/project.html). The sequenced charophyte green algal transcriptomes have been shown to be remarkably similar to land plants, which demonstrate the evolutionary significance of these studies. Currently about 60,000 expressed sequence tags (ESTs) are available in GenBank for *Klebsormidium* (NCBI, 22.07.2014), representing several different species and strains. However, so far no studies on the effect of desiccation on the transcriptome of a streptophyte green alga are available.

In the present study, we exposed laboratory-grown *Klebsormidium crenulatum* to a strong desiccation stress in order to study, at the transcriptome level, the key molecular processes required for desiccation tolerance of this alga. To corroborate our findings, we performed physiological measurements of CO_2 assimilation rates, effective quantum yield and antiradical power, and determined the cellular glutathione levels by immunoelectron microscopy. This enabled us to discuss the transcriptome findings in the context of the cellular strategies against desiccation.

Material and Methods

Algal cultures

Klebsormidium crenulatum (Kützing) Lokhorst [36] SAG 2415, previously isolated from an alpine soil crust [16], were cultivated in modified Bold's basal medium (3NMBBM (triple nitrate concentration [37]) or Waris-medium [38]) at 30–25 μmol photons $m^{-2}s^{-1}$ in a light:dark regime of 16:8 in an Intellus environmental controller (Percival Scientific, Perry, IA, USA) at 20°C as previously described [16,18]. The phylogenetic position of this strain was determined by *rbc*L sequences, deposited in GenBank under accession number JN190354 [18].

Desiccation experiment

Two-week-old cultures of *K. crenulatum* were concentrated by vacuum filtration onto S&S mixed cellulose membrane filters (Sigma Aldrich Z 612383, pore size 0.45 μm) in triplicate (n = 3). The experiment was carried out 2 h after onset of light. The filters were then desiccated on top of 4 glass columns holding a perforated metal grid inside a 200 mL polystyrol box over ~100 g silica gel (Silica Gel Orange, Carl Roth, Karlsruhe, Germany) for 2.5 h according to the methods of [39]. During this procedure, the temperature and the relative air humidity in the box were monitored with a PCE-MSR145S-TH mini data logger for air humidity and temperature (PCE Instruments, Meschede, Germany). The weight of the mixed cellulose membrane filters was determined prior to vacuum filtration. Then the weight of the fresh biomass (after filtration of the algae; cells and surrounding water film) and the weight of the desiccated biomass (after 2.5 h of desiccation over silica gel) were determined. The filter weight was subtracted to gain the net weight of the fresh or desiccated biomass, respectively. The reduction (%) of the water content of the fresh biomass was calculated following the equation:

$$Reduction(\%) = 100 - \left[\frac{desiccated\ biomass}{fresh\ biomass} \cdot 100\right]$$

The desiccated biomass was then dried at $100°C$ for 13 h. The relative water content (RWC; %) of the desiccated biomass was calculated following the equation:

$$RWC(\%) = 100 - \left[\frac{oven\ dried\ biomass}{desiccated\ biomass} \cdot 100\right]$$

Staining with FM 1–43

Control cells, cells desiccated for 2.5 h, and cells allowed to recover for 2 h in culture medium after desiccation were stained with a $40\ \mu M$ solution of FM1–43 (green biofilm cell stain, Invitrogen Ltd., Paisley, UK, prepared from a 20 mM stock solution in DMSO). Cells were observed directly in the staining solution in a Zeiss Pascal confocal laser scanning microscope (Carl Zeiss AG, Jena, Germany). Samples were excited with an argon laser beam (488 nm), and the emission was collected in two separate channels at a wavelength band between 505–550 nm (false-colored green) and at wavelengths longer than 560 nm (false-colored red).

Pulse amplitude fluorescence (PAM) and gas exchange measurements

The effective quantum yield ($\Delta F/Fm'$) of photosystem II (Yield II) was determined every 10 min during the dehydration period, using a pulse-amplitude modulated fluorimeter (PAM 2500, Heinz Walz GmbH, Effeltrich, Germany) according to the methods described by [39]. $\Delta F/Fm'$ was calculated as $(Fm'-F)/Fm'$ with F as the fluorescence yield of light-exposed algal cells (40 µmol photons $m^{-2}\ s^{-1}$) and Fm' as the maximum light-adapted fluorescence yield after employing an 800 ms saturation pulse, as described by Schreiber and Bilger [40]. The PAM light probe was positioned outside the desiccation box at a distance of 2 mm. The measurements were carried out on 3 individual filters (n = 3) containing algal suspension, placed in individual desiccation boxes. At each filter 3 measurement points were recorded every 10 min. The CO_2 assimilation rate of K. crenulatum was measured by a GFS-3000 portable gas exchange fluorescence system (Heinz Walz GmbH, Effeltrich, Germany) with a LED-Array/PAM-Fluorometer 3055-FL under strong desiccation stress at ~20% RH (n = 3) for 2.5 h. 100 mL homogeneous algal culture was filtered on a ME25 membrane filter (mixed cellulose ester, Whatman GmbH, Germany). Cells were removed from the filter areas that were not placed inside the measurement chamber. The assimilation rate was calculated based on the area (8 cm^2) of the measurement chamber. After a recovery period in culture medium overnight (21 h), the assimilation rate was monitored again. Standard conditions for all measurements were 20°C, 30 µmol

photons $m^{-2}\ s^{-1}$, 380 ppm absolute CO_2 content and a molar flow rate at the inlet of the cuvette of 750 µmol s^{-1}. Absolute water content was calculated with a Vaisala Humidity Calculator 1.3 (Vaisala, Helsinki, Finland) to adjust the required relative humidity in the GFS-3000 measurement chamber to 4919 ppm for 20% RH. Assimilation rate was measured at 1-min intervals, and in parallel Yield II was measured at 3-min interavls. After each measurement, the chlorophyll a content of the sample was determined. Chlorophyll a was extracted in dimethylformamide (DMF) and quantified according to Porra at al. [41]. Values of the initial 5 min of the desiccation period, of the desiccated sample, and of the initial 5 min of the recovery measurements were statistically analyzed with SPSS 15.0 for Windows (IBM Corporation, Somer, NY, USA), using one-way ANOVA followed by Tukeys post-hoc test ($p \leq 0.001$).

RNA isolation

RNA was isolated from K. crenulatum cells either directly concentrated on mixed cellulose filters or desiccated as described above. The filters were swiped with 450 µL of Life Guard Soil Preservation solution (MO BIO Laboratories, Inc., Carlsbad, CA, USA, Cat No. 12868-100). The cells were frozen in liquid nitrogen, then ground with a mortar and pestle, followed by RNA extraction using a PEQLAB Gold Plant RNA isolation kit (PEQLAB, Erlangen, Germany) according to the manufacturer's instructions. For DNA removal the RNA was treated with DNaseI (Thermo Scientific, DNaseI, RNase free). A final cleanup was performed, again using the PEQLAB RNA column.

Expression analysis

Random primed cDNA libraries were prepared by GATC Biotech AG (Konstanz, Germany). The library preparations were sequenced on an Illumina HiSeq 2000 as single-reads to 100 bp in the same flow cell (first sequencing run). All sequences were analyzed using the CASAVA v1.7 (Illumina, USA). To establish a reference transcriptome, the same amounts of RNA from the control and desiccated samples were pooled and a normalized random primed cDNA library was prepared from the pooled samples by GATC Biotech AG (Konstanz, Germany). The reference library was sequenced on a Roche GS FLX (1/2 plate), yielding 1,553,953 sequenced reads (636,002,000 sequenced bases). Assembly and basic expression analysis were performed at GATC Biotech AG. Briefly, the reference genome library was assembled using the De Novo Assembler (Newbler) 2.8 software from Roche. For expression analysis, sequence reads were mapped to the reference sequence (unigene.highcov.fa) using BWA with the default parameters. Only uniquely mapped reads were considered for further processing. PCR duplicates were removed using PICARD. The resulting high-quality sequence alignments were taken for coverage determination and variant detection. The alignments were processed to compute the read counts for each exon/CDS/transcript. Read counts were then normalized to 1 million reads (RPM) by considering the total reads mapped in each sample, to remove the uneven sequencing coverage over multiplexed samples. Finally, RPM counts were normalized to 1 Kbp transcript length (RPKM) to remove the uneven gene/transcript length bias. The final table included all the transcripts present in the reference employed. The entries were filtered based on defined threshold before doing the expression analysis. Determining counts, RPM and RPKM computations were done using in-house scripts by GATC-Biotech.

For differential expression analysis, the DeSeq software [42] was used, following the program's vignette (DESeq version 1.12.0, Last revision 2013-02-24). Pathways and modules were analyzed using the KEGG module and pathway databases at http://www.genome.jp/kegg/. Local Blast analyses were performed using the NCBI BLASTX program and the protein databases from the following species obtained at www.phytozome.net: *Chlamydomonas reinhardtii*, *Coccomyxa subellipsoidea*, *Micromonas pusilla*, *Ostreococcus lucimarinus*, *Physcomitrella patens*, *Selaginella italica*, *Zea mays*, *Oryza sativa*, *Aquilegia coerulea*, *Solanum tuberosum*, *Eucalyptus grandis*, *Theobroma cacao*, *Arabidopsis thaliana*, *Malus domesticus*, *Medicago trunculata* and *Populus trichocarpa* RNA-Seq data and the assembled contigs have been deposited in the NCBI's Gene Expression Omnibus (GEO) [43] and are accessible through the NCBI Sequence Read Archive (http://trace.ncbi.nlm.nih.gov/Traces/sra/) under the accession numbers SRR1514242 (assembled reference library) and SRR1563130 to SRR1563134 (transcriptomes).

Cytohistochemical detection of glutathione

Preparation of samples for transmission electron microscopy and immunogold labeling of glutathione was performed as previously described [44,45]. Briefly, algae were fixed in 2.5% paraformaldehyde and 0.5% glutardialdehyde in 0.06 M phosphate buffer (pH 7.2), rinsed in the buffer, dehydrated in increasing concentrations of acetone (50%, 70%, and 90%) and infiltrated with increasing concentrations of LR-White resin (30%, 60% and 100%; London Resin Company Ltd., Berkshire, UK). Samples were polymerized at 50°C. Ultrathin sections were blocked with 2% bovine serum albumin (BSA) in phosphate buffered saline (PBS, pH 7.2) and treated with the primary antibody (anti-glutathione rabbit polyclonal IgG, Millipore Corp., Billerica, MA, USA) diluted 1:50 in PBS containing 1% goat serum. After three brief rinses in PBS, sections were incubated with 10 nm gold-conjugated secondary antibodies (goat anti-rabbit IgG for glutathione, British BioCell International, Cardiff, UK) 1:50 in PBS and finally washed with distilled water. A minimum of 25 different cells were analyzed for gold particle density. The data obtained were presented as the number of gold particles per μm^2.

DPPH-Test

For determination of the antiradical power of fresh and 2.5 h desiccated *K. crenulatum* cells, a 2,2-diphenyl-1-picrylhydrazyl (DPPH)-test using 6-hydroxy-2,5,7,8-tetramethylchroman-2-carbonacid (Trolox, Sigma Aldrich, Germany) as standard was used according to the methods of [46]. *K. crenulatum* cultures were filtered according to the above-mentioned procedure onto Whatman GF/F glass fiber filters (Whatman, Dassel, Germany) and wrapped in aluminum foil, either immediately as a control (n = 3) or after desiccation for 2.5 h over silica gel (n = 3). Filters were then frozen in liquid nitrogen, and stored at −80°C prior to lyophilization. Lyophilized algae were then extracted in plastic vials containing 5 mL of 70% acetone for 24 h at 4°C under continuous shaking in darkness. 200 μL of a 150 μM DPPH-solution in 80% ethanol was transferred into a 96-well microtiter plate. The DPPH solution was complemented with 22 μL of the respective samples or Trolox standards in 70% acetone (7.8 μM-1 mM). After 25 min the absorption was determined for each sample at 516 nm. The statistical significance of means was tested by two-sample t-test.

Results

Experimental setup and physiological response to desiccation stress in *Klebsormidium crenulatum*

To identify the molecular processes involved in the desiccation response in *K. crenulatum*, filaments were transferred to filter membranes and incubated for 2.5 h in a desiccation chamber over silica gel. At the end of the experiment, the relative humidity in the desiccation chamber was ∼10% (Fig. 1). The weight of the fresh biomass (algae and surrounding water film) was reduced by 95.16±1.41% during the desiccation process. The experimentally desiccated algae had a RWC of 6.54±1.89%.

Within 40–50 min after exposure of the filters to silica gel, the effective quantum yield of *K. crenulatum* cells dropped to zero, suggesting complete inhibition of photosystem II (Fig. 1). When rehydrated with medium after the 2.5 h desiccation period, the effective quantum yield of photosystem II fluorescence did not recover for at least 2.5 h (not shown). Gas exchange measurements confirmed this observation (Fig. S1). The initial CO_2 assimilation rate was ∼0.5 μmol CO_2 m^{-2} s^{-1} at the beginning of the measurements (Fig. S1). CO_2 assimilation of photosynthesis was inhibited immediately when the RH dropped (Fig. S1). At the end of the desiccation period, a net respiration was found. Cells did not recover completely when the filters were rewetted, and the measurements continued; after a 21-h recovery period, respiration still predominated, although at a lower level (−0.05 to −0.1 CO_2 m^{-2} s^{-1}). In agreement with the measurements from the desiccation chamber (Fig. 1), and also in the gas-exchange measurements, the effective quantum yield of photosystem II (Yield II, Fig. S1) did not recover.

To assess the cellular damage resulting from the desiccation stress, a cell viability test using the FM1-43 dye was performed. In control cells of *K. crenulatum*, FM1-43 dye stained the cell membrane predominantly (Fig. 2 A); however in some cases, the membranes of the vacuoles and chloroplast were also stained if the cells were exposed to the dye for a longer period (not shown). In contrast, dead cells, which are always present in small numbers in a cell culture, showed a bright fluorescence in the cell lumen (Fig. 2 D).

The cytoplasm of cells exposed to desiccation stress for 2.5 h was stained in a granular pattern; however, the plasma membrane was still visible and no labeling of the plastid was observed (Fig. 2 B). After a recovery period of ∼2 h the plasma membrane was stained more prominently as in the desiccated cells, although weaker than in the control cells, and there was still some cytoplasmic staining (Fig. 2 C). However, we rarely observed the bright fluorescence of the total cell volume, indicative of dead cells (Fig. 2 D).

Molecular analysis

To identify desiccation stress-regulated cellular processes and functions in *K. crenulatum*, we first established a reference transcriptome database. Algae were transferred to filter membranes and RNA was isolated either directly or after applying desiccation stress (2.5 h desiccation over silica gel). RNA was isolated from control and desiccated samples (each in triplicate). Control samples contained more RNA than desiccated samples (Fig. S2, compare the amount of 18S and 25S between control and desiccated samples), indicating a considerable reduction in the amount of RNA during desiccation stress in *K. crenulatum*. In addition, in the electropherograms of the desiccated samples, the 18S and 25S rRNA peaks were reduced and an increased baseline noise and lower RNA integrity number (RIN between 5.7 to 5.8 in the control samples and between 3.8 to 4.2 in the desiccated samples) were observed, indicative of RNA degradation, which is nearly absent from the control samples (see also discussion).

For the reference library, all samples were pooled (same amount of RNA from each sample), and a normalized random primed cDNA library was prepared and sequenced using 454 sequencing. Random primed cDNA libraries were also prepared from the

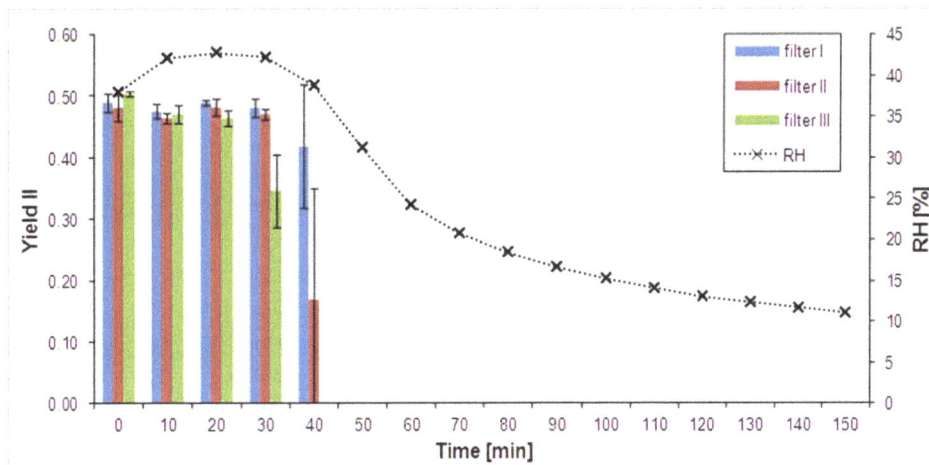

Figure 1. Effective quantum yield of photosystem II (Yield II) of *Klebsormidium crenulatum* **during desiccation (desiccated over 100 g of silica gel) monitored by PAM 2500.** Values represent means (n = 3), error bars show standard deviation (SD). The results of three filters containing similar amount of algal biomass are shown individually, to demonstrate that the decrease of Y II occurred within a short time frame (40–50 min). The relative air humidity (RH) inside the chamber (right y-axis) was monitored by a PCE sensor.

individual samples and sequenced using the Illumina technology. All control samples and two desiccated samples yielded sufficient reads for further analyses. In the following sections, we will first describe the reference library and then present our results from the expression analysis.

Reference library

454 sequencing yielded 1,553,953 sequenced reads (average read length 409 b, GC-content 52.17%, N 0.02%), giving more than 138,213 contigs with a mean length of 563 b. However, 83,276 of the contigs represented short singletons; therefore a high-coverage reference database was constructed, which contained only contigs containing at least 5 reads. This database contained 24,183 contigs with a mean sequence length of 1,327 b (N50 = 1,463 b). All further analyses were performed using this high-coverage assembly.

BLAST-analyses. To identify transcripts of interest, we compared the deduced peptides of the high-coverage assembly with the proteomes of selected green algae and embryophytes

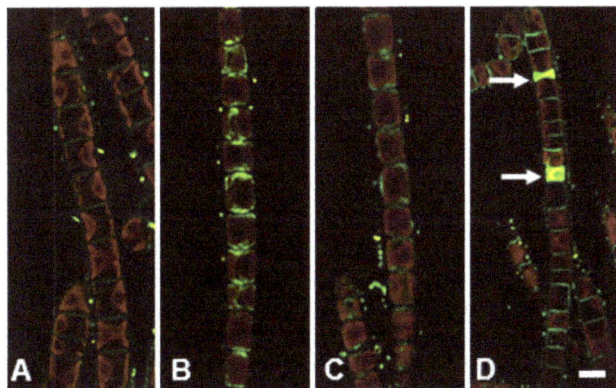

Figure 2. *Klebsormidium crenulatum* **stained with FM 1–43.** (A) Control cells. (B) Desiccated for 2.5 h. (C) 2 h recovery in culture medium after drying for 2.5 h. (D) Control filament with two dead cells (arrows). Bar 10 μm.

(Table 1) using BLASTX. A total of 13,316 contigs (55.1%) showed similarity to the deduced protein sequences in the genomes (for simplicity, henceforth called proteins) of the selected viridiplants (chlorophytes and streptophytes, e-value <exp -10). 10,505 *K. crenulatum* contigs = 43.4% showed similarity to chlorophyte protein sequences, whereas 12,596 K. *crenulatum* contigs = 52.1% showed similarity to streptophyte protein sequences. However, only 4,752, 5,949 and 7,701 *K. crenulatum* contigs showed similarity to proteins in all investigated viridiplants, chlorophytes or streptophytes respectively. Overall, more sequences showed similarity to streptophyte genomes than to chlorophyte genomes (Table 1) and generally the conservation of the amino acid sequences was higher when the deduced peptide sequences of *Klebsormidium* contigs were compared with streptophyte proteins than when comparing with chlorophyte proteins (not shown). These results highlight the genetic diversity of chlorophytes and the large number of streptophyte-specific contigs (2,791 *K. crenulatum* contigs).

KEGG-Modules and Pathway analyses. To investigate whether the reference genome covers most cellular functions, we searched the KEGG databases using the high-coverage contigs as a query. As shown in Table S1, most basic cellular processes (e.g. carbohydrate metabolism, fatty acid metabolism, nucleotide metabolism, amino acid metabolisms, photosynthesis, and respiration) were completely or nearly completely represented in the deduced peptide sequences.

Expression analysis

Illumina sequencing of random primed cDNA libraries yielded between 12 and 30 million reads per sample. The read statistics are given in Table S2. Only high-quality reads (unique or duplicates) were used for further analyses. Table 2 gives an overview of the number of *K. crenulatum* contigs showing changes in transcript expression level. 10,493 *K. crenulatum* contigs (43.4%) showed at least a 2-fold change in their expression level, thus the majority of the transcriptome is apparently unaffected. Only a few K. *crenulatum* contigs were expressed exclusively under desiccation stress or control conditions (Table 2) and more transcripts were down-regulated than up-regulated (Table 2). Interestingly, the percentage of up-regulated transcripts showing

Table 1. Similarity of *Klebsormidium crenulatum* contigs to other organisms.

Name	Growth type	Number of Blast hits	Up-regulated	Up-regulated (%)
Chlamydomonas reinhardtii	Unicellular Flagellate	8727	1545	17.70
Coccomyxa subellipsoidea	Unicellular Coccoid	8856	1624	18.34
Micromonas pusilla	Unicellular Flagellate	7759	1367	17.62
Ostreococcus lucimarinus	Unicellular Coccoid	7180	1273	17.73
Physcomitrella patens	Moss	11842	2095	17.69
Selaginella italica	Lycophyte	10781	1886	17.49
Zea mays	Grass	10771	1884	17.49
Oryza sativa	Grass	10590	1879	17.74
Aquilegia coerulea	Perennial Herb	10995	1924	17.51
Solanum tuberosum	Perennial Herb	10127	1870	18.47
Eucalyptus grandis	Tree	10897	1875	17.21
Theobroma cacao	Tree	11137	1945	17.46
Arabidopsis thaliana	Annual Herb	10995	1931	17.56
Malus domesticus	Tree	10670	1890	17.71
Medicago trunculata	Annual Herb	9552	1742	18.24
Populus trichocarpa	Tree	11070	1931	17.44
Ricinus communis	Shrub	11015	1907	17.31

24,183 contigs were blasted against the protein database of the selected organisms listed (obtained from Phytozome). The number of BLASTX hits (e-value <exp -10) is shown. The number of hits of contigs that were up-regulated under desiccation stress (twofold or more) and the percentage of the total hits are also shown.

similarity to viridiplant genomes is higher than the percentage of down-regulated transcripts showing similarity to viridiplant genomes. In Table 3 the 20 highest up-regulated *K. crenulatum* contigs are shown. The complete lists of all up- and down-regulated mRNAs including the most similar proteins in the 14 viridiplant proteomes searched, are presented in Table S3 and Table S4. Most of the highly up-regulated contigs do not show similarity to known viridiplant proteins and many show no similarity to any known protein at all. Thus a large number of regulated transcripts might represent specific adaptations of *K. crenulatum* to its environment. However, 100 of the 171 *K. crenulatum* contigs, up-regulated at least 10-fold, show similarity to proteins in viridiplant proteomes. Surprisingly, we could detect only 46 *Arabidopsis* proteins in this category. Interestingly, most of these *Arabidopsis* proteins are well conserved in embryophytes and 34 are also similar to proteins in chlorophytes, suggesting that these proteins represent basic cellular functions, generally up-regulated upon stress. In contrast, we could detect several additional *K. crenulatum* contigs (at least 10-fold up-regulated) that show similarity to other embryophytes. Overall there are 89 *K. crenulatum* contigs similar to proteins in *Medicago* and 61 to 71 *K. crenulatum* contigs similar to proteins from the trees investigated (except *Eucalyptus*, 49).

KEGG analysis of desiccation induced changes. To gain more insight into the molecular mechanisms of desiccation tolerance, we searched the KEGG database for pathways (Table 4) and modules (Table S1), using the complete, the up-regulated and the down-regulated datasets. *K. crenulatum* contigs could be assigned to 782 different KO functions using the pathway searching tool, and transcripts representing 229 KO functions were up-regulated and transcripts representing 186 KO functions were down-regulated, respectively. The majority of the deduced KEGG-pathways (for simplicity called pathways in the following sections) showed complex regulation patterns with several contigs up- and down-regulated (e.g. Figs. 3 and 4). The 15 pathways

which were overall most strongly up- and down-regulated are shown in Table 4 and Table 5. As is evident from Table 4, contigs involved in energy production (glycolysis, TCA cycle, respiration, photosynthesis etc.) were mainly up-regulated. Other important up-regulated pathways were galactose, ascorbate (Fig. 3 A) and glutathione (Fig. 3 B) metabolism; and plant hormone and calcium signaling. No contigs showed similarity to proteins of the auxin and gibberellin signal transduction pathways. In contrast, homologues for all components of the cytokinin signaling pathway (CRE1, down-regulated; AHP, up-regulated; B-ARR, down-regulated and A-AAR, up-regulated) and 3 of the 4 components of the abscisic acid signaling pathway (PP2C, SnRK2, ABF, all up-regulated) were found. Surprisingly, putative homologues of the ethylene and jasmonic acid receptors (ETR, JAR1) are also present. However, putative homologues of other components of these signaling pathways are missing, except for (CTR1 and EIN3) for which we also detected putative homologues.

Desiccation induced up-regulation of transcripts. Only for six of the top 20 up-regulated contigs it is possible to infer a putative function. Four of these transcripts are similar to plastidic proteins: two up-regulated transcripts are possibly involved in non-photochemical quenching (UN028420, ELIP1, 35.9-fold up-regulated; UN031973, NPQ4 (PSBS), 35.1-fold up-regulated); another interesting contig shows similarity to a plastidic glucose transporter (UN026579, GLT1, 30.1-fold up-regulated). Transcripts for sucrose phosphate synthase (2 contigs: UN036592, 3.6-fold; UN039619, 2.8-fold) and sucrose synthase (3 contigs: UN023940, 2.1-fold, UN038702 3.2-fold; UN039004 3.8-fold) are also up-regulated, whereas transcripts for cellulose synthase and some enzymes of the starch and trehalose metabolism are down-regulated. We also found a contig displaying strong similarity to AT2G35840, a sucrose-6-phosphate phosphohydrolase family protein (e-value: 3e-104). This transcript is 4-fold up-regulated; however, the pathway reconstruction tool at KEGG failed to recognize this transcript as a putative SPSP.

Table 2. Overview of changes in transcript expression during desiccation stress.

	Up-regulated	Down-regulated
At least 2-fold change	4129 (2264)	5898 (2343)
•Significant (P_{val}* <0.05):	3766	5011
•Significant (P_{adj}* <0.05):	3642	4584
•Number of KO-Terms	736	616
Expressed only in/upon controls/desiccation	31	435
•Significant (P_{val}* <0.05):	4	136
•Significant (P_{adj}* <0.05):	2	115

The number of *Klebsormidium crenulatum* contigs in each category is given except for Number of KO-terms, which gives the number of KO-Terms, obtained for up- and down-regulated *K. crenulatum* contigs with a P_{adj} <0.05. Number in brackets gives the percentage of *K. crenulatum* contigs that show similarity to viridiplant genomes.
*obtained with the DeSeq program. P_{val} = p value for the statistical significance of this change.
P_{adj} = p value adjusted for multiple testing with the Benjamini-Hochberg procedure, which controls the false discovery rate (FDR).

Several transcripts for putative enzymes of the galactinol/raffinose metabolism were found to be up-regulated, including a transcript for galactinol synthase (UN010053, BLASTX result: AT1G60450.1, AtGolS7, e-value 5 exp-30), the key enzyme for the biosynthesis of the raffinose family of oligosaccharides (RFO). Ascorbate is probably synthesized from GDP-Man as a transcript similar to GDP-D-mannose 3′, 5′-epimerase (UN025544, similar to AT5G28840.2, GME, e-value: 3E-49) is under desiccation stress 26-fold up-regulated, as are several contigs with similarity to enzymes using ascorbate as reductant. Similarly to ascorbate, glutathion plays an important role in reactive oxygen metabolism. Several transcripts coding for enzymes in the glutathion metabolism are up-regulated (Fig. 3 B). For example, contigs with

similarity to enzymes involved in the first step of glutathion biosynthesis, glutathione-S-transferase and glutathione peroxidase are up-regulated, while transcripts with similarity to glutathione-degrading enzymes and the glutathione-disulfide reductase are down-regulated. The importance of reactive oxygen species (ROS) protection is also highlighted by the up-regulation of a transcript similar to catalase 2 (UN029908, 29.4-fold up-regulation). To corroborate these findings, we investigated changes in the total cellular glutathione content, using an immune gold glutathione-labeling approach (Fig. 5), and the cellular antiradical power of *K. crenulatum*. The distribution of glutathione-specific gold labeling in *K. crenulatum* grown under control conditions revealed that the highest glutathione labeling density was found in mitochondria,

Table 3. *Klebsormidium crenulatum* contigs showing the largest changes upon desiccation stress.

K. crenulatum -ID	Fold Change	padj	Function
UN027167	Not expressed in controls	0.03141657	Plastid NADH:ubiquinone/plastoquinone oxidoreductase
UN034964	Not expressed in controls	0.00627352	
UN022359	91.4439431	0.00021721	
UN036634	60.5266278	8.9783E-16	Similar protein in some plants
UN043434	52.011157	7.675E-16	
UN036811	51.8868523	1.6101E-23	Similar protein in some plants
UN012067	51.0430016	1.3086E-11	Similar protein in some plants
UN034707	47.3497452	2.6678E-17	Similar protein in some plants
UN020446	45.7941202	1.3302E-21	Similar protein in some plants
UN000210	38.6089069	7.6031E-16	
UN026331	38.1383995	1.679E-07	Similar protein in some plants
UN028420	35.9012593	4.4486E-83	ELIP1 early light inducible protein
UN031973	35.1401051	1.6143E-65	NPQ4 (PSBS)
UN022671	33.789896	7.0903E-37	
UN035517	30.9168329	5.324E-104	Similar protein in some plants
UN052587	30.7842929	6.0114E-36	
UN028622	30.4695717	1.9817E-51	PHT6 phosphate transporter
UN026579	30.0812794	1.0935E-69	GLT1 plastid glucose transporter
UN030270	29.7741941	5.6407E-34	
UN029908	29.432577	2.1169E-81	Catalase 2

A tentative function is given when possible.

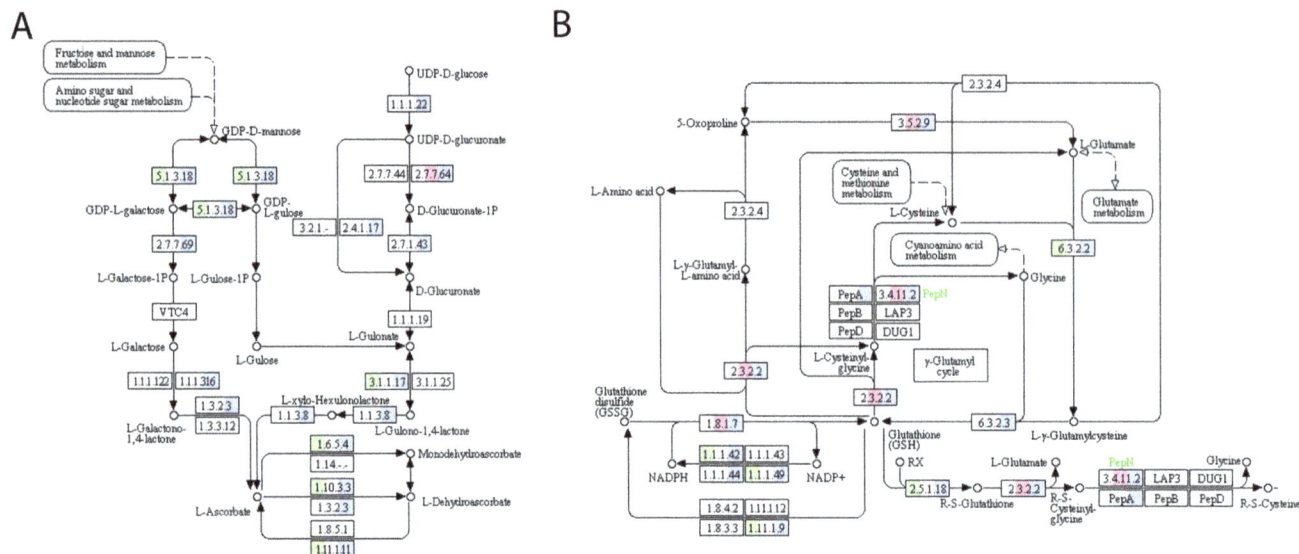

Figure 3. Ascorbate (A) and glutathione (B) metabolism in *Klebsormidium crenulatum*. (A) Biosynthesis of ascorbate via the D-galacturonate pathway is upregulated, as are ascorbate-consuming and recycling enzymes. (B) The biosynthesis of glutathione is up-regulated except for glutathione synthase. Glutathione degradation is down-regulated, whereas glutathione-consuming enzymes are up-regulated.

followed by nuclei, chloroplasts and the cytosol (Fig. 5 A and Table 6). Due to ultrastructural changes induced by desiccation, gold particles could not be clearly assigned to the individual cell compartments of desiccation-stressed *K. crenulatum* cells (Fig. 5 B). Thus, the total amount of glutathione labeling was calculated for both control and desiccation-stressed *K. crenulatum* cells (Table 7). The glutathione labeling density increased in desiccation-stressed *K. crenulatum* cells to about 177% of the value in control cells (Table 6). The total cellular antiradical power was determined using the DPPH-test (see material and methods). The cellular antiradical power was significantly higher (p <0.01) in 2.5 h desiccated samples than in control cells (Fig. 6).

Desiccation induced down-regulation of transcripts and complex regulation patterns.   Pathways, which were strongly down-regulated, were mainly involved in integrative cellular functions, such as cell division and DNA repair mechanisms. Purine-, pyrimidine-, most cofactors, and many amino acid biosyntheses were down-regulated as well. However, as mentioned above, most pathways showed complex regulation patterns. Fig. 4 illustrates this for the terpenoid backbone (Fig. 4 A) and carotenoid biosynthesis (Fig. 4 B). Contigs for the cytosolic mevalonate pathway are down-regulated, but contigs for the plastidic MEP/DOXP-pathway are not regulated, and many contigs of the carotenoid pathways are up-regulated and none down-regulated. In agreement with this, we also observed up-regulation of contigs coding for most light-harvesting complex proteins psbS and elip (see above), supporting the idea that non-photochemical quenching plays an important role in protection from ROS during desiccation stress.

Protein degradation using the ubiquitin/proteasome system is another interesting example with a complex regulation pattern; contigs for all core subunits are up-regulated, as well as some contigs for the enzymes of the ubiquitination mechanism. However, transcripts for some other components of the ubiquitination mechanism and some components of the regulatory proteasome lid and were down-regulated. In protein synthesis, contigs for nearly all ribosomal proteins were up-regulated, while contigs for some components of the nuclear ribosome processing

pathway and some aminoacyl-tRNA synthases were down-regulated.

Embryophyte desiccation/dehydration/drought stress-related proteins.   Several *K. crenulatum* contigs showed similarity to embryophyte proteins implicated in the desiccation response (Table 7). Generally, more than one contig showed similarity to a certain *Arabidopsis* gene, suggesting the presence of small protein families. Most contigs were significantly up-regulated, however a few contigs were not-regulated or down-regulated.

Discussion

In the present study we investigated the transcriptome and physiology of *K. crenulatum* under strong desiccation stress, including tests of cell viability. Cell viability in *K. crenulatum* was previously tested by using the amphiphilic styril dye FM 1–43 [17]. The present study corroborated the previous results, as we found the plasma membranes stained in control cells, whereas in desiccated cells, only slight uptake of the dye was observed. FM dyes only fluorescence under hydrophobic conditions, and mechanisms of uptake into embryophyte plant cells have been discussed [47,48]. However, even under the strong desiccation conditions used, the total cell volume was not stained, suggesting that the damage was not as extensive as reported in cells that were desiccated for 7 days [17]. In contrast, initially dead cells of control filaments show a bright FM 1–43 fluorescence throughout the whole cell lumen. We cannot fully exclude the presence of damage to the cell membrane or changes in the cell membrane composition caused by our desiccation treatment; however, these results indicate that cells survive strong desiccation over silica gel. To further analyze the physiological state of the desiccated *K. crenulatum* cells, photosynthetic parameters were measured. The CO_2 assimilation rate of *K. crenulatum* cells was determined for the first time in the present study. By gas exchange measurements it is possible to determine CO_2 assimilation rates of *K. crenulatum* under defined RH leading to desiccation under controlled conditions (temperature, CO_2 and H_2O content, light). Until

A

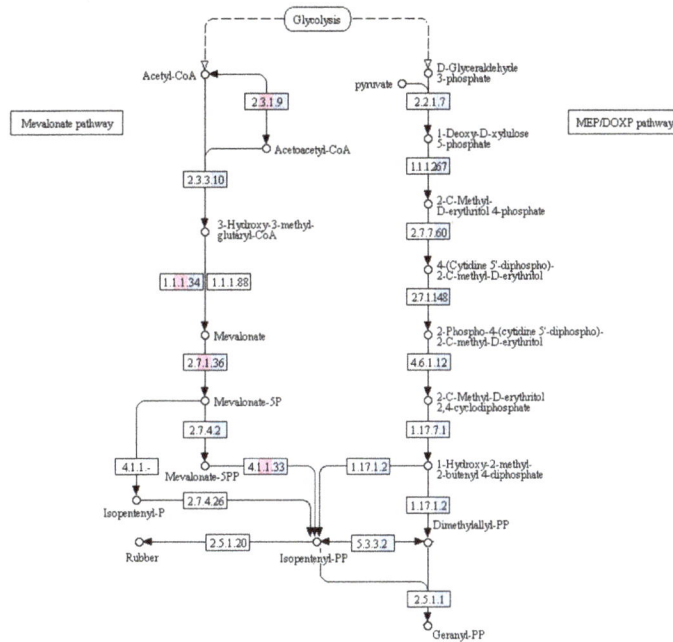

B

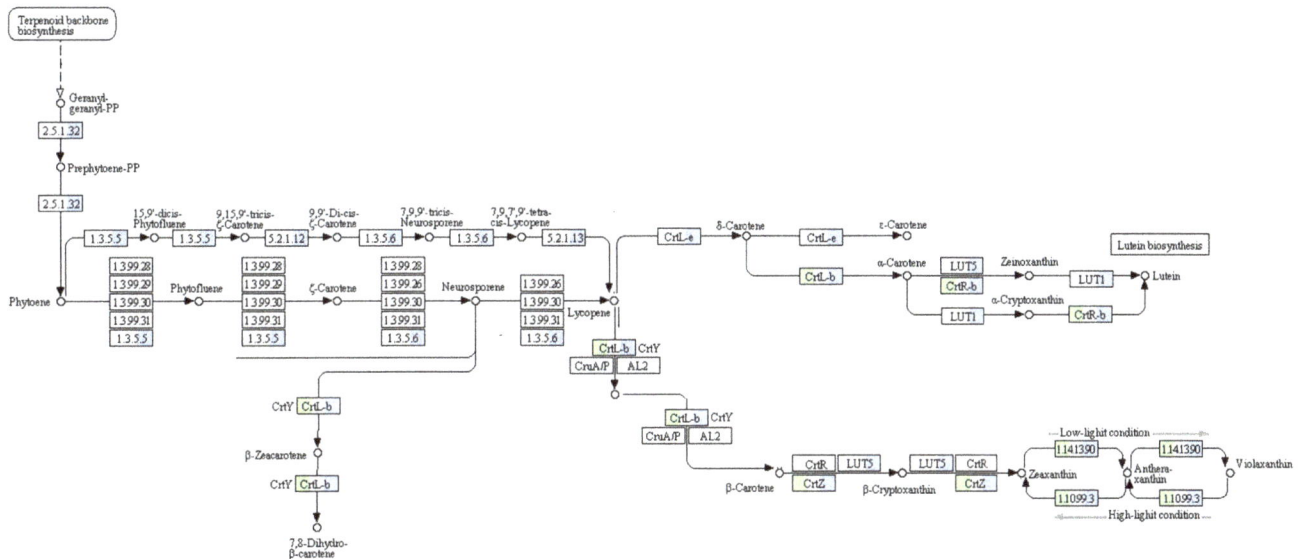

Figure 4. Biosynthesis of carotenoids in *Klebsormidium crenulatum*. (A) Terpenoid backbone synthesis, the cytosolic mevalonate pathway is inhibited. (B) Carotenoid biosynthesis is up-regulated.

now most measurements of photosynthetic parameters in green algae were performed in the fluid phase. Therefore, so far only chlorophyll fluorescence parameters could be determined in the desiccated state (e.g. [3]), but no information on assimilation rates was available.

During the desiccation period of 2.5 h under low light conditions (30 μmol photons) the assimilation rate dropped below zero, indicating that the cells shut down their photosynthetic activity; the still-measurable respiration rate might be attributed to failure to reach equilibrium to 20% RH in the chamber of the GFS-3000 (it is not possible to reach RHs as low as obtained in the desiccation box over silica gel); however after 21 h recovery, both

the assimilation rate and Yield II were near zero, indicating drastic perturbation of the cells. Suppression of photosynthesis in green algae during dehydration has been previously described [13,16,19,49]. In an earlier study using *K. crenulatum*, short-term desiccation for 3 hours at ambient room temperature and humidity was performed [16]. There, the F_v/F_m value recovered after 30 min rehydration to 48% of the initial value, and after 2 h the cells were fully recovered [16]. In a study using *K. dissectum* desiccated under different RH (100, 55 and 5%) for 1 or 3 weeks, the cells recovered within 1 or 2 weeks respectively, indicating that some cells have the ability to survive these harsh treatments [19].

Table 4. Effect of desiccation stress on transcript expression in *Klebsormidium crenulatum*: strongly up-regulated pathways.

KO number	Cellular Function	All contigs	Up-regulated contigs	% up-regulated	Down-regulated contigs	% down-regulated
ko00195	Photosynthesis	32	30	93.8	1	3.1
ko03010	Ribosome	124	114	91.9	4	3.2
ko00196	Photosynthesis - antenna proteins	12	9	75.0	1	8.3
ko00710	Carbon fixation in photosynthetic organisms	21	14	66.7	2	9.5
ko00190	Oxidative phosphorylation	79	48	60.8	5	6.3
ko04075	Plant hormone signal transduction	14	8	57.1	5	35.7
ko00030	Pentose phosphate pathway	18	10	55.6	4	22.2
ko00020	Citrate cycle (TCA cycle)	20	11	55.0	1	5.0
ko00630	Glyoxylate and dicarboxylate metabolism	29	15	51.7	4	13.8
ko00052	Galactose metabolism	15	7	46.7	4	26.7
ko00071	Fatty acid degradation	13	6	46.2	4	30.8
ko04020	Calcium signaling pathway	11	5	45.5	3	27.3
ko00010	Glycolysis/Gluconeogenesis	32	14	43.8	8	25.0
ko00053	Ascorbate and aldarate metabolism	15	6	40.0	2	13.3
ko00500	Starch and sucrose metabolism	36	14	38.9	15	41.7

The KEGG ontology (KO) numbers were retrieved using the KAAS annotation server. Using the KO numbers, pathway maps were constructed using the KEGG pathway reconstruction tool. The pathways with the largest percentage of up-regulated KO functions are shown. Human disease-related pathways and overview maps are not shown.

Transcriptome analysis

High-quality RNA is crucial for any transcriptome analysis. RNA agarose gels and Bioanalyzer electropherogram profiles indicated a reduced amount of RNA and some RNA degradation in our desiccated samples, but little or no degradation in our control samples (Fig. S2). We are convinced that the reduced RNA amount and decreased RNA integrity number are not exper-imental artifacts, but rather were caused by the applied stress conditions, for the following reasons: (1) The same protocol was used for all samples and samples were processed in parallel. (2) Similar results have been obtained for other systems, e.g. *Tortula ruralis* gametophytes [50], pea seeds [51] and yeast [52]. Particularly in the study by Chen et al. [51], a detailed analysis of the RNA integrity numbers (RIN, [53]) is given, and similar

Table 5. Effect of desiccation stress on transcript expression in *Klebsormidium crenulatum*: strongly down-regulated pathways.

KO number	Function	All contigs	Up-regulated contigs	% up-regulated	Down-regulated contigs	% down-regulated
ko03450	Non-homologous end joining	9	2	22.2	6	66.7
ko04140	Regulation of autophagy	11	1	9.1	7	63.6
ko00970	Aminoacyl-tRNA biosynthesis	27	3	11.1	15	55.6
ko03420	Nucleotide excision repair	35	5	14.3	19	54.3
ko03030	DNA replication	28	1	3.6	15	53.6
Ko03430	Mismatch repair	19	1	5.3	10	52.6
ko04142	Lysosome	43	7	16.3	21	48.8
ko00310	Lysine degradation	19	5	26.3	9	47.4
ko04150	mTOR signaling pathway	13	3	23.1	6	46.2
ko04111	Cell cycle - yeast	48	6	12.5	22	45.8
Ko00670	One carbon pool by folate	11	1	9.1	5	45.5
ko00513	Various types of N-glycan biosynthesis	20	2	10.0	9	45.0
ko04113	Meiosis - yeast	38	3	7.9	17	44.7
ko00640	Propanoate metabolism	23	6	26.1	10	43.5
ko00910	Nitrogen metabolism	14	4	28.6	6	42.9

The KEGG ontology (KO) numbers were retrieved using the KAAS annotation server. Using the KO numbers, pathway maps were constructed using the KEGG pathway reconstruction tool. The pathways with the largest percentage of down-regulated KO functions are shown. Human disease-related pathways and overview maps are not shown.

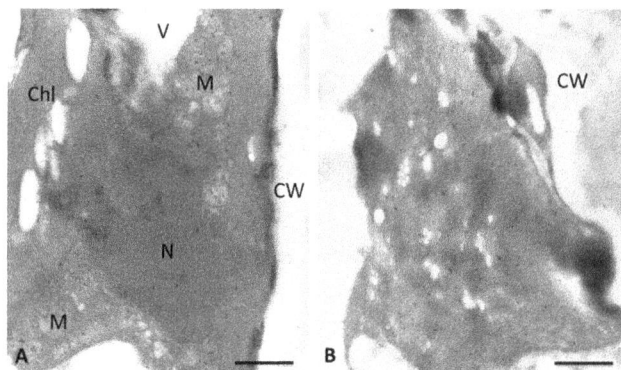

Figure 5. Immunogold labeling with 10 nm gold particles, of glutathione in transmission electron microscopic sections of *Klebsormidium crenulatum*. (A) Control cell. (B) Desiccated cell (2.5 h at ~ 10% RH). Bar 1 μm.

values to those obtained in this study were found in artificially aged pea seeds, which were considered as 'partially degraded' but still usable for the analysis.

To our knowledge this is the first broad-scale expression study of a *Klebsormidium* species in response to abiotic stress. ESTs have been sequenced for *Klebsormidium flaccidum* and *Klebsormidium subtile*, yielding about 60,000 ESTs (NCBI, 22.07.2014). As no reference data were available for *K. crenulatum*, we first established a reference transcriptome based on the pooled RNA samples. This reference transcriptome contains over 24,000 contigs. About 60% are similar to proteins in sequenced viridiplant genomes, with about 10% apparently present only in strepto-phytes, supporting the close relationship of *Klebsormidium* with embryophytes. However, currently the number of genomes from chlorophytes is limited. Therefore it seems possible that the addition of further genomes will increase the total number of *K. crenulatum* contigs that are similar to chlorophyte proteins. The established reference transcriptome covers all major cellular functions and metabolic pathways. However, for more than 30% of the contigs, no similar protein was found in any database, suggesting the existence of a large number of genus- or species-specific genes, which is very similar to the situation observed in other organisms [54].

Expression Analysis

Our expression analysis revealed that a large number of contigs are up- or down-regulated. Given the large number of contigs that show no similarity to any organism and which are strongly up-regulated in *K. crenulatum*, we did not perform any GO or KEGG-ontology enrichment study, but instead we focused on KEGG pathway analyses. There are three major responses to desiccation in *K. crenulatum*. First, contigs related to energy metabolism (glycolysis, TCA cycle, respiration) are up-regulated. The role of these pathways during desiccation tolerance is not clear. In actively growing roots and whole plants water stress inhibits mitochondrial respiration; however, in leaves the response is more variable, ranging from reduction in most cases over no change even to increase under severe water stress e.g. in *Arnica alpina*, *Triticum aestivum* or *Helianthus annuus* [55]. As the algal cell is more comparable with a plant leaf than with plant roots or seeds, we suggest that similar to the above mentioned examples, *Klebsormidium* cells try to increase respiration for energy production, which might also help to explain the high respiration

rate in the desiccated state, however it seems also possible that the cells just prepare for a fast energy production upon rehydration.

Photosynthesis is inhibited very rapidly upon desiccation (see above). *Klebsormidium* responds by up-regulation of genes involved in photosynthesis, obviously preparing for rehydration. Finally, the cells show strong protection against light damage and ROS stress, by increasing carotenoid and LHC biosynthesis as well as up-regulating glutathione and ascorbic-acid metabolism. The observed up-regulation of the pentose-phosphate cycle is probably required to provide the cells with the reducing agent NADPH. Overall these responses are very similar to typical embryophyte responses [56]. Interestingly, we found in *K. crenulatum* contigs that are similar to well-known plant proteins involved in the desiccation/drought response, e.g. LEA proteins [57] and genes involved in early response to drought [58] further emphasizing the similarities between the streptophyte algal and plant desiccation responses.

Osmolytes

It has been previously reported that in *Klebsormidium flaccidum*, sucrose is increased upon cold stress [59], and a similar reaction could be expected after desiccation stress. Indeed, we observed an up-regulation of sucrose synthase (K00695) and sucrose phosphate synthase (K00696). Nagao and Uemura [60] found that sucrose-phosphate phosphatase is present in *K. flaccidum*, but has a different structure, lacking an extensive C-terminal domain. In agreement with this, we observed a transcript similar to a plant sucrose-phosphate-phosphohydrolase family protein, which could not be mapped on the sucrose pathway by the reconstruct-pathway tool at KEGG. In addition to sucrose, members of the raffinose family of oligosaccharides (RFO) have been shown to be compatible solutes involved in tolerance of stresses such as desiccation and cold in plants [61]. Previously raffinose has been detected in *K. crenulatum* as one of the major soluble carbohydrates [18]. Contigs for several enzymes of the galactinol/raffinose metabolism were up-regulated in desiccated samples, suggesting that RFOs might function as compatible solutes in *K. crenulatum* as well.

Phytohormones

Several phytohormones have been implicated in signaling abiotic stress responses (recently reviewed [62]). ABA has been implicated in signaling of cold and desiccation responses [63] and we could detect contigs similar to all components of the ABA signaling pathway except the ABA receptor (PYR/PYL) in *K. crenulatum*, suggesting that this pathway might have evolved in streptophyte algae and facilitated colonization of the terrestrial habitat. It has been suggested that jasmonic acid acts even before ABA in the stress response [64]. Interestingly, a contig similar to JAR1 (the embryophyte jasmonic acid receptor) was also found, but no other contig showed similarity to other components of the jasmonic acid signaling pathway. This raises the possibility that the jasmonic acid receptor might function as an ABA receptor in *K. crenulatum*. Further work will be required to clarify this possibility.

Lipids

Air drying has recently been shown to increase the synthesis of triacylglycerol in the green alga *Chlorella kessleri* [65]. The authors subjected their samples to ambient air conditions on glass filters for 11 h, and obtained an increase of triacylglycerols from 0.3 to 15.3% (w/w), which corresponds to an increase from 4.7 to 70.3 mole% of fatty acids; the cell weight increased 2.7-fold in 96 h [65]. Our data also suggest that an increase in fatty acid production can be expected from the up-regulation of the

Table 6. Glutathione content of *Klebsormidium crenulatum* cells.

	Gold particles per μm^2	
	control	desiccated
Mitochondria	168 ± 8^a	n.d.
Chloroplasts	28 ± 1^d	n.d.
Nuclei	42 ± 2^c	n.d.
Cytosol	29 ± 1^d	n.d.
Whole cell	30 ± 1^e	53 ± 2^b
Cell without vacuole	32 ± 1^e	n.d.

Values are means with standard errors, and document the total amount of gold particles bound to glutathione per μm^2 in different cell compartments and whole cells of *Klebsormidium crenulatum* grown under control and desiccation conditions. n = 40 for the individual cell structures and n = 25 for the whole cell. Different lowercase letters indicate significant differences (P<0.05) analyzed with the Kruskal-Wallis test followed by post-hoc comparison according to Conover. n.d. = not determined.

Table 7. Plant desiccation/dehydration/drought stress-related contigs in *Klebsormidium crenulatum*.

Name	*Arabidopsis* Gene ID	Up/down-regulation	Padj	Similarity to the *Ath* gene
Late embryogenesis abundant (LEA) proteins				
LEA-related	AT5G60530.1	0.23991374	2.5624E-05	6E-18
LEA_2	**AT2G44060.2**	**1.86413136**	**0.00014204**	**1E-60**
LEA-related	AT1G54890.1	0.29400396	6.7547E-07	5E-13
LEA-related	AT5G54370.1	1.22790739	0.53229844	4E-16
LEA_4	**AT5G44310.2**	**3.76509689**	**1.5892E-16**	**1E-11**
LEA-related Hydroxyproline-rich glycoprotein	**AT3G44380.1**	**4.68751795**	**1.8879E-16**	**1E-17**
LEA_4	**AT5G44310.2**	**4.25267018**	**1.2762E-18**	**4E-13**
LEA_4	**AT5G44310.2**	**4.25267018**	**1.2762E-18**	**4E-13**
LEA-related	**AT3G62580.1**	**3.00106737**	**8.899E-13**	**8E-41**
LEA-related Hydroxyproline-rich glycoprotein	**AT2G01080.1**	**3.882294126**	**0.00435072**	**3E-11**
LEA-related Hydroxyproline-rich glycoprotein	**AT2G01080.1**	**4.659043737**	**4.6237E-05**	**2e-11**
LEA-related	AT3G62580.1	3.993800234	0.39220796	4E-17
Early responsive to dehydration stress (ERD) genes				
ERD-family protein	AT4G22120.6	0.40330981	0.00264642	eE-35
ERD-family PM protein	**AT4G04340.1**	**1.914357304**	**0.00119979**	**3E-29**
ERD-family PM protein	AT4G04340.1	0.719668846	0.54074033	4E-16
ERD4	**AT1G30360.1**	**3.326696704**	**0.09599311**	**3E-46**
ERD4	**AT1G30360.1**	**2.282664946**	**0.00333271**	**1E-148**
ERD4	**AT1G30360.1**	**4.305228158**	**7.4444E-15**	**4E-22**
ERD4	**AT1G30360.1**	**4.069239817**	**3.643E-21**	**1E-169**
ERD4	**AT1G30360.1**	**2.682142777**	**0.54503736**	**5E-61**
ERD-family PM protein	AT4G04340.1	1.200513952	0.40890632	1E-121
ERD4	AT1G30360.1	0.347378285	1.6875E-06	5E-166
ERD-family protein, localized in endomembrane system	AT4G02900.1	0.851420448	0.49554025	1E-100
Other				
Chloroplast drought-induced stress protein of 23 kDa, CIDSP32, thioredoxin	**AT1G76080.1**	**1.71514379**	**0.00189019**	**5E-74**
Drought-sensitive 1, DRS1,	AT1G80710.1	1.574535846	0.17300482	3E-64
Drought-sensitive 1, DRS1,	**AT1G80710.1**	**1.722093198**	**0.03356124**	**3e-63**

Significantly up-regulated genes are in bold.

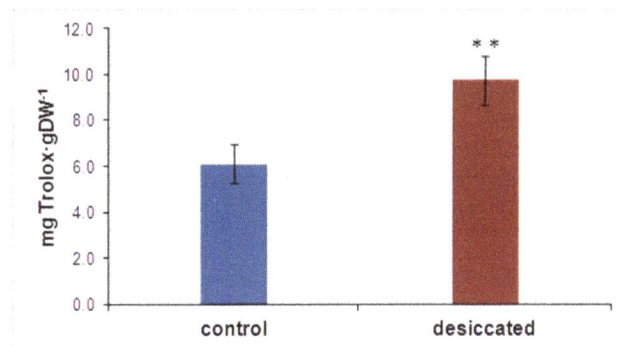

Figure 6. Antiradical power (Trolox equivalents g^{-1} dry weight (DW) of *Klebsormidium crenulatum* **control samples (n = 3) and samples (n = 3) desiccated above silica gel for 2.5 h.** The values are expressed in mg Trolox equivalent.g dry weight (DW)$^{-1}$. Means and standard deviations are shown. The statistical siginficance was thested by two sample *t*-test and demonstrated that the antiradical power of the desiccated samples is significantly higher (p <0.01).

respective enzymes. However, for acquiring desiccation tolerance, the composition of the biomembranes was found to be more crucial, as they are the initial targets in the desiccation process [66]. Moreover, chloroplast membranes are protected, as these authors reported the removal of the thylakoid lipid monogalacto-syldiacylglycerol (MGDG), which was hydrolyzed and converted into diacylglycerol (DAG). It was then further converted into phosphatidylinositol (PI) in the resurrection plant *Craterostigma plantagineum* [66]. Interestingly, we also find that in *K. crenulatum* the CDP-diacylglyerol-inositol 3-phosphatidyltransfer-ase (EC 2.7.8.11) is up-regulated, which is responsible for the conversion of CDP-diacylglycerol into phosphatidyl D myo-inositol. Moreover, enzymes in the diacylglycerol pathway such as digalactosyldiacyl-glycerol synthase (K09480) or enzymes leading to triacylglycerol [e.g. diacylglycerol O-acyltransferase (K00635) and phospholipid: diacylglycerol acyltransferase (K00679)] are up-regulated.

Antioxidants

Antioxidant protection has been studied extensively in lichens and photobionts of lichens [67–69]. It has been found that the green photobionts contain protective enzymes such as superoxide dismutase, catalase, peroxidases, glutathione reductase and ascorbate peroxidase, in combination with non-enzymatic substances such as glutathione, α-tocopherol and ascorbic acid (recently summarized [4]). Increased glutathione levels have been determined to be a consequence of supplemental UV irradiation in *Chlorococcum infusionum* and *Chlorogonium elongatum* [70]. In *Trebouxia excentrica*, the photobiont of the lichen *Cladonia vulcani*, the reduced form of glutathione (GSH) is progressively oxidized to glutathione disulfide (GSSG) during a desiccation period of 20 days [68].

Our transcriptome data showed that contigs with similarity to the first step of glutathion biosynthesis, glutathione-S-transferase and glutathione peroxidase are up-regulated. The latter are involved in the protection of the cells from oxidative damage by reducing lipid hydroperoxides to alcohols and free hydrogen peroxide to water. In contrast, contigs with similarity to glutathione-degrading enzymes are down-regulated. However, we currently cannot explain why the glutathione-disulfide reductase which is responsible for reduction of GSSG to GSH is also down-regulated. To corroborate these findings, we have tested

if elevated levels of glutathione could be found in *K. crenulatum*. We found a marked elevation (to 177% of the original value) as demonstrated by immunoelectron microscopy with polyclonal antibodies against glutathione [45]. It would have been interesting to determine the compartments which show the largest increases in glutathione levels. Unfortunately, the subcellular distribution, which showed the highest concentrations in mitochondria in control cells, was virtually impossible to determine in desiccated cells of *K. crenulatum*. The reason for this is probably that the cytoplasm appears condensed in desiccated *K. crenulatum* cells [17], as well as in other desiccated green algae (for a summary see [3]).

In concordance with the above findings, we measured a significant increase in the total antiradical power of *K. crenulatum* (this study, DPPH-test). This test is commonly used to determine the total antioxidant activity (e.g. [71,72]). For example, in *Saccharina latissima*, threefold higher level of antioxidants has been detected in generative (sorus) *versus* vegetative tissue [71]. In the case of the brown macroalga, this was attributed to a high phlorotannin content, which was considered responsible for protection of the reproductive tissue. In the present study we found evidence for a rapid adjustment of free-radical scavengers such as ascorbate and glutathione, induced by desiccation stress.

Additional potent antioxidants are pigments such as the secondary carotenoid astaxanthin, which has been found to be accumulated, e.g. in the green alga *Haematococcus pluvialis* [73]. However, based on our transcriptome data, *K. crenulatum* apparently cannot synthesize astaxanthin. Nevertheless we observed a strong up-regulation of enzymes for other carotenoids, which might also have antioxidant activity.

Conclusion

The present study revealed that a transcriptome approach is feasible for investigating a strong stress as imposed by the desiccation of *K. crenulatum*. Our results indicate that cells react to the desiccation stress, in attempting to re-establish their energy metabolism (up-regulation of contigs related to respiration and photosynthesis). Furthermore the response is surprisingly similar to that found in embryophytes (LEA proteins, ERD proteins, ROS protection, possibly accumulation of RFO osmolytes) and might also be regulated by the abscisic acid signaling pathway, highlighting the close relationship of streptophyte algae to land plants and supporting an important role for streptophyte algae in the adaptation process that eventually led to colonization of the terrestrial habitat.

Supporting Information

Figure S1 A CO$_2$ assimilation rate, filter temperature and effective quantum yield at different relative air humidity levels during a desiccation period of 2.5 h (RH 20%). Measurements were continued after a 21-h recovery period in culture medium. B CO$_2$ assimilation rate and effective quantum yield of photosystem II at the beginning of the desiccation period (n = 5, 5 min of initial desiccation period, blue bars), at the end of the desiccation period of 2.5 h (n = 5, purple bars) and after a recovery period of 21 h in culture medium (n = 5, green bars). Statistical analyses were carried out by one-way ANOVA with Tukey test (p≤0.001), C Effective quantum yield (Yield II) at the same time points as in B.

Figure S2   RNA integrity: A Gel electrophoretic separa-tion of isolated total RNA from *Klebsormidium crenula-*

tum **control cells (K1, K2, K3) and 2.5 h over silica gel-desiccated samples (T1, T2, T3). 18 S and 25 S RNA bands are marked.** The same percentage of the isolated total RNA was loaded for each sample. B Electropherograms of corresponding total RNA samples as shown in (A). The 18 S and 25 S bands were marked when applicable. RNA integrity (RIN) numbers were calculated with Agilent 2100 Expert software and ranged between RIN 5.7 and RIN 5.8 in the control samples and between RIN 3.8 and RIN 4.2 in the desiccated samples. RNA quality was considered appropriate when two distinct peaks were visible (K1 to K3); samples with elevated baselines (T1 to T3), where two peaks were still visible, were considered partially degraded, but were also used for the analysis. Although sample T3 had the highest RIN of the desiccated samples (RIN 4.2), no cDNA library could be constructed from this sample.

Table S1 Module reconstruction. KO identifiers were retrieved from the KEGG orthology database for the complete, up-regulated and down-regulated dataset used to reconstruct KEGG modules. Column A, B and C show the results for the complete (all), the up-regulated and the down-regulated data set, respectively. Column D displays the KO identifier and column E the name of the pathway or enzyme.

Table S2 Read statistics for NG-6357 K1 lib21278,NG-6357 K2 lib21279,NG-6357 K3 lib21280, NG-6357 T2 lib21282, NG-6357 T3 lib21283.

Table S3 *Klebsormidium crenulatum* **contigs up-regulated upon strong desiccation.** The white region displays the result of the expression analysis using the DeSeq program. The green and blue region give the accession number of the most similar protein in streptophytes (green) and chlorophytes (blue). Kcr-ID, sequence identifier; baseMean, mean normalized counts, averaged over all samples from both conditions; baseMeanA, mean normalized counts from condition A; baseMeanB, mean normalized counts from condition B; foldChange, fold change from condition A to B; log2FoldChange, the logarithm (to basis 2) of the fold change; pval, p value for the statistical significance of this change; padj; p value adjusted for multiple testing with the Benjamini-Hochberg procedure (see the R function p.adjust),

which controls false discovery rate (FDR). Ath-ID *Aradopsis thaliana*; Mtr-ID, *Medicago trunculata*; Aca-ID, *Aquilegia coerulea*; Stu-ID, *Solanum tuberosum*; Mdo-ID, *Malus domesticus*; Ptr-ID, *Populus trichocarpa*; Egr-ID, *Eucalyptos grandiflora*; Tca-ID, *Theobroma cacao*; Ppa-ID, *Physcomitrella patens*; Sit-ID *Selaginella italica*; Cre_ID, *Chlamydomonas reinhardtii*; Csu_ID, *Cocomyxa subellipsoidea*; Mpu_ID, *Micromonas pusilla*; Olu_ID; *Ostreococcus lucimarinus*.

Table S4 *Klebsormidium crenulatum* **contigs down-regulated upon strong desiccation.** The white region displays the result of the expression analysis using the DeSeq program. The green and blue region give the accession number of the most similar protein in streptophytes (green) and chlorophytes (blue). Kcr-ID, sequence identifier; baseMean, mean normalized counts, averaged over all samples from both conditions; baseMeanA, mean normalized counts from condition A; baseMeanB, mean normalized counts from condition B; foldChange, fold change from condition A to B; log2FoldChange, the logarithm (to basis 2) of the fold change; pval, p value for the statistical significance of this change; padj; p value adjusted for multiple testing with the Benjamini-Hochberg procedure (see the R function p.adjust), which controls false discovery rate (FDR). Ath-ID *Aradopsis thaliana*; Mtr-ID, *Medicago trunculata*; Aca-ID, *Aquilegia coerulea*; Stu-ID, *Solanum tuberosum*; Mdo-ID, *Malus domesticus*; Ptr-ID, *Populus trichocarpa*; Egr-ID, *Eucalyptos grandiflora*; Tca-ID, *Theobroma cacao*; Ppa-ID, *Physcomitrella patens*; Sit-ID *Selaginella italica*; Cre_ID, *Chlamydomonas reinhardtii*; Csu_ID, *Cocomyxa subellipsoidea*; Mpu_ID, *Micromonas pusilla*; Olu_ID; *Ostreococcus lucimarinus*.

Acknowledgments

We would like to thank O. Buchner, University of Innsbruck, for help in gas exchange measurements.

Author Contributions

Conceived and designed the experiments: AH BB. Performed the experiments: AH FK KB BZ KKB. Analyzed the data: AH FK BZ BB. Contributed reagents/materials/analysis tools: AH BB. Wrote the paper: AH FK BZ BB.

References

1. Alpert P (2006) Constraints of tolerance: why are desiccation tolerant organisms so small or rare? J Exp Bot 209: 1575–1584.
2. Oliver MJ, Cushman JC, Koster KL (2010) Dehydration tolerance in plants. In: Sunkar R, editor. Plant Stress Tolerance, Methods in Molecular Biology 639, (New York: Springer) pp. 3–24.
3. Holzinger A, Karsten U (2013) Desiccation stress and tolerance in green algae: Consequences for ultrastructure, physiological and molecular mechanisms. Frontiers in Plant Science, 4, 327, doi: 10.3389/fpls.2013.00327.
4. Agrawal SC (2012) Factors controlling induction of reproduction in green algae – review: the text. Folia Microbiol 57: 387–407.
5. Gasulla F, de Nova PG, Esteban-Carrasco A, Zapata JM, Barreno E, et al. (2009) Dehydration rate and time of desiccation affect recovery of the lichen alga *Trebouxia erici*: alternative and classical protective mechanisms. Planta 231: 195–208.
6. Lüttge U, Büdel B (2010) Resurrection kinetics of photosynthesis in desiccation-tolerant terrestrial green algae (Chlorophyta) on tree bark. Plant Biol 123: 437–444.
7. Becker B, Marin B (2009) Streptophyte algae and the origin of embryophytes. Ann Bot-London 103: 999–1004.
8. Wodniok S, Brinkmann H, Glöckner G, Heidel AJ, Philippe H, et al. (2011) Origin of land plants: Do conjugating green algae hold the key? BMC Evol Biol 11: 104. doi: 10.1186/1471-2148-11-104
9. Turmel M, Otis C, Lemieux C (2013) Tracing the evolution of streptophyte algae and their mitochondrial genome. Genome Biol Evol 5: 1817–1835.
10. Becker B (2013) Snowball earth and the split of Streptophyta and Chlorophyta. Trends Plant Sci 18: 180–183.
11. Zhong B, Xi Z, Goremykin VV, Fong R, McLenachan PA, et al. (2014) Streptophyte algae and the origin of land plants revisited using heterogeneous models with three new algal chloroplast genomes. Mol Biol Evol 31: 177–183.
12. Bertsch A (1966) CO₂ Gaswechsel der Grünalge *Apatococcus lobatus*. Planta 70: 46–72.
13. Gray DW, Lewis LA, Cardon ZG (2007) Photosynthetic recovery following desiccation of desert green algae (Chlorophyta) and their aquatic relatives. Plant Cell Environ 30: 1240–1255.
14. Wieners PC, Mudimu O, Bilger W (2012) Desiccation-induced non-radiative dissipation in isolated green lichen algae. Photosynth Res 113: 239–247.
15. Lunch CK, LaFountain AM, Thomas S, Frank HA, Lewis LA, et al. (2013) The xanthophyll cycle and NPQ in diverse desert and aquatic green algae. Photosynth Res 115: 139–151.
16. Karsten U, Lütz C, Holzinger A (2010) Ecophysiological performance of the aeroterrestrial green alga *Klebsormidium crenulatum* (Charophyceae, Strepto-phyta) isolated from an alpine soil crust with an emphasis on desiccation stress. J Phycol 46: 1187–1197.
17. Holzinger A, Lütz C, Karsten U (2011a) Desiccation stress causes structural and ultra-structural alterations in the aeroterrestrial green alga *Klebsormidium crenulatum* (Klebsormidiophyceae, Streptophyta) isolated from an alpine soil crust. J Phycol 47: 591–602.

18. Kaplan F, Lewis LA, Wastian J, Holzinger A (2012) Plasmolysis effects and osmotic potential of two phylogenetically distinct alpine strains of *Klebsormidium* (Streptophyta). Protoplasma 249: 789–804.

19. Karsten U, Holzinger A (2012) Light, temperature and desiccation effects on photosynthetic activity and drought-induced ultrastructural changes in the green alga *Klebsormidium dissectum* (Streptophyta) from a high alpine soil crust. Microb Ecol 63: 51–63.

20. Karsten U, Pröschold T, Mikhailyuk T, Holzinger A (2013) Photosynthetic performance of different genotypes of the green alga *Klebsormidium* sp. (Streptophyta) isolated from biological soil crusts of the Alps. Algol Stud 142: 45–62.

21. Škaloud P, Rindi F (2013) Ecological differentiation of cryptic species within an asexual protist morphospecies: A case study of filamentous green alga *Klebsormidium* (Streptophyta). J Eukar Microbiol 60: 350–362.

22. Ryšánek D, Hrcková K, P. škaloud (2014) Global ubiquity and local endemism of free-living terrestrial protists: phylogeographic assessment of the streptophyte alga *Klebsormidium*. Environm Microbiol doi:10.1111/1462–2920.12501

23. Rindi F, Guiry MD, López-Bautista JM (2008) Distribution, morphology, and phylogeny of *Klebsormidium* (Klebsormidiales, Charophyceae) in urban environments in Europe. J Phycol 44: 1529–1540.

24. Rindi F, Mikhailyuk TI, Sluiman HJ, Friedl T, López-Bautista JM (2011) Phylogenetic relationships in *Interfilum* and *Klebsormidium* (Klebsormidiophyceae, Streptophyta). Mol Phylogenet Evol 58: 218–231.

25. Gasulla F, Jain R, Barreno E, Guéra A, Balbuena TS, et al. (2013a) The response of *Asterochloris erici* (Ahmadjian) Skaloud et Peksa to desiccation: a proteomic approach. Plant Cell Environ 36: 1363–1378.

26. Nowrousian M (2010) Next-generation sequencing techniques for eukaryotic microorganisms: sequencing-based solutions to biological problems. Eukaryot Cell 9: 1300–1310.

27. Malone JH, Oliver B (2011) Microarrays, deep sequencing and the true measure of the transcriptome. BMC Biology 9: 34.

28. McIntyre LM, Lopiano KK, Morse AM, Amin V, Oberg AL, et al. (2011) RNA-seq: technical variability and sampling. BMC Genomics 12: 293.

29. Rodriguez MC, Edsgärd D, Hussain SS, Alquezar D, Rasmussen M, et al. (2010) Transcriptomes of the desiccation-tolerant resurrection plant *Craterostigma plantagineum*. Plant J 63: 212–228.

30. Usadel B, Fernie AR (2013) The plant transcriptome – from integrating observations to models. Front Plant Sci 4: 48.

31. Dinakar C, Bartels D (2013) Desiccation tolerance in resurrection plants: new insights from transcriptome, proteome, and metabolome analysis. Front Plant Sci 4: 482.

32. Guarnieri MT, Nag A, Smolinski SL, Darzins A, Seibert M, et al. (2011) Examination of triacylglycerol biosynthetic pathways *via de novo* transcriptomic and proteomic analyses in an unsequenced microalga. PLoS ONE 6: e25851.

33. Heinrich S, Valentin K, Frickenhaus S, John U, Wiencke C (2012) Transcriptomic analysis of acclimation to temperature and light stress in *Saccharina latissima* (Phaeophyceae). PLoS ONE 7: e44342.

34. Deng Y, Yao J, Wang X, Guo H, Duan D (2012) Transcriptome sequencing and comparative analysis of *Saccharina japonica* (Laminariales, Phaeophyceae) under blue light induction. PLoS ONE 7: e39704.

35. Timme RE, Delwiche CF (2010) Uncovering the evolutionary origin of plant molecular processes: comparison of *Coleochaete* (Coleochaetales) and *Spirogyra* (Zygnematales) transcriptomes. BMC Plant Biology 10: 96.

36. Lokhorst GM, Star W (1985) Ultrastructure of mitosis and cytokinesis in *Klebsormidium mucosum* nov. comb., formerly *Ulothrix verrucosa* (Chlorophyta). J Phycol 21: 466–476.

37. Starr RC, Zeikus JA (1993) UTEX – the culture collection of algae at the University of Texas at Austin 1993 List of cultures. J Phycol 29: 1–106.

38. McFadden GI, Melkonian M (1986) Use of HEPES buffer for microalgal culture media and fixation for electron microscopy. Phycologia 25: 551–557.

39. Karsten U, Herburger K, Holzinger A (2014) Dehydration, temperature and light tolerance in members of the aeroterrestrial green algal genus *Interfilum* (Streptophyta) from biogeographically different temperate soils. J Phycol DOI: 10.1111/jpy.12210.

40. Schreiber U, Bilger W (1993) Progress in chlorophyll fluorescence research: major developments during the past years in retrospect. Progr Bot 54: 151–173.

41. Porra RJ, Thompson WA, Kriedmann PE (1989) Determination of accurate extinction coefficients and simultaneous equations for assaying chlorophylls a and b extracted with four different solvents: verification of the concentration of chlorophyll standards by atomic absorption spectroscopy. Biochim Biophys Acta 975: 384–394.

42. Anders S, Huber W (2010) Differential expression analysis for sequence count data. Genome Biology 11: R106.

43. Edgar R, Domrachev M, Lash AE (2002) Gene Expression Omnibus: NCBI gene expression and hybridization array data repository. Nucleic Acids Res 30: 207–210.

44. Zechmann B, Müller M (2010) Subcellular compartmentation of glutathione in dicotyledonous plants. Protoplasma 246: 15–24.

45. Zechmann B, Koffler BE, Russell SD (2011) Glutathione synthesis is essential for pollen germination in vitro. BMC Plant Biol 11: 54.

46. Cruces E, Huovinen P, Gómez I (2012) Phlorotannin and antioxidant responses upon short-term exposure to UV radiation and elevated temperature in three South Pacific kelps. Photochem Photobiol 88: 58–66.

47. Emans N, Zimmermann S, Fischer R (2002) Uptake of a fluorescent marker in plant cells is sensitive to brefeldin A and wortmannin. Plant Cell 14: 71–86.

48. Bolte S, Talbot C, Boutte Y, Catrice C, Read ND, et al. (2004) FM dyes as experimental probes for dissecting vesicle trafficking in living plant cells. J Microsc Oxford 214: 159–173.

49. Häubner N, Schumann R, Karsten U (2006) Aeroterrestrial algae growing in biofilms on facades – response to temperature and water stress. Microb Ecol 51: 285–293.

50. Oliver MJ, Bewley JD (1984) Plant Desiccation and Protein Synthesis V. Stability of Poly (A)− and Poly (A)+ RNA during desiccation and their synthesis upon rehydration in the desiccation-tolerant moss *Tortula ruralis* and the intolerant moss *Cratoneuron filicinum*. Plant Physiol 74: 917–922.

51. Chen H, Osuna D, Colville L, Lorenzo O, Graeber K, et al. (2013) Transcriptome-wide mapping of pea seed ageing reveals a pivotal role for genes related to oxidative stress and programmed cell death. PLoS ONE 8: e78471.

52. Rapoport AI, Beker ME (1986) Ribonucleic acid degradation during dehydration of yeast cells. Microbiol 55: 689–691.

53. Schroeder A, Mueller O, Stocker S, Salowsky R, Leiber M, et al. (2006) The RIN: an RNA integrity number for assigning integrity values to RNA measurements. BMC Mol Biol. 7:3.

54. Adams M (2013) Open questions: genomics and how far we haven't come. BMC Biol 11: 109.

55. Atkin OK, Mahcerel D (2009) The crucial role of plant mitochondria in orchestrating drought tolerance. Ann Bot 103: 581–597.

56. Dinakar C, Djilianov D, Bartels D (2012) Photosynthesis in desiccation tolerant plants: energy metabolism and antioxidative stress defense. Plant Sci 182: 29–41.

57. Shi C-Y, Yang H, Wei C-L, Yu O, Zhang Z-Z, et al. (2011) Deep sequencing of the *Camellia sinensis* transcriptome revealed candidate genes for major metabolic pathways of tea-specific compounds. BMC Genomics 12: 131.

58. Shinozaki K, Yamaguchi-Shinozaki K (2000) Molecular responses to dehydration and low temperature: differences and cross-talk between two stress signalling pathways. Curr Opin Plant Biol 3: 217–223.

59. Nagao N, Matsui K, Uemura M (2008) *Klebsormidium flaccidum*, a charophycean green alga, exhibits cold acclimation that is closely associated with compatible solute accumulation and ultrastructural changes. Plant Cell Environ 31: 872–885.

60. Nagao M, Uemura M (2012) Sucrose phosphate phosphatase in the green alga *Klebsormidium flaccidum* (Streptophyta) lacks an extensive C-terminal domain and differs from that of land plants. Planta 235: 851–861.

61. ElSayed AI, Rafudeen MS, Golldack D (2014) Physiological aspects of raffinose family oligosaccharides in plants: protection against abiotic stress. Plant Biol 16: 1–8.

62. Ahmad P, Bhardwaj R, Tuteja N (2012) 'Plant Signaling Under Abiotic Stress Environment'. In: P Ahmad, MNV Prasad, editors. Environmental Adaptations and Stress Tolerance of Plants in the Era of Climate Change. Springer New York, pp. 297–323.

63. Osakabe Y, Osakabe K, Shinozaki K, Tran L-SP (2014) Response of plants to water stress. Front Plant Sci 5: 86. doi: 10.3389/fpls.2014.00086.

64. Djilianov D, Dobrev P, Moyankova D, Vaňková R, Georgieva D, et al. (2013) Dynamics of endogenous phytohormones during desiccation and recovery of the resurrection plant species *Haberlea rhodopensis*. J Plant Growth Reg 32: 564–574.

65. Shiratake T, Sato A, Minoda A, Tsuzuki M, Sato N (2013) Air-drying of cells, the novel conditions for simulated synthesis of triacylglycerol in green alga, *Chlorella kessleri*. PLoS ONE 8: e79630.

66. Gasulla F, Vom Dorp K, Dombrink I, Zähringer U, Gisch N, et al. (2013b) The role of lipid metabolism in the acquisition of desiccation tolerance in *Craterostigma plantagineum*: a comparative approach. Plant J 75: 726–741.

67. Kranner I, Birtic F (2005) A modulation role for antioxidants in desiccation tolerance. Integr Comp Biol 45: 734–740.

68. Kranner I, Cram WJ, Zorn M, Wornik S, Yoshimura I, et al. (2005) Antioxidants and photoprotection in a lichen as compared with its isolated symbiotic partners. Proc Natl Acad Sci USA 102: 3141–3146.

69. Weissman L, Garty J, Hochman A (2005) Characterization of enzymatic antioxidants in the lichen *Ramalina lacera* and their response to rehydration. Appl Environ Microbiol 71: 6508–6514.

70. Agrawal SC (1992) Effects of supplemental UV-B radiation on photosynthetic pigment, protein and glutathione contents in green algae. Environ Exp Bot 32: 137–143.

71. Holzinger A, Di Piazza L, Lütz C, Roleda MY (2011b). Sporogenic and vegetative tissues of *Saccharina latissima* (Laminariales, Phaeophyceae) exhibit distinctive sensitivity to experimentally enhanced ultraviolet radiation: photosynthetically active radiation ratio. Phycol Res 59: 221–235.

72. Plank DW, Szpylka J, Sapirstein H, Woollard D, Zapf CM, et al. (2012) Determination of antioxidant activity in foods and beverages by reaction with 2,2′-Diphenyl-1-Picrylhydrazyl (DPPH): Collaborative study first action 2012.04. J AOAC Internat 95: 1562–1569.

73. Borowitzka MA, Huisman JM, Osborn A (1991) Culture of the astaxanthin-producing green alga *Haematococcus pluvialis* 1. Effects of nutrients on growth and cell type. J Appl Phycol 3: 295–304.

16

Role of LARP6 and Nonmuscle Myosin in Partitioning of Collagen mRNAs to the ER Membrane

Hao Wang, Branko Stefanovic*

Department of Biomedical Sciences, College of Medicine, Florida State University, Tallahassee, Florida, United States of America

Abstract

Type I collagen is extracellular matrix protein composed of two α1(I) and one α2(I) polypeptides that fold into triple helix. Collagen polypeptides are translated in coordination to synchronize the rate of triple helix folding to the rate of posttranslational modifications of individual polypeptides. This is especially important in conditions of high collagen production, like fibrosis. It has been assumed that collagen mRNAs are targeted to the membrane of the endoplasmic reticulum (ER) after translation of the signal peptide and by signal peptide recognition particle (SRP). Here we show that collagen mRNAs associate with the ER membrane even when translation is inhibited. Knock down of LARP6, an RNA binding protein which binds 5′ stem-loop of collagen mRNAs, releases a small amount of collagen mRNAs from the membrane. Depolymerization of nonmuscle myosin filaments has a similar, but stronger effect. In the absence of LARP6 or nonmuscle myosin filaments collagen polypeptides become hypermodified, are poorly secreted and accumulate in the cytosol. This indicates lack of coordination of their synthesis and retro-translocation due to hypermodifications and misfolding. Depolymerization of nonmuscle myosin does not alter the secretory pathway through ER and Golgi, suggesting that the role of nonmuscle myosin is primarily to partition collagen mRNAs to the ER membrane. We postulate that collagen mRNAs directly partition to the ER membrane prior to synthesis of the signal peptide and that LARP6 and nonmuscle myosin filaments mediate this process. This allows coordinated initiation of translation on the membrane bound collagen α1(I) and α2(I) mRNAs, a necessary step for proper synthesis of type I collagen.

Editor: Alexander F. Palazzo, University of Toronto, Canada

Funding: The full funding for this manuscript came in part from the NIH grant R01 DK059466-07A2 to B.S. and in part from the College of Medicine, Florida State University. The funders had no role in study design, data collection and analysis, decision to publish, or preparation of the manuscript.

Competing Interests: The authors have declared that no competing interests exist.

* Email: branko.stefanovic@med.fsu.edu

Introduction

Type I collagen is the most abundant protein in human body, found predominantly in skin, bone, tendons and other connective tissues. The protein is composed of two α1(I) polypeptides and one α2 polypeptide. The polypeptides are co-translationally inserted into the lumen of the endoplasmic reticulum (ER), post-translationally modified by hydroxylations and glycosylations, folded into a triple helix and secreted into the extracellular environment [1,2]. After proteolytic processing, triple helices are polymerized and crosslinked into fibrils [3]. Evidence has been presented that processing and polymerization of collagen helices can take place even in the terminal secretory vesicles [4].

The abundance of type I collagen is primarily due to slow turnover of the protein, rather than due to high rate of synthesis. The half life of type I collagen is about 30–60 days, while its fractional synthesis rate (FSR), expressed as percentage per hour, is estimated to be $0.076 \pm 0.063\%$ per hour or about 2% per day [5]. This is substantially slower than, for example, the FSR of plasma proteins, which is about 0.5% per h [6,7,8]. However, in reparative [9] or reactive fibrosis [10,11,12,13] the rate of collagen synthesis increases several hundred fold. To enable such outburst of collagen production the cells employ a unique mechanism that is based on binding of LARP6 to the conserved stem loop (SL) found in the 5′ UTRs of mRNAs encoding type I collagen [14,15].

The collagen 5′SL binds LARP6 with high affinity and specificity and LARP6 recruits accessory factors to promote translation of collagen mRNAs. These factors include RNA helicase A (RHA), FKBP3 and STRAP [16,17,18]. RHA promotes loading of polysomes on collagen mRNA, most likely by unwinding the 5′SL [18]. FKBP3 is a chaperone involved in recycling of LARP6 [17], while STRAP is needed for coordination of translation of collagen α1(I) and α2(I) mRNAs [16]. Cells lacking STRAP produce predominantly homotrimers of type I collagen and hypermodified individual α1(I) and α2(I) polypeptides. In addition, LARP6 associates collagen mRNAs with vimentin intermediate filaments to prolong their half life [19]. These processes drive high collagen production in fibrosis.

Nonmuscle myosin is the motor protein which slides actin filaments [20]. The activity of nonmuscle myosin is needed for cell motility, kariokinesis and trafficking of intracellular vesicles and these functions of nonmuscle myosin have been extensively studied [21,22,23,24,25,26,27,28]. There are three isoforms of nonmuscle myosin, but isoforms IIA and IIB are the most abundant in cells. In fibrosis, quiescent fibroblasts and other cells capable of making type I collagen are activated and differentiate into myofibroblasts [29]. They highly upregulate nonmuscle myosin and acquire the ability to migrate [30,31,32]. At the same time they begin to produce large amount of type I collagen. It has been postulated

Table 1. Primers used for RT-PCR and sequences of siRNA.

GENE	FORWARD	REVERSE
COL1A1	AGAGGCGAAGGCAACAGTCG	GCAGGGCCAATGTCTAGTCC
COL1A2	CTTCGTGCCTAGCAACATGC	TCAACACCATCTCTGCCTCG
FIB	ACCAACCTACGGATGACTCG	GCTCATCATCTGGCCATTTT
ACT	GTGCGTGACATTAAGGAGAAG	GAAGGTAGTTTCGTGGATGCC
MMP12	ACACATTTCGCCTCTCTGCT	CCTTCAGCCAGAAGAACCTG
LARP6 siRNA	AGGACGUGCACGAGUUGGAUU	
CON siRNA	D-001210-01-05 (Dharmacon)	

that nonmuscle myosin filaments help secretion of vesicle containing already formed collagen fibers [4]. However, we were first to show that nonmuscle myosin filaments are also needed for synthesis and proper formation of the type I collagen [15,33].

In this manuscript we extend these observations and show that collagen mRNAs associate with the ER membrane in translation independent manner and that LARP6 and nonmuscle myosin filaments play a role in this process. Direct binding of collagen mRNAs to the ER membrane coordinates translation of collagen mRNAs, resulting in proper modifications of the polypeptides and effective secretion of heterotrimeric type I collagen.

Material and Methods

Chemicals

ML-7 (I2764), blebbistatin (B0560), puromycin (P7255), and digitonin (D41) were purchased from Sigma. Pateamine A was purchased from Southwestern Medical Center, Dallas, TX and iodixanol (Opti-prep) was purchased from Accurate Chemical & Scientific CORP (LYS3782).

Constructs

Dominant negative myosin light chain kinase (DN-MLCK) was a kind gift of Dr. P. Gallagher, Indiana University School of Medicine. The construct was described in [34]. The cDNA of DN-MLCK was recloned into pAdCMV-Track vector and recombinant adenovirus was constructed by recombination in E. coli, as described in [35]. For overexpression of DN-MLCK adenovirus was added at multiplicity of infection of 500 to human lung fibroblasts in culture.

Cells and transfections

Human lung fibroblasts (HLFs) immortalized by expression of telomerase reverse transcriptase have been described previously [14,36]. Scleroderma fibroblasts derived from the skin of a scleroderma patients was purchased from European collection of cell cultures (cell line BM0070). Human lung fibroblasts and scleroderma fibroblasts were grown under standard conditions in Dulbecco's modified Eagle's medium supplemented with 10% fetal bovine serum (Valley Biomedical) for up to 10 passages. HEK293T cells were grown in Dulbecco's modified Eagle's medium with 10% of FetalPlus serum for up to 10 passages. For analysis of proteins, ML-7 (40 µM) or Blebbistatin (100 µM) were added 16 hours before the analysis and for analysis of mRNAs they were added 3 hours before the experiment. Pateamine A (200 nM), puromycin (100 µg/ml) or cycloheximide (100 µg/ml) were added 3 hours before the analysis. HEK293T cells were transfected with 1 µg of plasmid per 35 mm dish by 293TransIT

reagent (Mirus). LARP6 specific siRNA with the sequence 5'-AGGACGUGCACGAGUUGGAUU-3' was purchased from Thermo Scientific (Stebo-000001). The non-targeting control siRNA was from Thermo Scientific (D-001210-01-20). HLFs were transfected with siRNA at the final concentration of 150 nM using Lipofectamine 2000 reagent (Invitrogen). The cells were harvested for analysis 48–72 hours after transfections.

RT-PCR analysis

Total RNA was isolated by Phenol-Chloroform sequential extraction and contaminated DNA was removed by treated with DNase I. 100 ng of total RNA or equivalent amounts (10%) of RNA extracted from cytosolic and membrane fractions were used in RT-PCR. Semi-quantitative RT-PCR reactions were performed using rTth reverse transcriptase (Boca Scientific, FL) and including radio-labeling of the PCR products with 32P-dATP [10,14,16,37,38]. The primers used are listed in Table 1. To maintain the reaction in the linear range, 23 cycles were used for collagen and fibronectin mRNA. The PCR products were resolved on sequencing gels and specific bands were detected by autoradiography.

Real time RT-PCR

RNA from soluble and membrane fractions was isolated by phenol/chloroform extraction and isopropanol precipitation and treated with DNase I to remove contaminating DNA. The first-strand cDNAs were synthesized using SuperScript II RT reverse transcriptase (Invitrogen) and oligo-dT primer using equal amounts (10%) of the soluble and membrane RNA. Quantitative real-time PCR analyses were performed on an IQ5 thermocycler (Bio-Rad) using SYBR green detection kit (Qiagen) and gene specific primers (Table 1). The signals for COL1A1, COL1A2 and MMP12 mRNAs were normalized to signal for actin mRNA. Actin mRNA was found to be equally distributed in soluble and membrane fractions and, therefore, was suitable for normalization. The normalized cytosolic and membrane mRNA signals were added and arbitrarily set as 1 for each gene and the fraction of mRNA in cytosol and membrane was calculated. The experiments were done in three replicates and data are presented as means and standard errors of the mean (SEM).

Western blots

Total cellular proteins were prepared by lysing cells in 150 mM NaCl, 10 mM MgCl2, 10 mM Tris 7.5 and 0.5% NP-40. The nuclei were removed by centrifugation and protein concentration measured by Bradford assay. Typically, 50 µg of total proteins was used for analysis. For analysis of proteins secreted into cellular medium, equal amount of cells were seeded in 35 mm dishes.

After reaching 80% confluency the cells were washed with serum free medium, 500 μl of serum free medium was added and incubation continued for 3 hours. The medium was collected and equal volumes were directly analyzed by Western blotting. The antibodies used were: anti-collagen α1(I) antibody (Rockland, PA); anti- collagen α2(I) antibody (Santa Cruz Biotech, Dallas, Texas); anti-fibronectin antibody (BD Transduction Laboratories, New Jersey); anti-LARP6 antibody (Abnova, CA); anti-tubulin antibody (Cell Signalling, Danvers, MA); anti-calnexin antibody (BD transduction, New Jersey); anti-actin antibody (Abcam, Cambridge, MA); anti-Golgin84 antibody (BD transduction, New Jersey), anti-myosin IIB antibody (Hybridoma bank, University of Iowa) and anti-GAPDH antibody (Santa Cruz Biotechnology). Anti-histone H4 antibody has been described in [39].

Fractionation of cells into cytosolic and membrane compartments

HLFs were grown to 90% confluency, washed with PBS, gently coated with permeabilization buffer (110 mM KOAc, 25 mM HEPES 7.2, 2.5 mM Mg(OAc)$_2$, 1 mM EGTA, 0.015% digitonin, 1 mM DTT, 1 mM PMSF) and rocked slowly on ice for 5 minutes. The soluble cytosolic material was collected and the cell remnants were washed with 110 mM KOAc, 25 mM HEPES 7.2, 2.5 mM Mg(OAc)$_2$, 1 mM EGTA, 0.004% digitonin, 1 mM DTT, 1 mM PMSF. The cell membranes were dissolved in 400 mM KOAc, 25 mM HEPES (pH 7.2), 15 mM Mg(OAc)$_2$, 1% NP-40, 0.5% DOC, 1 mM DTT, 1 mM PMSF by rocking on ice for 5 minutes and the extracted material was collected [40]. The samples were clarified by centrifugation at 7500 g for 10 minutes at 4°C and used for analysis by Western blot. RNA from the fractions was isolated by phenol-chloroform extraction and isopropanol precipitation. For separation of polysomes the soluble and membrane fractions were loaded onto 25% sucrose cushion and centrifuged at 34,200 rpm for 2 hours at 4°C [40]. The supernatant was collected as post-polysomal supernatant (PPS) and the pellets were dissolved in PBS as polysomes.

Separation of ER and Golgi compartments

HLFs grown to 80% confluency were treated with ML-7 or DMSO, washed in PBS and homogenized in buffer containing 0.25 M sucrose, 1 mM EDTA, 10 mM HEPES 7.4 using Dounce homogenizer. Nuclei were removed by centrifugation at 3000 g for 10 min and the supernatant was overlaid onto 5% to 25% iodixanol gradient and centrifuged at 34,200 rpm for 2 hours [41]. Sixteen 500 μl fractions were collected from the bottom of the gradient and 80 μl of each fraction was analyzed by Western blot.

Two-Dimensional Gel Electrophoresis

Proteins were precipitated in 10 volumes of 90% ethanol and protein pellets were solubilized and separated by iso-electric focusing on immobilized 7 cm long pH gradient strips with pH range of 3-10 (GE Healthcare) with total of 5,000 Vh using Ettan IPGphor 3 instrument. The strips were loaded of 7.5% SDS PAGE gels and after separation the gels were blotted and probed with anti-collagen antibody.

Results

Translation independent localization of collagen mRNAs to the ER membrane

Secreted proteins are translated at the membrane of the ER and co-translationally inserted into the lumen through SEC61 translocation channels [42]. It has been well accepted that the mRNAs encoding secreted proteins are targeting to the ER membrane by signal recognition particle (SRP) after translation of the signal peptide [43]. However, recently it has been shown that many mRNA can associate with the ER membrane in the translation independent manner and it has been postulated that RNA binding proteins mediate this process [44,45,46,47,48,49]. Since mRNAs encoding type I collagen bind LARP6 with high specificity [14], we investigated if collagen mRNAs can be targeted to the ER membrane in the absence of translation. To this goal we inhibited translation in human lung fibroblasts (HLF) using three drugs: pateamine A (inhibits translation initiation) [50,51], puromycin (dissociates polysomes) or cycloheximide (immobilizes polysomes). 3 h after inhibition of translation, we used selective detergent extraction to separate cytosolic RNA and membrane bound RNA [40] and analyzed collagen mRNAs distribution in these fractions. In untreated cells collagen α1(I) and α2(I) mRNAs were found entirely associated with the membrane fraction. The same was true for mRNAs encoding two other secreted proteins, fibronectin (FIB) and matrix metalloproteinase 12 (MMP12) (Fig 1A). Actin mRNA was equally distributed between the cytosol and membrane. Similar dual distribution of several mRNAs encoding cytosolic proteins has been observed before [52,40] and we used actin mRNA as additional control in our experiments. Real time RT-PCR measurement of the mRNA distribution is shown later. When the cells were treated with pateamine A (Fig 1B) or puromycin (Fig 1C), the membrane association of collagen α1(I) and α2(I) mRNAs remained unchanged, however, the localization of FIB and MMP12 mRNAs was dramatically altered. About 40% of MMP12 mRNA (for real time RT-PCR measurement see later) and 10–20% of fibronectin mRNA were released into the cytosol. Distribution of actin mRNA was unchanged.

Analysis of ribosomal RNA in the fractions revealed that in HLF about 80% of ribosomes are associated with the membrane (Fig 1A). This was expected, as HLF are specialized cells dedicated to secretion of extracellular matrix proteins. Similar, predominantly membrane distribution of ribosomal RNA, was found in other cell types that secrete large amounts of proteins, including plasmocytoma cells and mouse embryonic fibroblasts [53]. Most ribosomes were released from the membrane by treatment with pateamine A or puromycin (Fig 1B and C), as evidenced by the shift of ribosomal RNA into the cytosolic fraction. To verify that there was no cross-contamination of the fractions and no release of the nuclear material we analyzed the following proteins; calnexin as ER marker, GAPDH as cytosolyc marker and histone H4 nuclear marker (Fig 1A, B, C and D, bottom panels). The analysis demonstrated that the fractions were devoid of significant contamination. These results suggested that collagen mRNAs remain associated with the internal membranes in the absence of translation and that they may attach to the ER membrane differently than the mRNAs encoding other secreted proteins. Treatment with cycloheximide did not release FIB, MMP12 or collagen mRNAs from the membranes (Fig 1D). The analysis of ribosomal RNA revealed that most ribosomes remained attached to the membrane, suggesting that, when polysomes were immobilized, the membrane association of FIB and MMP12 mRNAs was retained.

A time course experiment (Fig 1E) revealed that collagen mRNAs remain associated with the membranes 12 h after the translation inhibition by Pat-A, with only tracing amounts of collagen α2(I) mRNA released after 12 h. FIB and MMP12 mRNA were partially released after 3 h and progressively increasing amounts were released at subsequent time points (Fig 1E). This suggested that mRNAs encoding secreted proteins

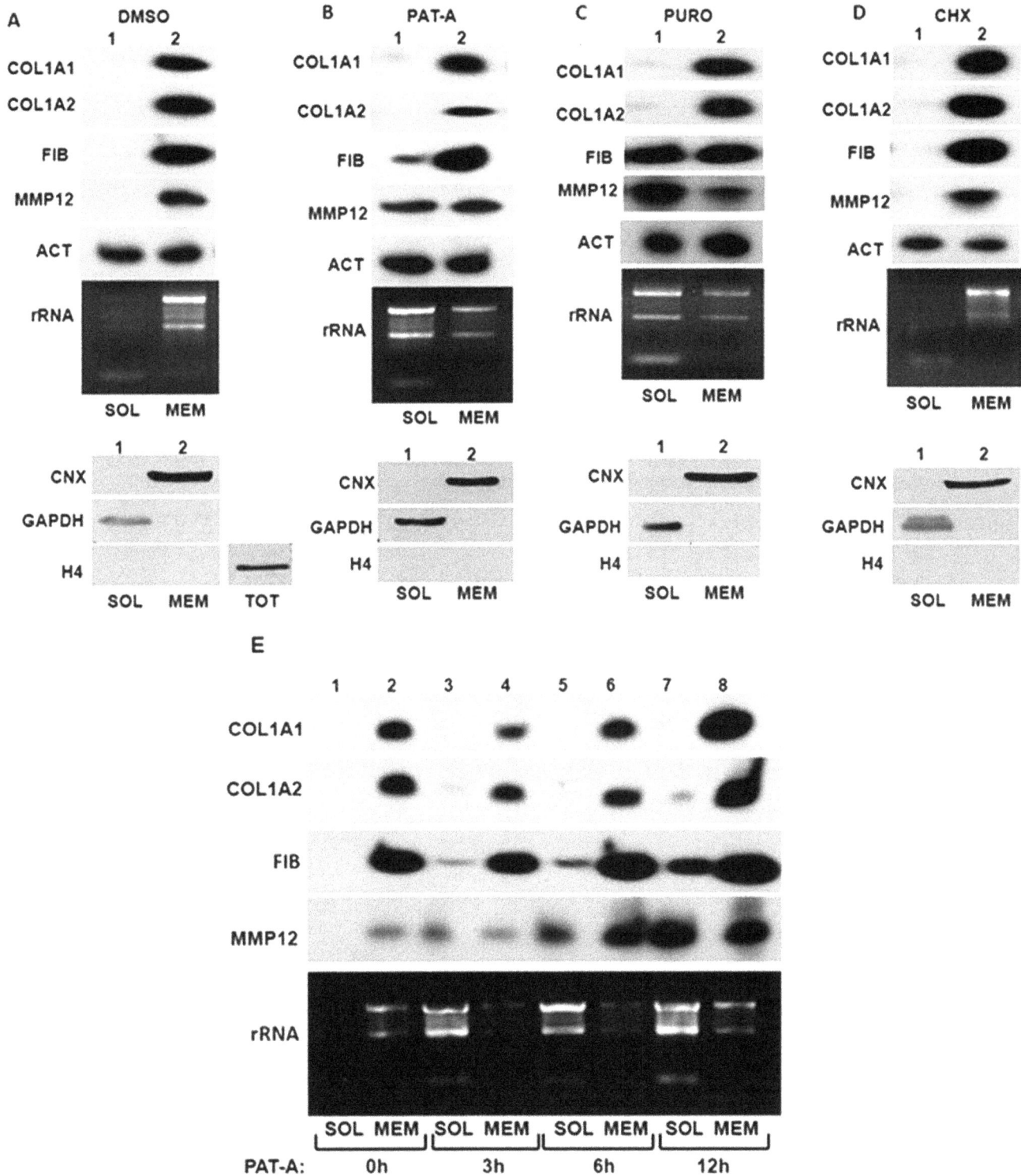

Figure 1. Translation is not required for membrane localization of collagen mRNAs. A. membrane partitioning of collagen mRNAs in control cells. Upper panel: HLFs were fractionated in cytosolic fraction (SOL, lane 1) or membrane fraction (MEM, lane 2) and collagen α1(I) (COL1A1), collagen α2(I) (COL1A2), fibronectin (FIB), matrix metalloproteinase 12 (MMP12) and actin (ACT) mRNAs were analyzed by RT-PCR. Distribution of ribosomal RNA (rRNA) is shown by agarose gel electrophoresis and ethidium bromide staining. Bottom panel: distribution of calnexin (CNX), GAPDH and histone H4 proteins in the fractions, analyzed by western blot. TOT; total cellular extract showing reactivity of the histone H4 antibody. B. Partitioning of collagen mRNAs after translation inhibition by pateamine A (PAT-A). Experiment as in A, except the cells were treated with PAT-A for 3 h prior to harvesting. C. Partitioning after translation inhibition by puromycin. D. Partitioning after translation inhibition by cycloheximide. E. Time course of mRNA redistribution after inhibition of translation. HLFs were treated with PAT-A for the indicated time periods and cells were fractionated into cytosolic (SOL) and membrane (MEM) fractions and the distribution of the mRNAs was analyzed by RT-PCR.

differ in the way they partition to the ER membrane and that collagen mRNAs stably associate with the internal membranes in the absence of translation.

Association of collagen mRNAs with the ER membrane is mediated by LARP6

To investigate if LARP6, as collagen mRNA specific RNA binding protein [14], is required for translation independent association of collagen mRNAs with the ER membrane, we knocked down LARP6 and analyzed the partitioning of collagen mRNAs. Using siRNA we were able to knock down LARP6 in lung fibroblasts by ~80% (Fig 2A). When LARP6 was knocked down, the expression of collagen α1(I) and α2(I) polypeptides was reduced when the peptides was measured either intracelullarly (Fig 2B) or secreted into the medium (Fig 2C). Thus, the cells with reduced amounts of LARP6 inefficiently synthesize type I collagen.

When partitioning of collagen mRNAs between soluble and membrane fractions was analyzed in control cells, collagen α1(I) and α2(I) mRNAs were found almost entirely in the membrane fraction, as before (Fig 2D). The purity of the fractions is shown at the bottom of Figs 2D and 2E. However, in LARP6 knock down cells, a significant amount of collagen α1(I) mRNA and lesser amounts of α2(I) mRNA were found free in the cytosol (Fig 2E). The partitioning of fibronectin mRNA and actin mRNA was not affected. For real time RT-PCR measurement of the mRNA distribution, including the MMP12 mRNA, see later. Thus, the reduced amounts of LARP6 altered the subcellular partitioning of collagen α1(I) mRNAs, with a marginal effect on collagen α2(I) mRNA. This suggested that the decreased production of collagen polypeptides in LARP6 knock down cells (Fig 2B and C) may be related to the impaired association of collagen mRNAs with the ER membrane.

To further corroborate that translation is dispensable for membrane association of collagen mRNAs we inhibited translation

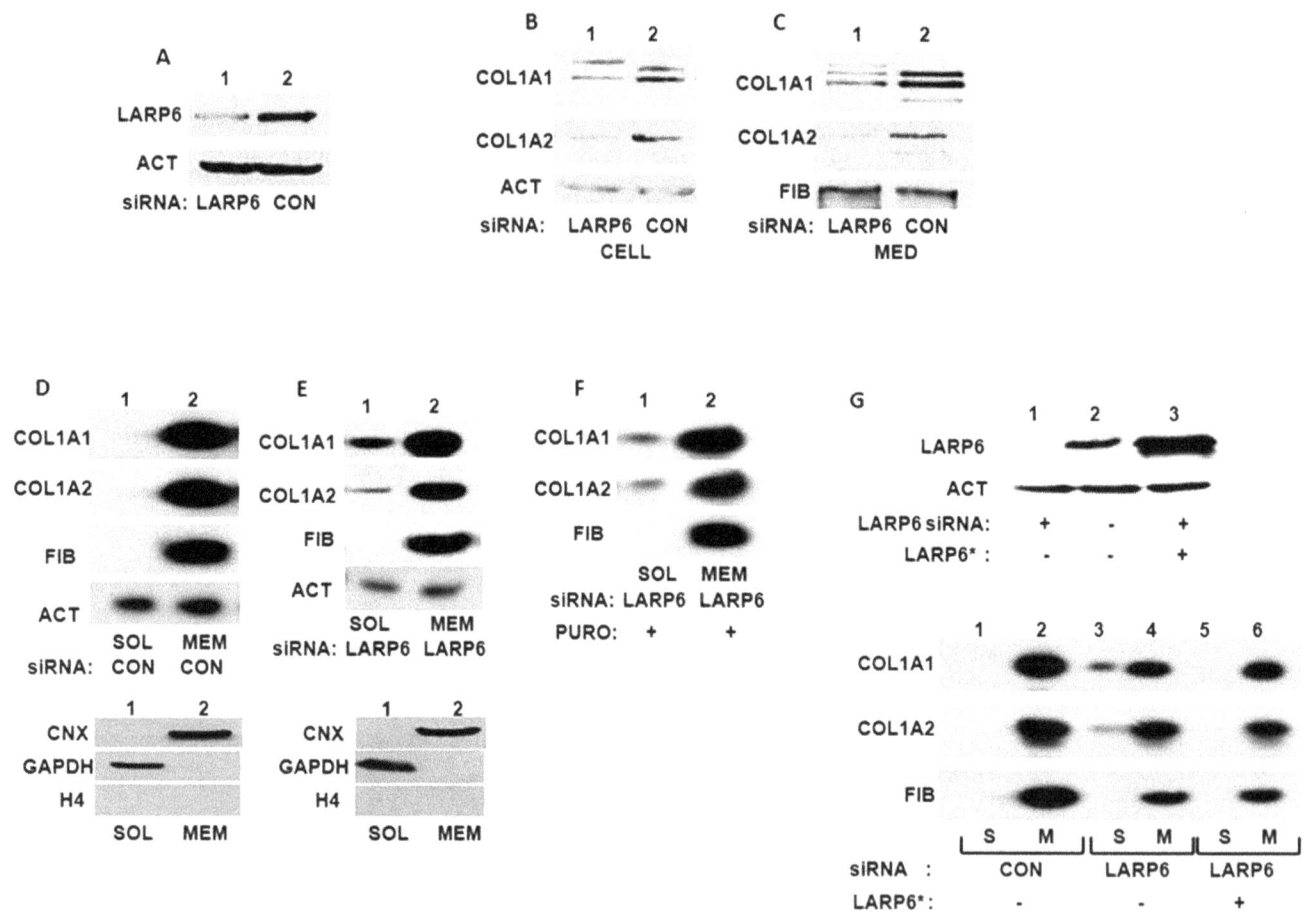

Figure 2. Knock down of LARP6 decreases expression of type I collagen. A. Knock down of LARP6. Expression of LARP6 was analyzed by western blot in HLFs after transfection of LARP6 siRNA (lane 1) and control siRNA (lane 2). Loading control: actin (ACT). B. Cellular level of collagen polypeptides after LARP6 knock down. Collagen α1(I) (COL1A1) and α2(I) (COL1A2) polypeptides were analyzed by western blot in LARP6 knock down HLFs (lane 1) and control HLFs (lane 2). C. Secretion of collagen polypeptides into the cellular medium. Cellular medium of cells in B was analyzed by western blot. Loading control: fibronectin (FIB). D. Partitioning of collagen mRNAs in control cells. Upper panel: HLFs transfected with control siRNA were fractionated into cytosolic (SOL) or membrane (MEM) fraction and analyzed for presence of collagen mRNAs and actin (ACT) and fibronectin (FIB) mRNAs by RT-PCR. Bottom panel: distribution of the marker proteins. E. Partitioning of collagen mRNAs in LARP6 knock down cells. Experiment as in D, except LARP6 siRNA was transfected. F. Partitioning of collagen mRNAs in LARP6 knock down cells after puromycin treatment. Experiment as in E, except the cells were treated with puromycin for 3 h. G. Rescue of partitioning of collagen mRNAs by supplementing LARP6. Upper panel: expression of LARP6 in LARP6 knock down cells (lane 1), in control cells (lane 2) and in LARP6 knocked down cells transduced with LARP6 adenovirus (lane 3). Loading control: actin (ACT). Bottom panel: Distribution of collagen mRNAs in the soluble (S) and membrane (M) fractions of control cells (lanes 1 and 2), LARP6 knock down cells (lanes 3 and 4) and LARP6 knock down cells supplemented with exogenous LARP6 (lanes 5 and 6).

by puromycin in LARP6 knock down cells (Fig 2F). No additive effect was observed and partitioning of collagen mRNAs was similar to that in LARP6 knock down cells in the presence of translation (compare Figs 2E and 2F).

Partitioning of collagen mRNAs was rescued when LARP6 was supplemented to the LARP6 knock down cells. Fig 2G, upper panel, shows knock down of LARP6 in HLF (lane 1) and the level of the protein in control cells (lane 2). Delivery of LARP6 by an adenovirus into the LARP6 knock down cells (LARP6*, lane 3) resulted in restoration of the LARP6 levels. This completely rescued the partitioning of collagen mRNAs (Fig 2G, bottom panel), which were now found exclusively in the membrane fraction (compare lanes 3 and 4 to lanes 5 and 6).

Altered modifications of collagen polypeptides in LARP6 knock down cells

Prior to folding into the triple helix collagen polypeptides must be modified by hydroxylations of selected prolines and lysines and glycosylation of hydroxy-lysines [1,2]. Hypermodifed collagen polypeptides appear when the rate of translation is not coupled to the rate of folding [54,55,56,57]. Therefore, we analyzed the modifications of collagen polypeptides present in soluble and membrane fractions after knocking down LARP6 using 2D SDS-PAGE and western blotting. The available antibody recognized only α2(I) polypeptide in 2D SDS-PAGE gels, so we could only show the results for this polypeptide. In cytosolic fraction of control cells collagen α2(I) polypeptide was found in small amounts and was presented as relatively homogeneous molecules, isoelectrically focused around pH 9.2. The knock down of LARP6 did not change their abundance or pI (lower panel) (Fig 3B, upper panels). In the membrane fraction of control cells α2(I) polypeptide was resolved as molecular species having isoelectric point from 8.8–9.1 with majority focusing at pH 9.1. In LARP6 knock down cells, an increased fraction of molecules focused between pH 7.5 and 8.8 (Fig 3B, lower panels, arrows), suggesting additional modifications that changed the pI. Although, we could not discern the nature of modifications causing the pI shift, it is known that hydroxylations of amino-acids shift the pI to a more acidic region [58,59]. Therefore, we assumed that the additional modifications represent excessive hydroxylations of prolines and lysines. This indicated that knock down of LARP6, not only decreases the total collagen synthesis, but also results in hypermodifications of individual polypeptides. Thus, direct targeting of collagen α1(I) mRNA to the ER membrane, and to a lesser extent of α2(I) mRNA, by LARP6 is critical for efficient synthesis of type I collagen, including proper posttranslational modifications.

Integrity of nonmuscle myosin filaments as requirement for collagen synthesis

Our previous work suggested that the integrity of nonmuscle myosin filaments is necessary for synthesis of type I collagen [15,33], while experiments in mice showed that cardiac fibrosis is found only in the regions of the heart re-expressing nonmuscle myosin [60]. Here we wanted to investigate if nonmuscle myosin filaments participate in the membrane targeting of collagen mRNAs. To this end we depolymerized the filaments using ML-7, a specific inhibitor of myosin light chain kinase (MLCK) [61,62]. Additionally, we inactivated the motor function of myosin by blebbistatin, an inhibitor of the myosin ATPase activity [63], to compare the requirement for the integrity of the filaments versus their motor function. Neither of the inhibitors changed the level of collagen mRNAs (Fig 4B and D), suggesting that transcription or stability of collagen mRNAs was not affected. Treatment with

ML-7 decreased the amount of α1(I) and α2(I) polypeptides secreted into the cellular medium, while the effect on the cellular level was minimal (Fig 4A), suggesting that a defect may be in the folding of collagen triple helix or its secretion into the medium. Treatment with blebbistatin had a weaker effect; the amount of collagen α1(I) and α2(I) polypeptides secreted into the cellular medium was affected less by this drug than by ML-7 (Fig 4C).

To distinguish if ML-7 treatment affects folding of collagen triple helix or its secretion into the medium we first analyzed distribution of collagen polypeptides throughout the secretory pathway. We separated ER from Golgi and early endosomes by centrifugation through iodixanol gradient [41] and analyzed the presence of collagen polypeptides in the fractions by western blot. We exposed the western blots to achieve similar signal intensity of control and ML-7 treated cells to directly compare the relative distribution of collagen polypeptides in the fractions. By analyzing calnexin (ER marker) [64] and Golgin84 (Golgi marker) [65], we assigned the fractions representing these compartments and assigned fractions containing early endosomes according to [41]. As shown in Fig 5A, the relative distribution of collagen α1(I) and α2(I) polypeptides throughout the secretory pathway was similar in cells treated with ML-7 and in control cells. The major difference was fraction 4, where control cells had significantly more collagen than ML-7 treated cells. However, fraction 4 also contained more fibronectin in ML-7 treated cells, which was normally secreted. Therefore, we concluded that the alteration of fraction 4 is not a likely cause of poor secretion of type I collagen. The relatively unperturbed distribution of collagen polypeptides throughout the secretory pathway suggested that collagen polypeptides that have entered into the pathway are transported normally between the compartments. However, we noticed that both collagen polypeptides were present in increased amount in the cytosol (Fig 5A, CYT fraction) in the ML-7 treated cells. This suggested that some collagen polypeptides accumulate in the cytosol when nonmuscle myosin filaments were disrupted.

To further investigate this phenomenon we fractionated cells in soluble and membrane fractions and analyzed the level of collagen polypeptides by western blot. In ML-7 treated HLFs, the level of both collagen polypeptides decreased by ~50% in the membrane fraction and increased by ~50% in the soluble fraction, compared to control cells (Fig 5B). ML-7 treatment did not change the distribution of calnexin, actin or GAPDH proteins. This verified that depolimerization of nonmuscle myosin filaments results in specific redistribution of type I collagen from the membrane compartment into the cytosol. Blebbistatin treatment did not increase the cytosolic retention of collagen polypeptides (Fig 5C), what is consistent with a minimal effect on their secretion (Fig 4C). Thus, integrity of the nonmuscle myosin filaments, rather than their motor function, appears to be critical for proper secretion of type I collagen.

To verify that a similar requirement holds for other collagen producing cells we analyzed human scleroderma skin fibroblasts (Fig 5D). About 50% of collagen polypeptides was shifted into the soluble fraction when scleroderma fibroblasts were treated with ML-7, suggesting that myosin dependent collagen production is a general characteristic of collagen producing cells.

To exclude the nonspecific effects of ML-7 we disrupted nonmuscle myosin by overexpressing dominant negative isoform of myosin light chain kinase (DN-MLCK) [33]. Myosin light chain kinase (MLCK) phosphorylates regulatory light chains of nonmuscle myosin to promote assembly of the filaments [66]. A dominant negative isoform of MLCK (DN-MLCK) was developed [34], which inhibits MLCK and which has been used in previous studies to disrupt nonmuscle myosin filaments [34,67,68].

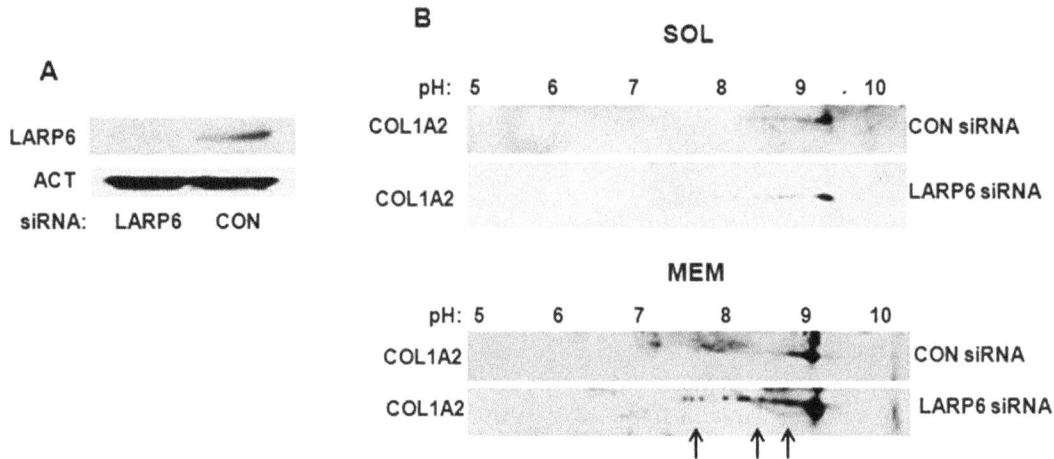

Figure 3. Collagen modifications in LARP6 knock down cells. A. Knock down of LARP6. Expression of LARP6 in HLFs transfected with LARP6 siRNA (lane 1) and control siRNA (lane 2). Loading control: actin (ACT). B. Modifications of collagen α2(I) polypeptides. HLFs transfected with control siRNA (CON) or LARP6 siRNA were fractionated into cytosolic (SOL, upper panels) and membrane (MEM, lower panels), the fractions were resolved by 2D SDS-PAGE gels and collagen α2(I) polypeptide (COL1A2) visualized by western blotting. Numbers indicate the pH range of the isoelectric focusing strips and excessive collagen modifications are indicated by arrows.

Overexpression of DN-MLCK in HLFs decreased the secretion and intarcellular level of collagen α2(I) polypeptide, while the secretion of collagen α1(I) polypeptide was unaltered (Fig 5E, lanes 1 and 3). A strong effect of DN-MLCK on α2(I) polypeptide and a minimal effect on α1(I) polypeptide was reported before for HLFs and was attributed to the ability of HLFs to excrete homotrimers of α1(I) polypeptides in myosin independent manner [33]. When the intracellular distribution of collagen polypeptides between soluble and membrane fraction was analyzed (Fig 5F), DN-MLCK caused predominantly cytosolic accumulation of α2(I) polypeptide and slightly increased cytosolic retention of α1(I) polypeptide (compare lanes 2 and 4). Overexpression of DN-MLCK in scleroderma fibroblasts resulted in decrease in expression of both collagen polypeptides (Fig 5G). These results corroborated the findings using ML-7 and supported the notion

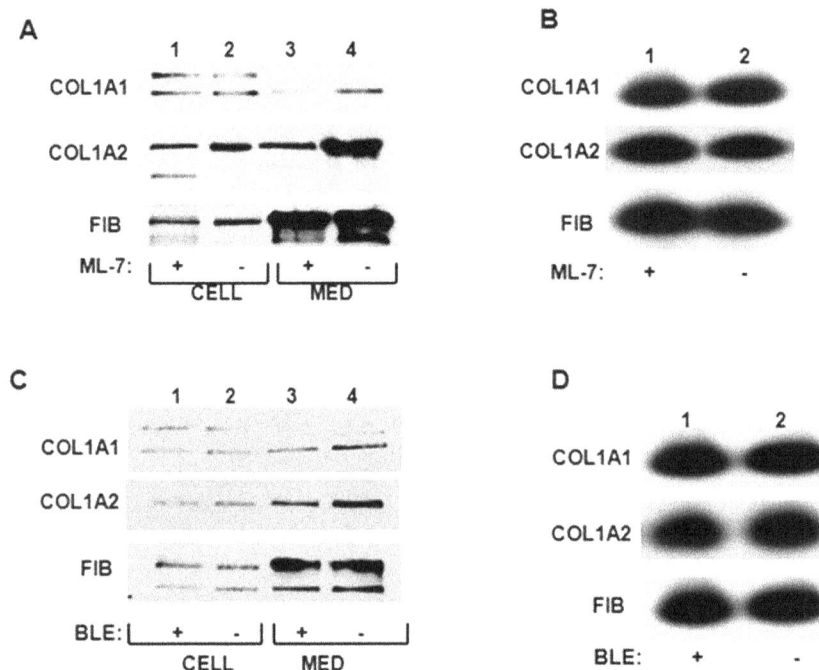

Figure 4. Disruption of nonmuscle myosin filaments decreases type I collagen production. A. Effect of ML-7. HLFs were treated with ML-7 and the level of collagen polypeptides intracellularly (lanes 1 and 2) or secreted into the culture medium (lanes 3 and 4) was analyzed by western blot. Loading control: fibronectin (FIB). B. ML-7 does not affect expression of collagen mRNAs. Total RNA was extracted from cells in A and analyzed for expression of collagen and fibronectin mRNAs by RT-PCR. C. Effect of blebbistatin. Experiment as in A, except HLFs were treated with blebbistatin. D. Expression of collagen mRNAs in cells treated with blebbistatin. Experiment as in C, except the cells were treated with blebbistatin.

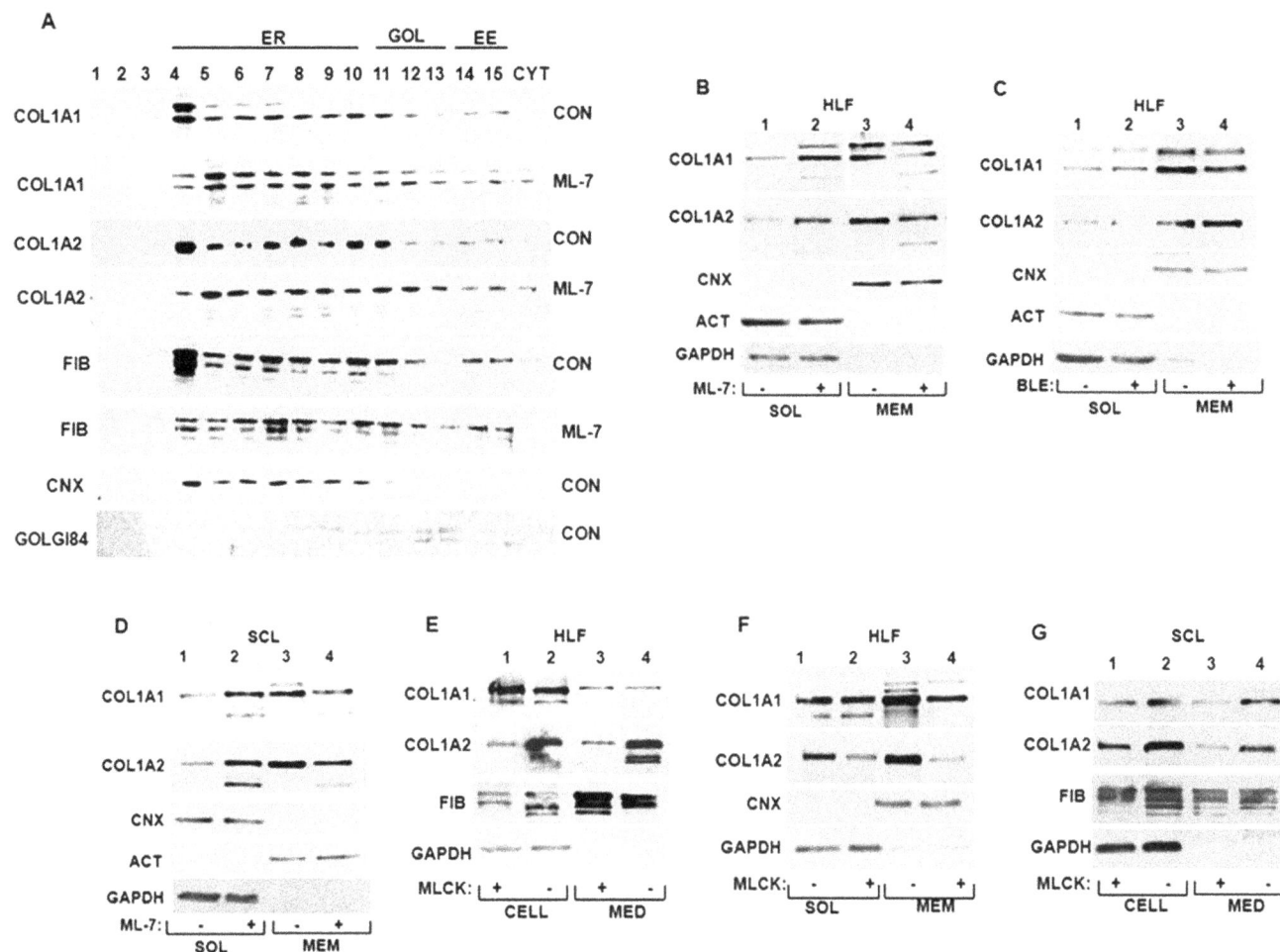

Figure 5. Disruption nonmuscle myosin results in cytosolic accumulation of collagen polypeptides. A. Distribution of collagen polypeptides in the secretory pathway compartments. Extracts of control and ML-7 treated HLFs were fractionated by density gradient centrifugation into fractions containing endoplasmic reticulum (ER), Golgi (GOL), early endosomes (EE) or cytosol (CYT) and the fractions were analyzed by western blot for presence of collagen α1(I) (COL1A1), collagen α2(I) (COL1A2) and fibronectin (FIB) polypeptides. Calnexin (CNX) was analyzed as a marker of the ER and golgin 84 (GOLGI84) as a marker of the Golgi complex. B. Cytosolic accumulation of collagen polypeptides after disruption of nonmuscle myosin in HLF. HLF were treated with ML-7 and fractionated into cytosolic (SOL, lanes 1 and 2) and membrane (MEM, lanes 3 and 4) fractions. The fractions were analyzed for collagen α1(I) (COL1A1), collagen α2(I) (COL1A2) polypeptides by western blot. Calnexin (CNX) was analyzed as a marker of membrane fraction and actin (ACT) and GAPDH as a markers of cytosolic fraction. C. Inhibition of the motor function of nonmuscle myosin has little effect on cytosolic accumulation of collagen polypeptides. Experiment as in B, except HLFs were treated with blebbistatin. D. Cytosolic accumulation of collagen polypeptides after disruption of nonmuscle myosin in scleroderma fibroblasts. Experiment as in B, except human scleroderma fibroblasts were used. E. Inhibition of collagen expression by DN-MLCK. DN-MLCK was overexpressed in HLFs and collagen α1(I) (COL1A1) and α2(I) (COL1A2) polypeptides were analyzed in cell extract (lanes 1 and 2) and in cellular medium (lanes 3 and 4). Loading controls; fibronectin (FIB) and GAPDH. F. Cytosolic accumulation of collagen polypeptides after disruption of nonmuscle myosin by DN-MLCK. DN-MLCK was overexpressed in HLF and cells were fractionated into cytosolic (SOL, lanes 1 and 2) and membrane (MEM, lanes 3 and 4) fractions. The fractions were analyzed for collagen α1(I) (COL1A1) and collagen α2(I) (COL1A2) polypeptides by western blot. Loading controls; calnexin (CNX) and GAPDH. G. Inhibition of collagen expression by DN-MLCK in scleroderma fibroblasts. DN-MLCK was overexpressed in scleroderma fibroblasts and collagen α1(I) (COL1A1) and α2(I) (COL1A2) polypeptides were analyzed in cell extract (lanes 1 and 2) and in cellular medium (lanes 3 and 4). Loading controls; fibronectin (FIB) and GAPDH.

that integrity of nonmuscle myosin filaments is a prerequisite for effective type I collagen synthesis.

Collagen polypeptides retained in the cytosol in absence of nonmuscle myosin filaments are hypermodified

Unfolded or grossly aberrant collagen polypeptides are retro-translocated from the ER into the cytosol for degradation [57,69,70]. Appearance of increased amount of collagen polypeptides in the cytosol upon inactivation of nonmuscle myosin suggested that these polypeptides may have been retro-translocated from the secretory pathway and that their abundance may have

saturated the degradation machinery. Therefore, we analyzed the posttranslational modifications of collagen α2(I) polypeptide found in the cytosol of ML-7 treated cells by 2D SDS-PAGE and western blotting (Fig 6A). Upper panels show cytosolic fractions and lower panels show membrane fractions of control cells and of cells treated with ML-7. For better qualitative comparison, the western blot of cytosolic fraction of control cells was exposed much longer to achieve a similar signal to that of ML-7 treated cells. α2(I) polypeptide found in the cytosol of ML-7 treated cells appeared massively hyper-modified, as >50% of molecules were shifted into the more acidic region between pH of 6.5 and 8 (arrows in

Fig 6A). α2(I) polypeptide in control cells had the pI around 9, as observed before (Fig 3). As hydroxylations typically change the pI towards more acidic region [58], and an acidic shift of hyper-hydroxylated collagen polypeptides has been reported in a patient with osteogenesis imperfecta [59], we again assumed that excessive hydroxylations of prolines and lysines represented the over-modifications. The isoelectric focusing of α2(I) polypeptide in the membrane fraction of ML-7 treated cells showed similar pattern to that in control cells (Fig 6A, lower panels). The most molecules in both cell types had a pI from 8 to 9, suggesting that almost all of hypermodified collagen polypeptides in the ML7 treated cells were translocated into the cytosol. We concluded that grossly abnormal collagen polypeptides are synthesized when nonmuscle myosin filaments are disrupted and that the hypermodified polypeptides are eliminated from the secretory pathway. This is in contrast with LARP6 knock down cells, where hypermodified collagen poly-peptides were primarily found in the membrane fraction. The analysis of scleroderma fibroblasts verified that ML-7 has a similar effect in cells of different origin (Fig 6C).

Treatment with blebbistatin also resulted in cytosolic accumu-lation of hypermodified α2(I) polypeptides, but the extent of modifications and the fraction of molecules affected was much smaller (Fig 6B), again suggesting that the motor function of

nonmuscle myosin is not as critical for collagen synthesis as the integrity of the filaments.

When nonmuscle myosin filaments were disrupted using DN-MLCK [34], hypermodified collagen α2(I) polypeptides also accumulated in the cytosol (Fig 6D), resembling the effects of ML-7. Thus, this result further confirmed that hypermodifications of collagen polypeptides are specifically due to disruption of nonmuscle myosin filaments and not to unanticipated effects of ML-7.

Although we could not analyze collagen α1(I) polypeptide due to lack of its recognition in 2D gels by the antibody available, we concluded that inactivation of nonmuscle myosin results in massive hyper-modifications of collagen polypeptides, retention of the hyper-modified molecules in the cytosol and diminished secretion of type I collagen and that these events were not associated with a general disruption of the secretory pathway.

Role of nonmuscle myosin in membrane partitioning of collagen mRNAs

We reasoned that hyper-modifications of α2(I) polypeptide caused by ML-7 treatment may also be due to perturbed partitioning of collagen mRNAs to the ER membrane. To test this we first analyzed if nonmuscle myosin is associated with the

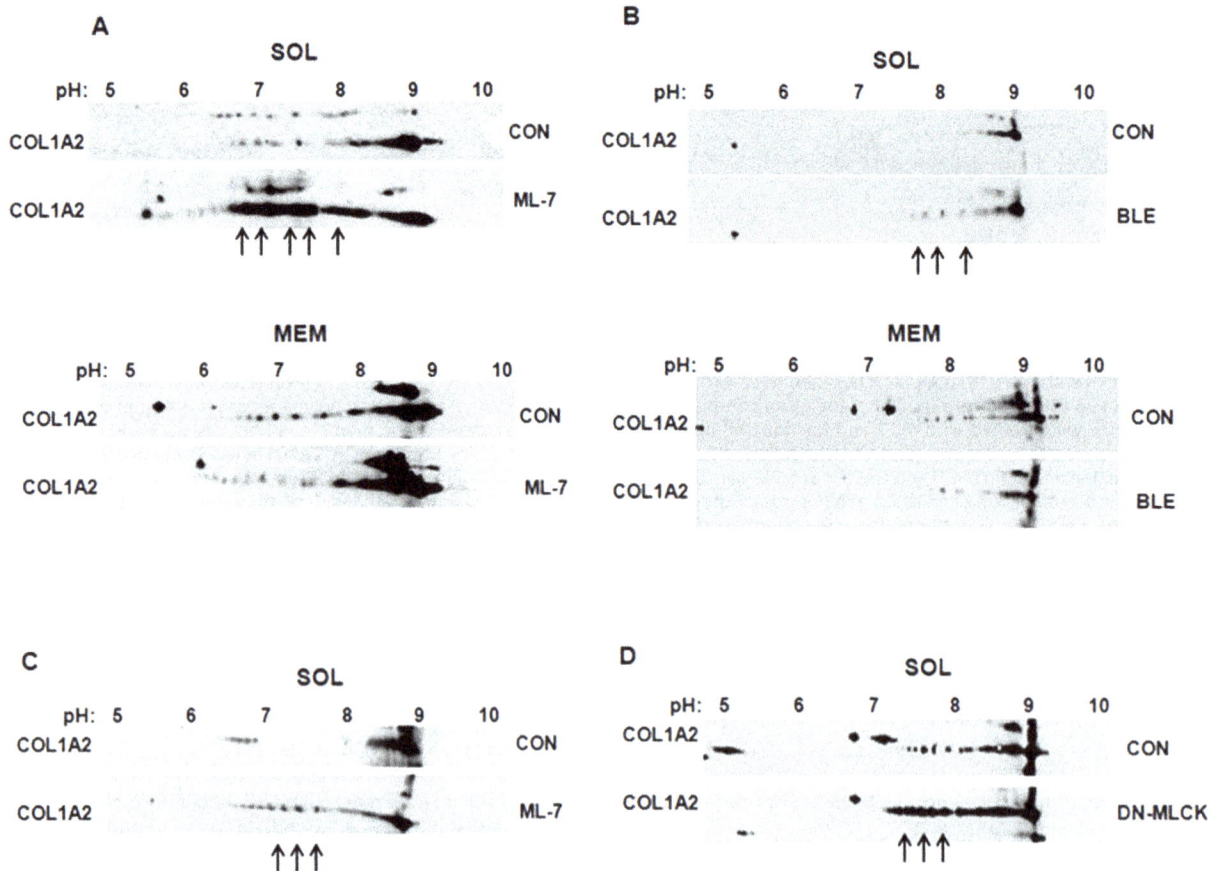

Figure 6. Hypermodifications of collagen α2(I) polypeptides accumulated in the cytosol after disruption of nonmuscle myosin. A. Modifications of collagen α2(I) polypeptides accumulated in the cytosol after ML-7 treatment. Control HLFs (CON) and HLF treated with ML-7 were fractionated into cytosolic (SOL) and membrane (MEM) fractions and the fractions were analyzed by 2D SDS-PAGE, followed by western blotting. The pH range of isoelectric focusing strips is shown on the top and excessive modifications of collagen α2(I) polypeptide are indicated by arrows. B. Same experiment as in A, except the cells were treated with blebbistatin. C. Experiment as in A, except scleroderma skin fibroblasts were used and only cytosolic (SOL) fraction is shown. D. DN-MLCK was overexpressed in HLFs and collagen α2(I) polypeptide was analyzed in the cytosolic fraction (SOL) of control and DN-MLCK expressing cells by 2D SDS-PAGE and western blotting. Arrows indicate the hypermodifications.

internal membranes. Using sequential detergent extraction we found that ~60% of nonmuscle myosin was extractable in the membrane fraction, while ~40% was found in the cytosol (Fig 7A, lanes 1 and 4). To assess if the membrane bound nonmuscle myosin supports polysomes formation, we further separated cytosolic and membrane fractions into polysomes and postpolysomal supernatant. In the cytosolic fraction nonmuscle myosin did not copurify with polysomes; all of it was found in the post polysomal supernatant (Fig 7A, lanes 2 and 3). A similar distribution was found in the membrane fraction, where all nonmuscle myosin was present in the post polysomal supernatant (Fig 7A, lanes 5 and 6). When cells were treated with ML-7, most of the nonmuscle myosin in the membrane fraction disappeared (Fig 7B, compare lanes 1 and 4), suggesting that depolymerization of the filaments causes detachment of the myosin from the membranes. Treatment with blebbistatin did not disrupt the association of nonmuscle myosin with the membranes (Fig 7C, lanes 4 and 5). About 50% of nonmuscle myosin was found associated with membranes in control and blebbistatin treated cells.

To exclude the possibility that depolymerization of nonmuscle myosin by ML-7 may have caused its sequestration into insoluble material, rather than its release from the membrane, we fractionated the cells as before. However, we also extracted the cell remnants with 1% SDS to solubilize the proteins that were insoluble in NP-40 used for the membrane extraction and analyzed all the fractions by western blot. Fig 7D shows that in untreated cells a significant fraction of nonmuscle myosin was found in the NP-40 insoluble fraction (lane 3). Importantly, ML-7 treatment did not increase the amount of insoluble myosin (lane 6), but decreased the amount associated with the membrane (lane 5). Pat-A treatment did not significantly alter the fractionation of nonmuscle myosin (lanes 7–9).

To assess if ML-7 causes retention of ribosomes in the insoluble fraction, we extracted total RNA from all fractions and analyzed it by agarose gel electrophoresis (Fig 7D, bottom panel). In control cells most ribosomal RNA was found in the membrane fraction (lane 2), but a significant portion was also found in the NP-40 insoluble fraction (lane 3). ML-7 did not change such distribution of ribosomal RNA. In contrast, in Pat-A treated cells most ribosomal RNA was found in the cytosol and it disappeared from the NP40 insoluble fraction. This suggested that ML-7 treatment did not cause significant sequestration of nonmuscle myosin and ribosomes into the insoluble material, but that inhibition of translation initiation releases polysomes from the membranes and from the NP-40 insoluble proteins.

Having established that nonmuscle myosin is bound to the internal membranes, but does not support polysomes, we assessed if it plays a role in partitioning of collagen mRNAs. When the cells were treated with ML-7 ~40% of collagen α1(I) mRNA and α2(I) mRNA were released from the membrane and were found free in the cytosol (Fig 8B), compared to control cells in which all collagen mRNAs were found associated with the membrane (Fig 8A). Partitioning of FIB, MMP12 and actin mRNAs was minimally altered, suggesting that mRNAs encoding other secreted proteins do not require intact nonmuscle myosin for their association with the ER membrane. Figure 8D summarizes the quantitative RT-PCR determinations of collagen mRNAs distributed between soluble and membrane fractions of cells treated with Pat-A, ML-7 or transfected with LARP6 siRNA. Since actin mRNA was used for normalization of the real time RT-PCR results, we verified that distribution of actin mRNA was unaltered upon ML-7, Pat-A and LARP6 siRNA treatment (Fig 8E). Most of the ribosomal RNA also remained bound to the membrane after ML-7 treatment,

indicating no major rearrangement of the translation machinery at the ER membrane, as predicted by the lack of association of nonmuscle myosin with polysomes. Blebbistatin did not have any effect of partitioning of collagen mRNAs (Fig 8C). This is consistent with a minimal effect of this drug on modifications of collagen polypeptides (Fig 6B), their secretion or cytosolic retention (Fig 5C).

From these experiments we concluded that the integrity of nonmuscle myosin filaments is required for direct membrane targeting of collagen mRNAs. In the absence of myosin dependent collagen mRNA partitioning the cells poorly secrete type I collagen and accumulate large amount of hyper-modified collagen polypeptides in the cytoplasm.

Discussion

Understanding excessive synthesis of type I collagen in fibrosis is a key to find a cure for this intractable condition. Random translation of collagen mRNAs and their targeting to the ER membrane by the signal recognition particle (SRP) can not account for the most features of type I collagen synthesis. In reparative and reactive fibrosis collagen producing cells upregulate collagen biosynthesis several hundred fold [11,12,13,71,72,73]. Heterotrimeric type I collagen is synthesized, without formation of the homotrimers of α1(I) polypeptides. Stable homotrimers of α1(I) polypeptides readily form when α2(I) polypeptide is not available [74,75,76,77], but their complete absence suggests tightly regulated shunting of α2(I) polypeptide into the biosynthetic pathway. If folding of collagen polypeptides into the heterotrimer is delayed, the polypeptides are hyper-modified and subjected to intracellular degradation [54,55,56]. This suggests that there is a kinetic equilibrium between translation, posttranslational modifications and folding and excludes random processes. Folding of collagen heterotrimer is highly concentration dependent, as mutual recognition of three molecules initiates the event. When collagen folding was mimicked in vitro using model peptides, the triple helix was formed at mM concentrations of the peptides [78,79]. It is not possible to achieve such high concentrations of collagen polypeptides within the lumen of the ER, unless the synthesis of α1(I) and α2(I) polypeptides is concentrated at discrete sub-regions. Molecular chaperones facilitate protein folding in vivo, but the only collagen specific chaperone is HSP48. HSP48 acts on all collagens and it has been demonstrated that HSP48 prevents lateral aggregation of the collagen trimers and does not facilitate their formation [70,80,81].

We postulate here the existence of a mechanism that facilitates type I collagen synthesis by directly targeting collagen mRNAs to the membrane of the ER. We show that: 1. mRNAs encoding type I collagen are associated with the ER membrane in the absence of translation. 2. LARP6, which specifically binds 5′SL of collagen mRNAs, is involved in this process. 3. Intact nonmuscle myosin filaments, rather than their motor function, are needed for association of collagen mRNAs with the ER membrane. 4. When partitioning of collagen mRNAs to the ER membrane is disrupted, collagen polypeptides become hyper-modified and are poorly secreted. Thus, to efficiently synthesize, modify and fold type I collagen it is critical to directly partition collagen mRNAs to the ER membrane.

Partitioning of mRNA to the ER membrane is believed to be mediated by RNA binding proteins [44,46,48]. Rrbp1 (ribosome binding protein 1, p180) is ribosomal receptor on the ER membrane [82]. Depletion of Rrbp1 results in decreased association of ribosomes with the ER membrane and poor secretion of type I collagen [83]. This suggests that Rrbp1

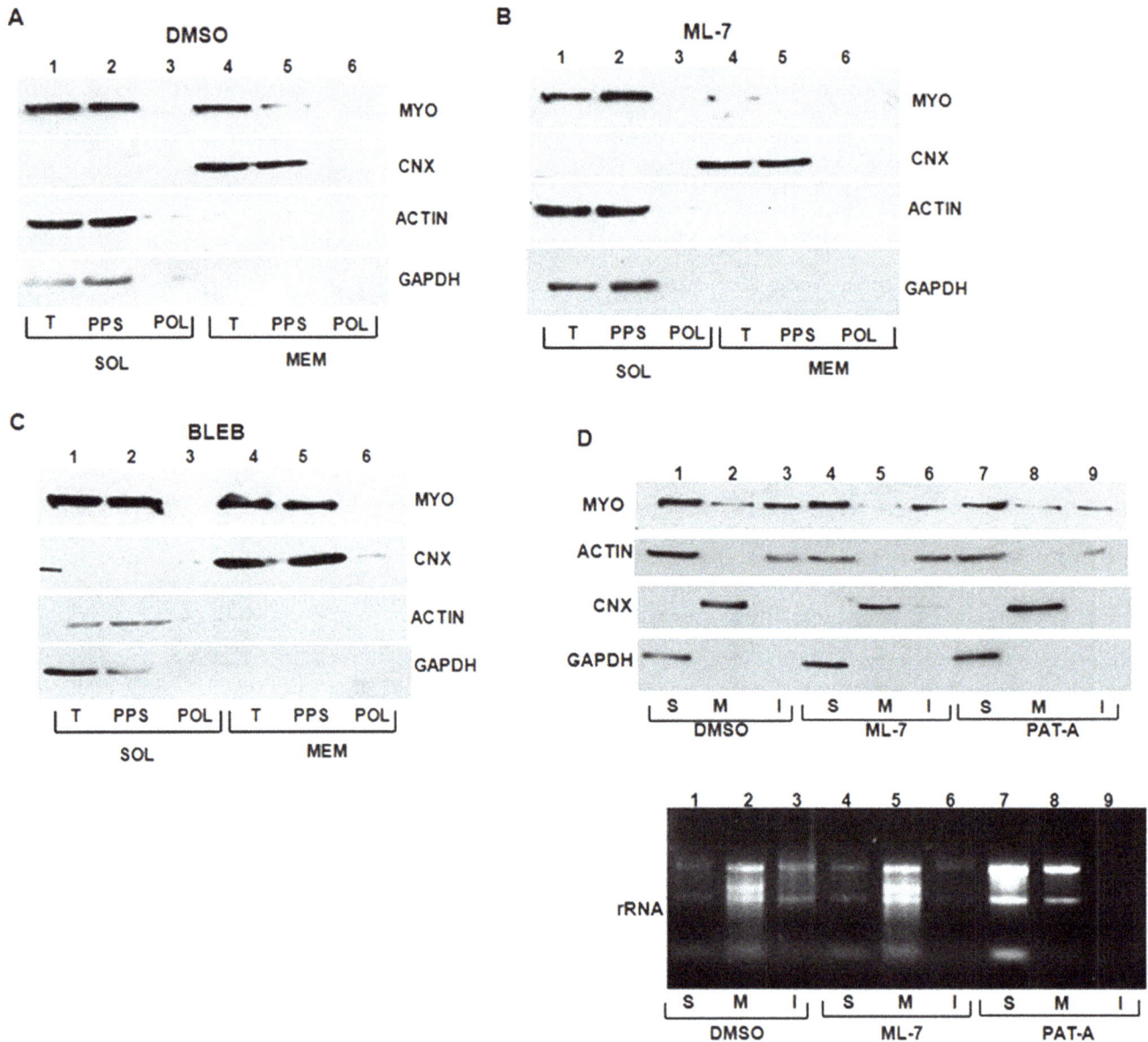

Figure 7. Association of nonmuscle myosin with internal membranes. A. Distribution of nonmuscle myosin in cytosolic and membrane fractions of HLFs. HLFs were fractionated into cytosolic (SOL) and membrane (MEM) fractions and the total fractions (T) were analyzed in lanes 1 and 4. Polysomes (POL, lanes 3 and 6) and post-polysomal supernatant (PPS, lanes 2 and 5) were furthered separated from the total fractions and the samples were analyzed for myosin IIb (MYO) by western blot. Calnexin (CNX) was analyzed as a marker of membrane fraction and actin (ACT) and GAPDH as a markers of cytosolic fraction. B. Disappearance of nonmuscle myosin from the membrane fraction after ML-7 treatment. Experiment as in A, except HLFs were treated with ML-7. C. Effect of blebbistatin on membrane association of myosin IIb. Experiment as in A, except HLFs were treated with blebbistatin. D. ML-7 treatment does not increase insoluble nonmuscle myosin. Upper panel: distribution of nonmuscle myosin (MYO), actin (ACT), calnexin (CNX) and GAPDH proteins in soluble (S), membrane (M) and NP-40 insoluble (I) fractions of control cells (lanes 1–3), ML-7 treated cells (lanes 4–6) and Pat-A treated cells (lanes 7–9). Bottom panel: distribution of ribosomal RNA.

facilitates translation at the ER membrane. It has been reported that Rrbp1 promotes translation independent localization of subset of mRNAs to the ER membrane by binding the RNA through its lysine rich domain, although Rrbp1 does not bind RNA in a sequence specific manner [48]. Ribosome independent maintanence of placental alkaline phosphatase mRNA and calreticulin mRNAs was mediated by Rrbp1 [84]. It is possible that Rrbp1 also participates in partitioning of collagen mRNAs, together with LARP6 and nonmuscle myosin, and this remains to be explored. Knock down of LARP6 resulted in dissociation of collagen α1(I) mRNA from the membrane, but the effect on α2(I)

mRNA was negligible (Figs 2 and 8D). However, in our experiments the knock down was not complete and our preliminary results suggests that binding of LARP6 to the 5′SL of collagen α2(I) mRNA is more stable than binding to the α1(I) 5′ SL. Thus, the residual LARP6 may have been sufficient to support membrane partitioning of α2(I) mRNA, while that of α1(I) mRNA was partially compromised. Making mutations in the 5′SL of collagen α1(I) and α2(I) mRNAs or creating total LARP6 knock cells is needed to fully understand the LARP6 independent partitioning. Nevertheless, this suggested that translation independent partitioning of collagen mRNAs to the ER membrane is

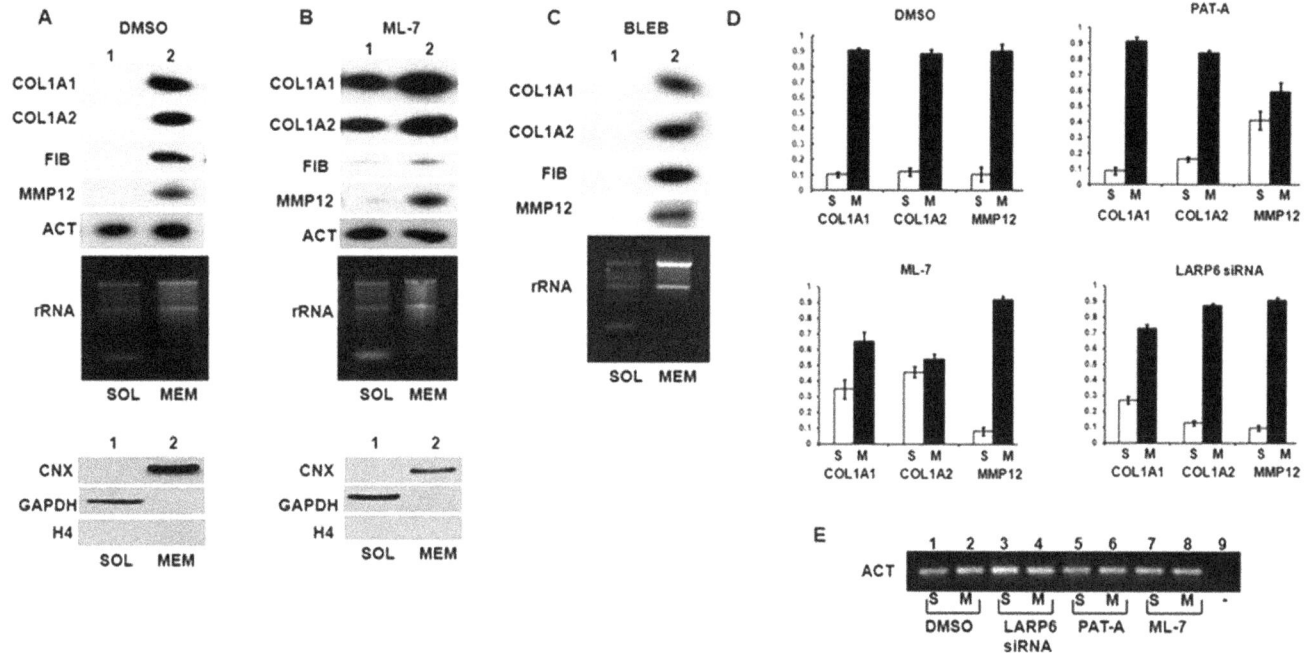

Figure 8. Membrane partitioning of collagen mRNAs after disruption of nonmuscle myosin. A. Partitioning of collagen mRNAs in control cells. Upper panel: HLFs were fractionated in cytosolic fraction (SOL, lane 1) or membrane fraction (MEM, lane 2) and collagen α1(I) (COL1A1), collagen α2(I) (COL1A2), fibronectin (FIB), matrix metalloproteinase 12 (MMP12) and actin (ACT) mRNAs were analyzed by RT-PCR. Distribution of ribosomal RNA (rRNA) is shown by agarose gel electrophoresis and ethidium bromide staining. Bottom panel: distribution of calnexin (CNX), GAPDH and histone H4 proteins in the fractions, analyzed by western blot. B. Partitioning of collagen mRNAs in ML-7 treated cells. Experiment as in A, except HLFs were treated with ML-7. C. Partitioning of collagen mRNAs in blebbistatin treated cells. Experiment as in A, except HLFs were treated with blebbistatin. D. Real time RT-PCR determination of collagen mRNAs. The treatment of HLFs is indicated above the panels. The cells were fractionated into cytosolic (S) and membrane (M) fractions, RNA was extracted and analyzed by real time RT-PCR for presence of collagen α1(I) (COL1A1), collagen α2(I) (COL1A2) and matrix metalloproteinase 12 (MMP12) mRNAs. The signals were normalized to actin mRNA, which was found in similar amounts in S and M fractions. The normalized expression of each mRNA in S and M fractions was summed up and arbitrarily set as 1. Bars represent the relative proportion of mRNAs in the fractions. Error bars: +- 1 SEM. E. Distribution of internal control actin mRNA. cDNA used in real time RT-PCR reactions in D was analyzed by PCR with primers specific for actin mRNA (ACT). Lanes 1 and 2, control cells, lanes 3 and 4, cells transfected with LARP6 siRNA, lanes 5 and 6, cells treated with Pat-A, lanes 7 and 8, cells treated with ML-7. Lane 9, control reaction without cDNA.

partially mediated by LARP6. LARP6 binds 5′ SL of collagen mRNAs with high affinity and the binding overlaps with the start codon. We hypothesize that LARP6 binding delays initiation of translation until collagen mRNAs are targeted to the ER. Overexpression of LARP6 inhibits translation of collagen mRNAs [14], indicating that LARP6 can act as a translational inhibitor. It has been reported that synthesis of type I collagen is not uniform throughout the ER, but concentrated at discrete loci [14,33]. Direct targeting to such specialized loci would increase the local concentration of α1(I) and α2(I) mRNAs, coordinate the synthesis of nascent collagen polypeptides, couple posttranslational modifications with folding and facilitate incorporation of the α2(I) polypeptide into the heterotrimer. The proteins which define the loci of collagen biosynthesis on the ER membrane are not completely characterized. Translocation associated membrane protein 2 (TRAM2) has been found associated with translocons engaged in synthesis of collagen polypeptides [85]. TRAM2 recruits the Ca++ pump of the ER, Serca2b, to these translocons and it has been postulated that the increased Ca++ concentrations stimulate molecular chaperones to fold collagen heterotrimers [85]. If there are other proteins besides TRAM2 that define the putative collagen specific translocons, remains to be discovered.

In cardiac fibrosis, the areas of fibrosis were found only in the regions of the heart re-expressing nonmuscle myosin [60] and we have reported nonmuscle myosin dependent collagen synthesis in fibroblasts [33]. The results shown here indicate that intact

filaments composed of nonmuscle myosin are necessary for the process of direct partitioning of collagen mRNAs. A large fraction of nonmuscle myosin can be extracted from the intracellular membranes (Fig 7). It has been reported that nonmuscle myosin associates with Golgi membranes [86] and that it also propels terminal vesicles loaded with pre-assembled collagen fibrils out of the cell [4]. Here we show that the secretory pathway through ER and Golgi is not affected by depolimerization of nonmuscle myosin (Fig 5). Instead, partitioning of collagen mRNAs to the ER membrane is perturbed, suggesting the role of nonmuscle myosin in this earliest step of the biosynthetic pathway (Fig 8). We have previously reported that LARP6 interacts with nonmuscle myosin [33], so it is possible that nonmuscle myosin filaments provide structural support for the components of the mechanism involved in partitioning of collagen mRNAs. Blebbistatin, an inhibitor of the motor function of myosin, has much smaller effect then ML-7, suggesting that the filament integrity, rather than their locomotor activity is essential.

The effect of disruption of nonmuscle myosin on collagen synthesis appears to be greater than knock down of LARP6. ML-7 treatment caused massive accumulation of hypermodified collagen polypeptides in the cytosol (Fig 6), a clear indication of dramatically perturbed collagen synthesis [56]. A similar result was obtained when nonmuscle myosin was disrupted by overexpression of DN-MLCK, excluding the nonspecific effects of ML-7 (Fig 5). When LARP6 was knocked down accumulation of hypermodified collagen was observed in the membrane compartment (Fig 3). The reason for this discrepancy is not clear and, although the residual LARP6 activity in the LARP6 knock down cells may have blunted the effect, it is likely that nonmuscle myosin may have additional roles in collagen biosynthesis.

Nonmuscle myosin has been profoundly studied in relation to cell motility and contractility [21,26,87,88]. In fibrosis, quiescent fibroblasts or other collagen producing cells become activated by profibrotic cytokines, such as TGFβ [89,90]. As part of the activation process the cells upregulate nonmuscle myosin expression and polymerize the filaments [30]; this allows their migration to the site of profibrotic injury. Our work implicates that formation of nonmuscle myosin filaments also allows the cells to partition collagen mRNAs to the ER membrane, as a prerequisite for productive synthesis of type I collagen. Thus, the ability of cells to migrate and the ability to produce type I collagen are two inseparable processes. Further characterization of their coupling will help understand the physiology of wound healing and pathogenesis of fibrosis.

Author Contributions

Conceived and designed the experiments: HW BS. Performed the experiments: HW BS. Analyzed the data: HW BS. Contributed reagents/materials/analysis tools: HW BS. Wrote the paper: HW BS.

References

1. Myllyharju J, Kivirikko KI (2004) Collagens, modifying enzymes and their mutations in humans, flies and worms. Trends Genet 20: 33–43.
2. Kivirikko KI (1998) Collagen biosynthesis: a mini-review cluster. Matrix Biol 16: 355–356.
3. Kadler KE, Hill A, Canty-Laird EG (2008) Collagen fibrillogenesis: fibronectin, integrins, and minor collagens as organizers and nucleators. Curr Opin Cell Biol 20: 495–501.
4. Kalson NS, Starborg T, Lu Y, Mironov A, Humphries SM, et al. (2013) Nonmuscle myosin II powered transport of newly formed collagen fibrils at the plasma membrane. Proc Natl Acad Sci U S A 110: E4743–4752.
5. el-Harake WA, Furman MA, Cook B, Nair KS, Kukowski J, et al. (1998) Measurement of dermal collagen synthesis rate in vivo in humans. Am J Physiol 274: E586–591.
6. Thalacker-Mercer AE, Campbell WW (2008) Dietary protein intake affects albumin fractional synthesis rate in younger and older adults equally. Nutr Rev 66: 91–95.
7. Bregendahl K, Yang X, Liu L, Yen JT, Rideout TC, et al. (2008) Fractional protein synthesis rates are similar when measured by intraperitoneal or intravenous flooding doses of L-[ring-2H5]phenylalanine in combination with a rapid regimen of sampling in piglets. J Nutr 138: 1976–1981.
8. Villa P, Arioli P, Guaitani A (1992) Protein turnover, synthesis and secretion of albumin in hepatocytes isolated from rats bearing Walker 256 carcinoma. In Vitro Cell Dev Biol 28A: 157–160.
9. Haukipuro K, Melkko J, Risteli L, Kairaluoma M, Risteli J (1991) Synthesis of type I collagen in healing wounds in humans. Ann Surg 213: 75–80.
10. Stefanovic B, Hellerbrand C, Holcik M, Briendl M, Aliebhaber S, et al. (1997) Posttranscriptional regulation of collagen alpha1(I) mRNA in hepatic stellate cells. Mol Cell Biol 17: 5201–5209.
11. Zeisberg M, Neilson EG (2010) Mechanisms of tubulointerstitial fibrosis. J Am Soc Nephrol 21: 1819–1834.
12. Pinzani M, Macias-Barragan J (2010) Update on the pathophysiology of liver fibrosis. Expert Rev Gastroenterol Hepatol 4: 459–472.
13. Wilson MS, Wynn TA (2009) Pulmonary fibrosis: pathogenesis, etiology and regulation. Mucosal Immunol 2: 103–121.
14. Cai L, Fritz D, Stefanovic L, Stefanovic B (2010) Binding of LARP6 to the conserved 5' stem-loop regulates translation of mRNAs encoding type I collagen. J Mol Biol 395: 309–326.
15. Cai L, Fritz D, Stefanovic L, Stefanovic B (2009) Coming together: liver fibrosis, collagen mRNAs and the RNA binding protein. Expert Rev Gastroenterol Hepatol 3: 1–3.
16. Vukmirovic M, Manojlovic Z, Stefanovic B (2013) Serine-threonine kinase receptor-associated protein (STRAP) regulates translation of type I collagen mRNAs. Mol Cell Biol 33: 3893–3906.
17. Manojlovic Z, Blackmon J, Stefanovic B (2013) Tacrolimus (FK506) prevents early stages of ethanol induced hepatic fibrosis by targeting LARP6 dependent mechanism of collagen synthesis. PLoS One 8: e65897.
18. Manojlovic Z, Stefanovic B (2012) A novel role of RNA helicase A in regulation of translation of type I collagen mRNAs. RNA 18: 321–334.
19. Challa AA, Stefanovic B (2011) A novel role of vimentin filaments: binding and stabilization of collagen mRNAs. Mol Cell Biol 31: 3773–3789.
20. Heissler SM, Manstein DJ (2013) Nonmuscle myosin-2: mix and match. Cell Mol Life Sci 70: 1–21.
21. Lofgren M, Ekblad E, Morano I, Arner A (2003) Nonmuscle Myosin motor of smooth muscle. J Gen Physiol 121: 301–310.
22. Bresnick AR (1999) Molecular mechanisms of nonmuscle myosin-II regulation. Curr Opin Cell Biol 11: 26–33.
23. Simerly C, Nowak G, de Lanerolle P, Schatten G (1998) Differential expression and functions of cortical myosin IIA and IIB isotypes during meiotic maturation, fertilization, and mitosis in mouse oocytes and embryos. Mol Biol Cell 9: 2509–2525.
24. Kelley CA (1997) Characterization of isoform diversity among smooth muscle and nonmuscle myosin heavy chains. Comp Biochem Physiol B Biochem Mol Biol 117: 39–49.
25. Ma X, Adelstein RS (2012) In vivo studies on nonmuscle myosin II expression and function in heart development. Front Biosci (Landmark Ed) 17: 545–555.
26. Ogut O, Yuen SL, Brozovich FV (2007) Regulation of the smooth muscle contractile phenotype by nonmuscle myosin. J Muscle Res Cell Motil 28: 409–414.
27. Even-Ram S, Doyle AD, Conti MA, Matsumoto K, Adelstein RS, et al. (2007) Myosin IIA regulates cell motility and actomyosin-microtubule crosstalk. Nat Cell Biol 9: 299–309.
28. Vicente-Manzanares M, Ma X, Adelstein RS, Horwitz AR (2009) Non-muscle myosin II takes centre stage in cell adhesion and migration. Nat Rev Mol Cell Biol 10: 778–790.
29. Hu B, Phan SH (2013) Myofibroblasts. Curr Opin Rheumatol 25: 71–77.
30. Moore CC, Lakner AM, Yengo CM, Schrum LW (2011) Nonmuscle myosin II regulates migration but not contraction in rat hepatic stellate cells. World J Hepatol 3: 184–197.
31. Liu Z, Van Rossen E, Timmermans JP, Geerts A, van Grunsven LA, et al. (2011) Distinct roles for non-muscle myosin II isoforms in mouse hepatic stellate cells. J Hepatol 54: 132–141.
32. Tangkijvanich P, Tam SP, Yee HF Jr (2001) Wound-induced migration of rat hepatic stellate cells is modulated by endothelin-1 through rho-kinase-mediated alterations in the acto-myosin cytoskeleton. Hepatology 33: 74–80.
33. Cai L, Fritz D, Stefanovic L, Stefanovic B (2010) Nonmuscle myosin-dependent synthesis of type I collagen. J Mol Biol 401: 564–578.
34. Jin Y, Atkinson SJ, Marrs JA, Gallagher PJ (2001) Myosin ii light chain phosphorylation regulates membrane localization and apoptotic signaling of tumor necrosis factor receptor-1. J Biol Chem 276: 30342–30349.
35. He TC, Zhou S, da Costa LT, Yu J, Kinzler KW, et al. (1998) A simplified system for generating recombinant adenoviruses. Proc Natl Acad Sci U S A 95: 2509–2514.
36. Yamada NA, Castro A, Farber RA (2003) Variation in the extent of microsatellite instability in human cell lines with defects in different mismatch repair genes. Mutagenesis 18: 277–282.
37. Jiang F, Parsons CJ, Stefanovic B (2006) Gene expression profile of quiescent and activated rat hepatic stellate cells implicates Wnt signaling pathway in activation. J Hepatol 45: 401–409.
38. Stefanovic L, Brenner DA, Stefanovic B (2005) Direct hepatotoxic effect of KC chemokine in the liver without infiltration of neutrophils. Exp Biol Med (Maywood) 230: 573–586.
39. Gunjan A, Verreault A (2003) A Rad53 kinase-dependent surveillance mechanism that regulates histone protein levels in S. cerevisiae. Cell 115: 537–549.
40. Stephens SB, Dodd RD, Lerner RS, Pyhtila BM, Nicchitta CV (2008) Analysis of mRNA partitioning between the cytosol and endoplasmic reticulum compartments of mammalian cells. Methods Mol Biol 419: 197–214.
41. Li X, Donowitz M (2008) Fractionation of subcellular membrane vesicles of epithelial and nonepithelial cells by OptiPrep density gradient ultracentrifugation. Methods Mol Biol 440: 97–110.
42. Mandon EC, Trueman SF, Gilmore R (2013) Protein translocation across the rough endoplasmic reticulum. Cold Spring Harb Perspect Biol 5.
43. Akopian D, Shen K, Zhang X, Shan SO (2013) Signal recognition particle: an essential protein-targeting machine. Annu Rev Biochem 82: 693–721.
44. Nicchitta CV, Lerner RS, Stephens SB, Dodd RD, Pyhtila B (2005) Pathways for compartmentalizing protein synthesis in eukaryotic cells: the template-partitioning model. Biochem Cell Biol 83: 687–695.
45. Nicchitta CV (2002) A platform for compartmentalized protein synthesis: protein translation and translocation in the ER. Curr Opin Cell Biol 14: 412–416.
46. Hermesh O, Jansen RP (2013) Take the (RN)A-train: localization of mRNA to the endoplasmic reticulum. Biochim Biophys Acta 1833: 2519–2525.
47. Kraut-Cohen J, Afanasieva E, Haim-Vilmovsky L, Slobodin B, Yosef I, et al. (2013) Translation- and SRP-independent mRNA targeting to the endoplasmic reticulum in the yeast Saccharomyces cerevisiae. Mol Biol Cell 24: 3069–3084.

48. Cui XA, Zhang H, Palazzo AF (2012) p180 promotes the ribosome-independent localization of a subset of mRNA to the endoplasmic reticulum. PLoS Biol 10: e1001336.

49. Pyhtila B, Zheng T, Lager PJ, Keene JD, Reedy MC, et al. (2008) Signal sequence- and translation-independent mRNA localization to the endoplasmic reticulum. Rna 14: 445–453.

50. Bordeleau ME, Cencic R, Lindqvist L, Oberer M, Northcote P, et al. (2006) RNA-mediated sequestration of the RNA helicase eIF4A by Pateamine A inhibits translation initiation. Chem Biol 13: 1287–1295.

51. Dang Y, Low WK, Xu J, Gehring NH, Dietz HC, et al. (2009) Inhibition of nonsense-mediated mRNA decay by the natural product pateamine A through eukaryotic initiation factor 4AIII. J Biol Chem 284: 23613–23621.

52. Lerner RS, Seiser RM, Zheng T, Lager PJ, Reedy MC, et al. (2003) Partitioning and translation of mRNAs encoding soluble proteins on membrane-bound ribosomes. Rna 9: 1123–1137.

53. Stephens SB, Dodd RD, Brewer JW, Lager PJ, Keene JD, et al. (2005) Stable ribosome binding to the endoplasmic reticulum enables compartment-specific regulation of mRNA translation. Mol Biol Cell 16: 5819–5831.

54. Pace JM, Wiese M, Drenguis AS, Kuznetsova N, Leikin S, et al. (2008) Defective C-propeptides of the proalpha2(I) chain of type I procollagen impede molecular assembly and result in osteogenesis imperfecta. J Biol Chem 283: 16061–16067.

55. Oliver JE, Thompson EM, Pope FM, Nicholls AC (1996) Mutation in the carboxy-terminal propeptide of the Pro alpha 1(I) chain of type I collagen in a child with severe osteogenesis imperfecta (OI type III): possible implications for protein folding. Hum Mutat 7: 318–326.

56. Tajima S, Takehana M, Azuma N (1994) Production of overmodified type I procollagen in a case of osteogenesis imperfecta. J Dermatol 21: 219–222.

57. Lamande SR, Chessler SD, Golub SB, Byers PH, Chan D, et al. (1995) Endoplasmic reticulum-mediated quality control of type I collagen production by cells from osteogenesis imperfecta patients with mutations in the pro alpha 1 (I) chain carboxyl-terminal propeptide which impair subunit assembly. J Biol Chem 270: 8642–8649.

58. Locke D, Koreen IV, Harris AL (2006) Isoelectric points and post-translational modifications of connexin26 and connexin32. FASEB J 20: 1221–1223.

59. Bateman JF, Mascara T, Chan D, Cole WG (1987) A structural mutation of the collagen alpha 1(I)CB7 peptide in lethal perinatal osteogenesis imperfecta. J Biol Chem 262: 4445–4451.

60. Pandya K, Kim HS, Smithies O (2006) Fibrosis, not cell size, delineates beta-myosin heavy chain reexpression during cardiac hypertrophy and normal aging in vivo. Proc Natl Acad Sci U S A 103: 16864–16869.

61. Bain J, McLauchlan H, Elliott M, Cohen P (2003) The specificities of protein kinase inhibitors: an update. Biochem J 371: 199–204.

62. Isemura M, Mita T, Satoh K, Narumi K, Motomiya M (1991) Myosin light chain kinase inhibitors ML-7 and ML-9 inhibit mouse lung carcinoma cell attachment to the fibronectin substratum. Cell Biol Int Rep 15: 965–972.

63. Limouze J, Straight AF, Mitchison T, Sellers JR (2004) Specificity of blebbistatin, an inhibitor of myosin II. J Muscle Res Cell Motil 25: 337–341.

64. Bergeron JJ, Brenner MB, Thomas DY, Williams DB (1994) Calnexin: a membrane-bound chaperone of the endoplasmic reticulum. Trends Biochem Sci 19: 124–128.

65. Bascom RA, Srinivasan S, Nussbaum RL (1999) Identification and character-ization of golgin-84, a novel Golgi integral membrane protein with a cytoplasmic coiled-coil domain. J Biol Chem 274: 2953–2962.

66. Somlyo AP, Somlyo AV (2003) Ca2+ sensitivity of smooth muscle and nonmuscle myosin II: modulated by G proteins, kinases, and myosin phosphatase. Physiol Rev 83: 1325–1358.

67. Connell LE, Helfman DM (2006) Myosin light chain kinase plays a role in the regulation of epithelial cell survival. J Cell Sci 119: 2269–2281.

68. Shi J, Takahashi S, Jin XH, Li YQ, Ito Y, et al. (2007) Myosin light chain kinase-independent inhibition by ML-9 of murine TRPC6 channels expressed in HEK293 cells. Br J Pharmacol 152: 122–131.

69. Fitzgerald J, Lamande SR, Bateman JF (1999) Proteasomal degradation of unassembled mutant type I collagen pro- alpha1(I) chains. J Biol Chem 274: 27392–27398.

70. Lamande SR, Bateman JF (1999) Procollagen folding and assembly: the role of endoplasmic reticulum enzymes and molecular chaperones. Semin Cell Dev Biol 10: 455–464.

71. Werner S, Grose R (2003) Regulation of wound healing by growth factors and cytokines. Physiol Rev 83: 835–870.

72. Shahbaz AU, Sun Y, Bhattacharya SK, Ahokas RA, Gerling IC, et al. (2010) Fibrosis in hypertensive heart disease: molecular pathways and cardioprotective strategies. J Hypertens 28 Suppl 1: S25–32.

73. Chen AY, Zirwas MJ, Heffernan MP (2010) Nephrogenic systemic fibrosis: a review. J Drugs Dermatol 9: 829–834.

74. Malfait F, Symoens S, Coucke P, Nunes L, De Almeida S, et al. (2006) Total absence of the alpha2(I) chain of collagen type I causes a rare form of Ehlers-Danlos syndrome with hypermobility and propensity to cardiac valvular problems. J Med Genet 43: e36.

75. Pope FM, Nicolls AC, Osse G, Lee KW (1986) Clinical features of homozygous alpha 2(I) collagen deficient osteogenesis imperfecta. J Med Genet 23: 377.

76. Miles CA, Sims TJ, Camacho NP, Bailey AJ (2002) The role of the alpha2 chain in the stabilization of the collagen type I heterotrimer: a study of the type I homotrimer in oim mouse tissues. J Mol Biol 321: 797–805.

77. Sims TJ, Miles CA, Bailey AJ, Camacho NP (2003) Properties of collagen in OIM mouse tissues. Connect Tissue Res 44 Suppl 1: 202–205.

78. Bachmann A, Kiefhaber T, Boudko S, Engel J, Bachinger HP (2005) Collagen triple-helix formation in all-trans chains proceeds by a nucleation/growth mechanism with a purely entropic barrier. Proc Natl Acad Sci U S A 102: 13897–13902.

79. Boudko S, Frank S, Kammerer RA, Stetefeld J, Schulthess T, et al. (2002) Nucleation and propagation of the collagen triple helix in single-chain and trimerized peptides: transition from third to first order kinetics. J Mol Biol 317: 459–470.

80. Koide T, Asada S, Nagata K (1999) Substrate recognition of collagen-specific molecular chaperone HSP47. Structural requirements and binding regulation. J Biol Chem 274: 34523–34526.

81. Nagata K (1996) Hsp47: a collagen-specific molecular chaperone. Trends Biochem Sci 21: 22–26.

82. Ueno T, Kaneko K, Sata T, Hattori S, Ogawa-Goto K (2012) Regulation of polysome assembly on the endoplasmic reticulum by a coiled-coil protein, p180. Nucleic Acids Res 40: 3006–3017.

83. Ueno T, Kaneko K, Katano H, Sato Y, Mazitschek R, et al. (2010) Expansion of the trans-Golgi network following activated collagen secretion is supported by a coiled-coil microtubule-bundling protein, p180, on the ER. Exp Cell Res 316: 329–340.

84. Cui XA, Zhang Y, Hong SJ, Palazzo AF (2013) Identification of a region within the placental alkaline phosphatase mRNA that mediates p180-dependent targeting to the endoplasmic reticulum. J Biol Chem 288: 29633–29641.

85. Stefanovic B, Stefanovic L, Schnabl B, Bataller R, Brenner DA (2004) TRAM2 protein interacts with endoplasmic reticulum Ca2+ pump Serca2b and is necessary for collagen type I synthesis. Mol Cell Biol 24: 1758–1768.

86. Duran JM, Valderrama F, Castel S, Magdalena J, Tomas M, et al. (2003) Myosin motors and not actin comets are mediators of the actin-based Golgi-to-endoplasmic reticulum protein transport. Mol Biol Cell 14: 445–459.

87. Conti MA, Adelstein RS (2008) Nonmuscle myosin II moves in new directions. J Cell Sci 121: 11–18.

88. Lo CM, Buxton DB, Chua GC, Dembo M, Adelstein RS, et al. (2004) Nonmuscle myosin IIb is involved in the guidance of fibroblast migration. Mol Biol Cell 15: 982–989.

89. Massague J (1996) TGFbeta signaling: receptors, transducers, and Mad proteins. Cell 85: 947–950.

90. Biernacka A, Dobaczewski M, Frangogiannis NG (2011) TGF-beta signaling in fibrosis. Growth Factors 29: 196–202.

Permissions

All chapters in this book were first published in PLOS ONE, by The Public Library of Science; hereby published with permission under the Creative Commons Attribution License or equivalent. Every chapter published in this book has been scrutinized by our experts. Their significance has been extensively debated. The topics covered herein carry significant findings which will fuel the growth of the discipline. They may even be implemented as practical applications or may be referred to as a beginning point for another development.

The contributors of this book come from diverse backgrounds, making this book a truly international effort. This book will bring forth new frontiers with its revolutionizing research information and detailed analysis of the nascent developments around the world.

We would like to thank all the contributing authors for lending their expertise to make the book truly unique. They have played a crucial role in the development of this book. Without their invaluable contributions this book wouldn't have been possible. They have made vital efforts to compile up to date information on the varied aspects of this subject to make this book a valuable addition to the collection of many professionals and students.

This book was conceptualized with the vision of imparting up-to-date information and advanced data in this field. To ensure the same, a matchless editorial board was set up. Every individual on the board went through rigorous rounds of assessment to prove their worth. After which they invested a large part of their time researching and compiling the most relevant data for our readers.

The editorial board has been involved in producing this book since its inception. They have spent rigorous hours researching and exploring the diverse topics which have resulted in the successful publishing of this book. They have passed on their knowledge of decades through this book. To expedite this challenging task, the publisher supported the team at every step. A small team of assistant editors was also appointed to further simplify the editing procedure and attain best results for the readers.

Apart from the editorial board, the designing team has also invested a significant amount of their time in understanding the subject and creating the most relevant covers. They scrutinized every image to scout for the most suitable representation of the subject and create an appropriate cover for the book.

The publishing team has been an ardent support to the editorial, designing and production team. Their endless efforts to recruit the best for this project, has resulted in the accomplishment of this book. They are a veteran in the field of academics and their pool of knowledge is as vast as their experience in printing. Their expertise and guidance has proved useful at every step. Their uncompromising quality standards have made this book an exceptional effort. Their encouragement from time to time has been an inspiration for everyone.

The publisher and the editorial board hope that this book will prove to be a valuable piece of knowledge for researchers, students, practitioners and scholars across the globe.

List of Contributors

Mattias O. Roth, Adam G. Wilkins, Georgina M. Cooke, David A. Raftos and Sham V. Nair
Department of Biological Sciences, Macquarie University, North Ryde, NSW, Australia

Malik N. Akhtar and Bruce R. Southey
Department of Animal Sciences, University of Illinois at Urbana-Champaign, Urbana, Illinois, United States of America

Per E. Andrén
Department of Pharmaceutical Biosciences, Uppsala University, Uppsala, Sweden

Jonathan V. Sweedler
Department of Chemistry, University of Illinois at Urbana-Champaign, Urbana, Illinois, United States of America

Sandra L. Rodriguez- Zas
Department of Animal Sciences, University of Illinois at Urbana-Champaign, Urbana, Illinois, United States of America
Department of Statistics, University of Illinois at Urbana-Champaign, Urbana, Illinois, United States of America
Institute for Genomic Biology, University of Illinois at Urbana-Champaign, Urbana, Illinois, United States of America

Yiguo Zhang, Lu Qiu, Shaojun Li, Yuancai Xiang and Jiayu Chen, Yonggang Ren
The Laboratory of Cell Biochemistry and Gene Regulation, College of Medical Bioengineering and Faculty of Life Sciences, Chongqing University, Shapingba District, Chongqing, China

Geqing Wang, Christopher A. MacRaild, Jamie S. Simpson and Raymond S. Norton
Medicinal Chemistry, Monash Institute of Pharmaceutical Sciences, Monash University, Parkville, Victoria, Australia

Biswaranjan Mohanty and Martin J. Scanlon
Medicinal Chemistry, Monash Institute of Pharmaceutical Sciences, Monash University, Parkville, Victoria, Australia
Australian Research Council Centre of Excellence for Coherent X-ray Science, Monash University, Parkville, Victoria, Australia

Mehdi Mobli
Centre for Advanced Imaging, University of Queensland, St Lucia, Queensland, Australia

Nathan P. Cowieson
Australian Synchrotron, Clayton, Victoria, Australia

Robin F. Anders
Department of Biochemistry, La Trobe University, Bundoora, Victoria, Australia

Sheena McGowan
Department of Biochemistry and Molecular Biology, Monash University, Clayton, Victoria, Australia

Michelle Davey, Ian C. Chute, Steven G. Griffiths, Simi Chacko, Nicolas Crapoulet, Sébastien Fournier, Andrew Joy, Michelle C. Caissie, Amanda D. Ferguson and Melissa Daigle
Atlantic Cancer Research Institute, Moncton, New Brunswick, Canada

Anirban Ghosh and Rodney J. Ouellette
Atlantic Cancer Research Institute, Moncton, New Brunswick, Canada
Department of Chemistry and Biochemistry, Université de Moncton, Moncton, New Brunswick Canada

Scott Lewis
New England Peptide Inc., Gardner, Massachusetts, United States of America

David Barnett
Atlantic Cancer Research Institute, Moncton, New Brunswick, Canada
Department of Chemistry and Biochemistry, Université de Moncton, Moncton, New Brunswick Canada
Department of Chemistry and Biochemistry, Mount Allison University, Sackville New Brunswick, Canada

M. Vicki Meli
Department of Chemistry and Biochemistry, Mount Allison University, Sackville New Brunswick, Canada

Stephen M. Lewis
Atlantic Cancer Research Institute, Moncton, New Brunswick, Canada
Department of Chemistry and Biochemistry, Université de Moncton, Moncton, New Brunswick Canada

Department of Microbiology and Immunology, Dalhousie University, Halifax, Nova Scotia, Canada
Department of Biology, University of New Brunswick, Saint John, New Brunswick, Canada

Duan Chen
Department of Mathematics and Statistics, University of North Carolina at Charlotte, Charlotte, North Carolina, United States of America

Andrey A. Bobko, Amy C. Gross, Randall Evans, Clay B. Marsh, Valery V. Khramtsov and Timothy D. Eubank
Division of Pulmonary, Allergy, Critical Care and Sleep Medicine, College of Medicine, The Ohio State University, Columbus, Ohio, United States of America

Avner Friedman
Mathematical Biosciences Institute, The Ohio State University, Columbus, Ohio, United States of America
Department of Mathematics, The Ohio State University, Columbus, Ohio, United States of America

Hubert Sytykiewicz, Grzegorz Chrzanowski, Paweł Czerniewicz, Iwona Sprawka, Iwona Łukasik, Sylwia Goławska and Cezary Sempruch
Siedlce University of Natural Sciences and Humanities, Department of Biochemistry and Molecular Biology, Siedlce, Poland

Irene Wuethrich., Janneke G. C. Peeters., Annet E. M. Blom, Christopher S. Theile, Zeyang Li, Eric Spooner, Hidde L. Ploegh and Carla P. Guimaraes
Whitehead Institute for Biomedical Research, Department of Biology, Massachusetts Institute of Technology, Cambridge, Massachusetts, United States of America

Rosa Mastrogiacomo, Paolo Pelosi and Andrea Serra
Department of Agriculture, Food and Environment, University of Pisa, Pisa, Italy

Chiara D'Ambrosio and Andrea Scaloni
Proteomics & Mass Spectrometry Laboratory, ISPAAM, National Research Council, Napoli, Italy

Alberto Niccolini and Angelo Gazzano
Department of Veterinary Sciences, University of Pisa, Pisa, Italy

Violette Frochaux and Michael W. Linscheid
Department of Chemistry, Humboldt-Universität zu Berlin, Berlin, Germany

Diana Hildebrand and Hartmut Schlüter
Department of Clinical Chemistry, University Medical Center Hamburg-Eppendorf Hamburg, Germany

Anja Talke
ProteaImmun GmbH, Berlin, Germany

Pradeep Kumar Naik
Department of Zoology, Guru Ghasidas Central University, Bilaspur, Chattishgarh, India

Iswar Baitharu
Department of Zoology, Guru Ghasidas Central University, Bilaspur, Chattishgarh, India
Department of Neurobiology, Defence Institute of Physiology and Allied Sciences, Defense Research Development Organisation, Timarpur, Delhi, India

Vishal Jain and Satya Narayan Deep
Department of Neurobiology, Defence Institute of Physiology and Allied Sciences, Defense Research Development Organisation, Timarpur, Delhi, India

Sabita Shroff
Department of Chemistry, Sambalpur University, Burla, India

Jayanta Kumar Sahu
Department of Life Science, National Institute of Technology, Rourkela, India

Govindasamy Ilavazhagan
Department of Research, Hindustan University, Chennai, Tamilnadu, India

Leila Priscila Peters, Giselle Carvalho, Paula Fabiane Martins, Manuella Nóbrega Dourado, Milca Bartz Vilhena and Ricardo Antunes Azevedo
Departamento de Genética, Escola Superior de Agricultura Luiz de Queiroz, Universidade de São Paulo, Piracicaba, Brazil

Marcos Pileggi
Departamento de Biologia Estrutural, Molecular e Genética, Universidade Estadual de Ponta Grossa, Ponta Grossa, Brazil

Weiti Cui, Kaikai Zhu, Qijiang Jin, Yanjie Xie, Jin Cui, Yan Xia, Jing Zhang and Wenbiao Shen
College of Life Sciences, Laboratory Center of Life Sciences, Nanjing Agricultural University, Jiangsu Province, Nanjing, China

Huiping Chen
Key Laboratory of Protection and Development Utilization of Tropical Crop Germplasm Resources, Hainan University, Haikou, China

Negar Seyed, Tahereh Taheri and Sima Rafati
Molecular Immunology and Vaccine Research Lab, Pasteur Institute of Iran, Tehran, Iran

Charline Vauchy, Magalie Dosset and Yann Godet
INSERM U1098, Unité Mixte de Recherche, Besançon, France
Etablissement Français du Sang de Bourgogne Franche-Comté , Besançon, France
Université de Franche-Comté , Besançon, France

Ali Eslamifar
Department of Electron Microscopy and Clinical Research, Pasteur Institute of Iran, Tehran, Iran

Iraj Sharifi
School of Medicine, Leishmaniasis Research Center, Kerman University of Medical Sciences, Kerman, Iran

Olivier Adotevi and Christophe Borg
INSERM U1098, Unité Mixte de Recherche, Besançon, France

Etablissement Français du Sang de Bourgogne Franche-Comté, Besançon, France
Université de Franche-Comté , Besançon, France
CHRU de Besançon, Service d'Oncologie, Besançon, France

Pierre Simon Rohrlich
INSERM U1098, Unité Mixte de Recherche, Besançon, France
Etablissement Français du Sang de Bourgogne Franche-Comté , Besançon, France
Université de Franche-Comté , Besançon, France
CHRU de Besançon, Service de pédiatrie, Besançon, France

Andreas Holzinger, Franziska Kaplan and Kathrin Blaas
University of Innsbruck, Functional Plant Biology, Innsbruck, Austria

Bernd Zechmann
Baylor University, Center for Microscopy and Imaging, Waco, Texas, United States of America

Karin Komsic-Buchmann and Burkhard Becker
University of Cologne, Botanical Institute, Biocenter, Cologne, Germany

Hao Wang and Branko Stefanovic
Department of Biomedical Sciences, College of Medicine, Florida State University, Tallahassee, Florida, United States of America

Index

www.ingramcontent.com/pod-product-compliance
Lightning Source LLC
Chambersburg PA
CBHW070152240326
41458CB00126B/4438